Deep Reinforcement Learning

Hao Dong • Zihan Ding • Shanghang Zhang
Editors

Deep Reinforcement Learning

Fundamentals, Research and Applications

Springer

Editors
Hao Dong
EECS
Peking University
Beijing, China

Zihan Ding
CS
Imperial College London
London, UK

Shanghang Zhang
EECS
University of California, Berkeley
Berkeley, USA

ISBN 978-981-15-4094-3 ISBN 978-981-15-4095-0 (eBook)
https://doi.org/10.1007/978-981-15-4095-0

Translation from the English language edition: Deep Reinforcement Learning by Hao Dong, Zihan Ding and Shanghang Zhang Copyright © Springer Nature Singapore Pte Ltd. 2020. All Rights Reserved.

This Springer imprint is published by the registered company Springer Nature Singapore Pte Ltd.
The registered company address is: 152 Beach Road, #21-01/04 Gateway East, Singapore 189721, Singapore

Foreword

I am impressed by the breadth of topics covered by this book. From fundamental underlying theory of deep reinforcement learning to technical implementation with elaborated code details, the authors devoted significant efforts to provide a comprehensive description. Such a style makes the book an ideal study material for novices and scholars. Embracing the open-source community is an indispensable reason for deep learning to have such a rapid development. I am glad that this book is accompanied by the open-source code. I believe that this book will be very useful for researchers who can learn from such a comprehensive overview of the field, as well as the engineers who can learn from scratch with hands-on practice using the open source code examples.

FREng MAE Director of Data Science Institute Yike Guo
Imperial College London
London, UK

This book provides the most reliable entry to deep reinforcement learning, bridging the gap between fundamentals and practices, featuring detailed explanation and demonstration of algorithmic implementation, offering tips and cheat sheet. The authors are researchers and practitioners from leading universities and open source community who conduct research on deep reinforcement learning or apply its new techniques in various applications. The book serves as an extremely useful resource for readers of diverse background and objectives.

Director of the Center on Frontiers of Computing Studies Baoquan Chen
Peking University
Beijing, China

This is a timely book in an important area—deep reinforcement learning (RL). The book presents a comprehensive set of tools in a clear and succinct fashion: covering the foundations and popular algorithms of deep RL, practical implementation details, as well as forward-looking research directions. It is ideally suited for anyone who would like to learn deep RL, to implement deep RL algorithms for their applications, or to begin fundamental research in the area of deep RL.

Princeton University Chi Jin
Princeton, NJ, USA

This is a book for pure fans of reinforcement learning, in particular deep reinforcement learning.

Deep reinforcement learning (DRL) has been changing our lives and the world since 2013 in many ways (e.g. autonomous cars, AlphaGo). It has showed the capability to comprehend the 'beauty of Go' better than professionals. The same idea is currently being implemented in technology, healthcare and finance. DRL explores the ultimate answer to one of the most fundamental questions: how do human beings learn from interaction with environment? This mechanism could be a silver bullet of avoiding the 'big data' trap, a necessary path towards 'Strong AI', as well as a virgin land that no human intelligence has touched before.

This book, written by a group of young researchers with full passion in machine learning, will show you the world of DRL and enhance your understanding by means of practical examples and experiences. Recommend to all learners who want to keep the key to future intelligence in their own pocket.

University College London Kezhi Li
London, UK

Preface

Deep reinforcement learning (DRL) combines deep learning (DL) with a reinforcement learning (RL) architecture. It has been able to perform a wide range of complex decision-making tasks that were previously intractable for a machine. Moreover, DRL has contributed to the recent great successes in artificial intelligence (AI) like AlphaGo and OpenAI Five. Indeed, DRL has opened up many exciting avenues to explore in a variety of domains such as healthcare, robotics, smart grids, and finance.

Divided into three main parts, this book provides a comprehensive and self-contained introduction to DRL. The first part introduces the foundations of DL, RL and widely used DRL methods and then discusses their implementations, which includes Chaps. 1–6. The second part covers selected DRL research topics in Chaps. 7–12, which are useful for those would like to specialize in DRL research. To help readers gain a deep understanding of DRL and quickly apply the techniques in practice, the third part including Chaps. 13–17 presents a rich set of applications, such as the AlphaZero and learning to run, with detailed descriptions.

The book is intended for computer science students, both undergraduate and postgraduate, who would like to learn DRL from scratch, practice its implementation, and explore the research topics. This book might also appeal to engineers and practitioners who do not have strong machine learning background but want to quickly understand how DRL works and use these techniques in their practical applications.

Beijing, China — Hao Dong

Acknowledgements

The authors would like to thank the people who provided feedback and suggestions on the contents of the book, including: Jie Fu from Mila, Jianhong Wang and Shikun Liu from Imperial College London, Kun Chen from Peking University, Meng Song from University of California, San Diego, Chen Ma, Chenjun Xiao and Jingcheng Mei from University of Alberta, Tong Yu from Samsung Research, Xu Luo from Fudan University, Dian Shi from University of Houston, Weipeng Zhang from Shanghai Jiaotong University, Yashu Kang from Georgia Institute of Technology, Chenxiao Zhao from East China Normal University, Tianlin Liu from Friedrich Miescher Institute, Gavin Ding from Borealis AI, Ruilong Su from Xiaohongshu Technology Co., Ltd., and Yingjun Pei from Chinese Academy of Sciences. We also want to thank Jared Sharp for the language proofread of most chapters in the book.

Many other people have contributed to this and the code base of the book—open-source contributors, such as Ruihai Wu, Luo Mai, Rundi Wu, Guo Li, Cheng Lai, and Jonathan Dekhtiar, who develop and maintain TensorLayer and the reinforcement learning examples, and colleagues who have provided important insights into the book design. To all these, we offer our thanks and gratitude. Hao Dong would especially like to thank the Center on Frontiers of Computing Studies of the Department of Computer Science at Peking University and Peng Cheng Laboratory for the strong support of developing and maintaining TensorLayer. Zihan Ding would like to thank Dr. Edward Johns for sharing his understandings and useful discussions.

Contents

Editors and Contributors

About the Editors

Hao Dong is currently an Assistant Professor at Peking University. He received his Ph.D. in Computing from Imperial College London in 2019, supervised by Prof. Yike Guo. Hao's research chiefly involves Deep Learning and Computer Vision, with the goal of reducing the amount of data required for learning intelligent systems. He is passionate about popularizing artificial intelligence technologies and established TensorLayer, a deep learning and reinforcement learning library for scientists and engineers, which won the Best Open Source Software Award at ACM Multimedia 2017.

Zihan Ding received his M.Sc. degree in Machine Learning with distinction from the Department of Computing, Imperial College London, supervised by Dr. Edward Johns. He holds double Bachelor degrees from the University of Science and Technology of China: in Photoelectric Information Science and Engineering (Physics) and in Computer Science and Technology. His research interests include deep reinforcement learning, robotics, computer vision, quantum computation and machine learning. He has published papers in ICRA, AAAI, NIPS, IJCAI, and Physical Review. He also contributed to the open-source projects TensorLayer RLzoo, TensorLet and Arena.

Shanghang Zhang is a postdoctoral research fellow in the Berkeley AI Research (BAIR) Lab, the Department of Electrical Engineering and Computer Sciences, UC Berkeley, USA. She received her Ph.D. from Carnegie Mellon University in 2018. Her research interests cover deep learning, computer vision, and reinforcement learning, as reflected in her numerous publications in top-tier journals and conference proceedings, including NeurIPS, CVPR, ICCV, and AAAI. Her research mainly focuses on machine learning with limited training data, including low-shot learning, domain adaptation, and meta-learning, which enables the learning system to automatically adapt to real-world variations and new environments. She was one of the “2018 Rising Stars in EECS”İ (a highly selective program launched at MIT in 2012, which has since been hosted at UC Berkeley, Carnegie Mellon, and Stanford annually). She has also been selected for the Adobe Academic Collaboration Fund, Qualcomm Innovation Fellowship (QInF) Finalist Award, and Chiang Chen Overseas Graduate Fellowship.

About the Authors

Hang Yuan is currently a Ph.D. candidate of Computer Science at the University of Oxford, specializing in AI Safety for Deep Learning and its applications in Healthcare AI. He conducted his master thesis at Swiss Federal Institute of Technology Lausanne (EPFL) with the Computer Vision Lab under Dr. Mathieu Salzmann and Dr. François Fleuret on the topic of delayed adversarial attack using recurrent neural networks for Deep Reinforcement Learning. Previously, he has also researched and studied at Carnegie Mellon University, Max Planck Institute for Intelligent Systems Empirical Inference Group and Imperial College London. He obtained his MSc degree at EPFL in Neuroscience and BSc at Jacobs University in Computer Science under the supervision of Prof. Herbert Jaeger.

Hongming Zhang is currently an engineer at the Institute of Automation, Chinese Academy of Sciences (CASIA). His research focuses on Reinforcement Learning and Game Theory. Before CASIA, he received his MSc degree in Statistics from Peking University, Bachelor degree in Mathematics from Beijing Normal University.

Jingqing Zhang is currently a Ph.D. candidate at Data Science Institute, Imperial College London under the supervision of Prof. Yike Guo. His research interest includes Deep Learning, Machine Learning, Text Mining, Data Mining and their applications. He received his BEng degree in Computer Science and Technology from Tsinghua University, 2016, and MRes degree with distinction in Computing from Imperial College London, 2017.

Yanhua Huang is currently a software engineer at Xiaohongshu Technology Co., Ltd., working on large-scale machine learning and reinforcement learning in recommender systems. He received his B.S. degree from the Department of Mathematics, East China Normal University in July 2016. Yanhua also contributed to some open-source projects, such as PyTorch, TensorFlow, and Ray.

Tianyang Yu is currently a MSc candidate of Computer Science at Nanchang University. Previously, he interned at the Institute of Automation, Chinese Academy of Sciences. Tianyang is interested in Reinforcement Learning and has strong experiences on applying Reinforcement Learning techniques into real-world applications.

Huaqing Zhang is currently a software engineer at Google LLC, exploring on the areas of multi-agent reinforcement learning and hierarchical game theory. He received the B.S. degree in Huazhong University of Science and Technology, Wuhan, China, in June 2013, and the Ph.D. degree in the department of electronic and computer engineering at University of Houston, Houston, TX, USA, in December 2017.

Ruitong Huang is currently a researcher at Borealis AI. His research interests broadly include topics such as online learning, convex optimization, adversarial learning, and reinforcement learning. Ruitong obtained his PhD in Statistical Machine Learning from the computing science department of University of Alberta. Before that, Ruitong spent four years at the University of Science and Technology of China for his Bachelor degree in Math and two years in the David R. Cheriton School of Computer Science at University of Waterloo for his Master's in Symbolic Computation.

Acronyms

AC	Actor-critic
ACKTR	Actor-critic using Kronecker-factored trust region
AGAIL	Action-guided adversarial imitation learning
AI	Artificial intelligence
AIRL	Adversarial inverse reinforcement learning
ANN	Artificial neural network
A2C	Advantage actor-critic
A3C	Asynchronous advantage actor-critic
BC	Behavioral cloning
BCO	Behavioral cloning from observation
BO	Bayesian optimization
BPTT	Backpropagation through time
CE	Cross entropy
CFD	Contrastive forward dynamics
CMA	Covariance matrix adaptation
CMA-ES	Covariance matrix adaptation evolution strategy
CNN	Convolutional neural network
CPU	Central processing unit
C51	Categorical 51
DAgger	Dataset aggreation
DDPG	Deep deterministic policy gradient
DDPGfD	Deep deterministic policy gradient from demonstration
DL	Deep learning
DMP	Dynamic movement primitives
DNN	Deep neural network
DP	Dynamic programming
DPG	Deterministic policy gradient
DQN	Deep Q-network
DQfD	Deep Q-learning from demonstrations
DRL	Deep reinforcement learning
EM	Expectation maximization

FAIL	Forward adversarial imitation learning
FC	Fully connected
FRL	Feudal reinforcement learning
FuN	Feudal network
GAN	Generative adversarial network
GAN-GCL	Generative adversarial network guided cost learning
GAIL	Generative adversarial imitation learning
GCL	Guided cost learning
GMM	Gaussian misture model
GMR	Gaussian mixture regression
GP	Gaussian process
GPU	Graphics processing unit
GPI	Generalized policy iteration
GPR	Gaussian process regression
HAM	Hierarchical abstract machine
HIRO	Hierarchical reinforcement learning with off-policy correction
HRL	Hierarchical reinforcement learning
IfO	Imitation learning from observation
IL	Imitation learning
ILPO	Imitating latent policies from observation
IMPALA	Importance weighted actor-learner architecture
InRL	Independent reinforcement learning
IRL	Inverse reinforcement learning
KL	Kullback-Leibler
KMP	Kernelized movement primitives
LQR	Linear quadratic regulators
LSTM	Long short-term memory
MARL	Multi-agent reinforcement learning
MaxEnt	Maximum entropy
MC	Monte Carlo
MCTS	Monte Carlo tree search
MDP	Markov decision process
ML	Machine learning
MLP	Multi-layer perceptron
MPO	Maximum a posteriori policy optimization
MRP	Markov reward process
MSE	Mean square error
NAC	Normalized actor-critic
OU	Ornstein-Uhlenbeck
PBT	Population based training
PER	Prioritized experience replay
PG	Policy gradient
POMDP	Partially observed Markov decision process
PPO	Proximal policy optimization
ProMP	Probabilistic movement primitives

QR-DQN	Quantile regression deep Q-network
RBF	Radial basis function
RCANs	Randomized-to-canonical adaptation networks
ReLU	Rectified linear unit
RIDM	Reinforced inverse dynamics modeling
RL	Reinforcement learning
RNN	Recurrent neural network
R2D2	Recurrent replay distributed DQN
SAC	Soft actor-critic
SEED	Scalable and efficient deep-RL
Sim2Real	Simulation to reality
SMDP	Semi-Markov decision process
SPG	Stochastic policy gradient
SRL	State representation learning
SVG	Stochastic value gradients
TCN	Time-contrastive networks
TD	Temporal difference
TD3	Twin delayed deep deterministic policy gradient
TRPO	Trust region policy optimization
UCB	Upper confidence bound
UCT	Upper confidence bounds applied to trees
VIME	Variational information maximizing exploration

Mathematical Notation

Jingqing Zhang, jingqing.zhang15@imperial.ac.uk.
We have tried to minimize the mathematical content of this book so as to minimize the requirements for understanding this field.

Fundamentals

x	A scalar
$\boldsymbol{x}$	A vector
$\boldsymbol{X}$	A matrix
$\mathbb{R}$	The set of real numbers
$\frac{\mathrm{d}y}{\mathrm{d}x}$	Derivative of y with respect to x
$\frac{\partial y}{\partial x}$	Partial derivative of y with respect to x
$\nabla_{\boldsymbol{x}} y$	Gradient of y with respect to $\boldsymbol{x}$
$\nabla_{\boldsymbol{X}} y$	Matrix derivatives of y with respect to $\boldsymbol{X}$
$P(X)$	A probability distribution over a discrete variable
$p(X)$	A probability distribution over a continuous variable, or over a variable whose type has not been specified
$X \sim p$	The random variable X has distribution p
$\mathbb{E}[X]$	Expectation of a random variable
$\mathrm{Var}[X]$	Variance of a random variable
$\mathrm{Cov}(X, Y)$	Covariance of two random variables
$D_{\mathrm{KL}}(P \| Q)$	Kullback-Leibler divergence of P and Q
$\mathcal{N}(\boldsymbol{x}; \boldsymbol{\mu}, \boldsymbol{\Sigma})$	Gaussian distribution over $\boldsymbol{x}$ with mean $\boldsymbol{\mu}$ and covariance $\boldsymbol{\Sigma}$

Deep Reinforcement Learning

s, s'	States
a	Action
r	Reward
R	Reward function
$\mathcal{S}$	Set of all non-terminal states
$\mathcal{S}^+$	Set of all states, including the terminal state
$\mathcal{A}$	Set of actions
$\mathcal{R}$	Set of all possible rewards
$\boldsymbol{P}$	Transition matrix
t	Discrete time step
T	Final time step of an episode
S_t	State at time t
A_t	Action at time t
R_t	Reward at time t, typically due, stochastically, to A_t and S_t
G_t	Return following time t
$G_t^{(n)}$	n-step return following time t
G_t^{λ}	λ-return following time t
π	Policy, decision-making rule
$\pi(s)$	Action taken in state s under *deterministic* policy π
$\pi(a\|s)$	Probability of taking action a in state s under *stochastic* policy π
$p(s', r\|s, a)$	Probability of transitioning to state s', with reward r, from state s and action a
$p(s'\|s, a)$	Probability of transitioning to state s', from state s taking action a
$v_\pi(s)$	Value of state s under policy π (expected return)
$v_*(s)$	Value of state s under the optimal policy
$q_\pi(s, a)$	Value of taking action a in state s under policy π
$q_*(s, a)$	Value of taking action a in state s under the optimal policy
V, V_t	Estimates of state-value function $v_\pi(s)$ or $v_*(s)$
Q, Q_t	Estimates of action-value function $q_\pi(s, a)$ or $q_*(s, a)$
τ	Trajectory, which is a sequence of states, actions and rewards, $\tau = (S_0, A_0, R_0, S_1, A_1, R_1, \dots)$
γ	Reward discount factor, $\gamma \in [0, 1]$
ϵ	Probability of taking a random action in ϵ-greedy policy
α, β	Step-size parameters
λ	Decay-rate parameter for eligibility traces

Introduction

Ever since the advent of the first computer in 1946, people have been striving to create more intelligent computers. Artificial Intelligence (AI) has benefited so much from the rapid development in the computing power and data volume that it can already outperform humans on many tasks, which were once considered intractable for machines such as board games like chess and Go, disease diagnosis, and video gaming. AI technology is also widely incorporated into other applications like drug discovery, weather prediction, advanced materials, recommended system, robotics perception and control, autonomous driving, human face recognition, speech recognition and dialog.

In the recent decade, not only do countries like China, the UK, the US, Japan and Germany have enacted concrete AI policies to support the development of AI but also tech giants like Google, Facebook, MicroSoft, Apple, Baidu, Huawei and Tencent have spent billions on AI research. AI is becoming almost omnipresent in our daily life, a few examples of which can be self-driving car, face ID, and chatbots. Without a doubt, AI is of paramount importance for the development of human society.

Before we dive into this book, we should first understand the relationships between various subdomains of AI, namely, machine learning (ML), deep learning (DL), reinforcement learning (RL), and the topic of this book—deep reinforcement learning (DRL). Figure 1 illustrates their relationships in a Venn diagram, and we will start to briefly introduce each of them in the following.

Artificial Intelligence

Since computers were first invented, scientists have endeavored to make the machines become more intelligent. However, the definition of intelligence even till today is still in an ongoing debate. So, without defining what intelligence is, Sir Alan Turing first introduced the Turing Test in his paper "Computing Machinery and Intelligence" at University of Manchester in 1950. The Turing test measures a

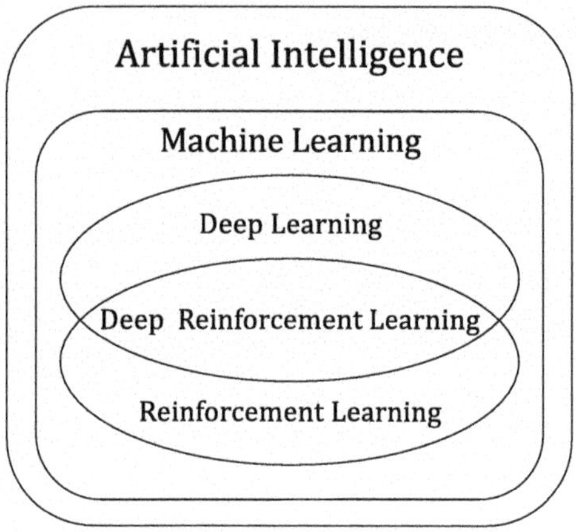

Fig. 1 Relationship of artificial intelligent, machine learning, deep learning, reinforcement learning, and deep reinforcement learning

machine's capability to imitate intelligent human behavior. Specifically, it describes an "imitation game", during which an interrogator asks a man and a computer in another room a series of questions, to determine which of the other two players is man, and which one is computer. The test is passed, if the computer can fool the interrogator.

AI was coined by John McCarthy in the famous Dartmouth conference in summer of 1956. This conference was seen as the starting point of AI being a field of computer science. In the early days of AI, the AI algorithms were mainly designed to solve problems that can be formulated by mathematical rules and logic rules.

Machine Learning

ML was coined in 1959 by Arthur Samuel (Bell Labs, IBM, Stanford). An AI system needs to has the ability to learn its own knowledge from the raw data. This capacity is known as ML. Many AI problems can be solved by designing a pattern recognition algorithm to extract features from raw data for that problem, and then providing these features to the ML algorithm.

For example, in the early days, to perform face recognition with a computer, we need specific facial feature extraction algorithms. The simplest way is to use Principal Component Analysis (PCA) to reduce the data dimension, and the feed these features into a classifier. Handcrafted feature engineering specific for face recognition is often required to improve the recognition performance. Nonetheless, it is fairly time-consuming to design the task-specific handcrafted feature extraction algorithms for different tasks, and let alone in many cases, it is extremely difficult to design a feature extraction algorithm. For example, the feature extraction of language translation requires the knowledge of grammar, which may require many language experts. A general algorithm is desired to extract features for different tasks, so as to reduce the reliance on prior knowledge from human.

Academics have invested lots of efforts in making ML learn the data representation automatically. Learning representation automatically is able to not only improve the performance but also rapidly reduce the cost to solve the AI problems.

Deep Learning

Deep Learning (DL) is a subset of ML algorithms based on artificial neural networks (ANN) Goodfellow et al. (2016). We call it neural network because it is inspired by biological neural networks. In 1943, Warren Sturgis McCulloch and Walter Pitts published "A Logical Calculus of the Ideas Immanent in Nervous Activity," McCulloch and Pitts (1943) which are deemed as the foundations for ANN. Since then, ANN shows the potential of automatic feature learning in which we do not need to design a specific feature learning algorithm for difficult input data, saving the development time of algorithms.

Deep Neural Network (DNN) is the "deep version" of ANN that consists of more neural network layers and can have greater data representation capacity as compared with the "shadow" neural networks. The difference between DL and non-DL methods is illustrated in Fig. 2, in which the DL methods free developers from hand-craft feature engineering to extracting and selecting useful features from input data for the final tasks. We also sometimes call this end-to-end learning as we only care about the input and the output and less on the feature. It is worth noting that this layer of abstraction is not always better as many people have spotted that DL methods tend to offer less transparency and interpretability.

Despite the promises DL has shown today, in the early step of DL history, due to the high computational cost of ANN, the hardware limitation of computers, and the black-box problem (we cannot explain what features the neural networks learned), DL was limited to use in practice and did not get much attention in academia.

This situation changed in 2012, mainly due to a neural network architecture called Alexnet Krizhevsky et al. (2012) which outperformed previous non-DL algorithms by more than 10% in image classification challenge event, ImageNet Russakovsky et al. (2015). DL starts to receive more attention and DL-based methods start to outperform many non-deep learning methods in different fields, such as computer vision Girshick (2015); Johnson et al. (2016); Ledig et al. (2017); Pathak et al. (2016); Vinyals et al. (2016) and natural language processing Bahdanau et al. (2015).

Fig. 2 Non-deep learning vs. deep learning algorithms

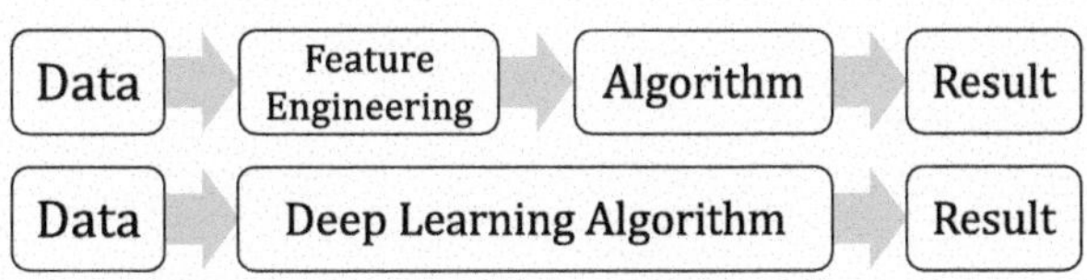

Reinforcement Learning

Even though, DL has a powerful data representation ability but it is not enough to build a smart AI system. This is because an AI system should not only able to learn from the provided data but also able to learn from interactions with the real world environment like a human. RL is a subset of ML that enables computers to learn by interacting with the real world environment.

In brief, RL separates the real world into two components—an environment and an agent. The agent interacts with the environment by performing specific actions and receives feedback from the environment. The feedback is usually termed as the "reward" in RL. The agent learns to perform "better" by trying to get more positive rewards from the environment. This learning process forms a feedback loop between the environment and agent, guiding the improvement of the agent with RL algorithms.

Deep Reinforcement Learning

DRL is to combine the advantages of DL and RL for building AI systems. The main reason to use DL in RL is to leverage the scalability of DNN in high-dimensional space, for example, the value function approximation utilizes the data representation of DNN to represent the highly compositional data distribution through end-to-end gradient-based optimization.

DeepMind, a research-oriented AI company established in London, plays an important role in the DRL history. In 2013, just one year after Alexnet, they published "Playing Atari with Deep Reinforcement Learning" which is the first successful DL model that learned how to play seven different Atari games using the raw pixels as the input without any adjustment of the model and learning algorithm. Different from the previous methods that relied on handcrafted features, DeepMind's method frees developer from feature engineering and outperforms all previous methods on six of the games and even surpasses a human expert on three of them.

In 2017, DeepMind's AlphaGo defeated the No.1 GO player Jie Ke in China, this event indicates that AI has the ability to perform better than human in a predefined environment via DRL algorithms. DRL is recognized as a subfield of ML that has the potential to achieve Artificial General Intelligence (AGI). However, there are still many challenges need to be addressed before we reach that point.

TensorLayer

Often, understanding the concepts is one thing and having to implement the mathematical formulae is a whole other thing. Therefore, at the end of many chapters of this book, we will also include a practical section in which we implement some of the key concepts in the corresponding chapter to better illustrate how different concepts are used in practice. Since DL is becoming increasingly popular, there exist many open-source frameworks, such as TensorFlow, Chainer, Theano, and Pytorch, to support automatic optimization for neural networks. In this book, we choose to adopt TensorLayer, a DL and DRL library designed specifically for researchers and engineers, which won the Best Open Source Software Award issued by ACM Multimedia in 2017. By the time we publish this book, TensorLayer supports TensorFlow as the computational backend, but with the continuous developing, TensorLayer may support more backends and the usage may be changed. Please refer to Github for more information https://github.com/tensorlayer/tensorlayer.

Beijing, China Hao Dong
Berkeley, USA Shanghang Zhang

References

Bahdanau D, Cho K, Bengio Y (2015) Neural machine translation by jointly learning to align and translate. In: Proceedings of the international conference on learning representations (ICLR)

Girshick R (2015) Fast R-CNN. In: Proceedings of the IEEE international conference on computer vision (ICCV), pp 1440–1448

Goodfellow I, Bengio Y, Courville A (2016) Deep learning. MIT Press, Cambridge. http://www.deeplearningbook.org

Johnson J, Alahi A, Fei-Fei L (2016) Perceptual losses for real-time style transfer and super-resolution. In: Proceedings of the European conference on computer vision (ECCV)

Krizhevsky A, Sutskever I, Hinton GE (2012) Imagenet classification with deep convolutional neural networks. In: Proceedings of the neural information processing systems. Advances in neural information processing systems, pp 1097–1105

Ledig C, Theis L, Huszar F, Caballero J, Cunningham A, Acosta A, Aitken A, Tejani A, Totz J, Wang Z, Shi W (2017) Photo-realistic single image super-resolution using a generative adversarial network. In: Proceedings of the IEEE conference on computer vision and pattern recognition (CVPR)

McCulloch WS, Pitts W (1943) A logical calculus of the ideas immanent in nervous activity. Bull Math Biophys 5(4):115–133

Pathak D, Krahenbuhl P, Donahue J, Darrell T, Efros AA (2016) Context encoders: feature learning by inpainting. In: Proceedings of the IEEE conference on computer vision and pattern recognition (CVPR), pp 2536–2544

Russakovsky O, Deng J, Su H, Krause J, Satheesh S, Ma S, Huang Z, Karpathy A, Khosla A, Bernstein M (2015) Imagenet large scale visual recognition challenge. Int J Comput Vis 115(3):211–252

Vinyals O, Toshev A, Bengio S, Erhan D (2016) Show and tell: lessons learned from the 2015 mscoco image captioning challenge. IEEE Trans Pattern Anal Mach Intell. arXiv:1609.06647v1

Part I
Fundamentals

Hao Dong
e-mail: hao.dong@pku.edu.cn

The first part of this book has six chapters to introduce the foundations of deep learning (DL), reinforcement learning (RL), widely used DRL algorithms and their implementations. Specifically, the first two chapters introduce the basic knowledge of DL and RL and the combination of the two, i.e. DRL, which are important for the readers to understand the rest of the book. You can skip these two chapters if you already have the related knowledge, but we highly recommend that you read through the second chapter to get familiar with the terminology and the mathematical formulas for the convenience of reading the following chapters.

The third chapter introduces the taxonomy of RL algorithms, which is intended to help readers to have an overview of modern DRL algorithms from different perspectives, such as model-based and model-free, policy-based and value-based, MC and TD methods, on-policy and off-policy, etc. We recommend that the readers go back to this chapter if there is any confusion about the categories and properties of specific algorithms when reading other chapters. For specific DRL algorithms, we introduce those which are most commonly applied, in detail, from the fourth to sixth chapters as well as providing the example codes to help the readers to understand the details of the algorithms and their implementations.

The related codes are released in the following link: https://github.com/tensorlayer/tensorlayer/tree/master/examples/reinforcement_learning.

Chapter 1
Introduction to Deep Learning

Jingqing Zhang, Hang Yuan, and Hao Dong

Abstract This chapter aims to briefly introduce the fundamentals for deep learning, which is the key component of deep reinforcement learning. We will start with a naive single-layer network and gradually progress to much more complex but powerful architectures such as convolutional neural networks (CNNs) and recurrent neural networks (RNNs). We will end this chapter with a couple of examples that demonstrate how to implement deep learning models in practice.

Keywords Deep learning · Convolutional neural networks · Recurrent neural networks

1.1 Introduction

This chapter introduces the basics of deep learning that will be used in deep reinforcement learning. For those who are already familiar with the fundamentals, please feel free to skip this chapter. This book's content is meant to be self-contained, but one might wish to refer to other books like Bishop (2006) and Goodfellow et al. (2016) to understand some of the topics in depth. Unlike classical reinforcement learning which uses analytical methods for function approximation, deep reinforcement learning relies on deep neural networks such that it can leverage the power of large data volume and increased computing resources. In general, there are two types of models.

J. Zhang
Imperial College London, London, UK
e-mail: jingqing.zhang15@imperial.ac.uk

H. Yuan
Oxford University, Oxford, UK
e-mail: hang.yuan@keble.ox.ac.uk

H. Dong (✉)
Peking University, Beijing, China
e-mail: hao.dong@pku.edu.cn

H. Dong et al. (eds.), *Deep Reinforcement Learning*,
https://doi.org/10.1007/978-981-15-4095-0_1

Discriminative Models study the conditional probability $p(\boldsymbol{y}|\boldsymbol{x})$ with input data $\boldsymbol{x}$ and a target label $\boldsymbol{y}$. In other words, discriminative models predict the label $\boldsymbol{y}$ given the input data $\boldsymbol{x}$. Discriminative models are mostly adopted in tasks such as classification and regression which require discriminative judgement. More specifically, in terms of **classification**, a model is designed to categorize the input data into specific classes from a set of given classes. The binary classification, as the most fundamental classification task, predicts one class from two candidates. For example, in the sentiment analysis (Maas et al. 2011), a piece of text is classified as either positive or negative. In contrast, in multi-label classification, the input data can be assigned with several classes at the same time. In some cases, instead of identifying the class directly, a classification model needs to calculate the probability distribution of classes. For example, the input data has a probability of 80% to be assigned with class A and a probability of 20% to be assigned with class B. This probabilistic representation is mostly needed during training for optimization purposes. Deep learning has achieved great success on classification tasks such as image classification (Krizhevsky et al. 2009) and text classification (Yang et al. 2019). Unlike classifications, which produce discrete class labels, a **regression** studies continuous values. An example of regression is to predict future traffic speed based on historical traffic data (Liao et al. 2018a,b). Regression models remain discriminative models as long as they are learning the conditional probability.

Generative Models are designed to study the joint probability $p(\boldsymbol{x}, \boldsymbol{y})$. Generative models are usually used to generate observed data by learning the distribution of the observed data. For example, the generative adversarial networks (GANs) (Goodfellow et al. 2014) are adopted to generate, reconstruct, and denoise images (Ledig et al. 2017; Yang et al. 2018). Nonetheless, techniques in deep learning like GANs have no explicit relationship with the distribution of the observed data but focus more on the similarity between generated samples and observed data. Meanwhile, generative models are also used for classification purposes like Naive Bayes (Ng and Jordan 2002; Rish et al. 2001). Although both generative models and discriminative models are used for classification, discriminative models only consider which label should be assigned given the observed data, while generative models try to learn the distribution of the observed data. For example, Naive Bayes studies the likelihood $p(\boldsymbol{x}|\boldsymbol{y})$, i.e. the probability of the observed data to be generated assuming a label.

Most deep neural networks that have been explored are discriminative models no matter whether they are initially designed for discriminative or generative problems. This is because many generative problems in practice can be simplified to classification or regression problems. For example, question answering (Devlin et al. 2019) selects which part of the provided context is the answer to the given question; abstractive summarization (Zhang et al. 2019b) selects words from vocabulary to assemble summaries based on the probability of each word. For both cases, they are trying to generate something but one uses a classification approach and the other uses a regression approach.

Concretely, this chapter covers the mechanical components and techniques such as the definitions of neurons, activation functions, and optimizers that can build up deep neural networks and deep learning applications. Fundamental deep neural networks such as multilayer perceptron (MLP), convolutional neural networks (CNNs), and recurrent neural networks (RNNs) are also within the scope of this chapter. Finally, Sect. 1.10 introduces examples of implementing deep neural networks by TensorFlow and TensorLayer. Please refer to Goodfellow et al. (2016) for a more detailed introduction to deep learning.

1.2 Perceptron

1.2.1 One Output

A neuron (node) is the basic unit of deep neural networks. Originally, the neuron was proposed to be an abstract representation of the real neuron in the brain, which receives electrical impulses from its dendrites. When this specific neuron is polarized enough, it will send an action potential spike via its axon to the other adjacent neurons. In a real biological system, these steps do not take place at once but at a more granular scale. Spiking neural networks are better suited in describing the underlying biological processes. At the moment, the deep learning community relies more on deep neural networks (DNNs), also known as artificial neural networks (ANNs). The neurons in deep neural networks are formalized with numerical inputs and outputs. A neuron can have many output neurons in the next layer and a neuron can also have many input neurons in the previous layer. This is a many-to-many relationship. A neuron in one layer aggregates the signals being passed through from its input neurons in the previous layer. This aggregated signal will then be passed through an activation function that will determine the neuronal behavior. Concretely, if the aggregated signal is strong enough, then the activation function will "activate" this neuron and pass forward a high value to the output neurons in the next layer. Otherwise, a low value will be passed forward instead (Fig. 1.1).

$$z = w_1x_1 + w_2x_2 + w_3x_3. \tag{1.1}$$

A neural network can have an arbitrary number of neurons with random connections among themselves, but for the ease of computation, the neurons are organized layer after layer. Typically, a single neuron will have at least two layers, namely the input and output layer as shown in Fig. 1.2. This network can be formalized by Eq. (1.1) and can help with simple decision-making. An example is helping a group of students decide whether or not they can play soccer on a day based on the weather condition. The decision may also rely on some other factors such as the expense of the soccer field and the students' availability. If the weather

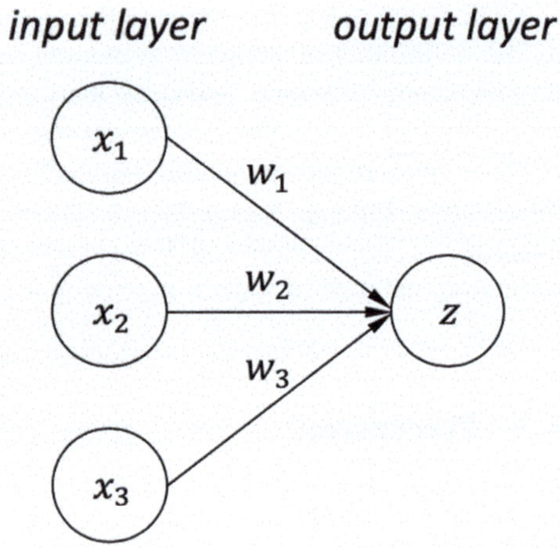

Fig. 1.1 A neural network with three input neurons and one output neuron

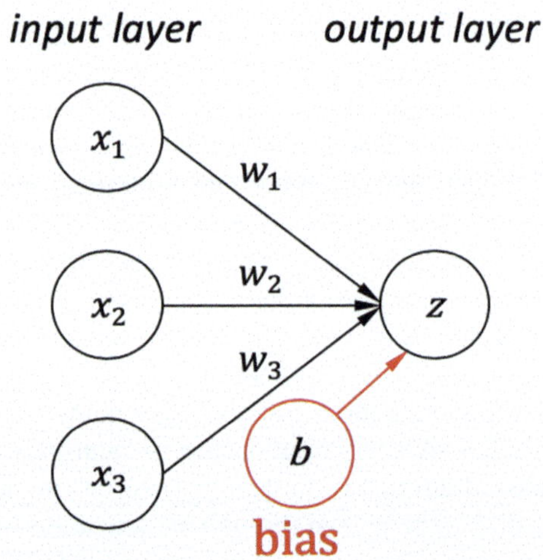

Fig. 1.2 A neural network with bias

condition has a higher impact on the decision, the corresponding weight (w) should have a greater absolute value. In contrast, factors of less importance should have weights with a lower absolute value. If a weight is set as zero, the corresponding input factor is discarded in the decision-making process. This kind of neural network is also called a single-layer neural network or **perceptron**.

1.2.2 Bias and Decision Boundary

A bias is an extra scalar that is added to the neuron to shift the value of the output. For example, Fig. 1.2 shows the single-layer neural network with a bias and it can be formalized as:

$$z = w_1 x_1 + w_2 x_2 + w_3 x_3 + b. \tag{1.2}$$

The bias can help a neural network to fit the data better. For example, let us define a binary classification problem, in which the label y is 1 if the input z is positive and 0 otherwise:

$$y = \begin{cases} 1 \text{ when } z > 0 \\ 0 \text{ otherwise} \end{cases} \tag{1.3}$$

Then the distribution of data samples is shown in Fig. 1.3 and we need to find out a set of weights and bias that can best fit the data. The decision boundary is defined to partition the data samples into the two classes for the binary classification. Formally, the decision boundary is $\{x_1, x_2, x_3 | w_1 x_1 + w_2 x_2 + w_3 x_3 + b = 0\}$.

Let us first simplify this problem by having only two inputs, i.e. $z = w_1 x_1 + w_2 x_2 + b$. As shown in the left-hand side of Fig. 1.3, without the bias component, i.e. $b = 0$, the decision boundary must cross the origin of the Cartesian coordinate as demonstrated by the blue line in the bottom-left corner. However, this apparently cannot fit the data distribution well enough as the data samples for both classes fall on the same side of the boundary. If the bias is non-zero, the decision boundary crosses both axes at $(0, -\frac{b}{w_2})$ and $(-\frac{b}{w_1}, 0)$, respectively, and this decision boundary can fit the data distribution better if the weights and bias are well chosen.

If we come back to the original setting of the problem where the neuron has three inputs, i.e. $z = w_1 x_1 + w_2 x_2 + w_3 x_3 + b$, the decision boundary will become a plane as shown in the right-hand side of Fig. 1.3. In a linear model like the single-layer neural networks defined in Eq. (1.2), the decision boundary is also called **hyperplane**.

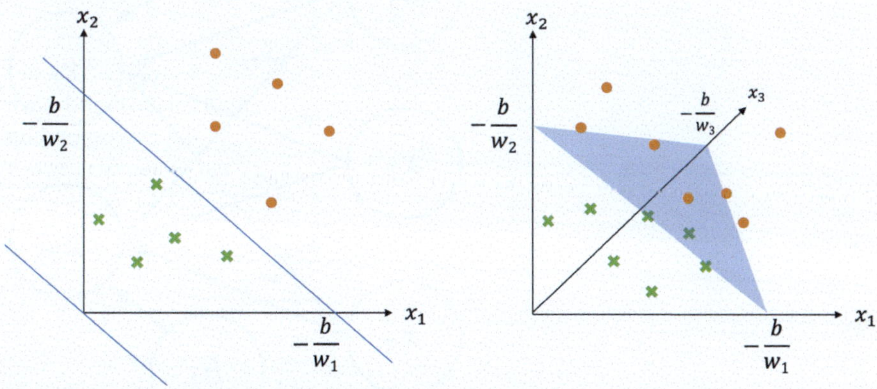

Fig. 1.3 Decision boundary of linear model with two and three inputs. Left: $z = w_1 x_1 + w_2 x_2 + b$, Right: $z = w_1 x_1 + w_2 x_2 + w_3 x_3 + b$

1.2.3 More Than One Output

The single-layer neural network can have multiple neurons. Figure 1.4 shows an example of a single-layer neural network with two outputs, which are computed by Eq. (1.4). Since each output is connected with all of the inputs, the output layer is also called the **dense layer**, or **fully connected (FC) layer**:

$$\begin{aligned} z_1 &= w_{11}x_1 + w_{12}x_2 + w_{13}x_3 + b_1 \\ z_2 &= w_{21}x_1 + w_{22}x_2 + w_{23}x_3 + b_2. \end{aligned} \tag{1.4}$$

In practice, the dense layer can be represented by matrix multiplication:

$$\boldsymbol{z} = \boldsymbol{W}\boldsymbol{x} + \boldsymbol{b} \tag{1.5}$$

where $\boldsymbol{W} \in \mathbb{R}^{m \times n}$ is a matrix to represent weights and $\boldsymbol{z} \in \mathbb{R}^m$, $\boldsymbol{x} \in \mathbb{R}^n$, $\boldsymbol{b} \in \mathbb{R}^m$ are column vectors to represent outputs, inputs, and biases, respectively. In the example by Eq. (1.4), $m = 2$ and $n = 3$.

$$\begin{bmatrix} z_1 \\ z_2 \end{bmatrix} = \begin{bmatrix} w_{11} & w_{12} & w_{13} \\ w_{21} & w_{22} & w_{23} \end{bmatrix} \begin{bmatrix} x_1 \\ x_2 \\ x_3 \end{bmatrix} + \begin{bmatrix} b_1 \\ b_2 \end{bmatrix} \tag{1.6}$$

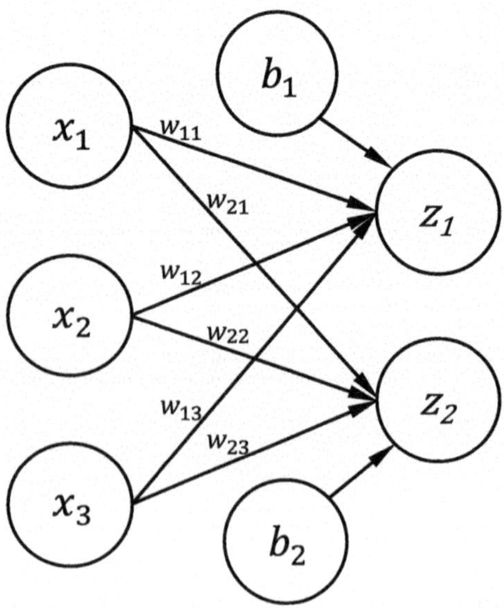

Fig. 1.4 The neural network with three input neurons and two output neurons

1.3 Multilayer Perceptron (MLP)

A multilayer perceptron (MLP) (Rosenblatt 1958; Ruck et al. 1990) stems from a single dense layer to have at least two dense layers. Figure 1.5 presents an MLP consisting of four dense layers. The three layers between the input and output layers are "hidden" because they are typically not accessible from outside the network, and we will refer them as the **hidden layers**. Compared with the network with a single dense layer, MLP can fit more complex data. In other words, MLP has a stronger learning capability than a single-layer neural network. However, more hidden layers in MLP do not necessarily lead to stronger learning capacity. The universal approximation theorem states that a feedforward network with one hidden layer (e.g., MLP with one hidden layer) and any squashing activation function (e.g., sigmoid or tanh) can approximate any Borel measurable function, given that the hidden layer has sufficient hidden units (Samuel 1959; Hornik et al. 1989; Goodfellow et al. 2016). However, in practice, such a network can be inflexible to train or hard to avoid overfitting if the hidden layer is extremely large. Therefore, deep neural networks including MLP typically have several hidden layers.

We start with the logic operations to demonstrate how a network approximates a function. The logic operations including AND, OR, NOR, NAND, XNOR, and XOR take two binary numbers and return either zero or one. For example, AND returns one if and only if the two binary numbers are both one. Simple logic operations can be easily approximated by the perceptron, which can be defined by Eq. (1.7).

$$f(\boldsymbol{x}) = \begin{cases} 1 & \text{if } z > 0 \\ 0 & \text{otherwise} \end{cases} \quad \text{where} \quad z = w_1 x_1 + w_2 x_2 + b \tag{1.7}$$

Figure 1.6 shows that hyperplanes defined by perceptron can be easily found to separate the points between zero and one for AND, OR, NOR, and NAND. However, it is not possible to do the same for XOR or XNOR.

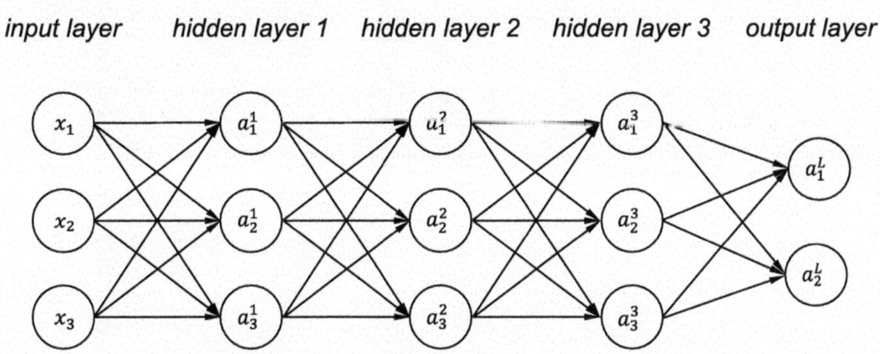

Fig. 1.5 An example of multilayer perceptron (MLP) with three hidden layers and one output layer. The neurons are represented by a_i^l, where l the layer index and i is the output index

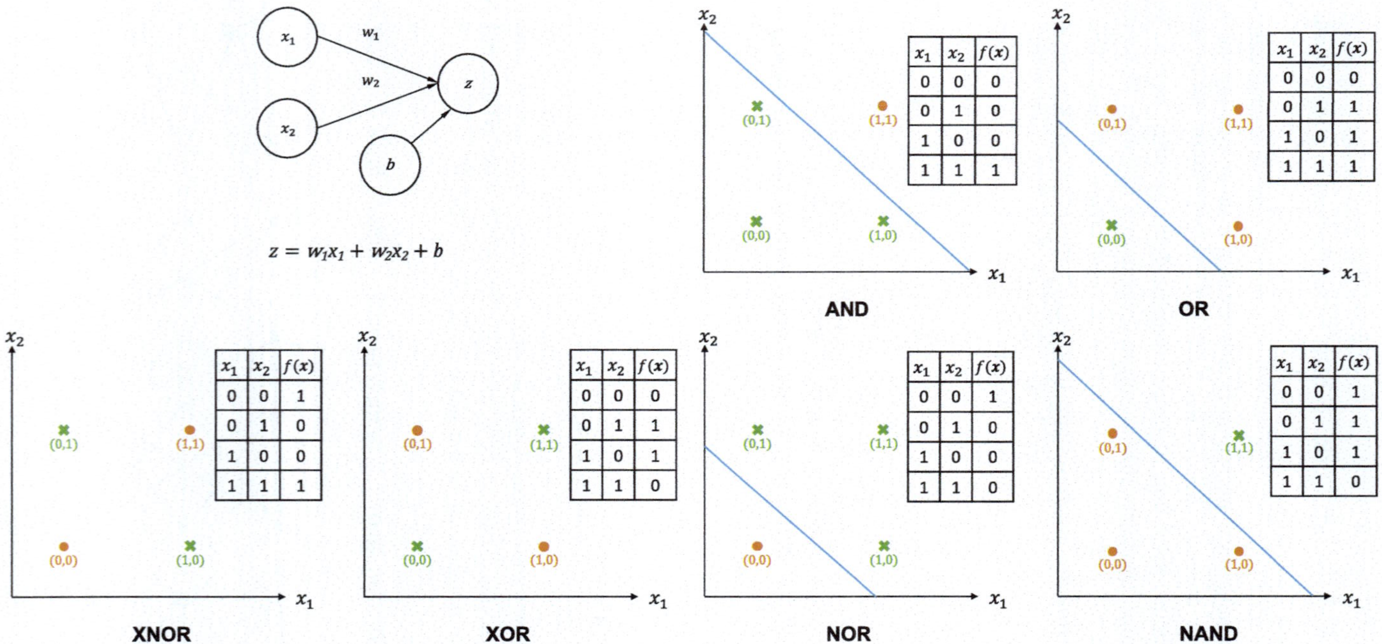

Fig. 1.6 Top left: The perceptron with two inputs and one output. The rest: Hyperplanes can be found to separate the points between zero (green) and one (orange) for AND, OR, NOR, NAND, but no hyperplane defined by perceptron can be found for XOR, XNOR

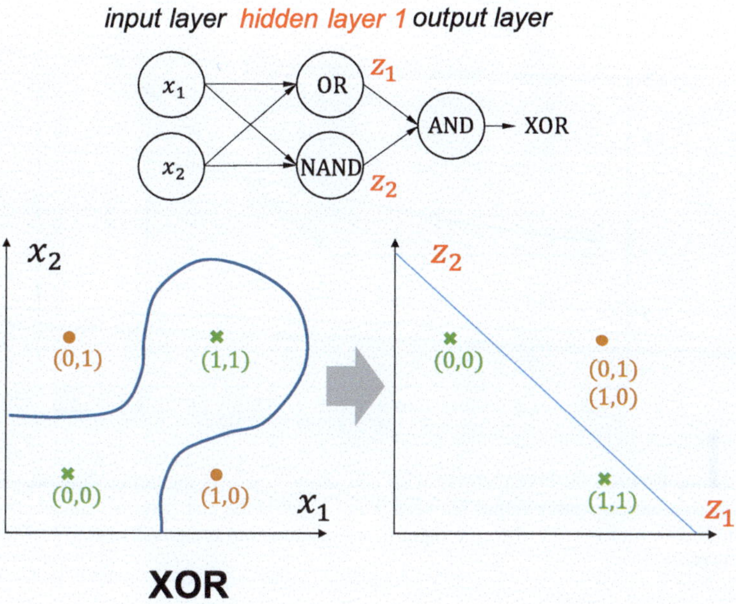

Fig. 1.7 Left: An MLP that approximates XOR. Mid and right: Transformation from the original data space to the feature space, where the data points are linearly separable

The XOR cannot be approximated by a linear model directly working on the original inputs x_1, x_2 like the perceptron, so we need to transform the inputs first. As an example, we use MLP with one hidden layer as shown in Fig. 1.7 to approximate XOR. This MLP first transforms the inputs x_1, x_2 into a new space by approximating the logic operations OR and NAND, and then, in the transformed space, the points are linearly separable by an approximation of AND. The transformed space is also named feature space and this example shows how learning features can improve the learning capacity of a model.

1.4 Activation Functions

Matrix addition and multiplication are both linear operators but the learning capability of a linear model is rather limited. For example, a linear model cannot easily approximate a cosine function. Most real-world problems that deep neural networks are applied to solve cannot be simplified as a linear transformation, so non-linearity is important for deep neural networks. In practice, the non-linearity of deep neural networks is introduced by activation functions, which are typically element-wise operations. In addition, the activation functions are necessary when a model needs to obtain probability vectors instead of vectors with arbitrary values. The choice of activation functions varies in different applications. Even though there

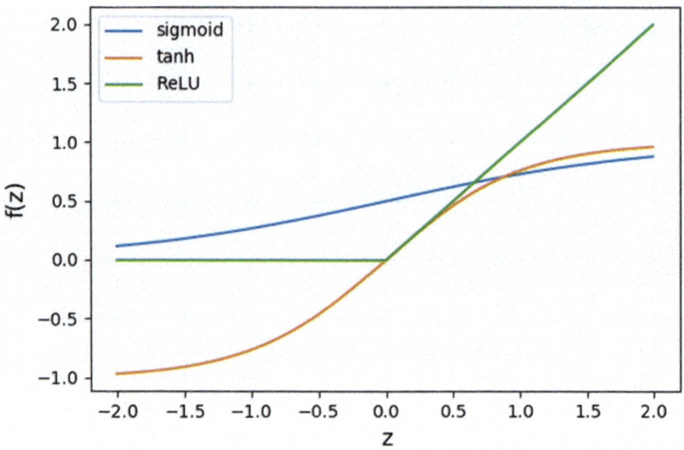

Fig. 1.8 Demonstration of three element-wise activation functions including sigmoid, tanh, and ReLU. The sigmoid constrains values between 0 and 1, while the tanh returns values between -1 and 1. The ReLU returns zero when the input is non-positive but is equivalent to $f(x) = x$ when the input is positive

exist some functions that work well in most deep learning applications, there might be other functions that have better performance on a case by case basis. Therefore, the design of activation functions remains an active research area. This section introduces four commonly used activation functions, namely sigmoid, tanh, ReLU, and softmax (Fig. 1.8).

The logistic **sigmoid** as an activation function has float output ranging between 0 and 1 as defined by Eq. (1.8). The sigmoid function can be used at the output layer for classification purpose. For example, a binary classifier with one output neuron uses sigmoid to constrain the output value between 0 and 1 and then converts it to a discrete class label (either 0 or 1) by using a threshold like 0.5.

$$f(z) = \frac{1}{1 + e^{-z}}. \tag{1.8}$$

Similar to the sigmoid function, the **hyperbolic tangent (tanh)** constrains output values to a limited range between -1 and 1 as defined by Eq. (1.9). The tanh function can be used in the hidden layers (Glorot et al. 2011) to provide non-linearity. It can also be used in the output layer, e.g. in the generation of images whose pixel values range between -1 and 1.

$$f(z) = \frac{e^{z} - e^{-z}}{e^{z} + e^{-z}}. \tag{1.9}$$

The **rectified linear unit (ReLU)**, also known as the rectifier, is defined by Eq. (1.10). The study by Glorot et al. (2011) shows that ReLU is more promising than sigmoid and tanh, and ReLU has also been widely adopted in recent works (He et al. 2016; Cao et al. 2017; Noh et al. 2015). The empirical advantages of ReLU are:

- Easier to implement and compute: in the implementation of ReLU, a simple comparison with zero is conducted first and then the activation is set to zero or z accordingly. Whereas in the sigmoid and tanh, the exponential function is harder to compute especially in the case of large networks.
- Easier for a network to optimize: ReLU function is close to being linear, consisting of two linear functions. This property makes the gradient large and consistent. The gradient of an active neuron by ReLU is always one, but the gradient of a neuron by sigmoid or tanh suffers from vanishing when the activated value approaches the limits (i.e., -1, 0, or 1).

$$f(z) = \begin{cases} 0 & \text{when } z <= 0 \\ z & \text{when } z > 0 \end{cases} \tag{1.10}$$

However, merely setting negative values to zero in ReLU can lead to information loss. Imagine, if a neuron constantly outputs zero, it will always output zero in the future and is unlikely to recover. This can happen because of an inappropriate learning rate or a negative bias. The work by Xu et al. (2015) proposes a solution to this with another activation function called leaky ReLU, which is defined in Eq. (1.11). The scalar α in this equation is a small positive value to control the slope (e.g., 0.01 or 0.02) so that a little information from the negative scope can be retained.

$$f(z) = \begin{cases} \alpha z & \text{when } z <= 0 \\ z & \text{when } z > 0 \end{cases} \tag{1.11}$$

The parametric ReLU (PReLU) (He et al. 2015) is similar to the leaky ReLU except that it considers α as a trainable parameter. There is no clear evidence to show which one of ReLU, leaky ReLU or PReLU is significantly better than the others since the choice varies in different scenarios.

Unlike the activation functions mentioned above, the **softmax** function, defined by Eq. (1.12), provides normalization based on all values from previous layer's outputs. The softmax function first computes the exponential function e^z and then normalizes each entry by dividing it.

$$f(z)_i = \frac{e^{z_i}}{\sum_{k=1}^{K} e^{z_k}} \tag{1.12}$$

In practice, the softmax function is typically only used in the output layer to normalize the output vector z into a probability vector, where each entry is non-negative and the entries are added to one. Therefore, the softmax function is widely used for classification.

1.5 Loss Functions

In deep learning, loss functions are defined to quantify an error, also known as the loss value or cost, between the prediction and target (i.e., ground truth, gold standard). The loss value is normally used as the objective to optimize the parameters of neural networks, such as the weights and biases. This section introduces some commonly used loss functions and Sect. 1.6 will introduce how to optimize the parameters based on the loss values.

1.5.1 Cross-Entropy Loss

Before we introduce the cross-entropy loss, we start with a similar concept named Kullback–Leibler (KL) divergence. The KL divergence measures the similarity between two distribution $P(x)$ and $Q(x)$:

$$D_{\mathrm{KL}}(P\|Q) = \mathbb{E}_{x\sim P}\left[\log \frac{P(x)}{Q(x)}\right] = \mathbb{E}_{x\sim P}[\log P(x) - \log Q(x)] \tag{1.13}$$

The KL divergence is non-negative and equals to 0 if and only if P and Q have the same distribution. Since the first term in KL divergence has no relation with Q, we introduce cross-entropy which can remove the first term.

$$H(P, Q) = -\mathbb{E}_{x\sim P} \log Q(x) \tag{1.14}$$

Therefore, minimizing the cross-entropy with respect to Q is equivalent to minimizing the KL divergence. As mentioned before, in some deep learning applications, e.g. classification, deep neural networks calculate a probability distribution of classes in practice instead of identifying the target class directly. Therefore, we can use the cross-entropy to measure how well the predicted distribution is and then update the network accordingly.

We start with binary classification as an example. In binary classification, for each input data sample x_i with target y_i (i.e., 0 or 1), a model needs to predict the probability of each candidate class $\hat{y}_{i,1}$, $\hat{y}_{i,2}$. Since $\hat{y}_{i,1} + \hat{y}_{i,2} = 1$, we can rewrite the prediction as $\hat{y}_i$ which represents the probability of one class, so the probability of the other class is $1 - \hat{y}_i$. Therefore, a neural network for binary classification typically has only one output neuron (with sigmoid) and following the definition of cross-entropy, we have:

$$\mathcal{L} = -\frac{1}{N}\sum_{i=1}^{N}\Big(y_i \log \hat{y}_i + (1 - y_i)\log(1 - \hat{y}_i)\Big), \tag{1.15}$$

where N represents the total number of data samples. Since y_i is either 0 or 1, only one of $y_i \log \hat{y}_i$ and $(1 - y_i)\log(1 - \hat{y}_i)$ is retained for each data sample. If $\forall i,\ y_i = \hat{y}_i$, the cross-entropy loss is zero.

In multinomial classification, where each data sample x_i is classified into one out of three or more candidate classes, a model predicts the probability of each class $\{\hat{y}_{i,1}, \hat{y}_{i,2}, \ldots, \hat{y}_{i,M}\}$, where $M \geq 3$ and $\sum_{j=1}^{M} \hat{y}_{i,j} = 1$. The target of each data sample is referred to as c_i, which is an integer between $[1, M]$, and it can be converted to a one-hot vector $\boldsymbol{y}_i = [y_{i,1}, y_{i,2}, \ldots, y_{i,M}]$, where only $y_{i,c_i} = 1$ and others are zero. Then, we can write the cross-entropy loss for the multinomial classification as below:

$$\begin{aligned}\mathcal{L} &= -\frac{1}{N}\sum_{i=1}^{N}\sum_{j=1}^{M} y_{i,j} \log \hat{y}_j = -\frac{1}{N}\sum_{i=1}^{N}(0 + \cdots + y_{i,c_i} \log \hat{y}_{c_i} + \cdots + 0) \\ &= -\frac{1}{N}\sum_{i=1}^{N} \log \hat{y}_{c_i}.\end{aligned} \tag{1.16}$$

1.5.2 $\mathcal{L}_p$ Norm

Given a vector $\boldsymbol{x}$, p-norm measures its scale such that a vector with larger values has a larger scale, and it is defined as follows, where p is an integer greater or equal to 1.

$$\begin{aligned}\|\boldsymbol{x}\|_p &= \left(\sum_{i=1}^{N} |x_i|^p\right)^{1/p} \\ \text{i.e.,} \|\boldsymbol{x}\|_p^p &= \sum_{i=1}^{N} |x_i|^p\end{aligned} \tag{1.17}$$

In deep learning, a p-norm can be used to measure the difference between two vectors written as $\mathcal{L}_p$, as in Eq. (1.18), where $\boldsymbol{y}$ and $\hat{\boldsymbol{y}}$ are the target and prediction, respectively.

$$\mathcal{L}_p = \|\boldsymbol{y} - \hat{\boldsymbol{y}}\|_p^p = \sum_{i=1}^{N} |y_i - \hat{y}_i|^p. \tag{1.18}$$

1.5.3 Mean Squared Error

The mean squared error (MSE) is the averaged $\mathcal{L}_2$ norm as defined by Eq. (1.19). The MSE can be used for regression problems in which the outputs of a neural network are continuous values. For example, the difference between two images can be measured by MSE between pixels of the two images.

$$\mathcal{L} = \frac{1}{N}\|\boldsymbol{y} - \hat{\boldsymbol{y}}\|_2^2 = \frac{1}{N}\sum_{i=1}^{N}(y_i - \hat{y}_i)^2, \tag{1.19}$$

where N is the number of data samples, and $\boldsymbol{y}$ and $\hat{\boldsymbol{y}}$ are the target and prediction, respectively.

1.5.4 Mean Absolute Error

Similar to MSE, the mean absolute error (MAE) can also be used for regression problems and is defined as the averaged $\mathcal{L}_1$ norm.

$$\mathcal{L} = \frac{1}{N}\sum_{i=1}^{N}|y_i - \hat{y}_i| \tag{1.20}$$

Both MSE and MAE minimize the difference between $\boldsymbol{y}$ and $\hat{\boldsymbol{y}}$. MSE offers a better mathematical property making it easier to compute the partial derivative which is required by gradient descent. In contrast, since the absolute term is not differentiable when $y_i = \hat{y}_i$, the partial derivative of MAE needs to walk around this case. In addition, when the difference between y_i and $\hat{y}_i$ is greater than 1, MSE has a larger error value compared to MAE (i.e., $(y_i - \hat{y}_i)^2$ vs $|y_i - \hat{y}_i|$) which can lead to a quicker optimization of a network.

1.6 Optimization

In this section we describe the optimization of deep neural networks, or in other words, how the parameters of deep neural networks are trained. This section covers back-propagation, gradient descent, stochastic gradient descent, and the selection of hyper-parameters.

1.6.1 Gradient Descent and Error Back-Propagation

Given a neural network and a loss function, the training of the neural network is formalized to learning its parameters $\boldsymbol{\theta}$ so that the loss $\mathcal{L}$ is minimized. Finding the minimum by searching $\boldsymbol{\theta}\ \ s.t.\ \ \nabla_{\boldsymbol{\theta}} \mathcal{L} = 0$ in a brute-force fashion is infeasible in practice, especially when the formula is as complex as that of a deep neural network. Therefore, we consider a process to approach the minimum by small steps and this technique is called **gradient descent**.

Figure 1.9 illustrates two examples of gradient descent. The learning process of gradient descent starts from a randomly picked point and the loss $\mathcal{L}$ decreases along with the update of parameters as denoted by the red dotted path. Similarly, in a neural network, its parameters are first randomly initialized and updated each step based on the partial derivative $\frac{\partial \mathcal{L}}{\partial \boldsymbol{\theta}}$. More specifically, the parameters are updated iteratively by $\boldsymbol{\theta} := \boldsymbol{\theta} - \alpha \frac{\partial \mathcal{L}}{\partial \boldsymbol{\theta}}$, where α the learning rate of each step and $\boldsymbol{\theta}$ mostly consists of weights $\boldsymbol{W}$ and biases $\boldsymbol{b}$ of each layer.

Back-Propagation (Rumelhart et al. 1986; LeCun et al. 2015) is a technique to compute the partial derivative $\frac{\partial \mathcal{L}}{\partial \boldsymbol{\theta}}$ in the network. To make the computation of $\frac{\partial \mathcal{L}}{\partial \boldsymbol{\theta}}$ clearer, we introduce an intermediate value $\boldsymbol{\delta} = \frac{\partial \mathcal{L}}{\partial z}$, which is the partial derivative of the loss $\mathcal{L}$ with respect to the layer's output z. Then, the partial derivatives of the loss $\mathcal{L}$ with respect to each parameter, which assemble $\frac{\partial \mathcal{L}}{\partial \boldsymbol{\theta}}$, are computed based on the intermediate value $\boldsymbol{\delta}$.

The layers are indexed as $l = 1, 2, \ldots L$, where L is the index of the output layer, each layer has an output z^l, an intermediate value $\boldsymbol{\delta}^l = \frac{\partial \mathcal{L}}{\partial z^l}$, and an activation output $\boldsymbol{a}^l = f(z^l)$ (where f is the activation function). We use an MLP with MSE loss and a sigmoid activation function as an example. Given $z^l = \boldsymbol{W}^l \boldsymbol{a}^{l-1} + \boldsymbol{b}^l$, $\boldsymbol{a}^l = f(z^l) = \frac{1}{1+e^{-z^l}}$, and $\mathcal{L} = \frac{1}{2} \| \boldsymbol{y} - \boldsymbol{a}^L \|_2^2$, we represent the partial derivative of

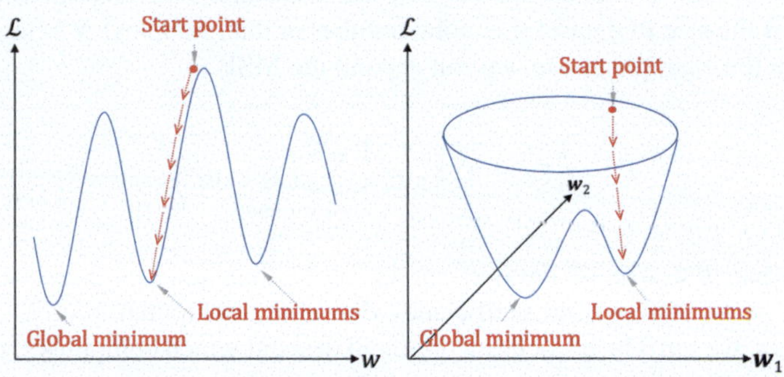

Fig. 1.9 Examples of gradient descent with a trainable parameter (Left) and two trainable parameters (Right). In gradient descent, the learning process starts from a ranomly picked point. With the parameters updates shown by the red arrows, the loss $\mathcal{L}$ gradually reachs a saddle point. Note that there is no guarantee the gradient descent can find the global minimum but in most cases a local minimum is approached

the activation output with respect to its original output as $\frac{\partial \boldsymbol{a}^l}{\partial z^l} = f'(z^l) = f(z^l)(1 - f(z^l)) = \boldsymbol{a}^l(1-\boldsymbol{a}^l)$ and the partial derivative of the loss with respect to the activation output as $\frac{\partial \mathcal{L}}{\partial \boldsymbol{a}^L} = (\boldsymbol{a}^L - \boldsymbol{y})$. To compute the partial derivative of the loss with respect to the output layer, we apply the chain rule as follows:

- $\boldsymbol{\delta}^L = \frac{\partial \mathcal{L}}{\partial z^L} = \frac{\partial \mathcal{L}}{\partial \boldsymbol{a}^L} \frac{\partial \boldsymbol{a}^L}{\partial z^L} = \left(\boldsymbol{a}^L - \boldsymbol{y}\right) \odot \left(\boldsymbol{a}^L \left(1 - \boldsymbol{a}^L\right)\right)$

Then, the partial derivative of the loss with respect to all the other layers' outputs can be computed recursively as follows, where $l = 1, 2, \ldots, L - 1$.

- Given $z^{l+1} = \boldsymbol{W}^{l+1}\boldsymbol{a}^l + \boldsymbol{b}^{l+1}$
- Then $\boldsymbol{\delta}^l = \frac{\partial \mathcal{L}}{\partial z^l} = \frac{\partial \mathcal{L}}{\partial z^{l+1}} \frac{\partial z^{l+1}}{\partial \boldsymbol{a}^l} \frac{\partial \boldsymbol{a}^l}{\partial z^l} = \left(\boldsymbol{W}^{l+1}\right)^T \boldsymbol{\delta}^{l+1} \odot \left(\boldsymbol{a}^l \left(1 - \boldsymbol{a}^l\right)\right)$

The second step of the back-propagation is to compute the partial derivative of the loss with respect to the parameters $\frac{\partial \mathcal{L}}{\partial \boldsymbol{W}^l}$ and $\frac{\partial \mathcal{L}}{\partial \boldsymbol{b}^l}$ of each layer based on the intermediate value $\boldsymbol{\delta}^l$.

- Given $z^l = \boldsymbol{W}^l\boldsymbol{a}^{l-1} + \boldsymbol{b}^l$, we have $\frac{\partial z^l}{\partial \boldsymbol{W}^l} = \boldsymbol{a}^{l-1}$ and $\frac{\partial z^l}{\partial \boldsymbol{b}^l} = 1$
- Then $\frac{\partial \mathcal{L}}{\partial \boldsymbol{W}^l} = \frac{\partial \mathcal{L}}{\partial z^l} \frac{\partial z^l}{\partial \boldsymbol{W}^l} = \boldsymbol{\delta}^l \left(\boldsymbol{a}^{l-1}\right)^T$, $\frac{\partial \mathcal{L}}{\partial \boldsymbol{b}^l} = \frac{\partial \mathcal{L}}{\partial z^l} \frac{\partial z^l}{\partial \boldsymbol{b}^l} = \boldsymbol{\delta}^l$

Finally, we use the $\frac{\partial \mathcal{L}}{\partial \boldsymbol{W}^l}$ and $\frac{\partial \mathcal{L}}{\partial \boldsymbol{b}^l}$ to update the parameters $\boldsymbol{W}^l$ and $\boldsymbol{b}^l$ as follows:

- $\boldsymbol{W}^l := \boldsymbol{W}^l - \alpha \frac{\partial \mathcal{L}}{\partial \boldsymbol{W}^l}$
- $\boldsymbol{b}^l := \boldsymbol{b}^l - \alpha \frac{\partial \mathcal{L}}{\partial \boldsymbol{b}^l}$

With the partial derivative $\frac{\partial \mathcal{L}}{\partial \boldsymbol{\theta}}$, gradient descent updates the parameter iteratively and converges to a minimum point of the loss function as in Fig. 1.9. In practice, the converged point is typically a local minimum rather than the global one. However, as deep neural networks offer a good representation capacity, the local minimums tend to be close to the global minimum (Goodfellow et al. 2016).

In gradient descent, the computation of the loss value $\mathcal{L}$ in each iteration can be expensive if the size of dataset (i.e., total number of data samples) N is large. Given the MSE in the example above, we can expand the MSE to:

$$\mathcal{L} = \frac{1}{2}\|\boldsymbol{y} - \boldsymbol{a}^L\|_2^2 = \frac{1}{2}\sum_{i=1}^{N}\left(y_i - a_i^L\right)^2 \tag{1.21}$$

In practice, the size of dataset can be more than tens of thousands so the gradient descent suffers from inefficiency due to the computation of $\mathcal{L}$. To tackle this problem, we introduce stochastic gradient descent which computes $\mathcal{L}$ of a small batch of data samples.

1.6.2 *Stochastic Gradient Descent and Adaptive Learning Rate*

Instead of computing the loss $\mathcal{L}$ of all training data in each iteration, the stochastic gradient descent (SGD) (Bottou and Bousquet 2007) randomly selects a small number of data samples from the training set. These small number of data samples are named as a **mini-batch**, and the quantity of data samples in the mini-batch is referred to as **batch size**. We can rewrite the Eq. (1.21) with batch size B and $B \ll N$ so that the computation of $\mathcal{L}$ is much more efficient.

$$\mathcal{L} = \frac{1}{2}\|\boldsymbol{y} - \boldsymbol{a}^L\|_2^2 = \frac{1}{2}\sum_{i=1}^{B}\left(y_i - a_i^L\right)^2 \tag{1.22}$$

The training process of stochastic gradient descent is outlined in Algorithm 1. If the parameters are updated with sufficient times (i.e., sufficient training steps/iterations), the mini-batches can cover the entire training set.

Algorithm 1 The training process of stochastic gradient descent (SGD)

Input: Parameters $\boldsymbol{\theta}$, learning rate α, number of training steps/iterations S
1: **for** $i = 0$ **to** S **do**
2: Compute $\mathcal{L}$ of a mini-batch;
3: Compute $\frac{\partial \mathcal{L}}{\partial \theta}$ by back-propagation;
4: $\nabla\boldsymbol{\theta} \leftarrow -\alpha * \frac{\partial \mathcal{L}}{\partial \boldsymbol{\theta}}$;
5: $\boldsymbol{\theta} \leftarrow \boldsymbol{\theta} + \nabla\boldsymbol{\theta}$; update the parameters
6: **end for**
7: **return** $\boldsymbol{\theta}$; return the trained parameters;

The learning rate controls the step size of each update in SGD. If the learning rate is too large, the SGD may fail to find the minimum as shown in Fig. 1.10. If the learning rate is too small, the SGD can be slow to converge (Fig 1.10) or become stuck in a local minimum which has high error (Fig 1.9). Therefore, it is difficult to determine a proper fixed learning rate. Recent studies proposed adaptive learning rates, such as Adam (Kingma and Ba 2014), RMSProp (Tieleman and Hinton 2017), and Adagrad (Duchi et al. 2011), which speed up the training process by automatically adapting the learning rate. Adam is one of the most frequently used algorithm. Instead of using the gradients to update the parameters directly, Adam computes the running average of the gradients and the second moment of the gradients to update the parameters as shown in Algorithm 2. The β_1 and β_2 terms are the forgetting factors, also known as momentum, for the gradients and the second moment of the gradients, respectively. By default, β_1 is 0.9 and β_2 is 0.999 (Kingma and Ba 2014).

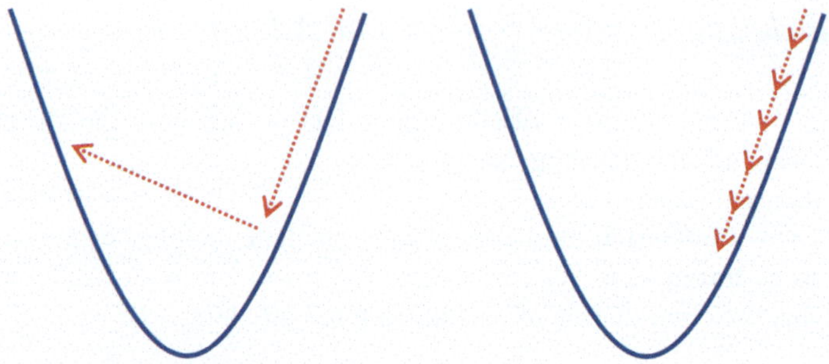

Fig. 1.10 A large learning rate may accelerate the training process but can also make it hard to train a model with ideal parameters. As shown in the left figure, which has a larger learning rate than the right figure, the loss value may increase after parameters update and it can be hard to approach the minimum. In contrast, in the right figure, which has a lower learning rate, the loss value decreases consistently but in a slower manner

Algorithm 2 The training process of Adam optimization

Input: parameters $\boldsymbol{\theta}$, learning rate α, number of training steps/iterations S, $\beta_1 = 0.9$, $\beta_2 = 0.999$, $\epsilon = 10^{-8}$

$\boldsymbol{m}_0 \leftarrow 0$; initialize the first moment vector
$\boldsymbol{v}_0 \leftarrow 0$; initialize the second moment vector
for $t = 1$ **to** S **do**
  $\frac{\partial \mathcal{L}}{\partial \boldsymbol{\theta}}$; compute the gradient using a random mini-batch
  $\boldsymbol{m}_t \leftarrow \beta_1 * \boldsymbol{m}_{t-1} + (1 - \beta_1) * \frac{\partial \mathcal{L}}{\partial \boldsymbol{\theta}}$; update the first moment
  $\boldsymbol{v}_t \leftarrow \beta_2 * \boldsymbol{v}_{t-1} + (1 - \beta_2) * (\frac{\partial \mathcal{L}}{\partial \boldsymbol{\theta}})^2$; update the second moment
  $\hat{\boldsymbol{m}}_t \leftarrow \frac{\boldsymbol{m}_t}{1-\beta_1^t}$; compute the running average of the first moment
  $\hat{\boldsymbol{v}}_t \leftarrow \frac{\boldsymbol{v}_t}{1-\beta_2^t}$; compute the running average of the second moment
  $\nabla \boldsymbol{\theta} \leftarrow -\alpha * \frac{\hat{\boldsymbol{m}}_t}{\sqrt{\hat{\boldsymbol{v}}_t + \epsilon}}$;
  $\boldsymbol{\theta} \leftarrow \boldsymbol{\theta} + \nabla \boldsymbol{\theta}$; update parameters
end for
return $\boldsymbol{\theta}$; return the trained parameters

1.6.3 Hyper-Parameter Selection

In deep learning, hyper-parameters refer to the settings of a model, such as the number of layers, and the settings of the training process, such as the number of steps, batch size, and learning rate. These settings can significantly affect the performance of a model, so selecting these hyper-parameters appropriately is essential to obtain an ideal model.

To evaluate the performance of different hyper-parameters, the data is usually split into training, validation, and testing sets. Then, multiple hyper-parameter settings are applied to the training set and evaluated on the validation set. Finally, the model with the best hyper-parameters that performs best on the validation set is selected for a final evaluation on the testing set.

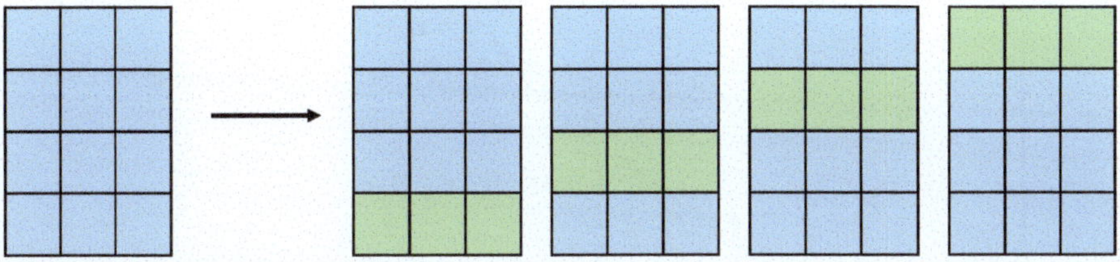

Fig. 1.11 An example of four-fold cross-validation. The dataset is split into four subsets (each row is a subset for demonstration purpose). In each trial, the blue subsets are the training set and the green subset is the testing set. The final evaluation result is the average of the four trials

Cross-Validation

For a small dataset, splitting the data into training and testing sets may be problematic. If the size of the training set is too small, the performance of a model can be harmed since there is no sufficient training data. On the other hand, if the testing set is too small, a model cannot be adequately evaluated. To tackle this problem, cross-validation is introduced.

In a k-fold cross-validation, a dataset is split into k non-overlapping subsets and each subset has the same size. The training/testing process is repeated for k times and, in each time, one of the subsets is selected for testing and the remainders for training. The final evaluation result is then averaged by the result across the k trials. Figure 1.11 illustrates an example of four-fold cross-validation.

1.7 Regularization

Regularization refers to a collection of methods which are designed to make sure a model not only works well on the training set but also on the testing data and new dataset. This section introduces the concept of overfitting and some regularization methods including weight decay, dropout, and batch normalization.

1.7.1 Overfitting

A machine learning model is optimized to minimize the training error (i.e., loss) but this cannot guarantee that the model can also perform well on the testing data. If the model is optimized "overly," the model may even have a significantly large testing error. This case is called overfitting. For example, in Fig. 1.12, the polynomial model represented by the dashed curve suffers from overfitting. This model fits the training data accurately but it fails to fit the testing data. Such a model with overfitting can be unreliable in real-world applications where there is always new data. In contrast,

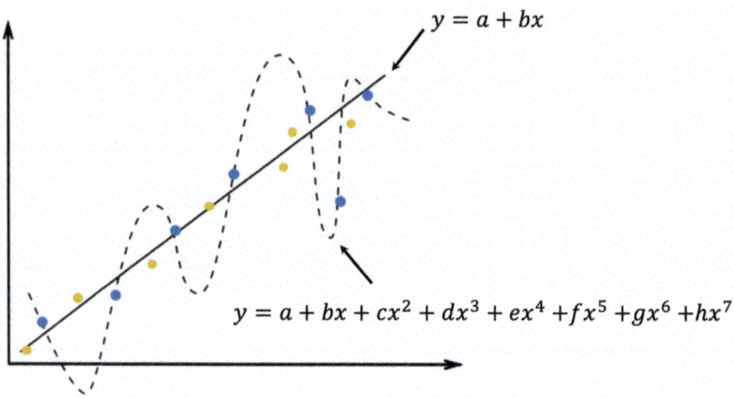

Fig. 1.12 A demonstration of overfitting. The blue dots represent training data, and the orange dots are testing data. Though the linear model represented by the solid line has a larger training error, it has much smaller testing error than the polynomial model represented by the dashed curve. Thus, we can say the polynomial model suffers from overfitting

the linear model represented by the solid straight line has fewer parameters while offering a better fit for the testing data.

Underfitting is opposite to overfitting, where the model cannot fit the training data, resulting in large error for both training and testing data. However, in practice, underfitting can be solved by using a larger model (more layers, more parameters, etc.), but solving overfitting is more challenging. The simplest way to alleviate overfitting is to use more training data which is not always possible since data acquisition and data labeling can be expensive.

1.7.2 Weight Decay

Weight decay is a simple but yet effective regularization method targeting the overfitting problem. It introduces a regularization term as a penalty to encourage $\boldsymbol{\theta}$ with smaller absolute values. For example, as Fig. 1.12 shows, if the parameters from c to h of the polynomial model have smaller absolute values, the model will have a lower swing range so that it can better fit the data. The loss function with the parameter norm penalty is defined as follows:

$$\mathcal{L}_{\text{total}} = \mathcal{L}(\boldsymbol{y}, \hat{\boldsymbol{y}}) + \lambda\Omega(\theta), \tag{1.23}$$

where $\mathcal{L}(\boldsymbol{y}, \hat{\boldsymbol{y}})$ is the original loss function computed from the target $\boldsymbol{y}$ and prediction $\hat{\boldsymbol{y}}$, Ω is the parameter norm penalty function and λ is a small value that controls the strength of the regularization. Two of the most commonly used parameter norm penalty functions are $\mathcal{L}_1 = \|\boldsymbol{W}\|$ and $\mathcal{L}_2 = \|\boldsymbol{W}\|_2^2$. The parameters

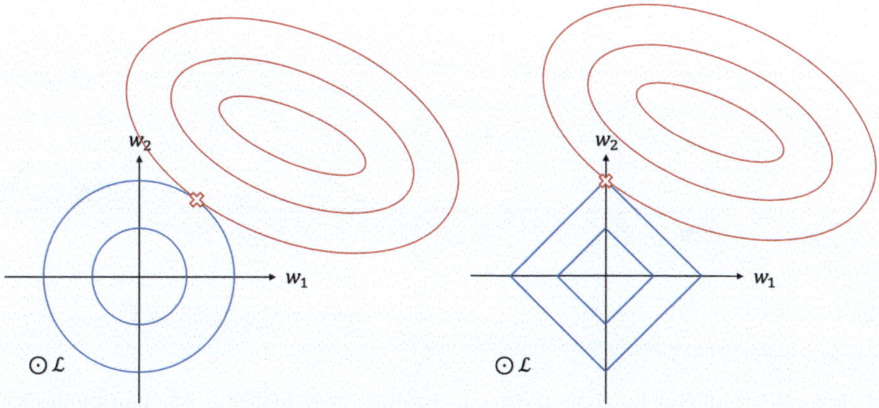

Fig. 1.13 Left: A demonstration of contour lines of the original loss (red) and $\mathcal{L}_2$ (blue). Right: A demonstration of contour lines of the original loss (red) and $\mathcal{L}_1$ (blue). The interaction points (red crosses) of the two contour lines in each sub-figure indicate that $\mathcal{L}_1$ may tend to produce parameters valued zero and $\mathcal{L}_1$ may produce parameters with similar absolute values

of deep neural networks often have absolute values smaller than 1, so $\mathcal{L}_1$ can lead to a large penalty than $\mathcal{L}_2$ since $|w| > w^2$ when $|w| < 1$. Therefore, the loss function with $\mathcal{L}_1$ has the property which encourages the parameters of a network to have rather small values, or even zeros. This enables the network to implicitly perform feature selection, i.e. discarding some input features by setting the corresponding parameters to zero or some small values. As Fig. 1.13 shows, given two parameters w_1, w_2, in the coordinate system, ${w_1}^2 + {w_2}^2 = r^2$ is a circle with radius of r and $|w_1| + |w_2| = r$ is a square with diagonal length of $2r$, both of which are demonstrated by the blue contour lines. The red contour lines indicate the original loss $\mathcal{L}(\boldsymbol{y}, \hat{\boldsymbol{y}})$. The intersection points, represented by the red crosses, of the parameter norm penalties and the original loss, indicate that $\mathcal{L}_1$ is more likely to produce parameters valued zero than $\mathcal{L}_2$, while $\mathcal{L}_2$ may produce parameters with similar absolute values.

1.7.3 Dropout

Deep neural networks with large numbers of neurons can suffer from the co-adaptation of neurons which can result in overfitting. The co-adaptation of neurons means that the neurons are dependent on each other. If one of the neurons fails, all dependent neurons may also fail and this can lead to the failure of the entire neural network. Dropout (Hinton et al. 2012; Srivastava et al. 2014) is a popular technique to address this problem by preventing the co-adaptation of neurons (i.e., parameters). To prevent the co-adaptation of parameters, during training, the hidden outputs are randomly set to zero, which resembles a random disconnection

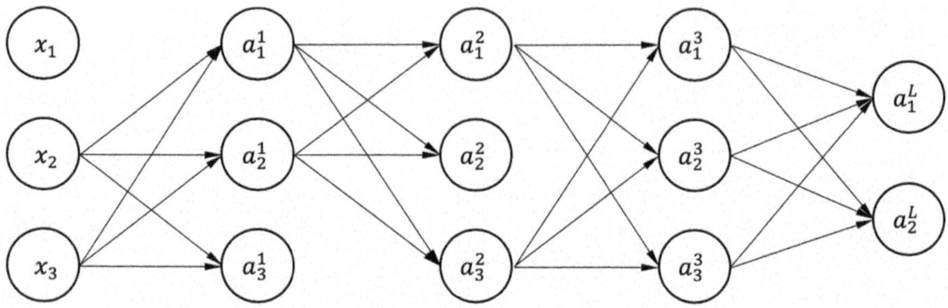

Fig. 1.14 Applying dropout to MLP where some neurons are randomly deactivated

of neurons from one layer to the next, as illustrated in Fig. 1.14. During back-propagation, with a zero-valued output $\boldsymbol{a}$, the corresponding partial derivative of the loss with respect to the layer output $\boldsymbol{\delta}$ will be zero. In other words, only the remaining connected neurons are updated. Therefore, the dropout method can train different sub-networks while allowing all of them to share the same parameters (Hinton et al. 2012). During testing, dropout is disabled, and no outputs are set to zero. This means that all sub-networks work together to predict the final result (i.e., ensemble learning (Hara et al. 2016)). The theoretical proof of dropout was not presented in the original work by Hinton et al. (2012), but more recent studies proved its effectiveness in ensemble learning (Hara et al. 2016) and Bayesian approximation (Gal and Ghahramani 2016).

1.7.4 Batch Normalization

Batch normalization (Ioffe and Szegedy 2015) normalizes the inputs of a layer to have a mean of 0 and a variance of 1 and can improve the performance of a neural network and its training stability. Specifically, during training, the batch normalization layer estimates the mean and variance of the batch inputs using a moving average. Then, the moving mean and variance are updated to normalize the batch inputs. During testing, the moving mean and variance are fixed and applied to normalize the inputs.

Besides improving the performance and stability, batch normalization provides regularization. Similar to the dropout process that adds a random factor to the hidden values, the moving mean and variance of batch normalization introduce randomness as they are updated in each iteration according to the random mini-batch. Therefore, a neural network is encouraged during training to be robust enough to deal with the variation (Fig. 1.15).

Fig. 1.15 An example of image data augmentation. The top-left image is the original image and the others are obtained by random flip, rotate, shear, shift, and zoom on the original image

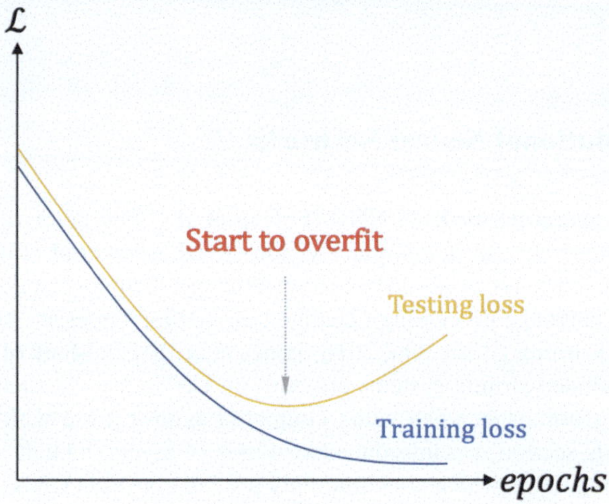

Fig. 1.16 A demonstration of where the overfitting starts. The early stopping can be applied so that the training process is terminated before the overfitting starts

1.7.5 Other Methods for Alleviating Overfitting

There are many other methods designed to prevent overfitting, such as early stopping and data augmentation. Early stopping allows early termination of the training process once it matches an empirical criterion, such as a threshold of accuracy on the validation set. Figure 1.16 shows that the testing loss may start to increase during training (i.e., the overfitting starts) and early stopping can be applied so that the training process is terminated before the overfitting starts. Data augmentation increases the size of training data by augmenting the existing training data. For example, image data can be augmented by simply flipping, rotating, shifting, and zooming. Data augmentation methods that generate arbitrary but reasonable data can reduce overfitting and improve the performance of a model (Simonyan and

Zisserman 2015; He et al. 2016; Howard et al. 2017; Dong et al. 2017b). As with an image, the audio can be augmented by adding noise or perturbation. A recent study by Ko et al. (2015) showed that audio data augmentation with speed perturbation can improve the performance of speech recognition algorithms.

However, it is not applicable to use similar augmenting transformations on textual data since the order of words provides specific meaning. For example, "people like dogs" is not semantically equivalent to "dogs like people." A practical way to augment textual data can be rephrasing sentences by replacing words with pre-defined synonyms (Zhang et al. 2015). Moreover, instead of augmenting the raw textual data, another study (Reed et al. 2016) interpolates the text embeddings of two random sentences so that the model is aware of the gaps in the text latent space.

1.8 Convolutional Neural Networks

Convolutional neural networks (CNNs) (LeCun et al. 1989) are a variant of MLP and are particularly useful in computer vision (Krizhevsky et al. 2012; Simonyan and Zisserman 2015; He et al. 2016), time series prediction (van den Oord et al. 2016), natural language processing (Zhang et al. 2019a; Yin et al. 2017), and also reinforcement learning (Rusu et al. 2016; James et al. 2019). Many of the deployed real-world machine learning systems are built on CNNs, which often demonstrate far superior performances when being compared against those with conventional methods. In this section, we introduce two kinds of layers, namely convolutional layer and pooling layer, which are commonly used to construct CNNs.

Convolutional Layer The convolutional layer has the most distinguishable feature of CNNs. The idea of its design stems from the study of the human brain again where we have an array of nearby neurons processing a subset of the visual input. Concretely, as Fig. 1.17 has shown, the convolution volume uses four different neurons to process the same region from the input image. Different neurons could be responsible for different tasks such as edge, color, or angle detection. The neuron in the convolution input is locally connected rather than being connected to all units from the previous layer. Convolutional layers can also be stacked one by one, which means a convolutional layer can be applied to the output from another convolutional layer. The benefit of a convolutional layer is that it has far fewer connections to the previous layer than a dense layer so the convolutional layer typically can be trained more quickly. Figure 1.17 also shows that each neuron in a convolutional layer contains all the information of a small region and across all channels. For example, if the input layer is the RGB image input layer, then a neuron in the convolutional layer has the information after the filter is applied to a small region of the image across all the RGB image channels.

Regarding the convolution operation inside the convolutional layer, it uses filters to extract various important features. A layer has an input of height/width W. When

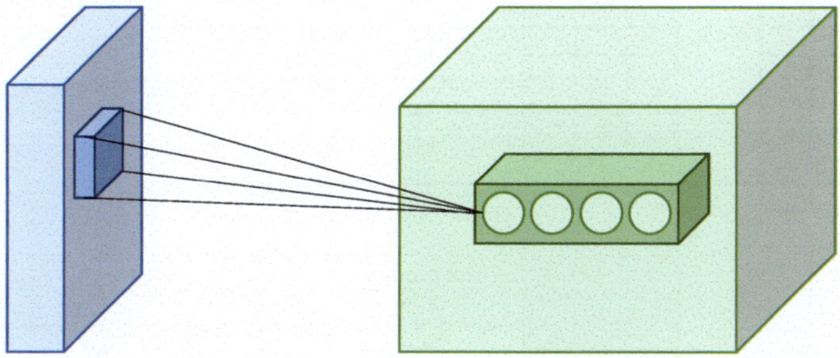

Fig. 1.17 Computation of the convolution volume from a sample image. There are four neurons applied to the same region in this example

we convolve an input with a filter of size F, we simply compute a dot product between the input and the filter values in a sliding window fashion. Then we move to apply the filter to the next block. The stride S describes how far each input block is away from each other. For instance, with the stride of two ($S = 2$), the filter is applied to the block that is one element away, skipping one row/column essentially. Lastly, sometimes in order to ensure that boundary values are well-considered, we have to add zeros on the edge, namely padding. We let the padding size be P. The output volume size of a convolutional layer can be computed by

$$\left\lfloor \frac{W - F + 2P}{S} + 1 \right\rfloor \tag{1.24}$$

The output volume has the same depth (number of output channels) as the number of filters. Figure 1.18 shows a concrete example of the convolution operation. In this example, there is an image of size 4×4 (height $\times$ width) with 3 input channels (RGB), and 1 filter sized $3 \times 3 \times 3$ (filter height $\times$ filter width $\times$ input channels) with a stride $S = 1$ and a padding $P = 0$. According to Eq. (1.24), the output height/width is $(4 - 3 + 0)/1 + 1 = 2$. The depth of the output (number of output channels) is 1 since there is 1 filter. To compute the top-left value in each channel, we first compute the dot products between the input image and the filter, which generate three values, and then sum up the three values to produce the top-left value. The convolution operation is a special case of $\sum_i w_i x_i$, where w_i is non-zero in a much smaller set. The output can then be passed through an activation function which introduces non-linearity.

Pooling Layer Pooling takes advantage of the fact that, for images, neighboring pixels are similar. So it is assumed that proper down-sampling, such as only retaining the maximum or the average of a small region, is beneficial for modeling. There are typically two types of pooling layers to reduce the dimensions, namely max-pooling and average-pooling. In Fig. 1.19, we are showing examples of max-pooling and average-pooling on a 4×4 input with a stride of 2. The pooling layer

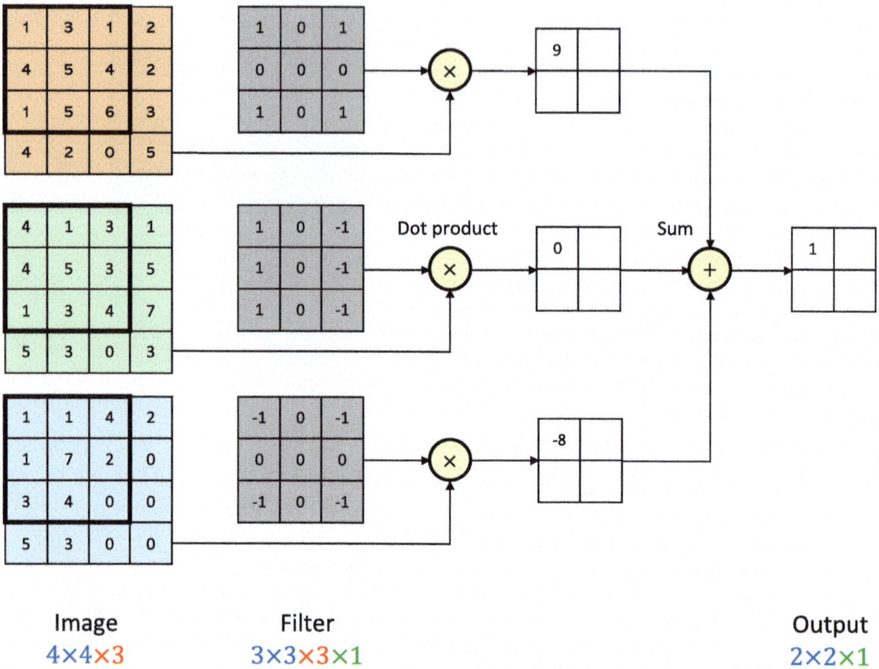

Fig. 1.18 Illustration of the convolution operation. In this example, 1 filter with size 3×3×3 (filter height × filter width × input channels) is applied on an image sized 4 × 4 (height × width) with 3 input channels (RGB). The dot products between the image and the filter are computed across the channels. The values obtained from the dot products are summed up to produce the top-left value of the output

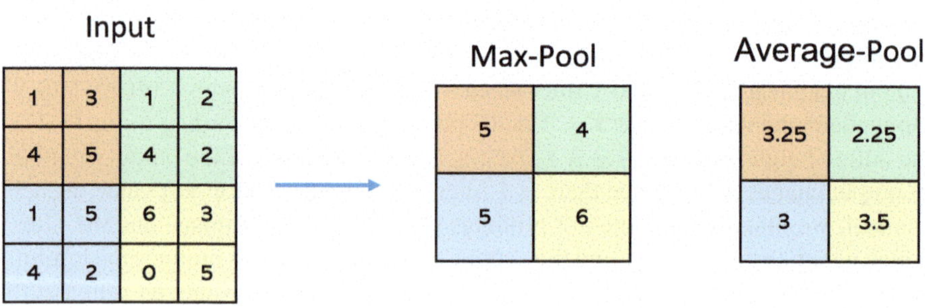

Fig. 1.19 2 × 2 max-pooling and average-pooling examples with a stride of 2 on a 4 × 4 input

reduces the dimensions of the output significantly, which makes computation in the following layers more efficient. For example, there can be hundreds of channels after a convolutional layer. Before the output is passed to a dense layer, reducing the dimensions of the output by pooling is preferred so that the successive dense layer has less computation workload.

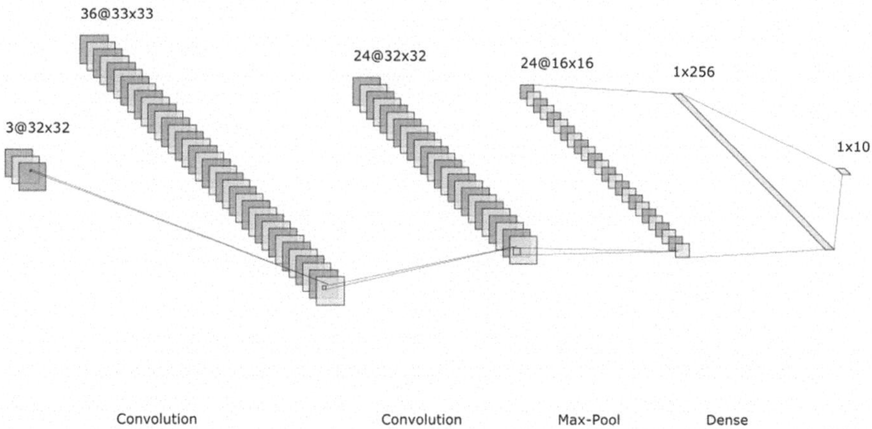

Fig. 1.20 A example of CNN with two convolutional layers, a max-pooling layer, and a dense layer. Figure created by NN-SVG[1]

Overall, the convolutional layer and pooling layer together with the dense layer are the basic components to construct CNNs. Figure 1.20 demonstrates a CNN with two convolutional layers, a max-pooling layer, and a dense layer. Note that activation functions can be applied to the output of the convolutional layers in the same way as the dense layer.

CNNs adopt the idea of **parameter sharing** which is different from MLP. The parameter sharing across different parts of a model makes the model more efficient (fewer parameters and less memory) and possible to handle variable data forms (different lengths and sizes). Recall that, in a dense layer, there is a weight matrix whose element w_{ij} denotes the connectivity between the i-th neuron in the previous layer and the j-th neuron in the current layer. However, in a convolutional layer, the filters are essentially weights, which are used repeatedly when the output values are being computed. The repeated usage of filters reduces the number of parameters needed in a convolutional layer and this is why a convolutional layer typically has far fewer parameters than a dense layer given similar sizes of input and output.

Batch normalization (batch-norm layers) (Ioffe and Szegedy 2015) can be integrated with CNNs to accelerate the training due to the internal covariate shift. As mentioned above, the input of a batch-norm layer is normalized by a mean and a variance, which are independent of other layers. Therefore, intuitively, the batch normalization simplifies the interactions between layers in the gradient update and allows larger learning rates which accelerate the training.

LeNet (LeCun et al. 1998), AlexNet (Krizhevsky et al. 2012), and VGGnet (Simonyan and Zisserman 2015) are some popular CNNs. How to design the architecture of CNNs for a specific task or a general scenario is still an on-going

[1] http://alexlenail.me/NN-SVG/LeNet.html.

research topic. The design can be an empirical driven exercise and requires lots of trials. However, recent works in neural architecture search seem to have provided more insights (Zoph and Le 2016; Zoph et al. 2018).

1.9 Recurrent Neural Networks

Recurrent neural networks (RNNs) (Rumelhart et al. 1986) is another class of deep learning architectures and it is designed to process sequential data. Unlike the images which can be represented by a grid of values, the sequential data refers to a sequence of values $\{x_1, x_2, \ldots, x_n\}$, which is also a common data format. For example, a document is composed of a sequence of words, and the values of a stock can be represented by a sequence of stock prices.

An important feature of the sequential data is the interaction among elements within the sequence. For example, provided with a snippet of text, a human reader may easily infer the content that would come next by only reading the beginning. However, the modeling of such interaction within the sequence can be more challenging if the sequence is longer. Therefore, RNNs should be able to effectively accumulate information provided by the sequential data and adequately consider the impact of earlier values on later ones in the sequence.

The design of RNNs, like that of CNNs, also adopts parameter sharing. The use of parameter sharing allows the same weight to be utilized repeatedly across multiple locations in the input sequential data. For example, RNNs should be able to learn that the sentences "Deep learning has been popular since the 2010s." and "Since the 2010s, deep learning has been popular." express the same meaning even though the positions of words are different. Similarly, when the CNNs are used to classify an image of a cat, the position of the cat in the image should not change the decision made by the CNNs (Fig. 1.21).

Simple Cell Similar to the CNNs which can process images of variable sizes, the RNNs can also easily be adjusted to process sequences with variable lengths. The

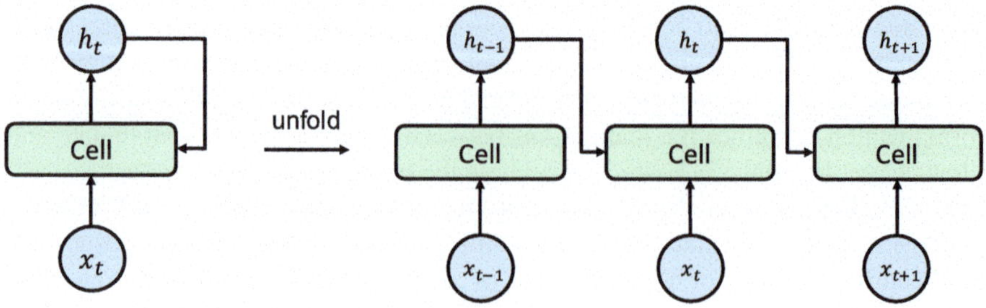

Fig. 1.21 An illustration of RNN architecture. The cell ingests the value $\boldsymbol{x}_t$ and the previous hidden state $\boldsymbol{h}_{t-1}$, and then outputs the new hidden state $\boldsymbol{h}_t$

idea of RNNs is to define a computation unit, referred to as a cell, and the cell is repeatedly computed given each value in the sequence one by one. The cell has a state which accumulates the information so far. When the cell is computed, it takes a value from the sequence and the previous state of the cell as inputs, and then generates a new state, which will be used in the next computation round. The simplest RNN cell applies a linear transformation which can be defined as follows:

$$\boldsymbol{h}_t = \boldsymbol{W}[\boldsymbol{x}_t; \boldsymbol{h}_{t-1}] + \boldsymbol{b} \tag{1.25}$$

In this equation, the previous state of the cell $\boldsymbol{h}_{t-1}$ is concatenated with the value $\boldsymbol{x}_t$ and then multiplied by the linear kernel $\boldsymbol{W}$. A bias $\boldsymbol{b}$ can also be added to the state. An RNN constructs a deep computational graph as the linear kernel is repeatedly multiplied. Such a deep computational graph may cause the exploding of gradients if the eigenvalues of $\boldsymbol{W}$ are greater than 1 in magnitude or vanishing of gradients if the eigenvalues are less than 1 in magnitude. The exploding of gradients can make the learning process volatile while the vanishing of gradients can make the optimization of objectives (cost or loss) less effective. The RNNs with the simple cell may suffer from either problem if the input sequence is lengthy.

LSTM The long short-term memory networks or LSTMs (Hochreiter et al. 1997) are more sophisticated RNNs to handle the long-term dependencies in long sequences, and the LSTM computation can serve as a cell in RNNs.

Unlike the simple cell, the LSTM cell has two states: cell state $\boldsymbol{C}_t$ and hidden state $\boldsymbol{h}_t$. The update process of the cell state forms an information highway (the orange line in Fig. 1.22) which runs across the entire sequence with simple computations. This feature allows an easier flow of information throughout the sequence so that the dependency between two values that are located far away from each other in the sequence (i.e., long-term dependency) can be properly considered. Meanwhile, the hidden state is involved with gated computations. The gate controls

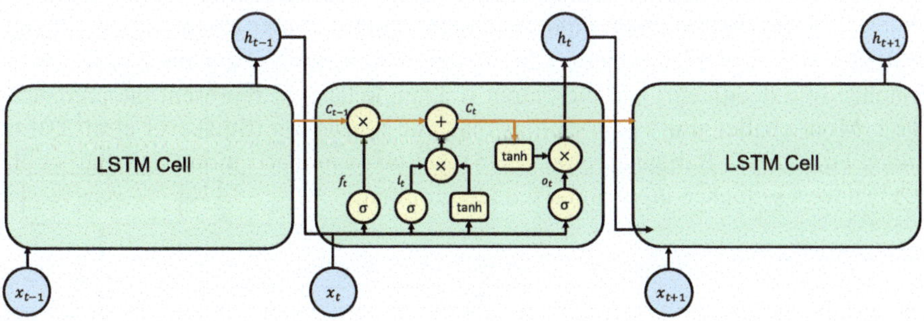

Fig. 1.22 An illustration of RNN with the LSTM cell. There are two states in the LSTM which are the cell state $\boldsymbol{C}_t$ and the hidden state $\boldsymbol{h}_t$. In addition, the three gates control whether information should be removed or added. Figure reproduced based on Olah (2015)

whether to forget or add information to the flow and is implemented by the sigmoid function. The output of the sigmoid function is restricted between 0 and 1. In other words, when the sigmoid function outputs 1, the corresponding information should be totally kept. In contrast, the corresponding information should be totally forgotten if the sigmoid function outputs 0.

There are three kinds of gates in an LSTM cell: the forget gate, input gate, and output gate. The forget state first determines whether certain information should be removed from the cell state based on the new input. In addition, the input gate controls whether the new input should be added into the cell state for longer storage and also for a replacement to any information which has been forgotten. Then finally, the output state decides what the cell should output based on the new cell state. The three gates and the computation within the LSTM cell can be formally defined as follows. Note that σ represents the sigmoid function.

$$\begin{aligned}
\text{Forget gate:}\quad \boldsymbol{f}_t &= \sigma(\boldsymbol{W}_f[\boldsymbol{h}_{t-1}; \boldsymbol{x}_t] + \boldsymbol{b}_f)\\
\text{Input gate:}\quad \boldsymbol{i}_t &= \sigma(\boldsymbol{W}_i[\boldsymbol{h}_{t-1}; \boldsymbol{x}_t] + \boldsymbol{b}_i)\\
\text{Output gate:}\quad \boldsymbol{o}_t &= \sigma(\boldsymbol{W}_o[\boldsymbol{h}_{t-1}; \boldsymbol{x}_t] + \boldsymbol{b}_o)\\
\text{Update cell state:}\quad \boldsymbol{C}_t &= \boldsymbol{f}_t \times \boldsymbol{C}_{t-1} + \boldsymbol{i}_t \times \tanh(\boldsymbol{W}_C[\boldsymbol{h}_{t-1}; \boldsymbol{x}_t] + \boldsymbol{b}_C)\\
\text{Update hidden state:}\quad \boldsymbol{h}_t &= \boldsymbol{o}_t \times \tanh(\boldsymbol{C}_t)
\end{aligned} \tag{1.26}$$

There is a family of gated RNNs that uses gated recurrent units (or GRUs) and the LSTM is a member of this family. Recent works have investigated different RNN architectures but it is still unclear which one is clearly better than others (Cho et al. 2014; Jozefowicz et al. 2015).

RNNs are widely adopted in deep learning to process sequential data like natural language and time series (Liao et al. 2018b; Chung et al. 2014; Mikolov et al. 2010) and also applied to solve reinforcement learning problems (Peng et al. 2018; Wierstra et al. 2010). Based on the relations between inputs and outputs, the architecture of RNNs can be modified in different scenarios. For example, a typical example of sequence input and single output is text classification (Zhang et al. 2019a; Lee and Dernoncourt 2016) where the input is a sequence of words (a sentence or a document) and the output is a single label to represent the predicted class. More challenging tasks such as machine translation (Sutskever et al. 2014; Luong et al. 2015; Bahdanau et al. 2015) and text summarization (Nallapati et al. 2017) have a sequence input and a sequence output.

1.10 Deep Learning Examples

This section introduces examples of how to implement deep learning models in TensorFlow[2] and TensorLayer.[3] TensorFlow (Abadi et al. 2016) by Google is an open-source library that enables researchers and engineers to develop deep learning models, while TensorLayer (Dong et al. 2017a) provides a moderate abstraction over TensorFlow to make such development easier and more flexible. The content of this section is validated on Python 3, TensorFlow 2.0, and TensorLayer 2.0 or later. In the future, TensorLayer will support different computational backend not only TensorFlow.

1.10.1 Tensor and Gradients

The tensor is the most fundamental computation unit in TensorFlow and it is used to represent outputs of an operation. A tensor can be created by operations such as `tf.constant`, `tf.matmul`, etc. Tensor does not store the values of the operation's outputs but provides access to the computation of those values in a TensorFlow session. In TensorFlow 2.0, there is no need to run a session manually, as in eager execution, graphs and sessions are designed to stay in the backend. For examples, in the matrix multiplication as shown below, matrices can be created by `tf.constant` and the multiplication is computed by `tf.matmul` whose output is another matrix.

Matrix multiplication in TensorFlow by Tensor.

```
>>> import tensorflow as tf
>>> a = tf.constant([[1, 2], [1, 2]])
# tf.Tensor(
# [[1 2]
#  [1 2]], shape=(2, 2), dtype=int32)
>>> b = tf.constant([[1], [2]])
# tf.Tensor(
# [[1]
#  [2]], shape=(2, 1), dtype=int32)
>>> c = tf.matmul(a, b)
# tf.Tensor(
# [[5]
#  [5]], shape=(2, 1), dtype=int32)
```

In the forward propagation of deep neural networks, the tensors are automatically connected by each other as a graph. Based on the graph and the automatic

[2]https://github.com/tensorflow/tensorflow.

[3]https://github.com/tensorlayer/tensorlayer.

differentiation technique provided by TensorFlow, gradients can be computed in the back-propagation. TensorFlow 2.0 provides `tf.GradientTape` to compute gradients of recorded operations with respect to its input variables. For example, the code below shows an example of computing gradients in back-propagation. The forward propagation and the computation of loss are within the scope of `tf.GradientTape`, while the back-propagation and the update of weights are outside the scope. The `tf.GradientTape` records all operations that are executed within the scope onto a `tape`. Then the gradients associated with each recorded operation and its input variables are computed by reverse-mode automatic differentiation. Once the function `tape.gradient()` is called, the resources held by `tf.GradientTape` are released.

Gradients computation in TensorFlow and TensorLayer.

```
import tensorflow as tf
import tensorlayer as tl
def train(model, dataset, optimizer):
   # given a model which is an instance of Model by TensorLayer
   # traverse the dataset where x is input and y is target output
   for x, y in dataset:
      # create the scope of gradient tape
      with tf.GradientTape() as tape:
         prediction = model(x) # forward propagation
         loss = loss_fn(prediction, y) # loss function
      # back-propagation and computing gradients, then the
          resources held by the GradientTape are released
      gradients = tape.gradient(loss, model.trainable_weights)
      # apply the gradients to weights and update the weights by
          the optimizer
      optimizer.apply_gradients(zip(gradients,
          model.trainable_weights))
```

1.10.2 Define a Model

In TensorLayer 2.0, `Model` is an entity that consists of multiple `Layers` and defines the propagation between the `Layers`. TensorLayer 2.0 provides two sets of APIs to define a model. Static model APIs allow users to build a model fluently and dynamic model APIs provide more flexibility in the forward propagation. A static model requires users to manually construct a graph and compile it. Once the model is compiled, the forward propagation cannot be changed. Unlike the static model, the dynamic model can be executed eagerly like Python normally does and the forward propagation is mutable.

In the implementation of models, as shown in the examples below, the difference between a static model and a dynamic model can be summarized in two aspects. First, when layers in a static model are declared, the connection between layers (i.e., the forward propagation) is defined explicitly at the same time. Based on the connection, for each layer, TensorLayer can automatically infer the size of input variables from previous layers and then construct weights. When the `Model` is finally instanced, only inputs and outputs need to be specified and TensorLayer automatically builds a graph based on the connection. However, when a dynamic model is initialized, the forward propagation is still unknown until it is defined in the function `forward` later. Thus, the size of input variables cannot be automatically inferred and it has to been manually provided via the argument `in_channels`.

Second, the forward propagation of a static model is fixed once the model is constructed, so it is easier to accelerate the computation of a static model. TensorFlow 2.0 provides a new feature called `tf.function` which can be used as a decorator and accelerate the computation. Unlike the static model, the forward propagation in a dynamic model can be more flexible. For example, the forward flow can be controlled based on input values or arguments specified by users. Users are also allowed to use or abandon any layer in the forward propagation of a dynamic model.

An example of a static model: multilayer perceptron (MLP)

```
import tensorflow as tf
from tensorlayer.layers import Input, Dense
from tensorlayer.models import Model

# a multilayer perceptron (MLP) model with three dense layers
def get_mlp_model(inputs_shape):
    ni = Input(inputs_shape)
    # since the connection between layers is explicitly defined
    # in_channels of each layer is automatically inferred
    nn = Dense(n_units=800, act=tf.nn.relu)(ni)
    nn = Dense(n_units=800, act=tf.nn.relu)(nn)
    nn = Dense(n_units=10, act=tf.nn.relu)(nn)
    # automatic build a model based on the connection between
        layers
    M = Model(inputs=ni, outputs=nn)
    return M

MLP = get_mlp_model([None, 784])
# switch to evaluation mode
MLP.eval()
# ingest data into the model
# the computation can be accelerated by using @tf.function in
    TensorFlow 2.0
outputs = MLP(data)
```

An example of a dynamic model: multilayer perceptron (MLP)

```
import tensorflow as tf
from tensorlayer.layers import Input, Dense
from tensorlayer.models import Model
```

```
class MLPModel(Model):
    def __init__(self):
        super(MLPModel, self).__init__()
        # since the connection between layers is unknown so far,
            in_channels has to be manually provided
        # assume the input data is size 784
        self.dense1 = Dense(n_units=800, act=tf.nn.relu,
            in_channels=784)
        self.dense2 = Dense(n_units=800, act=tf.nn.relu,
            in_channels=800)
        self.dense3 = Dense(n_units=10, act=tf.nn.relu,
            in_channels=800)

    def forward(self, x, foo=False):
        # define the forward propagation
        z = self.dense1(z)
        z = self.dense2(z)
        out = self.dense3(z)
        # control the forward flow in a dynamic model
        if foo:
            out = tf.nn.softmax(out)
        return out

MLP = MLPModel()
# switch to evaluation mode
MLP.eval()
# ingest data into the model
# the argument foo controls the forward flow
outputs_1 = MLP(data, foo=True) # with softmax
outputs_2 = MLP(data, foo=False) # without softmax
```

1.10.3 Customized Layers

TensorLayer 2.0 provides more than a hundred layers for users, and at the same time, TensorLayer 2.0 also supports `Lambda Layer` so that users can easily customize layers. The simplest example is to pass a lambda function into a `Lambda Layer` as shown below. Users may also define a customized function with arguments and the arguments can be passed by `fn_args` when the `Lambda Layer` is initialized or called.

```
import tensorlayer as tl
x = tl.layers.Input([8, 3], name='input')
y = tl.layers.Lambda(lambda x: 2*x)(x) # this layer has no
    trainable weights.

def customize_fn(input, foo): # arguments can be set by fn_args
    in Lambda Layer.
    return foo * input
```

```
z = tl.layers.Lambda(customize_fn, fn_args={'foo': 42})(x) #
    this layer has no weights.
```

The Lambda Layer can also have trainable weights. The example below shows that the weight is defined outside the customized function and it should be passed into the Lambda Layer by fn_weights.

```
import tensorflow as tf
import tensorlayer as tl
a = tf.Variable(1.0) # weight which is defined outside the scope
    of the customized function.
def customize_fn(x):
   return x + a
x = tl.layers.Input([8, 3], name='input')
y = tl.layers.Lambda(customize_fn, fn_weights=[a])(x) # weights
    are passed by fn_weights, which should be a list.
```

Moreover, the Lambda Layer enables the compatibility of Keras in TensorLayer. Users may define a Keras model and pass the model into a Lambda Layer as a function since the Keras model is callable. The trainable weights of the Keras model need to be fetched and then passed into the Lambda Layer so that the Keras model can be updated together with the customized model.

```
import tensorflow as tf
import tensorlayer as tl
# define a Keras model
layers = [
   tf.keras.layers.Dense(10, activation=tf.nn.relu),
   tf.keras.layers.Dense(5, activation=tf.nn.sigmoid),
   tf.keras.layers.Dense(1, activation=tf.identity)
]
perceptron = tf.keras.Sequential(layers)
# in order to get trainable_variables of keras
_ = perceptron(np.random.random([100, 5]).astype(np.float32))

class CustomizeModel(tl.models.Model):
   def __init__(self):
      super(CustomizeModel, self).__init__()
      self.dense = tl.layers.Dense(in_channels=1, n_units=5)
      self.lambdalayer = tl.layers.Lambda(perceptron,
          perceptron.trainable_variables) # pass the trainable
          weights of the model into the Lambda layer.

   def forward(self, x):
      z = self.dense(x)
      z = self.lambdalayer(z)
      return z
```

1.10.4 MLP: Image Classification on MNIST

With the `Models`, `Layers`, and other supportive APIs provided by TensorLayer 2.0, users can design and implement their own deep learning models in a straightforward and flexible manner. To help readers have a better understanding of how to write a deep learning model by TensorLayer, let us start from an MLP to classify images on the MNIST dataset (LeCun et al. 1998), which collects 70,000 images of handwritten digits. The implementation of a deep learning example typically has five steps including data loading, building a model, training, testing, and saving the model.

TensorLayer provides APIs in the submodule `tl.files` to load various popular datasets including MNIST, CIFAR10, PTB, CelebA, etc. For example, the MNIST dataset can be loaded by `tl.files.load_mnist_dataset` with a specific shape. The datasets are typically split into three subsets: the training set, validation set, and testing test.

```
# Loading the MNIST dataset by TensorLayer
X_train, y_train, X_val, y_val, X_test, y_test =
    tl.files.load_mnist_dataset(shape=(-1, 784)) # each image in
    MNIST is originally sized 28x28, i.e. has 784 pixels.
```

As introduced in the Sect. 1.10.2, an MLP model can be implemented as a either static or dynamic model in TensorLayer 2.0. In this example, the MLP model is designed to have three `Dense` layers and is implemented as a static model. But unlike a conventional MLP, the MLP model in this example also has three `Dropout` layers, which are used to prevent overfitting.

```
# build the model
ni = tl.layers.Input([None, 784]) # the input is aligned with
    the shape of data
# the layers of the MLP is connected one by one
nn = tl.layers.Dropout(keep=0.8)(ni)
nn = tl.layers.Dense(n_units=800, act=tf.nn.relu)(nn)
nn = tl.layers.Dropout(keep=0.5)(nn)
nn = tl.layers.Dense(n_units=800, act=tf.nn.relu)(nn)
nn = tl.layers.Dropout(keep=0.5)(nn)
nn = tl.layers.Dense(n_units=10, act=None)(nn)
# create the model with specified inputs and outputs
network = tl.models.Model(inputs=ni, outputs=nn, name="mlp")
```

The training of the MLP model on the MNIST dataset is to learn the weights of the model. Users can trigger the training process by simply calling the function `tl.utils.fit`. In addition, the testing step is to validate if the model has properly learned from the data and can be triggered by `tl.utils.test`.

```
# Define a metric to evaluate the accuracy of the model.
# Different from the loss function, the metric is NOT used to
    backpropagate or update the model.
```

```
def acc(_logits, y_batch):
    return tf.reduce_mean(
        tf.cast(
            tf.equal(
                tf.argmax(_logits, 1),
                tf.convert_to_tensor(y_batch, tf.int64)),
            tf.float32),
        name='accuracy'
    )

# Training
tl.utils.fit(
    network, # the model
    train_op=tf.optimizers.Adam(learning_rate=0.0001), # the
        optimizer
    cost=tl.cost.cross_entropy, # the loss function
    X_train=X_train, y_train=y_train, # training set
    acc=acc, # the metrics to evaluate the accuracy of a model
    batch_size=256, # the size of mini-batch
    n_epoch=20, # number of epoch to train
    X_val=X_val, y_val=y_val, eval_train=True, # validation set
)

# Testing
tl.utils.test(
    network, # the model just trained
    acc=acc, # the metrics to evaluate the accuracy of a model
    X_test=X_test, y_test=y_test, # testing set
    batch_size=None, # the size of mini-batch. If None, the whole
        testing set is fed into the network together, so only set
        it None when the testing set is small.
    cost=tl.cost.cross_entropy # the loss function
)
```

Finally, the weights of the trained MLP model can be saved to a local file so that the model can be restored later for inference.[4]

```
# save network weights to a file
network.save_weights('model.h5')
```

1.10.5 CNN: Image Classification on CIFAR10

The CIFAR-10 dataset (Krizhevsky et al. 2009) was a challenging and popular benchmark for image classification. It collects images from 10 classes and each

[4] The full code of the MLP example is available at https://github.com/tensorlayer/tensorlayer/tree/master/examples/basic_tutorials.

class has 6000 images. The images are sized 32 × 32 with RGB color and each image exclusively focuses on one single object (class) such as a dog, airplane, ship, etc. In TensorLayer 2.0, CIFAR-10 can be easily loaded and augmented by using `Dataset` and `Dataloader` APIs.

```
# pre-defined data augmentation
def _fn_train(img, target):
    # 1. Randomly crop a [height, width] section of the image.
    img = tl.prepro.crop(img, 24, 24, False)
    # 2. Randomly flip the image horizontally.
    img = tl.prepro.flip_axis(img, is_random=True)
    # 3. Subtract off the mean and divide by the variance of the
        pixels.
    img = tl.prepro.samplewise_norm(img)
    target = np.reshape(target, ())
    return img, target

# loading the training set
train_ds = tl.data.CIFAR10(train_or_test='train', shape=(-1, 32,
    32, 3))
# feed the dataset into a dataloader, which integrates data
    augmentation
train_dl = tl.data.Dataloader(train_ds, transforms=[_fn_train],
    shuffle=True, batch_size=batch_size,
    output_types=(np.float32, np.int32))

# loading the testing set
test_ds = tl.data.CIFAR10(train_or_test='test', shape=(-1, 32,
    32, 3))
# feed the dataset into a dataloader
test_dl = tl.data.Dataloader(test_ds, batch_size=batch_size)

# the images can be accessed by iteration
for X_batch, y_batch in train_dl:
    # code to train/test a model
```

In this example, a CNN model with batch normalization (Ioffe and Szegedy 2015) is trained to classify the images from CIFAR-10. The model has two convolution blocks, each of which contains a batch normalization layer, and the blocks are followed by three dense layers.[5]

```
# a static CNN model with BatchNorm
def get_model_batchnorm(inputs_shape):
    # customized initialization
    W_init = tl.initializers.truncated_normal(stddev=5e-2)
    W_init2 = tl.initializers.truncated_normal(stddev=0.04)
    b_init2 = tl.initializers.constant(value=0.1)
```

[5]The full source code of the CNN example is available at https://github.com/tensorlayer/tensorlayer/tree/master/examples/basic_tutorials.

```
# start from a input layer
ni = Input(inputs_shape)

# the first convolution block with a Conv2d, a BatchNorm and
    a MaxPool.
nn = Conv2d(64, (5, 5), (1, 1), padding='SAME',
    W_init=W_init, b_init=None)(ni)
nn = BatchNorm2d(decay=0.99, act=tf.nn.relu)(nn)
nn = MaxPool2d((3, 3), (2, 2), padding='SAME')(nn)

# the second convolution block with a Conv2d, a BatchNorm and
    a MaxPool.
nn = Conv2d(64, (5, 5), (1, 1), padding='SAME',
    W_init=W_init, b_init=None)(nn)
nn = BatchNorm2d(decay=0.99, act=tf.nn.relu)(nn)
nn = MaxPool2d((3, 3), (2, 2), padding='SAME')(nn)

# the outputs of the convolution blocks are finally fed into
    three Dense layers
nn = Flatten()(nn) # reshape the tensor
nn = Dense(384, act=tf.nn.relu, W_init=W_init2,
    b_init=b_init2)(nn)
nn = Dense(192, act=tf.nn.relu, W_init=W_init2,
    b_init=b_init2)(nn)
nn = Dense(10, act=None, W_init=W_init2)(nn)

# create the model given the inputs and outputs
M = Model(inputs=ni, outputs=nn, name='cnn')
return M
```

1.10.6 RNN and Seq2seq: Chatbot

Chatbots are designed to conduct conversation by audio and text in general. In this example, we simplify the chatbot which takes text as inputs and responses in text. In this sense, the seq2seq by (Sutskever et al. 2014) can be a good fit for the chatbot. The seq2seq model has a sequence input and a sequence output. For example, both the input and output can be a sentence, which is a sequence of words. In chatbot, the seq2seq model takes a sentence as input and is trained to respond properly with another sentence. The seq2seq was originally proposed for machine translation but has potentials on many other sequence-to-sequence scenarios such as traffic prediction (Liao et al. 2018b) and text summarization (Liu et al. 2018). In practice, the seq2seq model consists of two RNNs: one encoder and one decoder. The encoder RNN learns the representation of the input sequence and the decoder RNN generates the response against the input. TensorLayer provides APIs to build a seq2seq model with one line of code.

```
# Seq2seq model
model_ = Seq2seq(
    decoder_seq_length=decoder_seq_length, # the upper limit of
        the sequence length in the decoding
    cell_enc=tf.keras.layers.GRUCell, # the cell for the encoder
        (RNN)
    cell_dec=tf.keras.layers.GRUCell, # the cell for the decoder
        (RNN)
    n_layer=3, # number of RNN layers for the encoder and decoder
    n_units=256, # number of hidden units in RNN layers
    embedding_layer=tl.layers.Embedding(vocabulary_size=vocabulary
        _size, embedding_size=emb_dim), # the embedding layer of
        the encoder
)
```

An example output of the seq2seq based chatbot model[6] is demonstrated below. The model ingests the input query which is a sentence and outputs several candidate responses.

```
Query > happy birthday have a nice day
 > thank you so much
 > thank babe
 > thank bro
 > thanks so much
 > thank babe i appreciate it
```

References

Abadi M, Barham P, Chen J, Davis A, Dean J, Devin M, Geoffrey S, Irving G, Devin M, Kudlur M, Manjunath J, Monga R, Moore S, Murray DG, Derek B, Tucker P, Vasudevan V, Warden P, Wicke M, Yu Y, Zheng X (2016) TensorFlow: a system for large-scale machine learning. In: USENIX Symposium on Operating Systems Design and Implementation. USENIX Association, Berkeley

Bahdanau D, Cho K, Bengio Y (2015) Neural machine translation by jointly learning to align and translate. In: Proceedings of the international conference on learning representations (ICLR)

Bishop CM (2006) Pattern recognition and machine learning. Springer, Berlin

Bottou L, Bousquet O (2007) The tradeoffs of large scale learning. In: Proceedings of the 20th international conference on neural information processing systems. Advances in neural information processing systems, vol 20, pp 161–168

Cao Z, Simon Z, Wei SE, Sheikh SE (2017) Realtime multi-person 2D pose estimation using part affinity fields. In: Proceedings of the IEEE conference on computer vision and pattern recognition (CVPR)

Cho K, Van Merriënboer B, Gulcehre C, Bahdanau D, Bougares F, Schwenk H, Bengio Y (2014) Learning phrase representations using RNN encoder-decoder for statistical machine translation.

[6]The full source code of chatbot is available at https://github.com/tensorlayer/seq2seq-chatbot.

In: Proceedings of the 2014 conference on empirical methods in natural language processing (EMNLP)

Chung J, Gulcehre C, Cho K, Bengio Y (2014) Empirical evaluation of gated recurrent neural networks on sequence modeling. Preprint. arXiv:14123555

Devlin J, Chang MW, Lee K, Toutanova K (2019) BERT: pre-training of deep bidirectional transformers for language understanding. In: Proceedings of the 2019 conference of the North American chapter of the association for computational linguistics: human language technologies, vol 1 (long and short papers). Association for Computational Linguistics, Minneapolis, pp 4171–4186. https://doi.org/10.18653/v1/N19-1423

Dong H, Supratak A, Mai L, Liu F, Oehmichen A, Yu S, Guo Y (2017a) TensorLayer: a versatile library for efficient deep learning development. In: Proceedings of the ACM Multimedia (MM). http://tensorlayer.org

Dong H, Zhang J, McIlwraith D, Guo Y (2017b) I2t2i: learning text to image synthesis with textual data augmentation. In: Proceedings of the IEEE international conference on image processing (ICIP)

Duchi J, Hazan E, Singer Y (2011) Adaptive subgradient methods for online learning and stochastic optimization. J Mach Learn Res 12:2121–2159

Gal Y, Ghahramani Z (2016) Dropout as a Bayesian approximation: representing model uncertainty in deep learning. In: Proceedings of the international conference on machine learning (ICML), pp 1050–1059

Glorot X, Bordes A, Bengio Y (2011) Deep sparse rectifier neural networks. In: Proceedings of the international conference on artificial intelligence and statistics (AISTATS), pp 315–323

Goodfellow I, Pouget-Abadie J, Mirza M, Xu B, Warde-Farley D, Ozair S, Courville A, Bengio Y (2014) Generative adversarial nets. In: Proceedings of the neural information processing systems conference. Advances in neural information processing systems

Goodfellow I, Bengio Y, Courville A (2016) Deep learning. MIT Press, Cambridge. http://www.deeplearningbook.org

Hara K, Saitoh D, Shouno H (2016) Analysis of dropout learning regarded as ensemble learning. In: Proceedings of the international conference on artificial neural networks (ICANN). Springer, Berlin, pp 72–79

He K, Zhang X, Ren S, Sun J (2015) Delving deep into rectifiers: surpassing human-level performance on ImageNet classification. In: Proceedings of the IEEE international conference on computer vision, pp 1026–1034

He K, Zhang X, Ren S, Sun J (2016) Deep residual learning for image recognition. In: Proceedings of the IEEE conference on computer vision and pattern recognition (CVPR)

Hinton GE, Srivastava N, Krizhevsky A, Sutskever I, Salakhutdinov RR (2012) Improving neural networks by preventing co-adaptation of feature detectors. Preprint. arXiv:12070580

Hochreiter S, Hochreiter S, Schmidhuber J, Schmidhuber J (1997) Long short-term memory. Neural Comput 9(8):1735–1780

Hornik K, Stinchcombe M, White H (1989) Multilayer feedforward networks are universal approximators. Neural Netw 2(5):359–366

Howard AG, Zhu M, Chen B, Kalenichenko D, Wang W, Weyand T, Andreetto M, Adam H (2017) MobileNets: efficient convolutional neural networks for mobile vision applications. Preprint. arXiv:170404861

Ioffe S, Szegedy C (2015) Batch normalization: accelerating deep network training by reducing internal covariate shift. Preprint. arXiv:150203167

James S, Wohlhart P, Kalakrishnan M, Kalashnikov D, Irpan A, Ibarz J, Levine S, Hadsell R, Bousmalis K (2019) Sim-to-real via sim-to-sim: data-efficient robotic grasping via randomized-to-canonical adaptation networks. In: Proceedings of the IEEE conference on computer vision and pattern recognition, pp 12627–12637

Jozefowicz R, Zaremba W, Sutskever I (2015) An empirical exploration of recurrent network architectures. In: International conference on machine learning, pp 2342–2350

Kingma D, Ba J (2014) Adam: a method for stochastic optimization. In: Proceedings of the international conference on learning representations (ICLR)

Ko T, Peddinti V, Povey D, Khudanpur S (2015) Audio augmentation for speech recognition. In: Annual conference of the international speech communication association

Krizhevsky A, Hinton G et al (2009) Learning multiple layers of features from tiny images. Technical Report. Citeseer

Krizhevsky A, Sutskever I, Hinton GE (2012) ImageNet classification with deep convolutional neural networks. In: Advances in neural information processing systems, pp 1097–1105

LeCun Y, Boser B, Denker JS, Henderson D, Howard RE, Hubbard W, Jackel LD (1989) Backpropagation applied to handwritten zip code recognition. Neural Comput 1(4):541–551

LeCun Y, Bottou L, Bengio Y, Haffner P et al (1998) Gradient-based learning applied to document recognition. Proc IEEE 86(11):2278–2324

LeCun Y, Bengio Y, Hinton G (2015) Deep learning. Nature 521(7553):436

Ledig C, Theis L, Huszar F, Caballero J, Cunningham A, Acosta A, Aitken A, Tejani A, Totz J, Wang Z, Shi W (2017) Photo-realistic single image super-resolution using a generative adversarial network. In: Proceedings of the IEEE conference on computer vision and pattern recognition (CVPR)

Lee JY, Dernoncourt F (2016) Sequential short-text classification with recurrent and convolutional neural networks. In: Proceedings of the 2016 conference of the North American chapter of the association for computational linguistics: human language technologies. Association for Computational Linguistics, San Diego, pp 515–520. https://doi.org/10.18653/v1/N16-1062

Liao B, Zhang J, Cai M, Tang S, Gao Y, Wu C, Yang S, Zhu W, Guo Y, Wu F (2018a) Dest-ResNet: a deep spatiotemporal residual network for hotspot traffic speed prediction. In: 2018 ACM multimedia conference on multimedia conference. ACM, New York, pp 1883–1891

Liao B, Zhang J, Wu C, McIlwraith D, Chen T, Yang S, Guo Y, Wu F (2018b) Deep sequence learning with auxiliary information for traffic prediction. In: Proceedings of the 24th ACM SIGKDD international conference on knowledge discovery & data mining. ACM, New York, pp 537–546

Liu PJ, Saleh M, Pot E, Goodrich B, Sepassi R, Kaiser L, Shazeer N (2018) Generating wikipedia by summarizing long sequences. In: International conference on learning representations. https://openreview.net/forum?id=Hyg0vbWC-

Luong T, Pham H, Manning CD (2015) Effective approaches to attention-based neural machine translation. In: Proceedings of the 2015 conference on empirical methods in natural language processing. Association for Computational Linguistics, Lisbon, pp 1412–1421. https://doi.org/10.18653/v1/D15-1166

Maas AL, Daly RE, Pham PT, Huang D, Ng AY, Potts C (2011) Learning word vectors for sentiment analysis. In: Proceedings of the 49th annual meeting of the association for computational linguistics: human language technologies, HLT '11, vol 1. Association for Computational Linguistics, Stroudsburg, pp 142–150. http://dl.acm.org/citation.cfm?id=2002472.2002491

Mikolov T, Karafiát M, Burget L, Cernocky J, Khudanpur S (2010) Recurrent neural network based language model. In: INTERSPEECH 2010, 11th annual conference of the international speech communication association, Makuhari

Nallapati R, Zhai F, Zhou B (2017) SummaRuNNer: a recurrent neural network based sequence model for extractive summarization of documents. In: Proceedings of the thirty-first AAAI conference on artificial intelligence, AAAI'17. AAAI Press, Palo Alto, pp 3075–3081

Ng AY, Jordan MI (2002) On discriminative vs. generative classifiers: a comparison of logistic regression and naive Bayes. In: Proceedings of the neural information processing systems. Advances in neural information processing systems. Conference, pp 841–848

Noh H, Hong S, Han B (2015) Learning deconvolution network for semantic segmentation. In: Proceedings of the international conference on computer vision (ICCV), pp 1520–1528

Olah C (2015) Understanding LSTM networks. https://colah.github.io/posts/2015-08-Understanding-LSTMs/

Peng XB, Andrychowicz M, Zaremba W, Abbeel P (2018) Sim-to-real transfer of robotic control with dynamics randomization. In: 2018 IEEE international conference on robotics and automation (ICRA). IEEE, Piscataway, pp 1–8

Reed S, Akata Z, Yan X, Logeswaran L, Schiele B, Lee H (2016) Generative adversarial text to image synthesis. In: Proceedings of the international conference on machine learning (ICML)
Rish I et al (2001) An empirical study of the naive Bayes classifier. In: International joint conference on artificial intelligence 2001 workshop on empirical methods in artificial intelligence. vol 3, pp 41–46
Rosenblatt F (1958) The perceptron: a probabilistic model for information storage and organization in the brain. Psychol Rev 65(6):386
Ruck DW, Rogers SK, Kabrisky M, Oxley ME, Suter BW (1990) The multilayer perceptron as an approximation to a Bayes optimal discriminant function. IEEE Trans Neural Netw 1(4):296–298
Rumelhart DE, Hinton GE, Williams RJ (1986) Learning representations by back-propagating errors. Nature 323(6088):533
Rusu AA, Rabinowitz NC, Desjardins G, Soyer H, Kirkpatrick J, Kavukcuoglu K, Pascanu R, Hadsell R (2016) Progressive neural networks. Preprint. arXiv:160604671
Samuel A (1959) Some studies in machine learning using the game of checkers. IBM J Res Dev 3:210–219
Simonyan K, Zisserman A (2015) Very deep convolutional networks for large-scale image recognition. In: Proceedings of the international conference on learning representations (ICLR)
Srivastava N, Hinton G, Krizhevsky A, Sutskever I, Salakhutdinov R (2014) Dropout: a simple way to prevent neural networks from overfitting. J Mach Learn Res 15(1):1929–1958
Sutskever I, Vinyals O, Le QV (2014) Sequence to sequence learning with neural networks. In: Proceedings of the neural information processing systems. Advances in neural information processing systems. Conference, pp 3104–3112
Tieleman T, Hinton G (2017) Divide the gradient by a running average of its recent magnitude. COURSERA: neural networks for machine learning. Technical Report
van den Oord A, Dieleman S, Zen H, Simonyan K, Vinyals O, Graves A, Kalchbrenner N, Senior A, Kavukcuoglu K (2016) WaveNet: a generative model for raw audio. In: Arxiv. https://arxiv.org/abs/1609.03499
Wierstra D, Förster A, Peters J, Schmidhuber J (2010) Recurrent policy gradients. Log J IGPL 18(5):620–634
Xu B, Wang N, Chen T, Li M (2015) Empirical evaluation of rectified activations in convolutional network. In: Proceedings of the international conference on machine learning (ICML) workshop
Yang G, Yu S, Dong H, Slaubaugh, GG, Dragotti PL, Ye X, Liu F, Arridge SR, Keegan J, Guo Y, Firmin DN (2018) DAGAN: deep de-aliasing generative adversarial networks for fast compressed sensing MRI reconstruction. IEEE Trans Med Imaging 37(6):1310–1321
Yang Z, Dai Z, Yang Y, Carbonell J, Salakhutdinov RR, Le QV (2019) XLNet: generalized autoregressive pretraining for language understanding. In: Advances in neural information processing systems, pp 5754–5764
Yin W, Kann K, Yu M, Schütze H (2017) Comparative study of CNN and RNN for natural language processing. Preprint. arXiv:170201923
Zhang X, Zhao J, LeCun Y (2015) Character-level convolutional networks for text classification. In: Advances in neural information processing systems, pp 649–657
Zhang J, Lertvittayakumjorn P, Guo Y (2019a) Integrating semantic knowledge to tackle zero-shot text classification. In: Proceedings of the 2019 conference of the North American chapter of the association for computational linguistics: human language technologies, vol 1 (long and short papers). Association for Computational Linguistics, Minneapolis, pp 1031–1040. https://doi.org/10.18653/v1/N19-1108
Zhang J, Zhao Y, Saleh M, Liu PJ (2019b) PEGASUS: Pre-training with extracted gap-sentences for abstractive summarization. Preprint. arXiv:191208777

Zoph B, Le QV (2016) Neural architecture search with reinforcement learning. Preprint. arXiv:161101578

Zoph B, Vasudevan V, Shlens J, Le QV (2018) Learning transferable architectures for scalable image recognition. In: Proceedings of the IEEE conference on computer vision and pattern recognition, pp 8697–8710

Chapter 2
Introduction to Reinforcement Learning

Zihan Ding, Yanhua Huang, Hang Yuan, and Hao Dong

Abstract In this chapter, we introduce the fundamentals of classical reinforcement learning and provide a general overview of deep reinforcement learning. We first start with the basic definitions and concepts of reinforcement learning, including the agent, environment, action, and state, as well as the reward function. Then, we describe a classical reinforcement learning problem, the bandit problem, to provide the readers with a basic understanding of the underlying mechanism of traditional reinforcement learning. Next, we introduce the Markov process, together with the Markov reward process and the Markov decision process. These notions are the cornerstones in formulating reinforcement learning tasks. The combination of the Markov reward process and value function estimation produces the core results used in most reinforcement learning methods: the Bellman equations. The optimal value functions and optimal policy can be derived through solving the Bellman equations. Three main approaches for solving the Bellman equations are then introduced: dynamic programming, Monte Carlo method, and temporal difference learning. We further introduce deep reinforcement learning for both policy and value function approximation in policy optimization. The contents in policy optimization are introduced in two main categories: value-based optimization and policy-based optimization. In value-based optimization, the gradient-based methods are introduced for leveraging deep neural networks, like Deep Q-Networks. In policy-based optimization, the deterministic policy gradient and stochastic policy gradient are

Z. Ding (✉)
Imperial College London, London, UK
e-mail: zhding@mail.ustc.edu.cn

Y. Huang
Xiaohongshu Technology Co., Ltd., Shanghai, China

H. Yuan
Oxford University, Oxford, UK
e-mail: hang.yuan@keble.ox.ac.uk

H. Dong
Peking University, Beijing, China
e-mail: hao.dong@pku.edu.cn

H. Dong et al. (eds.), *Deep Reinforcement Learning*,
https://doi.org/10.1007/978-981-15-4095-0_2

introduced in detail with sufficient mathematical proofs. The combination of value-based and policy-based optimization produces the popular actor-critic structure, which leads to a large number of advanced deep reinforcement learning algorithms. This chapter will lay a foundation for the rest of the book, as well as providing the readers with a general overview of deep reinforcement learning.

Keywords Reinforcement learning · Multi-armed bandit · Markov process · Bellman equation · Dynamic programming · Monte Carlo method · Temporal difference learning · Value-based optimization · Deterministic policy gradient · Stochastic policy gradient

2.1 Introduction

This chapter introduces the basic knowledge of reinforcement learning (RL) as well as deep reinforcement learning (DRL), including the definitions and explanations of basic concepts, as well as the theoretical proofs of some theorems in reinforcement learning domain, which are the fundamentals of advanced topics in (deep) reinforcement learning. Therefore, we encourage the readers to read through and understand well about the contents in this chapter before moving on to the following chapters. We will start with the basic concepts in reinforcement learning.

The **agent** and **environment** are the basic components of reinforcement learning, as shown in Fig. 2.1. The environment is an entity that the agent can interact with. For example, an environment can be a Pong game, which is shown on the right-hand side of Fig. 2.2. The agent controls the paddle to hit the ball back and forth. An agent can "interact" with the environment by using a predefined **action set** $\mathcal{A} = \{A_1, A_2 \ldots\}$. The action set describes all possible actions. In Pong, the action set can be {moveUp, moveDown}. The goal of reinforcement learning algorithms is to teach the agent how to interact "well" with the environment so that the agent is able to obtain a good score under a predefined evaluation metric. In Pong, the metric could just be the score that a player gets. An agent will receive a **reward** r of 1 when the ball hits the wall on the opposite side. In other words, if the agent misses

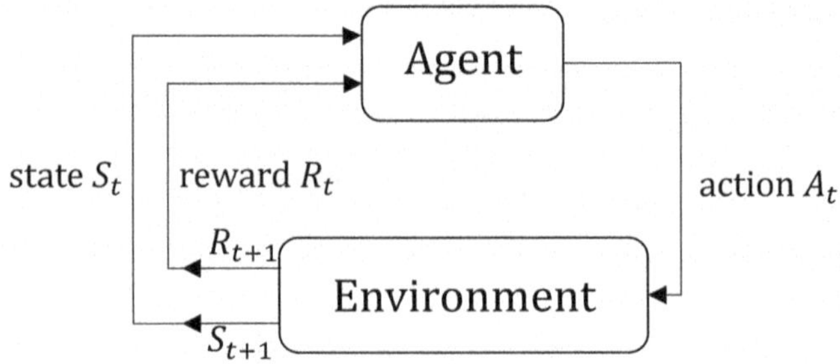

Fig. 2.1 Agent and environment

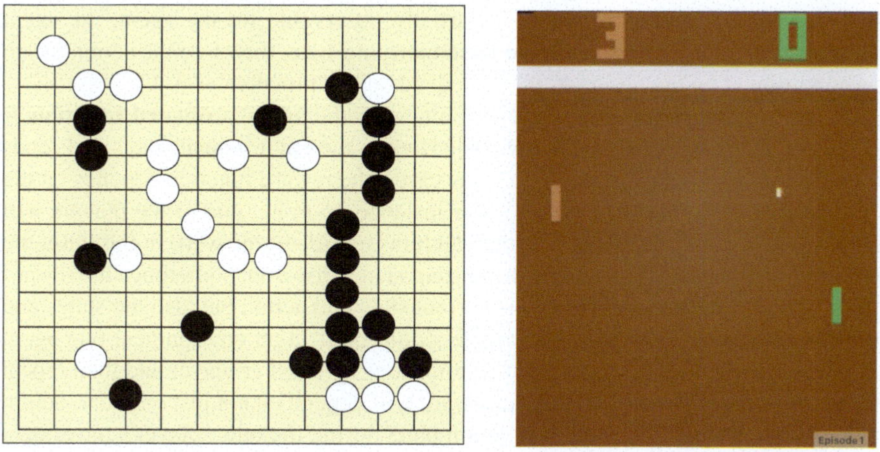

Fig. 2.2 Two types of gaming environments: on the left is the game of Go, where the observation contains the complete state of the environment, and thus the environment is fully observable. On the right is the Atari Pong game, where the observation of a single frame does not contain the velocity of the ball, and thus the environment is partially observable

the ball and lets the ball hit the wall on its side, then its opponent will receive a reward of 1.

Now, let us take a closer look at the relationship between the agent and the environment as depicted in Fig. 2.1. At an arbitrary time step, t, the agent first observes the current state of the environment, S_t, and the corresponding reward value, R_t. The agent then decides what to do next based on the state and reward information. The action the agent intends to perform, A_t, gets fed back into the environment such that we can obtain the new state S_{t+1} and reward R_{t+1}. The observation of the environment state s (s is a general representation of state regardless of time step s) from the agent's perspective does not always contain all the information about the environment. If the observation only contains partial state information, the environment is **partially observable**. Nevertheless, if the observation contains the complete state information of the environment, the environment is **fully observable**. In practice, the observation is usually a function of the state, which makes it difficult to differentiate the observations that contain all the state information from the ones that do not. A better understanding would be from the information perspective that a fully observable environment should not miss any information in the function from the underlying state of the whole environment to the observation of the agent.

The board game Go, shown on the left-hand side of Fig. 2.2, is a typical example of a fully observable environment, where all the placement information of Go pieces is fully observable for both of the players. The Atari game of Pong with a single frame as observation is a partially observable environment, where the velocity of the ball is important for the decision but not available from the stationary frame.

In the literature of reinforcement learning, the action a (a is a general representation of action regardless of time step t) is usually conditioned on the state s to represent the behavior of the agent, under the assumption of fully observable environments. If the environment is partially observable for the agent, the agent usually cannot have direct access to the underlying state, therefore the action has to be conditioned on the observation, without advanced process.

To provide feedback from the environment to the agent, a **reward function** R generates an **immediate reward** R_t according to the environment status and sends it to the agent at every time step. In some cases, the reward function depends on the current state only, i.e., $R_t = R(S_t)$. For instance, in Pong, one of the players will receive the reward immediately, if the ball hits one side of the wall. In this case, the reward function only depends on the current state. However, sometimes the reward function can depend on not only the current state and action but also the states and actions at other time steps. An example would be if in an environment one agent is asked to memorize a sequence of actions done by another player and then repeat the same sequence of actions. So the reward will depend on not just one state-action pair but also the sequence of state-action pairs during the other player's movement and this player's movement. A reward function based on the current state, or even the current state and action will not be indicative for the agent when mimicking the whole sequence.

In reinforcement learning, a **trajectory** is a sequence of states, actions, and rewards:

$$\tau = (S_0, A_0, R_0, S_1, A_1, R_1, \ldots)$$

which records how the agent interacts with the environment. The initial state in a trajectory, S_0, is randomly sampled from the **start-state distribution**, denoted by ρ_0, in which:

$$S_0 \sim \rho_0(\cdot) \tag{2.1}$$

For example, the Atari Pong game always starts with a ball in the center of the screen and the game of GO usually starts with a chess piece on a random location of the chessboard.

The transition from a state to the next state can be either a **deterministic transition process** or a **stochastic transition process**. For the deterministic transition, the next state S_{t+1} is governed by a deterministic function:

$$S_{t+1} = f(S_t, A_t), \tag{2.2}$$

where a unique next state S_{t+1} can be found. For the stochastic transition process, the next state S_{t+1} is described as a probabilistic distribution:

$$S_{t+1} \sim p(S_{t+1}|S_t, A_t) \tag{2.3}$$

A trajectory, being referred to also as an **episode**, is a sequence that goes from the initial state to the terminal state (for finite cases). For example, playing one entire game can be considered as an episode. The terminal state is reached when the agent wins or loses the game. Sometimes, one episode can have several sub-games rather than only one. For example, an episode can contain 21 sub-games for the Gym Pong game.

Finally, we shall discuss two important concepts before the end of the section, exploitation and exploration, as well as the well-known exploration-exploitation trade-off. **Exploitation** means maximizing the agent performance using the existing knowledge, and its performance is usually evaluated by the expected reward. For example, a gold digger now has an ore providing him two grams of gold per day, and he knows that the largest gold ore can give him five grams of gold per day. However, he also knows that finding a new ore will not only force him to stop exploiting the current ore but also costs him extra time with a risk of not finding anything at all in the end. Having these in mind he decides to dig the current ore until it is exhausted to maximize the reward (gold in this case) via exploitation and give up exploration, given the large risks of exploration based on his current knowledge. The policy he took here is the **greedy** policy, which means the agent constantly performs the action that yields the highest expected reward based on current information, rather than taking risky trials which may lead to lower expected rewards.

Exploration means increasing existing knowledge by taking actions and interacting with the environment. Back to the example of the gold digger, he wishes to spend some time to find new ore, and if he finds a bigger gold ore, he can get more reward per day. To have a larger long-term return, the short-term return may be sacrificed. The gold digger is always facing the exploitation and exploration dilemma as he needs to decide how much the yield a gold mine has for him to stay and how little the yield a gold mine has for him to keep exploring. Maybe, he also wants to see enough ores before he can make a well-informed decision. From the above descriptions, the readers may already have a primary understanding about the exploration-exploitation trade-off. The **exploration-exploitation trade-off** describes the balance between how much efforts the agent makes on exploration and exploitation, respectively. The trade-off between exploration and exploitation is a central theme of reinforcement learning research and reinforcement learning algorithm development. We will explain it with the following bandit problem.

2.2 Bandits

Single-Armed Bandit is a simple gambling machine as shown on the left-hand side of Fig. 2.3. The agent (player) interacts with the environment (machine) by pulling a single arm down, and receives a reward when the machine hits the jackpot. In a casino, there will usually be many bandits lining up in a row. The agent can choose to pull an arm of any of these slot machines. The distributions of the reward values r conditioned on the actions a, $P(r|a)$, for different bandits are different but fixed. The agent, however, does not know the distributions in the beginning, and the knowledge is acquired through the trials of the agents. The goal is to maximize

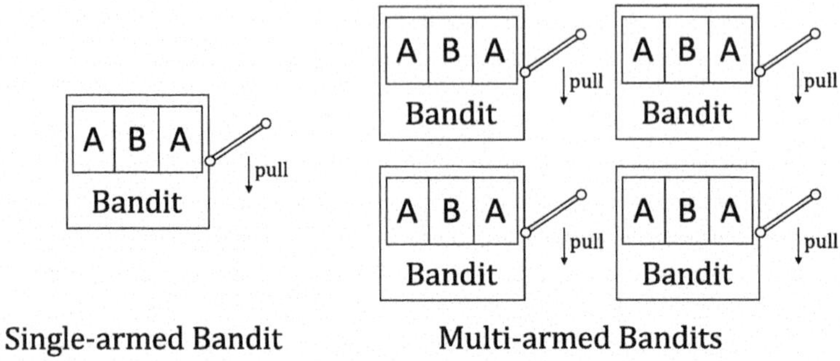

Fig. 2.3 Single-armed bandit (left) and multi-armed bandits (right)

the payoffs after some number of selections. The agent will have to choose among various slot machines at each time step and we refer this game as **multi-armed bandit** (MAB) which is shown on the right-hand side of Fig. 2.3. MAB gives the agent the freedom to make strategic choices of which arm to pull.

We try to solve the MAB problem with standard reinforcement learning methods. The action a of the agent is to choose which arm to pull. A reward will be given right after the action is conducted. Formally, at time step t, we are trying to maximize the expected action value defined as follows:

$$q(a) = \mathbb{E}[R_t | A_t = a]$$

If we already know the true value $q(a)$ of each action a, then the problem will be trivial to solve because we can always choose the action that would yield the best q value. However, in reality, we typically need to estimate the q value and we denote the estimate as $Q(a)$, which should be as close to $q(a)$ as possible.

The MAB problem is an excellent example to illustrate the exploration-exploitation trade-off. After one has estimated the q values for some states, if the agent is always going to take the action that has the greatest Q value, then this agent is considered to be greedy and is exploiting the already estimated q values. If the agent takes on any action that does not have the best Q value, then this agent is considered to be exploring different options. Neither doing only exploration or exploitation is a good way to improve the policy of the agent in most cases.

A simple action-value based method is to estimate $Q_t(a)$ using the ratio between the total rewards received by choosing one action by time t and the total number of times that this specific action has been chosen:

$$Q_t(a) = \frac{\text{sum of the rewards by choosing } a \text{ before } t}{\text{number of times } a \text{ was chosen before } t} = \frac{\sum_{i=0}^{t-1} R_i \cdot \mathbb{1}_{A_i=a}}{\sum_{i=0}^{t-1} \mathbb{1}_{A_i=a}}$$

$\mathbb{1}_{\text{predicate}}$ is one when the predicate is true otherwise it is zero. The greedy approach can be thus written as

$$A_t = \arg\max_a Q_t(a) \tag{2.4}$$

We can, however, easily convert this greedy method into one that also explores other states with probability ϵ. We call this method ϵ-greedy as it randomly chooses an action with probability ϵ and most of the time behaves in a greedy fashion. If we have an infinite number of time steps, we are guaranteed to have $Q_t(a)$ converge to $q(a)$. Moreover, the simple action-value based method is also an online learning approach which we will explain in detail in the next section.

2.2.1 *Online Prediction and Online Learning*

Online prediction problems are the class of problems where the agent has to make predictions about the future. For instance, imagine you have been in Hawaii for a week, and are asked to predict whether it will rain in the next days. Another example can be predicting afternoon oil prices based on the observed fluctuations in oil prices in the morning. Online prediction problems need to be solved with online methods. Online learning is distinguished from traditional statistic learning in the following aspects:

- The sample data is presented in an ordered sequence instead of an unordered batch.
- We will often have to consider the worst case rather than the average case because we do not want things to go out of control in the learning stage.
- The learning objective can be different as online learning often tries to minimize regret whereas statistical learning tries to minimize empirical risk. We will explain what regret is later.

Let us take look at a trivial example in the context of the MAB problem. Let us say, we observe a reward R_t at each time step t. An easy solution to find out what the best action is to update the estimate of the q value using R_t and A_t. A traditional way to compute the mean is to sum up all the previous rewards when A_t has been selected and divide that sum by the count of A_t. This approach is more like batch learning as it involves recomputing every time using a batch of data points. The online alternative would be to use a running average by doing the new estimate using the previous estimate: $Q_i(A_t) = Q_i(A_t) - Q_i(A_t)/N$; $Q_{i+1}(A_t) = Q_i(A_t) + R_t/N$. Q_i is the q value after A_t has been selected i times, and N is the number of times that A_t has been selected.

2.2.2 *Stochastic Multi-Armed Bandit*

Concretely, if we have $K \geq 2$ arms, we will need to select an arm to pull for each time step $t = 1, 2, \cdots, T$. At each step t, we can observe a reward R_t^i by selecting the ith arm.

Algorithm 1 Multi-armed bandit learning

Initialize K arms;
Number of time steps T;
Each arm is associated with $v_i \in [0, 1]$. The reward being returned is drawn i.i.d from v_i
for $t = 1, 2, \ldots, T$ **do**
  The agent selects $A_t = i$ from the K arms.
  The environment returns the reward vector $R_t = (R_t^1, R_t^2, \cdots, R_t^K)$.
  The agent observes reward R_t^i.
end for

In a traditional sense, one often tends to maximize the rewards. However, for a stochastic MAB, we will focus on another metric, regret. The regret after n steps is defined as:

$$RE_n = \max_{j=1,2,\ldots,K} \sum_{t=1}^{n} R_t^j - \sum_{t=1}^{n} R_t^i$$

The first term in the subtraction is the total reward that we accumulate until time n for always receiving the maximized rewards and the second term is the actual accumulated rewards in a trial that has gone through n time steps.

In order to select the best action, we should try to minimize the expected regret because of the stochasticity introduced by our actions and rewards. We will differentiate two different types of regret, the expected regret and pseudo-regret. The expected regret is defined as:

$$\mathbb{E}[RE_n] = \mathbb{E}\left[\max_{j=1,2,\ldots,T} \sum_{t=1}^{n} R_t^j - \sum_{t=1}^{n} R_t^i\right] \tag{2.5}$$

The pseudo-regret is defined as:

$$\overline{RE}_n = \max_{j=1,2,\ldots,T} \mathbb{E}\left[\sum_{t=1}^{n} R_t^j - \sum_{t=1}^{n} R_t^i\right] \tag{2.6}$$

The key distinction between the above two regrets is the order of the maximization and expectation. The expected regret is harder to compute. This is because for the pseudo-regret we only need to find the action that optimizes the regret in expectation, however, for the expected regret, we will have to find the expected regret that is optimal over actions across different trials. Concretely, we have $\mathbb{E}[RE_n] \geq \overline{RE}_n$.

Let μ_i be the mean of v_i, where v_i is the reward value of the $i-$th arm, $\mu^* = \max_{i=1,2,\ldots,T} \mu_i$. In the stochastic setting, we can rewrite Eq. (2.6) as:

$$\overline{RE}_n = n\mu^* - \mathbb{E}\left[\sum_{t=1}^{n} R_t^i\right] \tag{2.7}$$

One way to minimize the pseudo-regret is to select the best arm to pull given the observed sample pulls using ϵ-greedy which we already talked about before. A more sophisticated method is called Upper Confidence Bound (UCB) algorithm. UCB makes use of Hoeffding's lemma to derive an upper confidence bound estimate and chooses the arm whose sample mean has been the greatest so far.

We now introduce the concept of UCB strategy. The exact treatment of using UCB on stochastic MAB for regret optimization can be found in Bubeck et al. (2012). We will explain UCB for the situation when we optimize the policy with respect to the reward. In stochastic MAB, even though the rewards are drawn from a distribution, the reward function is still stationary over time. Let us refer back to the ϵ-greedy method. The ϵ-greedy method explores non-optimal states with a probability ϵ, but the issue is that it considers all the non-optimal states the same and does not make any differentiation. If we want to thoroughly visit every state, we should certainly prioritize the states that have not been visited yet or the states with fewer visits. UCB helps resolve this issue by rewriting Eq. (2.4) into:

$$A_t = \arg\max_a \left[Q_t(a) + c\sqrt{\frac{\ln t}{N_t(a)}}\right] \tag{2.8}$$

$N_t(a)$ is the number of times action a has been selected till time t and c is a positive real number that determines how much exploration needs to be done. Eq. (2.8) is how we would select an action if we have a non-stationary reward function. When $N_t(a)$ is zero, we consider action a to have the max value. To understand how UCB works, let us focus on the square-root term, which reflects the amount of uncertainty we have for the q-value estimate for a. As the number of times that a is selected increases, the uncertainty decreases as the denominator decreases. Also as more actions other than a is being selected, the uncertainty increases because the $\ln t$ increases but $N_t(a)$ does not. The log operator on t is saying that the impact of incoming new time steps decays as we have more time steps in total. UCB gives some form of an upper bound of the q value for a, and c is the confidence level.

2.2.3 Adversarial Multi-Armed Bandit

Stochastic MAB's reward functions are determined by the probabilistic distributions that are usually not changed. In reality, this might not be the case. Imagine that if in a casino, a player wants to make a large profit by playing a group of slot machines and he has found out which machines are more likely to yield better returns, the casino owner will change the behavior of the machines so that the casino does not lose money. This is exactly why adversarial MAB modeling is needed when the rewards are no longer governed by a stationary probabilistic distributions but arbitrarily determined by some adversary. Formally, in adversarial MAB, the reward for the ith arm at time t will be denoted by $R_t^i \in [0, 1]$, at the same time, the player will select an arm at time t, which is denoted by $I_t \in \{1, \ldots, K\}$.

One might wonder what if an adversary simply sets all the rewards to zero. This situation could happen, but there would be little point in playing this game if the player can get nothing in return. As a matter of fact, even when the opponent can freely decide what the rewards are going to look like, he would not do so because if all the rewards are zeros, no one will want to play the game anymore. His job is actually to give the player enough rewards just to trick the player in believing that he might have a chance of winning in the long term.

Algorithm 2 Adversarial multi-armed bandit

Initialize K arms;
for $t = 1, 2, \ldots, T$ **do**
  The agent selects I_t from the K arms.
  The adversary selects a reward vector $\boldsymbol{R}_t = (R_t^1, R_t^2, \ldots, R_t^K) \in [0, 1]^K$
  The agent observes reward $R_t^{I_t}$ and maybe also observes the rest of the reward vector depending on the specific problem set up.
end for

Algorithm 2 describes the general setup for adversarial MAB. At each time step, the agent will choose an arm I_t to pull and the adversary will decide the reward vector $\boldsymbol{R}_t$ for this time step. The agent might only be able to observe the reward for the arm he selects $R_t^{I_t}$ or the possible rewards for all the machines, $\boldsymbol{R}_t(\cdot)$. We still need two more pieces of information for the problem formulation. The first one is how much the adversary knows about the player's past behavior. It matters because, for some casino owners, they might adapt their reward strategies based on the player's behavior for more benefit. We will call the adversary who sets the rewards independent of the past history an **oblivious adversary** and the one that sets the rewards based on the past history a **non-oblivious adversary**. The second piece of information we have to specify is how much the player knows about the reward vector. We call the game in which a player has full knowledge of the reward vector a **full-information game** and the game with the knowledge of the reward for the action being played a **partial-information game**.

The difference between the oblivious and non-oblivious adversaries only starts to matter when we have a non-deterministic player. If we have a deterministic player, a player whose game strategy does not change, it is fairly easy to show that an adversary can always lower the regret to $\overline{RE} \geq n/2$ where n is the number of pulls the player takes. Therefore, let us focus on a non-deterministic player with full information in the first place. We can make use of the hedge algorithm to tackle this problem. In Algorithm 3, we first set G function to be

Algorithm 3 Hedge for adversarial multi-armed bandit

Initialize K arms;
$G_i(0)$ for $i = 1, 2, \ldots, K$;
for $t = 1, 2, \ldots, T$ **do**
The agent selects $A_t = i_t$ from the distribution $p(t)$, where

$$p_i(t) = \frac{exp(\eta G_i(t-1))}{\sum_j^K exp(\eta G_j(t-1))}$$

The agent observes reward vector g_t.
Let $G_i(t) = G(t-1) + g_t^i, \ \forall i \in [1, K]$.
end for

zero for all the arms, and use the softmax function to obtain the probability density function for the new action. η is a parameter greater than zero to control the temperature. The G function is updated by adding all the new rewards received for all the arms to allow the arm that has received the greatest reward being the most likely one to be selected. We refer this algorithm as Hedge. Hedge is also a building block for the method used under a partial-information game. If we want to limit the agent's observation to only R_t^i, then we will need to expand our reward scalar to a vector such that it can be passed to hedge. Exponential-weight algorithm for Exploration and Exploitation (Exp3) is the method that builds on Hedge for the partial information game. It further utilizes a blending of $p(t)$ and a uniform distribution to ensure that all the machines will get selected and hence the name exploration and exploitation. Auer et al. (1995) have more details on how Exp3 can be used and offer analysis on the confidence bound for regret.

2.2.4 *Contextual Bandits*

Contextual bandits are also sometimes called associative search tasks. The associative search tasks are best explained in comparison with the non-associative search tasks, the MAB tasks that we just described. When the reward function for the task is stationary, we only need to find the best action, otherwise, when the task is non-stationary we will try to keep track of the changes. This is the case for the non-associative search tasks, but reinforcement learning problems can become a lot more complicated. For instance, if we have several MAB tasks to play, and we will have to choose one at each time step. Even though we can still estimate the general expected reward, the performance is unlikely to be optimal. For cases like this, it would be useful to associate certain features of the slot machine with the learned expected reward. Imagine, for each slot machine, there is an LED light shining different colors at different times. Let us say if the machines with the red light always yield greater reward than the ones with the blue light, then we will be able to associate that information with our action selection policy, i.e. selecting the machines with red lights more.

Contextual bandit tasks are an intermediate between MAB tasks and the full reinforcement learning problems. They are similar to the MAB tasks because, for both situations, their actions only impact the immediate rewards. It is also similar to the full reinforcement learning setting because both require learning a policy function. To convert contextual bandits tasks to full reinforcement learning tasks, we will need to allow the actions to influence not just the intermediate rewards but also the future environment states.

2.3 Markov Decision Process

2.3.1 Markov Process

A Markov process (MP) is a discrete stochastic process with Markov property, which simplifies the simulation of the world in continuous space. Figure 2.4 shows an example of MP. Each circle represents a state and each edge represents a state transition. This graph simulates how a person works on two tasks and goes to bed in the end. To understand how this diagram works, let us look at this example together. Imagine, we are currently doing "Task1", and then with a probability of 0.7 we continue to execute "Task2", after which if we manage to pass with a probability of 0.6, we will pass the exam and then go straight to bed.

Figure 2.5 shows the **probabilistic graphical model** of MP in a probabilistic inference view, which will be frequently mentioned in later sections. In probabilistic graphical models, specifically the ones that we use in this book, a circle indicates a variable, and the arrow with a single direction indicates the relationship between two variables. For example, "$a \rightarrow b$" indicates that variable b is dependent on variable a. The variable in a circle in white denotes a normal variable, while the variable in a circle with a shade of gray denotes an observed variable (shown in later figures of Sect. 2.7), which provides information for taking an inference process of other normal variables. A solid black square with variables inside indicates those variables are iterative, which will be shown in later figures as well. The probabilistic graphical model can help us to have a more intuitive sense of the relationships between

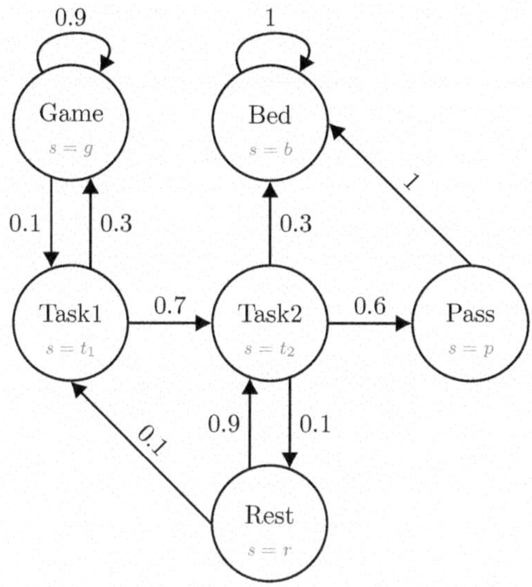

Fig. 2.4 A Markov process example. s denotes the current state and the values on the edges denote the probabilities of moving from one state to another

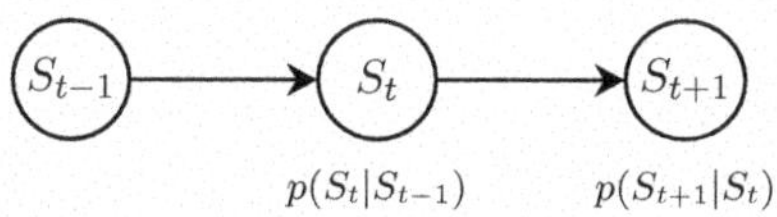

Fig. 2.5 Graphical model of Markov process: a finite representation with t indicating the time step and $p(S_{t+1}|S_t)$ indicating the state transition probability

variables in reinforcement learning, as well as providing rigorous references when we derive the gradients with respect to different variables along the MP chains.

MP follows the assumption of **Markov chain** where the next state S_{t+1} is only dependent on the current state S_t, with the probability of a state jumping to the next state described as follows:

$$p(S_{t+1}|S_t) = p(S_{t+1}|S_0, S_1, S_2, \ldots, S_t) \tag{2.9}$$

This formula describes the "memoryless" property, i.e. **Markov property**, of the Markov chain. Also, if $p(S_{t+2} = s'|S_{t+1} = s) = p(S_{t+1} = s'|S_t = s)$ holds for any time step t and for all possible states, then it is a stationary transition function along the time axis, which is called the **time-homogeneous** property, and the corresponding Markov chain is time-homogeneous Markov chain.

We also frequently use s' to represent the next state, in which the probability that state s at time t will lead to state s' at time $t + 1$ is as following in a time-homogeneous Markov chain:

$$p(s'|s) = p(S_{t+1} = s'|S_t = s) \tag{2.10}$$

The time-homogeneous property is a basic assumption for most of the derivations in the book, and we will not mention it but follow it as default in most cases. However, in practice, the time-homogeneous may not always hold, especially for non-stationary environments, multi-agent reinforcement learning, etc., which concerns with time-inhomogeneous/non-homogeneous cases.

Given a finite **state set** $\mathcal{S}$, we can have a **state transition matrix** $\boldsymbol{P}$. The $\boldsymbol{P}$ for Fig. 2.4 is as follows:

$$\boldsymbol{P} = \begin{array}{c} \begin{array}{cccccc} g & t_1 & t_2 & r & p & b \end{array} \\ \left[\begin{array}{cccccc} 0.9 & 0.1 & 0 & 0 & 0 & 0 \\ 0.3 & 0 & 0.7 & 0 & 0 & 0 \\ 0 & 0 & 0 & 0.1 & 0.6 & 0.3 \\ 0 & 0.1 & 0.9 & 0 & 0 & 0 \\ 0 & 0 & 0 & 0 & 0 & 1 \\ 0 & 0 & 0 & 0 & 0 & 1 \end{array}\right] \end{array} \begin{array}{c} g \\ t_1 \\ t_2 \\ r \\ p \\ b \end{array}$$

where $\boldsymbol{P}_{i,j}$ represents the probability of transferring the current state S_i to the next state S_j. For example, in Fig. 2.4, state $s = r$ will jump to state $s = t_1$ with a probability of 0.1, and to state $s = t_2$ with 0.9. The sum of each row is equal to 1 and the $\boldsymbol{P}$ is always a square matrix. These probabilities indicate the whole process is stochastic. Markov process can be represented by a tuple of $< \mathcal{S}, \boldsymbol{P} >$. Many simple processes in our world can be represented by this random process as an approximation, which is also a foundation of reinforcement learning methods. Mathematically, the next state is sampled from $\boldsymbol{P}$ as follows:

$$S_{t+1} \sim \boldsymbol{P}_{S_t,\cdot} \tag{2.11}$$

where symbol $\sim$ represents the next state S_{t+1} is randomly sampled according to the categorical distribution of $\boldsymbol{P}_{S_t,\cdot}$.

For infinite state set or continuous case, a finite matrix cannot be used to represent the transition relationship anymore. Therefore the transition function $p(s'|s)$ is applied as before, with a corresponding relationship $p(s'|s) = \boldsymbol{P}_{s,s'}$ for finite cases.

2.3.2 Markov Reward Process

Even though the agent can interact with the environment via the state transition matrix $\boldsymbol{P}_{s,s'} = p(s'|s)$, there is no way for MP to provide reward feedback from the environment to the agent. To provide the feedback, Markov reward process (MRP) extends MP from $< \mathcal{S}, \boldsymbol{P} >$ to $< \mathcal{S}, \boldsymbol{P}, R, \gamma >$. The R and γ represent the **reward function** and **reward discount factor**, respectively. An example of MRP is shown in Fig. 2.6. Figure 2.7 shows the graphical model of MRP in a probabilistic inference

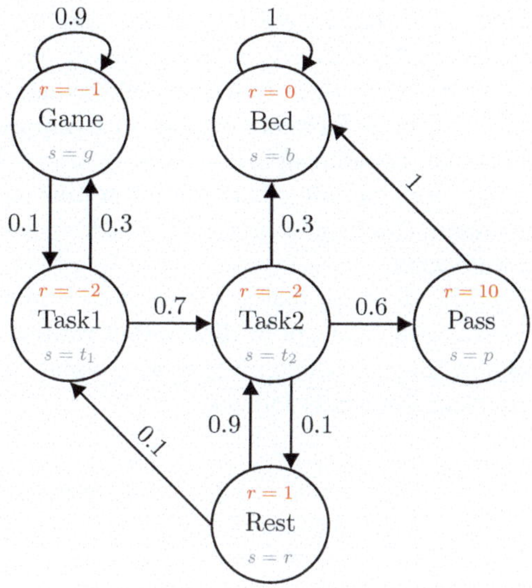

Fig. 2.6 A Markov reward process example. The s denotes the current state and the r denotes the immediate reward for each state. The values on the edges denote the probabilities of moving from one state to the next state

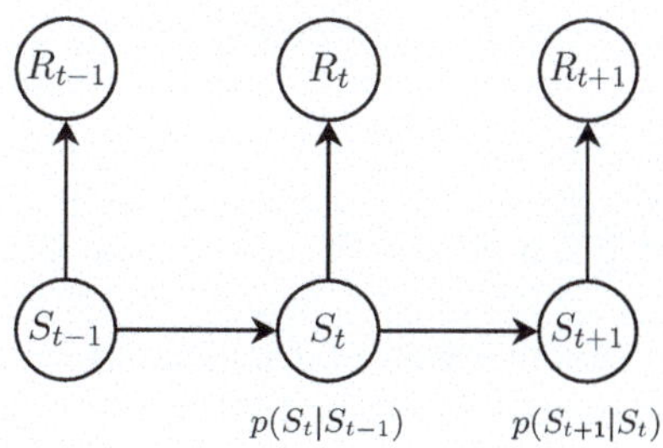

Fig. 2.7 Graphical model of a Markov reward process: a finite representation with t indicating the time step

view. The reward function depends on the current state:

$$R_t = R(S_t) \tag{2.12}$$

But the reward is also a result of the previous action based on the previous state. To better understand the reward as a function of the state, let us take a look at this example. If the agent passes the exam, the agent can obtain an immediate reward of ten, and taking a rest can obtain a reward of one, but if the agent works on a task, the reward of two will be lost. Given a list of immediate reward r for each time step in a single trajectory τ, **a return is the cumulative reward of a trajectory**, in which the **undiscounted return** of finite process with T time steps (not counting the initial one) is as follows:

$$G_{t=0:T} = R(\tau) = \sum_{t=0}^{T} R_t \tag{2.13}$$

where R_t is the immediate reward at time step t, and T represents the time step of the terminal state, or the total number of steps in a finite episode. For example, the trajectory (g, t_1, t_2, p, b) has an undiscounted return of $5 = -1 - 2 - 2 + 10$. Note that some other literature may use G to represent the return, and R to represent the immediate reward, but in this book, we use R as the reward function, therefore $R_t = R(S_t)$ gives the immediate reward at time step t, while $R(\tau) = G_{t=0}^{(T)}$ represents the return along the trajectory $\tau_{0:T}$, and r as a general representation of immediate reward value.

Often, the steps that are closer have a greater impact than the distant ones. We introduce the concept of **discounted return**. The discounted return is a weighted sum of rewards which gives more weights to the closer time steps. We define the discounted return as follows:

$$G_{t=0:T} = R(\tau) = \sum_{t=0}^{T} \gamma^t R_t. \tag{2.14}$$

where a **reward discount factor** $\gamma \in [0, 1]$ is used to reduce the weights as the time step increases. For example in Fig. 2.6, given $\gamma = 0.9$, the trajectory (g, t_1, t_2, p, b) has a return of $2.87 = -1 - 2 \times 0.9 - 2 \times 0.9^2 + 10 \times 0.9^3$. If $\gamma = 0$, the return is only related to the current immediate reward; if $\gamma = 1$, it is the undiscounted return. The discounted factor is especially critical when handling with infinite MRP cases, as it can prevent the return from going to infinite as the time step goes to infinite. Therefore it makes the infinite MRP process evaluative.

Another view of discount factor γ: for conciseness, the reward discount factor γ is sometimes omitted in literature (Levine 2018) in a discrete-time finite-horizon MRP. The discount factor can also be incorporated into the process by simply modifying the transition dynamics, such that any action produces a transition into an absorbing state with probability $1 - \gamma$, and all standard transition probabilities are multiplied by γ.

The **value function** $V(s)$ represents the **expected return from the state** s. For example, if there are two different next states S_1 and S_2, the values estimated with the current policy are $V^\pi(S_1)$ and $V^\pi(S_2)$. The agent policy usually selects the next state with higher value. If the agent acts according to the policy π, we denote the value function as $V^\pi(s)$:

$$V(s) = \mathbb{E}[R_t | S_0 = s] \tag{2.15}$$

A simple way to estimate the $V(s)$ is **Monte Carlo method**, we can randomly sample a large number of trajectories starting from state s according to the given state transition matrix $\boldsymbol{P}$. Take Fig. 2.6 as an example, given $\gamma = 0.9$ and $\boldsymbol{P}$, to estimate $V^\pi(s = t_2)$, we can randomly sample four trajectories as follows and compute the returns of all trajectories individually (Note that, in practice, the number of trajectories is usually far larger than four, but for demonstration purposes

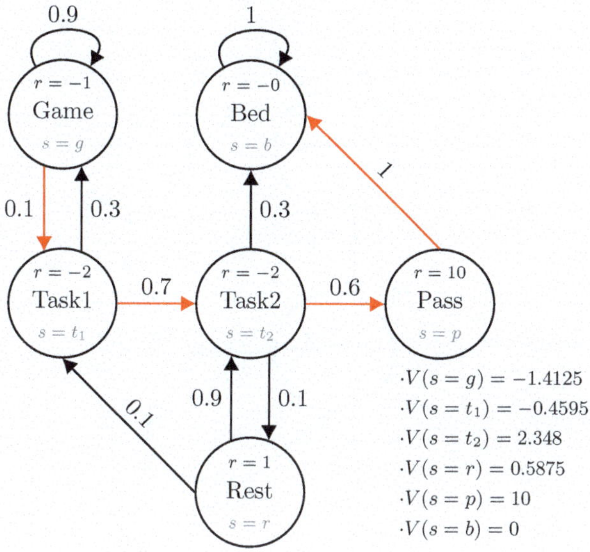

Fig. 2.8 Markov reward process and the estimated value function $V(s)$ by randomly choosing four trajectories for each state i.e., Monte Carlo method. The red edges indicate the learned policy

we simply sample four trajectories here.):

- $s = (t_2, b)$, $R = -2 + 0 \times 0.9 = -2$
- $s = (t_2, p, b)$, $R = -2 + 10 \times 0.9 + 0 \times 0.9^2 = 7$
- $s = (t_2, r, t_2, p, b)$, $R = -2 + 1 \times 0.9 - 2 \times 0.9^2 + 10 \times 0.9^3 + 0 \times 0.9^4 = 4.57$
- $s = (t_2, r, t_1, t_2, b)$, $R = -2 + 1 \times 0.9 - 2 \times 0.9^2 - 2 \times 0.9^3 + 0 \times 0.9^4 = -0.178$

Given the returns of all trajectories, the estimated expected return under state $s = t_2$ is $V(s = t_2) = (-2 + 7 + 4.57 - 0.178)/4 = 2.348$. Figure 2.8 shows all estimated expected returns for all states. Given the expected returns under all states, the simplest policy for the agent is to jump to the next state that has the highest expected return. The actions that can maximize the expected return are highlighted by red in Fig. 2.8. Apart from Monte Carlo methods, there are many other methods to compute $V(s)$, such as Bellman expectation equation and inverse matrix method, etc., which will be introduced later.

2.3.3 Markov Decision Process

Markov decision processes (MDPs) have been studied since the 1950s and have been widely used in modeling disciplines such as economics, control theory, and robotics. To model the process of sequential decision making, MDP is better than MR and MRP. As Fig. 2.9 shows, different from MRP that the immediate rewards are conditioned on the state only (reward values on nodes), the immediate rewards of MDP are associated with the action and state (reward values on edges). Likewise,

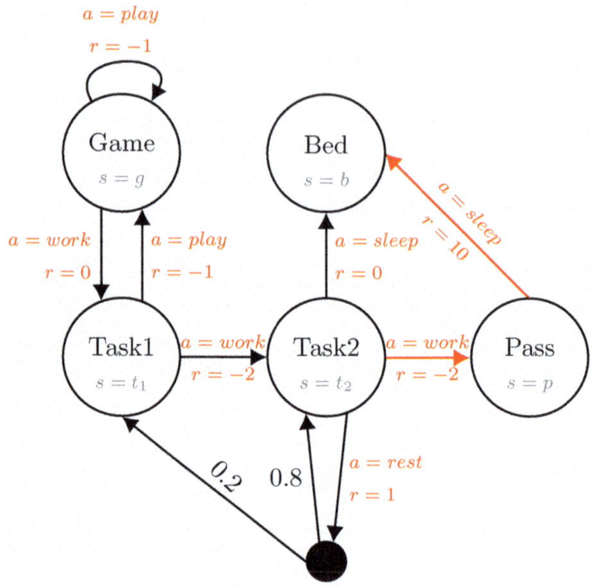

Fig. 2.9 A Markov decision process example, different from MRP in which the immediate rewards are associated with the state only. The immediate rewards of MDP is associated with the current state and the action just taken. The black solid node is an initial state

given an action under a state, the next state is not fixed. For example, if the agent acts "rest" under state $s = t_2$, the next state can be either $s = t_1$ or t_2. Fig. 2.10 shows the graphical model of MDP in a probabilistic inference view.

As mentioned above, MP can be defined as the tuple $< \mathcal{S}, \boldsymbol{P} >$, and MRP is defined as the tuple $< \mathcal{S}, \boldsymbol{P}, R, \gamma >$, where the element of state transition matrix is $\boldsymbol{P}_{s,s'} = p(s'|s)$. This representation extends the finite-dimension state transition matrix to an infinite-dimension probability function. Here, MDP is defined as a tuple of $< \mathcal{S}, \mathcal{A}, \boldsymbol{P}, R, \gamma >$. The element of state transition matrix becomes:

$$p(s'|s, a) = p(S_{t+1} = s'|S_t = s, A_t = a) \tag{2.16}$$

For instance, most of the edges in Fig. 2.9 have a state transition probability of one, e.g., $p(s' = t_2|s = t1, a = work) = 1$, except that $p(s'|s = t_2, a = rest) = [0.2, 0.8]$ which means if the agent performs action a = rest at state $s = t_2$, it has 0.2 probability will transit to state $s' = t_1$, and 0.8 probability will keep the current state. The non-existing edges have a state transition probability of zero e.g., $p(s' = t_2|s = t_1, a = rest) = 0$.

$\mathcal{A}$ represents the **finite action set** $\{a_1, a_2, \ldots\}$, and the immediate reward becomes:

$$R_t = R(S_t, A_t) \tag{2.17}$$

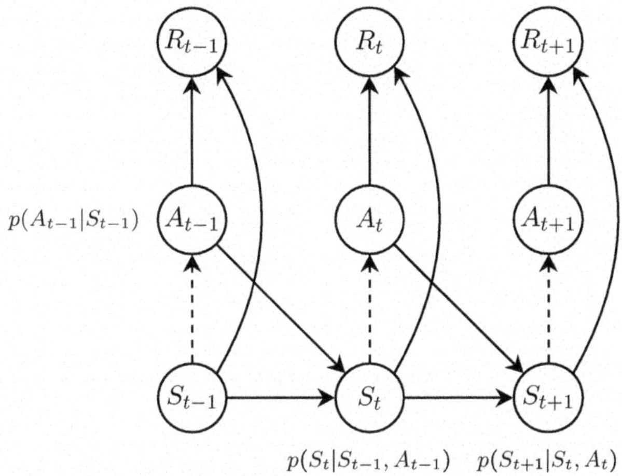

Fig. 2.10 Graphical model of Markov decision process: a finite representation with t indicating the time step, $p(A_t|S_t)$ as the action choice based on current state, and $p(S_{t+1}|S_t, A_t)$ as the state transition probability based on current state and action. The dashed lines indicate the decision process made by the agent

A **policy** π represents the way in which the agent behaves based on its observations of the environment. Specifically, the policy is a mapping from the each state $s \in \mathcal{S}$ and action $a \in \mathcal{A}$ to the probability distribution $\pi(a|s)$ for taking action a in state s, where the distribution is:

$$\pi(a|s) = p(A_t = a|S_t = s), \exists t \tag{2.18}$$

Expected return is the expectation of returns over all possible trajectories under a policy. Therefore, **the goal of reinforcement learning is to find the higher expected return by optimizing the policy**. Mathematically, given the start-state distribution ρ_0 and the policy π, the probability of a T-step trajectory for MDP is:

$$p(\tau|\pi) = \rho_0(S_0) \prod_{t=0}^{T-1} p(S_{t+1}|S_t, A_t)\pi(A_t|S_t) \tag{2.19}$$

Given the reward function R and all possible trajectories τ, the **expected return** $J(\pi)$ is defined as follows:

$$J(\pi) = \int_{\tau} p(\tau|\pi)R(\tau) = \mathbb{E}_{\tau\sim\pi}[R(\tau)] \tag{2.20}$$

where p here means that the trajectory with higher probability will have a higher weight to the expected return. The **RL optimization problem** is to improve the policy for maximizing the expected return with optimization methods. The **optimal policy** π^*can be expressed as:

$$\pi^* = \arg\max_{\pi} J(\pi) \tag{2.21}$$

where the $*$ symbol means "optimal" for the rest of the book.

Given policy π, the **value function** $V(s)$, the expected return under the state, can be defined as:

$$\begin{aligned} V^{\pi}(s) =& \mathbb{E}_{\tau\sim\pi}[R(\tau)|S_0 = s] \\ =& \mathbb{E}_{A_t\sim\pi(\cdot|S_t)}\left[\sum_{t=0}^{\infty} \gamma^t R(S_t, A_t)|S_0 = s\right] \end{aligned} \tag{2.22}$$

where $\tau \sim \pi$ means the trajectories τ are sampled given the policy π, $A_t \sim \pi(\cdot|S_t)$ means the action under a state is sampled from the policy, the next state is determined by the state transition matrix $\boldsymbol{P}$ given state s and action a.

In MDP, given an action, we have the **action-value function**, which depends on both the state and the action just taken. It gives an expected return under a state and an action. If the agent acts according to a policy π, we denote it as $Q^{\pi}(s, a)$, which is defined as:

$$\begin{aligned} Q^{\pi}(s, a) &= \mathbb{E}_{\tau\sim\pi}[R(\tau)|S_0 = s, A_0 = a] \\ &= \mathbb{E}_{A_t\sim\pi(\cdot|S_t)}\left[\sum_{t=0}^{\infty} \gamma^t R(S_t, A_t)|S_0 = s, A_0 = a\right] \end{aligned} \tag{2.23}$$

We need to keep in mind that the $Q^{\pi}(s, a)$ depends on π, as the estimation of the value is an expectation over the trajectories by the policy π. This also indicates if the π changes, the corresponding $Q^{\pi}(s, a)$ will also change accordingly. We therefore usually call the value function estimated with a specific policy the **on-policy value function**, for the distinction from the **optimal value function** estimated with the optimal policy.

We can observe the relation between $v_{\pi}(s)$ and $q_{\pi}(s, a)$:

$$q_{\pi}(s, a) = \mathbb{E}_{\tau\sim\pi}[R(\tau)|S_0 = s, A_0 = a] \tag{2.24}$$

$$v_{\pi}(s) = \mathbb{E}_{a\sim\pi}[q_{\pi}(s, a)] \tag{2.25}$$

There are two simple ways to compute the value function $v_{\pi}(s)$ and action-value function $q_{\pi}(s, a)$: The first is the **exhaustive method** follows Eq. (2.19), it first computes the probabilities of all possible trajectories that start from a state, and then follows Eqs. (2.22) and (2.23) to compute the $V^{\pi}(s)$ and $Q^{\pi}(s, a)$ for this state.

The exhaustive method computes the $V^\pi(s)$ for each state individually. However, in practice, the number of possible trajectories would be large and even infinite. Instead of using all possible trajectories, we can use **Monte Carlo method** as described in the previous MRP section to estimate the $V^\pi(s)$ by randomly sampling a large number of trajectories. In reality, the estimation formulas of value functions can be simplified, by leveraging the Markov property in MRP, which leads to the Bellman equations in the next section.

2.3.4 Bellman Equation and Optimality

Bellman Equation

The Bellman equation, also known as the Bellman expectation equation, is used to compute the expectation of value function given policy π, over the sampled trajectories guided by the policy. We call this "on-policy" manner as in reinforcement learning the policy is usually changing, and the value function is conditioned on or estimated by current policy.

Recall that the definitions of a value function or an action-value function are $v_\pi(s) = \mathbb{E}_{\tau\sim\pi}[R(\tau)|S_0 = s]$ and $q_\pi(s,a) = \mathbb{E}_{\tau\sim\pi}[R(\tau)|S_0 = s, A_0 = a]$. We can derive the Bellman equation for on-policy state-value function in a recursive relationship:

$$\begin{aligned}
v_\pi(s) &= \mathbb{E}_{a\sim\pi(\cdot|s),s'\sim p(\cdot|s,a)}[R(\tau_{t:T})|S_t = s] \\
&= \mathbb{E}_{a\sim\pi(\cdot|s),s'\sim p(\cdot|s,a)}[R_t + \gamma R_{t+1} + \gamma^2 R_{t+2} + \ldots + \gamma^T R_T|S_t = s] \\
&= \mathbb{E}_{a\sim\pi(\cdot|s),s'\sim p(\cdot|s,a)}[R_t + \gamma(R_{t+1} + \gamma R_{t+2} + \ldots + \gamma^{T-1} R_T)|S_0 = s] \\
&= \mathbb{E}_{a\sim\pi(\cdot|s),s'\sim p(\cdot|s,a)}[R_t + \gamma R_{\tau_{t+1:T}}|S_t = s] \\
&= \mathbb{E}_{A_t\sim\pi(\cdot|S_t),S_{t+1}\sim p(\cdot|S_t,A_t)}[R_t + \gamma\mathbb{E}_{a\sim\pi(\cdot|s),s'\sim p(\cdot|s,a)}[R_{\tau_{t+1:T}}]|S_t = s] \\
&= \mathbb{E}_{A_t\sim\pi(\cdot|S_t),S_{t+1}\sim p(\cdot|S_t,A_t)}[R_t + \gamma v_\pi(S_{t+1})|S_t = s] \\
&= \mathbb{E}_{a\sim\pi(\cdot|s),s'\sim p(\cdot|s,a)}[r + \gamma v_\pi(s')]
\end{aligned} \tag{2.26}$$

The final formula above holds because s, a are general representations of states and actions, while S_t, A_t are state and action at time step t only. S_t, A_t are sometimes separated from the general representations s, a to show more clearly over whom the expectation is taken over in some of above formulas.

Note that in the above derivation we show the Bellman equation for MDP process, however, the Bellman equation for MRP can be derived by simply removing the action from it:

$$v(s) = \mathbb{E}_{s'\sim p(\cdot|s)}[r + \gamma v(s')] \tag{2.27}$$

There is also Bellman equation for on-policy action-value function: $q_\pi(s,a) = \mathbb{E}_{s'\sim p(\cdot|s,a)}[R(s,a) + \gamma \mathbb{E}_{a'\sim\pi(\cdot|s')}[q_\pi(s',a')]]$, which can be derived as follows:

$$
\begin{aligned}
&q_\pi(s,a)\\
&= \mathbb{E}_{a\sim\pi(\cdot|s),s'\sim p(\cdot|s,a)}[R(\tau_{t:T})|S_t = s, A_t = a]\\
&= \mathbb{E}_{a\sim\pi(\cdot|s),s'\sim p(\cdot|s,a)}[R_t + \gamma R_{t+1} + \gamma^2 R_{t+2} + \cdots + \gamma^T R_T|S_t = s, A_t = a]\\
&= \mathbb{E}_{a\sim\pi(\cdot|s),s'\sim p(\cdot|s,a)}[R_t + \gamma(R_{t+1} + \gamma R_{t+2} + \cdots + \gamma^{T-1} R_T)|S_t = s, A_t = a]\\
&= \mathbb{E}_{S_{t+1}\sim p(\cdot|S_t,A_t)}[R_t + \gamma \mathbb{E}_{a\sim\pi(\cdot|s),s'\sim p(\cdot|s,a)}[R_{\tau_{t+1:T}}]|S_t = s]\\
&= \mathbb{E}_{S_{t+1}\sim p(\cdot|S_t,A_t)}[R_t + \gamma \mathbb{E}_{A_{t+1}\sim\pi(\cdot|S_{t+1})}[q_\pi(S_{t+1}, A_{t+1})]|S_t = s]\\
&= \mathbb{E}_{s'\sim p(\cdot|s,a)}[R(s,a) + \gamma \mathbb{E}_{a'\sim\pi(\cdot|s')}[q_\pi(s',a')]]
\end{aligned}
$$

The above derivation is based on the finite MDP with maximal length of T, however, these formulas still hold when in the infinite MDP, simply with T replaced by "∞". The two Bellman equations also do not depend on the formats of policy, which means they work for both stochastic policies $\pi(\cdot|s)$ and deterministic policies $\pi(s)$. The usage of $\pi(\cdot|s)$ is for simplicity here. Also, in deterministic transition processes, we have $p(s'|s,a) = 1$.

Solutions of Bellman Equation

The Bellman equation for MRP as in Eq. (2.27) can be solved directly if the transition function/matrix is known, which is called the **inverse matrix method**. We rewrite Eq. (2.27) in a vector form for cases with discrete and finite state space as:

$$\boldsymbol{v} = \boldsymbol{r} + \gamma \boldsymbol{P}\boldsymbol{v} \tag{2.28}$$

where $\boldsymbol{v}$ and $\boldsymbol{r}$ are vectors with their elements $v(s)$ and $R(s)$ for all $s \in \mathcal{S}$, and $\boldsymbol{P}$ is the transition probability matrix with elements $p(s'|s)$ for all $s, s' \in \mathcal{S}$.

Given $\boldsymbol{v} = \boldsymbol{r} + \gamma \boldsymbol{P}\boldsymbol{v}$, we can directly solve it with:

$$\boldsymbol{v} = (\boldsymbol{1} - \gamma \boldsymbol{P})^{-1}\boldsymbol{r} \tag{2.29}$$

the complexity of the solution is $O(n^3)$, where n is the number of states. Therefore this method does not work for a large number of states, meaning it may not be feasible for large-scale or continuous-valued problems. Fortunately, there are some iterative methods for solving the large-scale MRP problems in practice, like dynamic programming, Monte Carlo estimation, and temporal-difference learning, which will be introduced in detail in later sections.

Optimal Value Functions

Since on-policy value functions are estimated with respect to the policy itself, different policies will lead to different value functions, even for the same set of states and actions. Among all those different value functions, we define the optimal value function as:

$$v_*(s) = \max_{\pi} v_\pi(s), \forall s \in \mathcal{S}, \tag{2.30}$$

which is actually the **optimal state-value function**. We also have the **optimal action-value function** as:

$$q_*(s, a) = \max_{\pi} q_\pi(s, a), \forall s \in \mathcal{S}, a \in \mathcal{A}, \tag{2.31}$$

And they have the relationship:

$$q_*(s, a) = \mathbb{E}[R_t + \gamma v_*(S_{t+1})|S_t = s, A_t = a], \tag{2.32}$$

which can be derived easily by taking the maximization in the last formula of Eq. (2.26) and plugging in Eqs. (2.25) and (2.30):

$$\begin{aligned} q_*(s, a) &= \mathbb{E}\left[R(s, a) + \gamma \max_{\pi} \mathbb{E}\left[q_\pi\left(s', a'\right)\right]\right] \\ &= \mathbb{E}\left[R(s, a) + \gamma \max_{\pi} v_\pi\left(s'\right)\right] \\ &= \mathbb{E}[R_t + \gamma v_*(S_{t+1})|S_t = s, A_t = a]. \end{aligned} \tag{2.33}$$

Another relationship between the two is:

$$v_*(s) = \max_{a \sim \mathcal{A}} q_*(s, a) \tag{2.34}$$

which is obvious by simply maximizing the two sides of Eq. (2.25).

Bellman Optimality Equation

In the above sections we introduced on-policy Bellman equations for normal value functions, as well as the definitions of optimal value functions. So we can apply the Bellman equation on the pre-defined optimal value functions, which gives us the **Bellman optimality equation**, or called Bellman equation for optimal value functions, as follows.

The Bellman equation for optimal state-value function is:

$$v_*(s) = \max_a \mathbb{E}_{s' \sim p(\cdot|s,a)}[R(s,a) + \gamma v_*(s')], \tag{2.35}$$

which can be derived as follows:

$$\begin{aligned}
v_*(s) &= \max_a \mathbb{E}_{\pi^*, s' \sim p(\cdot|s,a)}[R(\tau_{t:T})|S_t = s] \\
&= \max_a \mathbb{E}_{\pi^*, s' \sim p(\cdot|s,a)}\Big[R_t + \gamma R_{t+1} + \gamma^2 R_{t+2} + \cdots + \gamma^T R_T|S_t = s\Big] \\
&= \max_a \mathbb{E}_{\pi^*, s' \sim p(\cdot|s,a)}[R_t + \gamma R_{\tau_{t+1:T}}|S_t = s] \\
&= \max_a \mathbb{E}_{s' \sim p(\cdot|s,a)}\Big[R_t + \gamma \max_{a'} \mathbb{E}_{\pi^*, s' \sim p(\cdot|s,a)}\left[R_{\tau_{t+1:T}}\right]|S_t = s\Big] \\
&= \max_a \mathbb{E}_{s' \sim p(\cdot|s,a)}[R_t + \gamma v_*(S_{t+1})|S_t = s] \\
&= \max_a \mathbb{E}_{s' \sim p(\cdot|s,a)}[R(s,a) + \gamma v_*(s')]
\end{aligned} \tag{2.36}$$

Bellman equation for optimal action-value function is:

$$q_*(s,a) = \mathbb{E}_{s' \sim p(\cdot|s,a)}[R(s,a) + \gamma \max_{a'} q_*(s',a')], \tag{2.37}$$

which can be derived similarly. Readers can take a practice by finishing the proof.

2.3.5 *Other Important Concepts*

Deterministic and Stochastic Policies

In the previous sections, the policy is represented as a probability distribution as $\pi(a|s) = p(A_t = a|S_t = s)$, where the action of the agent is sampled from the distribution. A policy with action sampled from the probability distribution is actually called the **stochastic policy distribution**, with the action:

$$a = \pi(\cdot|s) \tag{2.38}$$

However, if we reduce the variance of the probability distribution of a stochastic policy and narrow down its range to the limit, we will get a Dirac delta function (δ function) as a distribution, which is the **deterministic policy** $\pi(s)$. Deterministic policy $\pi(s)$ also means given a state there is only one unique action as follows:

$$a \sim \pi(s) \tag{2.39}$$

Note that the deterministic policy is no longer a mapping from a state and action to the conditional probability distribution, but rather a mapping from a state to an action directly. This slight difference will lead to some different derivations in the policy gradient method introduced in later sections. More detailed categories of policies in reinforcement learning, especially for deep reinforcement learning with parameterized policies are introduced in Sect. 2.7.3.

Partially Observed Markov Decision Process

As mentioned in previous sections, when the state in reinforcement learning environment is not fully represented by the observation for the agent, the environment is partially observable. For a Markov decision process, it is called the partially observed Markov decision process (POMDP), which forms a challenge for improving the policy without complete information of the environment states.

2.3.6 *Summary of Terminology in Reinforcement Learning*

Apart from the terminology in mathematics notations at the beginning of the book, the summary of terminology for common contents in reinforcement learning is provided as follows:

- R the reward function, $R_t = R(S_t)$ as the reward of state S_t for MRP, $R_t = R(S_t, A_t)$ for MDP, $S_t \in \mathcal{S}$.
- $R(\tau)$ the γ-discounted return of a trajectory τ, $R(\tau) = \sum_{t=0}^{\infty} \gamma^t R_t$.
- $p(\tau)$ the probability of a trajectory:
  - $p(\tau) = \rho_0(S_0) \prod_{t=0}^{T-1} p(S_{t+1}|S_t)$ for MP and MRP, $\rho_0(S_0)$ as start-state distribution.
  - $p(\tau|\pi) = \rho_0(S_0) \prod_{t=0}^{T-1} p(S_{t+1}|S_t, A_t)\pi(A_t|S_t)$ for MDP, $\rho_0(S_0)$ as start-state distribution.
- $J(\pi)$ the expected return of policy π, $J(\pi) = \int_\tau p(\tau|\pi)R(\tau) = \mathbb{E}_{\tau\sim\pi}[R(\tau)]$
- π^* optimal policy: $\pi^* = \arg\max_\pi J(\pi)$
- $v_\pi(s)$ value of state s under policy π (expected return)
- $v_*(s)$ value of state s under the optimal policy
- $q_\pi(s, a)$ value of taking action a in state s under policy π
- $q_*(s, a)$ value of taking action a in state s under the optimal policy
- $V(s)$ the estimates of state-value function for MRP starting from state s:

- $V^{\pi}(s)$ the estimates of on-policy state-value function for MDP, given a policy π, we have expected return:
  - $V^{\pi}(s) \approx v_{\pi}(s) = \mathbb{E}_{\tau\sim\pi}[R(\tau)|S_0 = s]$
- $Q^{\pi}(s, a)$ the estimates of on-policy action-value function for MDP, given a policy π, we have expected return:
  - $Q^{\pi}(s, a) \approx q_{\pi}(s, a) = \mathbb{E}_{\tau\sim\pi}[R(\tau)|S_0 = s, A_0 = a]$
- $V^*(s)$ the estimates of optimal state-value function for MDP, we have expected return according to the optimal policy:
  - $V^*(s) \approx v_*(s) = \max_{\pi} \mathbb{E}_{\tau\sim\pi}[R(\tau)|S_0 = s]$
- $Q^*(s, a)$ the estimates of on-policy action-value function for MDP, we have expected return according to the optimal policy:
  - $Q^*(s, a) \approx q_*(s, a) = \max_{\pi} \mathbb{E}_{\tau\sim\pi}[R(\tau)|S_0 = s, A_0 = a]$
- $A^{\pi}(s, a)$ the estimated advantage function of state s and action a:
  - $A^{\pi}(s, a) = Q^{\pi}(s, a) - V^{\pi}(s)$
- Relationship of on-policy state-value function $v_{\pi}(s)$ and on-policy action-value function $q_{\pi}(s, a)$:
  - $v_{\pi}(s) = \mathbb{E}_{a\sim\pi}[q_{\pi}(s, a)]$
- Relationship of optimal state-value function $v_{\pi}(s)$ and optimal action-value function $q_{\pi}(s, a)$:
  - $v_*(s) = \max_a q_*(s, a)$
- $a_*(s)$ the optimal action for state s according to optimal action-value function:
  - $a_*(s) = \arg\max_a q_*(s, a)$
- Bellman equations for on-policy state-value function:
  - $v_{\pi}(s) = \mathbb{E}_{a\sim\pi(\cdot|s), s'\sim p(\cdot|s,a)}[R(s, a) + \gamma v_{\pi}(s')]$
- Bellman equations for on-policy action-value function:
  - $q_{\pi}(s, a) = \mathbb{E}_{s'\sim p(\cdot|s,a)}[R(s, a) + \gamma \mathbb{E}_{a'\sim\pi(\cdot|s')}[q_{\pi}(s', a')]]$
- Bellman equations for optimal state-value function:
  - $v_*(s) = \max_a \mathbb{E}_{s'\sim p(\cdot|s,a)}[R(s, a) + \gamma v_*(s')]$
- Bellman equations for optimal action-value function:
  - $q_*(s, a) = \mathbb{E}_{s'\sim p(\cdot|s,a)}[R(s, a) + \gamma \max_{a'} q_*(s', a')]$

2.4 Dynamic Programming

Dynamic Programming (DP) was first introduced by Richard E. Bellman in the 1950s (Bellman et al. 1954) and has been successfully applied to a range of challenging fields. In this term, "dynamic" means that the problem has sequential or temporal components, and "programming" refers to an optimizing policy. DP provides a general framework for complex dynamic problems by breaking them down into sub-problems. For example, each number in the Fibonacci sequence is the sum of the two preceding ones, starting from 0 and 1. One can calculate the 4th number $F_4 = F_3 + F_2$ by $F_4 = (F_2 + F_1) + F_2$ by reusing the solution of the preceding sub-problem $F_2 = F_1 + F_0$. However, DP requires full knowledge of the environment, such as the reward model and the transition model, of which we often have limited knowledge in reinforcement learning. Nonetheless, it does provide a fundamental framework for learning to interact with the MDP incrementally, as most of the reinforcement learning algorithms attempt to achieve.

There are two properties that a problem must have for DP to be applicable: **optimal substructure** and **overlapping sub-problems**. Optimal substructure means that the optimal solution of a given problem can be decomposed into solutions to its sub-problems. Overlapping sub-problems implies that the number of sub-problems is finite and sub-problems occur recursively so that the sub-solutions can be cached and reused. While MDPs with finite actions and states satisfy both properties, the Bellman equation gives the recursive decomposition and value functions store the optimal solution of sub-problems. So in this section, we assume that state set and action set are both finite, and a perfect model of the environment is given.

2.4.1 Policy Iteration

Policy Iteration aims to manipulate the policy directly. Starting from arbitrary policy π, we can evaluate it by applying the Bellman equation recursively:

$$v_\pi(s) = \mathbb{E}_\pi[R_t + \gamma v_\pi(S_{t+1}) | S_t = s] \tag{2.40}$$

where the expectation is over all possible transitions based on full knowledge of the environment. A natural idea to obtain a better policy is acting greedily with respect to v_π:

$$\pi'(s) = \text{greedy}(v_\pi) = \arg\max_{a \in \mathcal{A}} q_\pi(s, a). \tag{2.41}$$

The improvement can be proved by:

$$
\begin{aligned}
v_\pi(s) &= q_\pi(s, \pi(s)) \\
&\leq q_\pi(s, \pi'(s)) \\
&= \mathbb{E}_{\pi'}[R_t + \gamma v_\pi(S_{t+1}) \,|\, S_t = s] \\
&\leq \mathbb{E}_{\pi'}[R_t + \gamma q_\pi(S_{t+1}, \pi'(S_{t+1})) \,|\, S_t = s] \\
&\leq \mathbb{E}_{\pi'}[R_t + \gamma R_{t+1} + \gamma^2 q_\pi(S_{t+2}, \pi'(S_{t+2})) \,|\, S_t = s] \\
&\leq \mathbb{E}_{\pi'}[R_t + \gamma R_{t+1} + \gamma^2 R_{t+2} + \ldots \,|\, S_t = s] = v_{\pi'}(s).
\end{aligned}
\tag{2.42}
$$

Apply policy evaluation and greedy improvement above successively until $\pi = \pi'$ forms the policy iteration. Generally, the procedure of policy iteration can be summarized as follows. Given an arbitrary policy π_t, for each state s in each iteration t, we first evaluate $v_{\pi_t}(s)$ and then find a better policy π_{t+1}. We call the former stage **policy evaluation** and the later stage **policy improvement**. Furthermore, we use the term **generalized policy iteration (GPI)** to refer to the general interaction of policy evaluation and policy improvement, as shown in Fig. 2.11.

One fundamental question is whether the process of policy iteration converges on the optimal value v_*. At each iteration in policy evaluation, for fixed and deterministic policy π, the value function update can be rewritten by the **Bellman expectation backup operator** $\mathcal{T}^\pi$:

$$
(\mathcal{T}^\pi V)(s) = (\mathcal{R}^\pi + \gamma \mathcal{P}^\pi V)(s) = \sum_{r,s'} (r + \gamma V(s')) P(r, s'|s, \pi(s)). \tag{2.43}
$$

Fig. 2.11 Generalized policy iteration

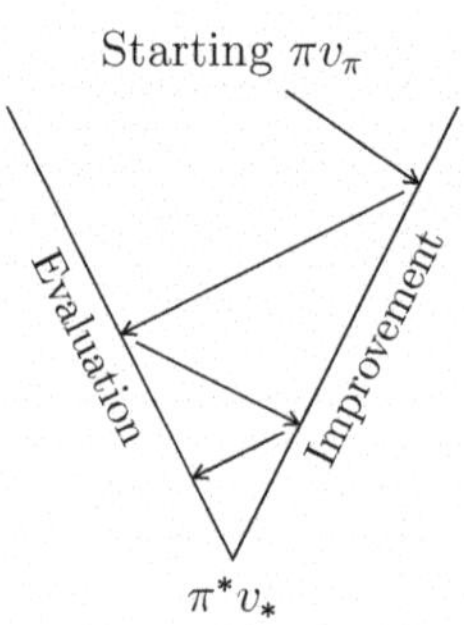

Then for arbitrary value functions V and V', we have the following contraction proof for $\mathcal{T}^{\pi}$:

$$
\begin{aligned}
|\mathcal{T}^{\pi}V(s) - \mathcal{T}^{\pi}V'(s)| &= |\sum_{r,s'}(r + \gamma V(s'))P(r, s'|s, \pi(s)) \\
&\quad - \sum_{r,s'}(r + \gamma V'(s'))P(r, s'|s, \pi(s))| \\
&= |\sum_{r,s'}\gamma(V(s') - V'(s'))P(r, s'|s, \pi(s))| \\
&\leq \sum_{r,s'}\gamma|V(s') - V'(s')|P(r, s'|s, \pi(s)) \\
&\leq \sum_{r,s'}\gamma\|V - V'\|_{\infty}P(r, s'|s, \pi(s)) \\
&= \gamma\|V - V'\|_{\infty},
\end{aligned}
\tag{2.44}
$$

where $\|V - V'\|_{\infty}$ is the ∞-norm. By contraction mapping theorem (also known as the Banach fixed-point theorem), iterative policy evaluation will converge on the unique fixed point of $\mathcal{T}^{\pi}$. Since $\mathcal{T}^{\pi}v_{\pi} = v_{\pi}$ is a fixed point, so that iterative policy evaluation converges on v_{π}. Note that the policy improvement is monotonic, and there is only a finite number of greedy policies with respect to value functions in finite MDP. The policy improvement will stop after a finite number of steps, i.e., the policy iteration will converge on v_*.

2.4.2 *Value Iteration*

The theoretical basis of **value iteration** is the **principle of optimality** which tells us that π is the optimal policy on one state if and only if π achieves the optimal value for any reachable successor state. So if we know the solution to sub-problems $v_*(s')$, we can find the solution of any initial state s by one-step full backups:

$$
v_*(s) = \max_{a \in \mathcal{A}} R(s, a) + \gamma \sum_{s' \in \mathcal{S}} P(s'|s, a)v_*(s'). \tag{2.45}
$$

The procedure of value iteration is to apply the updates above from the final state and backward successively. Similar to the convergence proof in policy iteration, the **Bellman optimality operator** $\mathcal{T}^*$:

$$
(\mathcal{T}^*V)(s) = (\max_{a \in \mathcal{A}} \mathcal{R}^a + \gamma\mathcal{P}^a V)(s) = \max_{a \in \mathcal{A}} R(s, a) + \gamma \sum_{s' \in \mathcal{S}} P(s'|s, a)V(s') \tag{2.46}
$$

Algorithm 4 Policy iteration

Initialize V and π for all states
repeat
  // Do policy evaluation
  repeat
    $\delta \leftarrow 0$
    for $s \in \mathcal{S}$ **do**
      $v \leftarrow V(s)$
      $V(s) \leftarrow \sum_{r,s'}(r + \gamma V(s'))P(r, s'|s, \pi(s))$
      $\delta \leftarrow \max(\delta, |v - V(s)|)$
    end for
  until δ is smaller than a positive threshold
  // Do policy improvement
  $stable \leftarrow true$
  for $s \in \mathcal{S}$ **do**
    $a \leftarrow \pi(s)$
    $\pi(s) \leftarrow \arg\max_a \sum_{r,s'}(r + \gamma V(s'))P(r, s'|s, a)$
    if $a \neq \pi(s)$ **then**
      $stable \leftarrow false$
    end if
  end for
until $stable = true$
return policy π

is also a contraction mapping for arbitrary value functions V and V'

$$
\begin{aligned}
|\mathcal{T}^* V(s) - \mathcal{T}^* V'(s)| = |\max_{a \in \mathcal{A}} & \left[R(s, a) + \gamma \sum_{s' \in \mathcal{S}} P(s'|s, a)V(s') \right] \\
& - \max_{a \in \mathcal{A}} \left[R(s, a) + \gamma \sum_{s' \in \mathcal{S}} P(s'|s, a)V'(s') \right] | \\
& \leq \max_{a \in \mathcal{A}} |R(s, a) + \gamma \sum_{s' \in \mathcal{S}} P(s'|s, a)V(s') - R(s, a) \\
& - \gamma \sum_{s' \in \mathcal{S}} P(s'|s, a)V'(s')| \\
& = \max_{a \in \mathcal{A}} |\gamma \sum_{s' \in \mathcal{S}} P(s'|s, a)(V(s') - V'(s'))| \\
& \leq \max_{a \in \mathcal{A}} \gamma \sum_{s' \in \mathcal{S}} P(s'|s, a)|V(s') - V'(s')|
\end{aligned}
\tag{2.47}
$$

$$
\begin{aligned}
&\leq \max_{a \in \mathcal{A}} \gamma \sum_{s' \in \mathcal{S}} P(s'|s,a) \|V - V'\|_\infty \\
&= \gamma \|V - V'\|_\infty \max_{a \in \mathcal{A}} \sum_{s' \in \mathcal{S}} P(s'|s,a) \\
&= \gamma \|V - V'\|_\infty.
\end{aligned}
$$

Since v_* is a fixed point of $\mathcal{T}^*$, the value iteration converges on the optimal value v_*. Note that in value iteration, only the actual value of successor states are known. In other words, the values are not complete so we use value function V instead of value v in the proof above.

It is not obvious when to stop the value iteration algorithm. Williams and Baird III (1993) gives a sufficient stopping criterion in theory that if the maximum difference between two successive value functions is less than ϵ, then the value of the greedy policy differs from the value function of the optimal policy by no more than $\frac{2\epsilon\gamma}{1-\gamma}$ at any state.

Algorithm 5 Value iteration

Initialize V for all states
repeat
  $\delta \leftarrow 0$
  for $s \in \mathcal{S}$ **do**
    $u \leftarrow V(s)$
    $V(s) \leftarrow \max_a \sum_{r,s'} P(r, s'|s, a)(r + \gamma V(s'))$
    $\delta \leftarrow \max(\delta, |u - V(s)|)$
  end for
until δ is smaller than a positive threshold
Output greedily policy $\pi(s) = \arg\max_a \sum_{r,s'} P(r, s'|s, a)(r + \gamma V(s'))$

2.4.3 *Other DPs: Asynchronous DP, Approximate DP, Real-Time DP*

DP methods described so far use synchronous backups, i.e., the value of each state is backed up on the basis of systematic sweeps. One of the efficient variants is asynchronous updates, which is a trade-off between speed and accuracy. Asynchronous DP is also available for reinforcement learning settings and is guaranteed to converge if all states continue to be selected. There are three simple ideas behind asynchronous DP:

1. In-Place Update

   Synchronous value iteration stores two copies of value function $V_{t+1}(\cdot)$ and $V_t(\cdot)$:

$$V_{t+1}(s) \leftarrow \max_{a\in\mathcal{A}} R(s,a) + \gamma \sum_{s'\in\mathcal{S}} P(s'|s,a)V_t(s'). \tag{2.48}$$

   In-place value iteration only stores one copy of value function:

$$V(s) \leftarrow \max_{a\in\mathcal{A}} R(s,a) + \gamma \sum_{s'\in\mathcal{S}} P(s'|s,a)V(s'). \tag{2.49}$$

2. Prioritized Sweeping

   In asynchronous DP, one more thing that needs to be considered is the update order. Given a transition (s, a, s'), prioritized sweeping views the absolute value of its Bellman error as its magnitude:

$$|V(s) - \max_{a\in\mathcal{A}}(R(s,a) + \gamma \sum_{s'\in\mathcal{S}} P(s'|s,a)V(s'))|. \tag{2.50}$$

   It can be implemented efficiently by maintaining a priority queue where the Bellman error of each state is stored or updated after each backup.

3. Real-Time Update

   After each time step t, no matter which action is taken, real-time update will only back up the current state S_t by:

$$V(S_t) \leftarrow \max_{a\in\mathcal{A}} R(S_t,a) + \gamma \sum_{s'\in\mathcal{S}} P(s'|S_t,a)V(s'). \tag{2.51}$$

   It can be viewed as selecting the states to update by the guide of the agent's experience.

Both synchronous and asynchronous DP back up over the full state set to estimate the expected return of the next state. Under the perspective of probability, a biased but efficient choice is using sampled data. We will discuss this topic extensively in the next section.

2.5 Monte Carlo

Unlike DP, Monte Carlo (MC) methods do not require perfect knowledge of the environment. MC only needs experience for learning. MC is also a class of sampling-based methods. MC can obtain good performance by learning from experience with little prior knowledge about the environment. "Monte Carlo" refers to the class of algorithms that have a large component of randomness. Indeed so, when using MC in reinforcement learning, we will average the rewards for each state-action pair from different episodes. One example can be with the contextual bandit problem that we talked about earlier in this chapter. If there is an LED light on different slot machines, the player can gradually learn from the association between the lighting information with the relevant reward. We will consider a particular arrangement of the lights for our state, and the corresponding possible reward is the value for this state. Initially, we might not have a good estimate for the state-value, but gradually as we play more, the average state-value pairs should be closer to the real ones. In this section, we will investigate how to do this estimation properly and then how to make the best use of this information. Also, we assume that the problem is episodic and an episode will always terminate regardless of the actions taken.

2.5.1 Monte Carlo Prediction

State-Value Prediction To start with, we will look at the case when using MC methods to estimate the state-value function for a given policy π. The most intuitive way to do this is to estimate the state-value function from experience by simply averaging the return from a particular policy. More specifically, let the function $v_\pi(s)$ be the state-value function under policy π. We then collect a pool of episodes that pass through s. We call each appearance of state s in an episode a visit to state s. There are two types of estimations, **first-visit MC** and **every-visit MC**. The first-visit MC only considers the return of the first visit to state s in the whole episode, however, every-visit MC considers every visit to state s in the episode. These two methods share lots of similarities but have a few theoretical differences. In Algorithm 6, we are showing exactly how $v_\pi(s)$ is computed with first-visit MC estimation. It is simple to convert the first-visit MC prediction to the every-visit MC prediction by removing the check for a state being the first state. Note that both types of methods will converge to $v_\pi(s)$ if we take the number of visits to state s to infinity.

MC methods can estimate different states independently from each other. Unlike DP, MC does not use bootstrapping, estimating the value of the current step with the estimation from other steps (e.g. the next step). This unique feature will enable one to estimate the state-value directly from the true sampled returns, which can be of less bias but higher variances.

The state-value function will be handy if a model is given as we can easily select the optimal action for an arbitrary state by looking at the combined average of the state-value for a specific action, as in DP. When a model is not known, we will have to estimate the state-action value instead. Each state-action pair has to be estimated separately. Now, one learning objective has become the estimation of $q_\pi(s, a)$, the expected return at state s by performing action a, under policy π. This is essentially the same as the estimation of the state-value function as we can just take the average return at state s and when action a is taken. The only issue is that there might exist

Algorithm 6 First-visit MC prediction

Input: Initialize policy π
Initialize $V(s)$ for all states
Initialize a list of returns: $Returns(s)$ for all states
repeat
  Generate an episode under π: $S_0, A_0, R_0, S_1, \cdots, S_{T-1}, A_{T-1}, R_t$
  $G \leftarrow 0$
  $t \leftarrow T - 1$
  for $t >= 0$ **do**
    $G \leftarrow \gamma G + R_{t+1}$
    if $S_0, S_1, \cdots, S_{t-1}$ does not have S_t **then**
      $Returns(S_t)$.append(G)
      $V(S_t) \leftarrow$ mean($Returns(S_t)$)
    end if
    $t \leftarrow t - 1$
  end for
until convergence

states that could never be visited, and thus have zero return. To choose the optimal strategy, we must fully explore all states. A naive solution to this issue is to directly specify the starting state-action pair for each episode and each state-action pair has a non-zero probability of getting selected. In this way, we can ensure that all state-action pairs can be visited if we have enough episodes. We refer this assumption as exploring starts.

2.5.2 *Monte Carlo Control*

Now, we shall adapt GPI to MC to see how it is used in control. Recall that GPI consists of two stages: policy evaluation and policy improvement. Policy evaluation is the same as that of DP as introduced in the previous section and therefore, we will discuss more about policy improvement. We will make use of a greedy policy for the action-value, we do not need to have a model in this case. The greedy policy will always choose the action that has maximal value for a given state:

$$\pi(s) = \arg\max_a q(s, a) \tag{2.52}$$

We will go from policy evaluation to policy improvement. For each policy improvement, we will need to construct π_{t+1} based on q_{π_t}. We will show how the policy improvement theorem is applicable here:

$$q_{\pi_t}(s, \pi_{t+1}(s)) = q_{\pi_t}(s, \arg\max_a q_{\pi_t}(s, a)) \tag{2.53}$$

$$= \max_a q_{\pi_t}(s, a) \tag{2.54}$$

$$\geq q_{\pi_t}(s, \pi_t(s)) \tag{2.55}$$

$$\geq v_{\pi_t}(s) \tag{2.56}$$

The above proves that π_{t+1} will be no worse than π_t, and thus eventually the optimal policy can be found. This means that we can use MC for control without much knowledge about the environment but only the sampled episodes. Here, we have two assumptions that we need to resolve. The first is the exploring starts and the second is that we have an infinite number of episodes. We will keep the exploring starts for now but focus on the second assumption. An easy way to relax this assumption is to avoid the infinite number of episodes needed for policy evaluation by directly alternating between evaluation and improvement for single states.

Algorithm 7 MC exploring starts

```
Initialize π(s) for all states
Initialize Q(s, a) and Returns(s, a) for all state-action pairs
repeat
    Randomly select S_0 and A_0 s.t. all state-action pairs' probabilities are nonzero.
    Generate an episode from S_0, A_0 under π: S_0, A_0, R_0, S_1, ··· , S_{T-1}, A_{T-1}, R_t
    G ← 0
    t ← T − 1
    for t >= 0 do
        G ← γG + R_{t+1}
        if S_0, A_0, S_1, A_1 ··· , S_{t-1}, A_{t-1} does not have S_t, A_t then
            Returns(S_t, A_t).append(G)
            Q(S_t, A_t) ← mean(Returns(S_t, A_t))
            π(S_t) ← arg max_a Q(S_t, a)
        end if
        t ← t − 1
    end for
until convergence
```

2.5.3 *Incremental Monte Carlo*

As we have seen in both Algorithm 6 and Algorithm 7, we have to take the averages of the lists of observed rewards, the state values and the state-action values respectively. There exists a more efficient computational method that allows us to get rid of the lists of observed returns and simplify the mean calculation step. We will thus do the update in an episode by episode way. We let the $Q(S_t, A_t)$ be the estimation of the state-action value after it has been selected for $t-1$ times, which can be then rewritten as:

$$Q(S_t, A_t) = \frac{G_1 + G_2 + \cdots + G_{t-1}}{t-1} \tag{2.57}$$

The naive implementation of this is to keep a record of all the returned G values, and then divide their sum by the visit times. However, we can also compute the same value by the following:

$$Q_{t+1} = \frac{1}{t}\sum_{i=1}^{t} G_i \tag{2.58}$$

$$= \frac{1}{t}\left(G_t + \sum_{i=1}^{t-1} G_i\right) \tag{2.59}$$

$$= \frac{1}{t}\left(G_t + (t-1)\frac{1}{t-1}\sum_{i=1}^{t-1} Q_i\right) \tag{2.60}$$

$$= \frac{1}{t}(G_t + (t-1)Q_t) \tag{2.61}$$

$$= Q_t + \frac{1}{t}(G_t - Q_t) \tag{2.62}$$

The formulation will give us a much easier time when it comes to the return computation. This can also appear in a more general form as:

$$\text{NewEstimate} \leftarrow \text{OldEstimate} + \text{StepSize} \cdot (\text{Target} - \text{OldEstimate}) \tag{2.63}$$

The "StepSize" is a parameter that controls how fast the estimate is being updated.

2.6 Temporal Difference Learning

Temporal difference (TD) learning describes another class of algorithm that is at the core of reinforcement learning by combining the ideas both from DP and MC. Similar to DP, TD uses bootstrapping in the estimation, however, like MC, it does not require full knowledge of the environment in the learning process, but applies a sampling-based optimization approach. In this chapter, we will first introduce how TD can be used in policy evaluation and then elaborate on the differences and commonalities between MC, TD, and DP. Lastly, we will end this chapter with Q-learning, an extremely useful and powerful learning algorithm in both classical reinforcement learning and deep reinforcement learning.

2.6.1 TD Prediction

As its name suggests, TD utilizes the error, the difference between the target value and the estimated value, at different time steps to learn. That reason why it is also using bootstrapping is that TD forms the target from the observed return and an estimated state value for the next state. More precisely, the most basic TD method makes the update using:

$$V(S_t) \leftarrow V(S_t) + \alpha[R_{t+1} + \gamma V(S_{t+1}) - V(S_t)] \tag{2.64}$$

This method is also called TD(0) or one-step TD for looking one-step ahead. N-step TD can also be developed easily by extending the target value with discounted rewards in the N-step future and the estimated state value at the N-th step. If we observe carefully, the target value during update for MC is G_t which is known only after one episode, whereas for TD the target value is $R_{t+1} + \gamma V(S_{t+1})$ which can be computed step by step. In Algorithm 8, we are showing how TD(0) can be used to do policy evaluation.

Algorithm 8 TD(0) for state-value estimation

Input policy π
Initialize $V(s)$ and step size $\alpha \in (0, 1]$
for each episode **do**
  Initialize S_0
  for Each step S_t in the current episode **do**
    $A_t \leftarrow \pi(S_t)$
    $R_{t+1}, S_{t+1} \leftarrow Env(S_t, A_t)$
    $V(S_t) \leftarrow V(S_t) + \alpha[R_{t+1} + \gamma V(S_{t+1}) - V(S_t)]$
  end for
end for

Before we move on, it is worthwhile spending time to take a closer look at what DP, MC, and TD have in common and how they differ from one another. These three types of algorithms sit at the heart of reinforcement learning and often their usage is combined together in modern reinforcement learning application. Even though all of them can be used for policy evaluation and policy improvement, their subtle differences can contribute to major performance variations in deep reinforcement learning.

Some forms of GPI are being used by these three methods. The main difference lies in their policy evaluation schemes. The most obvious difference is that both DP and TD use bootstrapping but MC does not, and DP requires full knowledge of the model but MC and TD do not. Furthermore, let us dive deeper into how the learning objectives differ among these three.

$$v_\pi(s) = \mathbb{E}_\pi[G_t|S_t = s] \tag{2.65}$$
$$= \mathbb{E}_\pi[R_{t+1} + \gamma G_{t+1}|S_t = s] \tag{2.66}$$
$$= \mathbb{E}_\pi[R_{t+1} + \gamma v_\pi(S_{t+1})|S_t = s] \tag{2.67}$$

Equation (2.65) stands for the state value estimation for MC methods and Eq. (2.67) represents the same for DP methods. Both of them are only estimation but not the true values. TD combines both the MC sampling and DP bootstrapping. We will now explain briefly how TD can be better than either DP or MC.

First of all, TD does not need a model to learn which DP requires. When TD is being compared with MC, TD is using an online learning approach meaning it can learn at every step, however, in order for MC methods to learn, it will have to wait until one episode is finished which can be tricky to deal with if the task has very long episodes. There are also problems that are continuous and cannot be learned in an episodic fashion. Moreover, TD can be faster because it can learn from transitions disregarding the actions being taken. MC cannot do this. Both TD and MC methods will eventually converge to $v_\pi(s)$ asymptotically, nonetheless, we do not have a proof to show which converges faster but that TD methods converge faster empirically.

Before we move on, it is also worth discussing the **variance and bias trade-off** between TD and MC methods. We know that in a supervised setting a large bias means that the model is underfitting for the data distribution and a large variance means that the model is overfitting the data. The bias of an estimator is the difference between the estimation and the true value. In the case of state value estimation, bias can be defined as $\mathbb{E}[V(S_t)] - V(S_t)$. The variance of an estimator describes how noisy the estimator is. Again for state-value estimation, variance can be defined as $\mathbb{E}[(\mathbb{E}[V(S_t)] - V(S_t))^2]$. In prediction, regardless of whether it is for the state or state-action approximation, both TD and MC are doing the update of the form:

$$V(S_t) \leftarrow V(S_t) + \alpha[\text{Target Value} - V(S_t)]$$

Essentially, we are doing a weighted average across different episodes. TD and MC differ in their ways of handling the target value. MC methods directly estimate the accumulative rewards until the end of an episode, which is exactly how the state value is defined. They will have no bias, however TD has a greater bias because its target is estimated with bootstrapping method, $R_{t+1} + \gamma v_\pi(S_{t+1})$. Now, let us see why MC tends to have a larger variance. The accumulated reward it has is computed until the end of each episode, which can vary a lot as different episodes can have vastly different outcomes. TD resolves this issue by just looking at the target value locally depending on the current reward and the estimated reward for the next state/action. Naturally, TD should have less variance.

TD(λ)

DP and MC have lots of similarities and perhaps there is a mid-ground between these two paradigms which could be more efficient in solving the problem at hand. Indeed, TD (λ) is the mid-ground between DP and MC, but we will need to introduce the concept of **eligibility traces** and λ return first.

To put it in a simple way, an eligibility trace can provide us with various computational advantages. To see this, we need to talk about semi-gradient methods quickly and then explain how eligibility traces can be used. For a detailed treatment of the policy-gradient method, please refer to Sect. 2.7. Here, we are simply using some concepts from the gradient-based methods to explain what eligibility traces can do. Imagine if our state-value function is not in a tabular form, but in a functional form parameterized by a weight vector $\boldsymbol{w} \in \mathbb{R}^n$. $\boldsymbol{w}$ can be, for instance, be the weight for a neural network. We aim to have $v(s, \boldsymbol{w}) \approx v_\pi(s)$. To do this, we can use stochastic gradient update to reduce the quadratic loss between our approximation and the true value function. The update rule w.r.t. the weight vector can be written as the following:

$$\boldsymbol{w}_{t+1} = \boldsymbol{w}_t - \frac{1}{2}\alpha \nabla_{\boldsymbol{w}_t}[v_\pi(S_t) - V(S_t, \boldsymbol{w}_t)]^2 \tag{2.68}$$

$$= \boldsymbol{w}_t + \alpha[v_\pi(S_t) - V(S_t, \boldsymbol{w}_t)]\nabla_{\boldsymbol{w}_t} V(S_t, \boldsymbol{w}_t) \tag{2.69}$$

where α is a positive step size.

An eligibility trace is a vector $z_t \in \mathbb{R}^n$, which is used in such a way that, during learning, whenever a component of w_t is used for estimation, the corresponding component value in z_t also increases and then starts to fade away. The learning will take place if there is a TD error happening before the value in the trace falls back to zero. We first initialize all values using zero and then increase the trace using the gradient. The decay rate is $\gamma\lambda$:

$$z_{-1} = 0 \tag{2.70}$$

$$z_t = \gamma\lambda z_{t-1} + \nabla_{\boldsymbol{w}_t} V(S_t, \boldsymbol{w}_t) \tag{2.71}$$

It becomes easy to see that when $\lambda = 1$, the former sum becomes zero and the return is the same as that of an MC method. When $\lambda = 0$, it essentially becomes a one-step TD method. This is because the trace will always only contain the gradient of the one-step TD error. An eligibility trace is thus a great way to combine MC and TD methods.

Moving along, a λ-return is an estimated return value over the next n steps. λ-returns are a combination of n discounted returns with an existing estimate at the last step. Formally, it can be written as:

$$G_{t:t+n} = R_{t+1} + \gamma R_{t+2} + \cdots + \gamma^{n-1} R_{t+n} + \gamma^n v(S_{t+n}, \boldsymbol{w}_{t+n-1}) \tag{2.72}$$

t here is a nonzero scalar and is also less than or equal to $T - n$. We can make use of a weighted return in estimation as long as their weights sum up to one. TD(λ) makes use of this weighted averaging in its update with $\lambda \in [0, 1]$:

$$G_t^{\lambda} = (1 - \lambda) \sum_{n=1}^{\infty} \lambda^{n-1} G_{t:t+n} \tag{2.73}$$

Intuitively, what this means is that the very next step return has the largest weight $1 - \lambda$, the two-step return has a weight of $(1 - \lambda)\lambda$, and the weight decays at each step with a rate of λ. To have a clear picture, let us have a terminal state at time T, then the above can be rewritten as

$$G_t^{\lambda} = (1 - \lambda) \sum_{n=1}^{T-t-1} \lambda^{n-1} G_{t:t+n} + \lambda^{T-t-1} G_t \tag{2.74}$$

The TD error δ_t can be defined as:

$$\delta_t = R_{t+1} + \gamma V(S_{t+1}, \boldsymbol{w}_t) - V(S_t, \boldsymbol{w}_t) \tag{2.75}$$

The update rule is based on the proportion of the TD error and trace. See Algorithm 9 for details.

Algorithm 9 Semi-gradient TD(λ) for state-value

```
Input: policy π
Initialize a differentiable state function v, step size α and value function weight w
for each episode do
  Initialize S_0
  z ← 0
  for Each step S_t in the current episode do
    Select A_t using policy that is based on π
    R_{t+1}, S_{t+1} ← Env(S_t, A_t)
    z ← γλz + ∇V(S_t, w_t)
    δ ← R_{t+1} + γV(S_{t+1}, w_t) − V(S_t, w_t)
    w ← w + αδz
  end for
end for
```

2.6.2 *Sarsa: On-Policy TD Control*

For TD control, the methodology is similar to that of the prediction task except that we will transition from the state-to-state alternation to state-action pair alternation.

The update rule can, therefore, be framed as:

$$Q(S_t, A_t) \leftarrow Q(S_t, A_t) + \alpha[R_{t+1} + \gamma Q(S_{t+1}, A_{t+1}) - Q(S_t, A_t)] \quad (2.76)$$

When S_t is the terminal state, the Q value for the next state-action pair will be zero. It is referred to by the acronym Sarsa because we have this chain of being in a state, taking an action, receiving a reward and being in a new state to take another action. The chain allows us to do a simple update step. The state value gets updated for each transition and the updated state value is influencing the policy being used to determine the behavior, so it is also an on-policy method. On-policy methods generally describe the class of methods that have an update policy which is the same as its behavior policy, whereas, for the off-policy methods, these two are different. An example of an off-policy method is Q-learning which we will talk about later. It assumes a greedy approach while doing the update of its q-value function, whereas, in fact, for its behavior it is using other policies such as ϵ-greedy. We now entail the steps for Sarsa in Algorithm 10. We will have the convergence for both the optimal policy and action-state values as long as each state-action pair is visited an infinite number of times.

Algorithm 10 Sarsa (on-policy TD control)

Initialize $Q(s, a)$ for all state-action pairs.
for each episode **do**
  Initialize S_0
  Select A_0 using policy that is based on Q
  for Each step S_t in the current episode **do**
    Select A_t from S_t using policy that is based on Q
    $R_{t+1}, S_{t+1} \leftarrow Env(S_t, A_t)$
    Select A_{t+1} from S_{t+1} using policy that is based on Q
    $Q(S_t, A_t) \leftarrow Q(S_t, A_t) + \alpha[R_{t+1} + \gamma Q(S_{t+1}, A_{t+1}) - Q(S_t, A_t)]$
  end for
end for

What we have shown only has one step time horizon meaning that the approximation only involves the state-action value of the next step. We can call this 1-step Sarsa or Sarsa(0), although we can easily extend the bootstrapped target value to include future steps down the road to, for instance, reduce our bias. As shown by the backup tree in Fig. 2.12, we see an array of Sarsa variants' state-action spectrum starting from the most basic 1-step Sarsa all the way up to the infinite step Sarsa which is an equivalent of the MC because its target value accounts for the accumulated rewards until the terminal state. To incorporate this change, we rewrite the discounted returns as the following:

$$G_{t:t+n} = R_{t+1} + \gamma R_{t+2} + \cdots + \gamma^{n-1} R_{t+n} + \gamma^n Q_{t+n-1}(S_{t+n}, A_{t+n}) \quad (2.77)$$

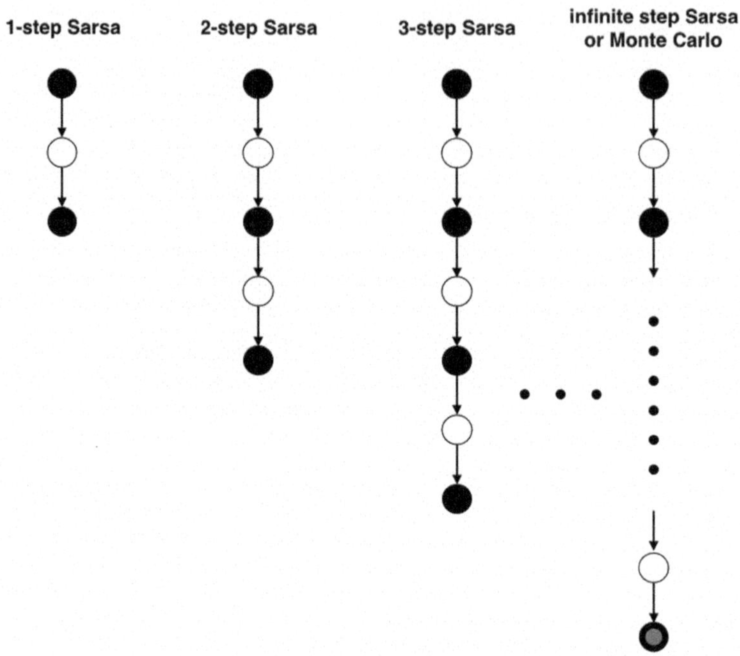

Fig. 2.12 The backup tree for the coverage of the n-step Sarsa methods. Each black circle represents a state and each white circle represents an action. The last state of the infinite step Sarsa is a terminal state

The n-step Sarsa is described in Algorithm 11. The major difference it has from the one-step version is that it has to go back in time to do the update, whereas the one-step version can do the update as it goes.

Convergence of Sarsa

Now we will discuss the convergence theory of Sarsa algorithm for finite action space (discrete cases), which requires some additional conditions described below.

Definition 2.1 A learning policy is defined as **Greedy in the Limit with Infinite Exploration (GLIE)** if it satisfies the following two properties:

1. If a state is visited infinitely often, then each possible action in that state is chosen infinitely often, i.e., $\lim_{k\to\infty} N_k(s, a) = \infty, \forall a$, if $\lim_{k\to\infty} N_k(s) = \infty$.
2. The policy converges on a greedy policy with respect to the learned Q-function in the limit (as $t \to \infty$), i.e., $\lim_{k\to\infty} \pi_k(s, a) = \mathbb{1}(a == \arg\max_{a'\in\mathcal{A}} Q_k(s, a'))$, where the "==" is a comparison operator and $\mathbb{1}(a == b)$ is 1 if true and 0 otherwise.

Algorithm 11 n-step Sarsa

Initialize $Q(s, a)$ for all state-action pairs.
Initialize step-size $\alpha \in (0, 1]$.
Determine a fixed policy π or use ϵ-greedy.
for each episode **do**
  Initialize S_0
  Select A_0 using $\pi(S_0, A)$
  $T \leftarrow$ INTMAX (the length of an episode)
  $\gamma \leftarrow 0$
  for $t \leftarrow 0, 1, 2, \ldots$ until $\gamma - T - 1$ **do**
    if $t < T$ **then**
      $R_{t+1}, S_{t+1} \leftarrow Env(S_t, A_t)$
      if S_{t+1} is terminal **then**
        $T \leftarrow t + 1$
      else
        Select A_{t+1} using $\pi(S_t, A)$
      end if
    end if
    $\tau \leftarrow t - n + 1$ (the time step to update. This is an n-step Sarsa, so we will only update the estimate that is $n + 1$ steps ago and we will continue to do so until all the eligible states have been updated.
    if $\tau \geq 0$ **then**
      $G \leftarrow \sum_{i=\tau+1}^{min(r+n,T)} \gamma^{i-\gamma-1} R_i$
      if $\gamma + n < T$ **then**
        $G \leftarrow G + \gamma^n Q(S_{t+n}, A_{\gamma+n})$
      end if
      $Q(S_\gamma, A_\gamma) \leftarrow Q(S_\gamma, A_\gamma) + \alpha[G - Q(S_\gamma, A_\gamma)]$
    end if
  end for
end for

The GLIE is a condition for the convergence of the learning policies, for any reinforcement learning algorithm that converges to the optimal value function and whose estimates are always bounded. For example, we can derive a GLIE policy with ϵ-greedy strategy as follows:

Lemma 2.1 *The ϵ-greedy policy is GLIE if ϵ reduces to zero with $\epsilon_k = \frac{1}{k}$.*

We can therefore have the convergence theorem of Sarsa algorithm.

Theorem 2.1 *For a finite state-action MDP and a GLIE learning policy, with the action-value function Q estimated with Sarsa (1-step) by Q_t for time step t. Then Q_t converges to Q^* and the learning policy π_t converges to an optimal policy π^*, if the following conditions are satisfied:*

1. *The Q values are stored in a lookup table;*
2. *The learning rate $\alpha_t(s, a)$ associated with the state-action pair (s, a) at time t satisfies $0 \leq \alpha_t(s, a) \leq 1$, $\sum_t \alpha_t(s, a) = \infty$ and $\sum_t \alpha_t^2(s, a) < \infty$ and $\alpha_t(s, a) = 0$ unless $(s, a) = (S_t, A_t)$;*
3. *$Var[R(s, a)] < \infty$.*

A typical sequence of the learning rate as the second condition required is $\alpha_t(S_t, A_t) = \frac{1}{t}$. The proofs of above theorems are not introduced here, but interested readers can refer to the paper by Singh et al. (2000).

2.6.3 *Q-Learning: Off-Policy TD Control*

Q-learning is an off-policy TD method that is very similar to Sarsa and plays an important role in deep reinforcement learning application such as the deep Q-network, which we will discuss in the next chapter. As shown in Eq. (2.78), the main difference that Q-learning has from Sarsa is that the target value now is no longer dependent on the policy being used but only on the state-action function.

$$Q(S_t, A_t) \leftarrow Q(S_t, A_t) + \alpha[R_{t+1} + \gamma \max_a Q(S_{t+1}, a) - Q(S_t, A_t)] \quad (2.78)$$

Algorithm 12 Q-learning (off-policy TD control)

Initialize $Q(s, a)$ for all state-action pairs and step size $\alpha \in (0, 1]$
for each episode **do**
  Initialize S_0
  for Each step S_t in the current episode **do**
    Select A_t using policy that is based on Q
    $R_{t+1}, S_{t+1} \leftarrow Env(S_t, A_t)$
    $Q(S_t, A_t) \leftarrow Q(S_t, A_t) + \alpha[R_{t+1} + \gamma \max_a Q(S_{t+1}, a) - Q(S_t, A_t)]$
  end for
end for

In Algorithm 12, we have shown how Q-learning can be used for TD control. It is easy to convert Q-learning to Sarsa by first choosing the action using the state and return, and second changing the target value in the update step to be the estimated action value for the next step instead. This is also a one-step version. We can adapt the Q-learning into a n-step version by adapting the target value in Eq. (2.78) to include the discounted returns for the future steps.

Convergence of Q-Learning

The convergence of Q-learning follows similar conditions as the Sarsa algorithm. Apart from the GLIE condition for the policy, the convergence of Q function in Q-learning also requires the same requirements on its learning rate and the bounded reward values, which will not be duplicated here. Details and proofs are available in the papers (Szepesvári 1998; Watkins and Dayan 1992).

2.7 Policy Optimization

2.7.1 Overview

In reinforcement learning, the ultimate goal of the agent is to improve its policy to acquire better rewards. Policy improvement in the optimization domain is called policy optimization (Fig. 2.13). For deep reinforcement learning, the policy and value functions are usually parameterized by variables in deep neural networks, and therefore enable the gradient-based optimization methods to be applied. For example, Fig. 2.14 shows the graphical model of MDP with the policy parameterized by variables θ, on a discrete finite time horizon $t = 0, \ldots, N-1$. The reward function follows $R_t = R(S_t, A_t)$ and action $A_t \sim \pi(\cdot|S_t; \theta)$. The dependencies among variables in the graphical models can help us to understand the underlying relationships of the MDP for estimation, and it can be useful when we take derivatives on the final objective to optimize variables on the dependency graphs, so we will display all those graphical models in this chapter to help understand the deduction process, especially for differential process. Recently, Levine (2018), Fu et al. (2018) proposed the method of **control as inference**, which uses a graphical model with additional variables indicating optimality on the MDP to incorporate the probabilistic/variational inference framework into maximum entropy reinforcement learning with the same objective. This method enables the inference tools to be applied in the reinforcement learning policy optimization process. But the details of those methods are beyond the scope of the book here.

Apart from some linear methods, the parameterization of value functions with deep neural networks is one way of achieving **value function approximation**, and it's the most popular way in the modern deep reinforcement learning domain. Value function approximation is useful because we cannot always acquire the true value function easily, and actually we cannot get the true function for most cases in practice. Figure 2.15 shows the model of MDP with both parameterized policy

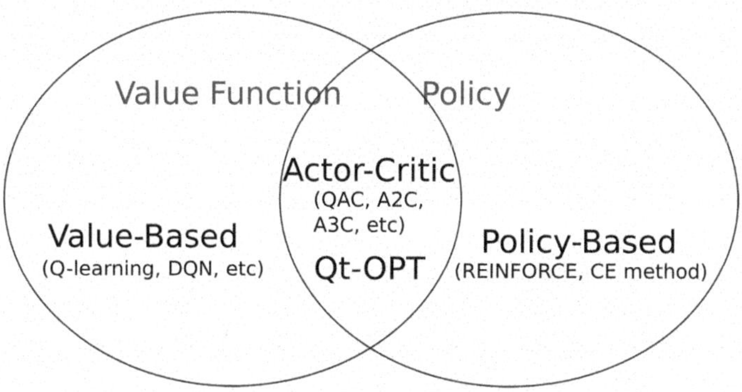

Fig. 2.13 Overview of policy optimization in reinforcement learning

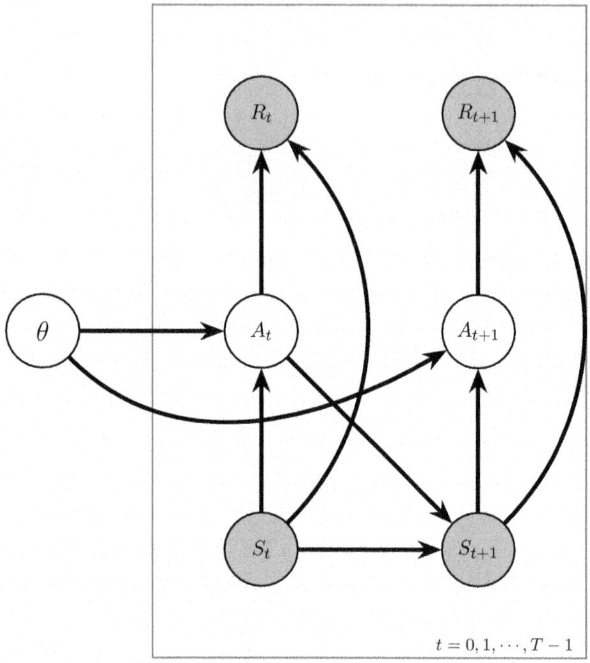

Fig. 2.14 Graphical model of MDP with parameterized policy

π_θ and parameterized value function $V_w^\pi(S_t)$, via parameters θ and w respectively. Figure 2.16 shows the model with parameterized policy π_θ and Q-value functions $Q_w^\pi(S_t, A_t)$. The gradient-based optimization methods can be used for improving parameterized policies, usually through the method called **policy gradient** in reinforcement learning terminology. However, there are also non-gradient-based methods for optimizing less complicated policies, like the cross-entropy (CE) method and so on.

As shown in Fig. 2.13, the methods in policy optimization fall into two main categories: (1) **value-based optimization** methods like Q-learning, DQN, etc., which optimize the action-value function to obtain the preferences for the action choice, and (2) **policy-based optimization** methods like REINFORCE, the cross-entropy method, etc., which directly optimize the policy according to the sampled reward values. A combination of these two categories was found to be a more effective approach by people (Sutton et al. 2000; Peters and Schaal 2008; Kalashnikov et al. 2018), which forms one of the most widely used architecture in model-free reinforcement learning called **actor-critic**. Actor-critic methods employ the optimization of value function as the guidance of policy improvement. The typical algorithms in the combined category include actor-critic-based algorithms and other algorithms built upon that, which will be described in detail in later this chapter and the following chapters.

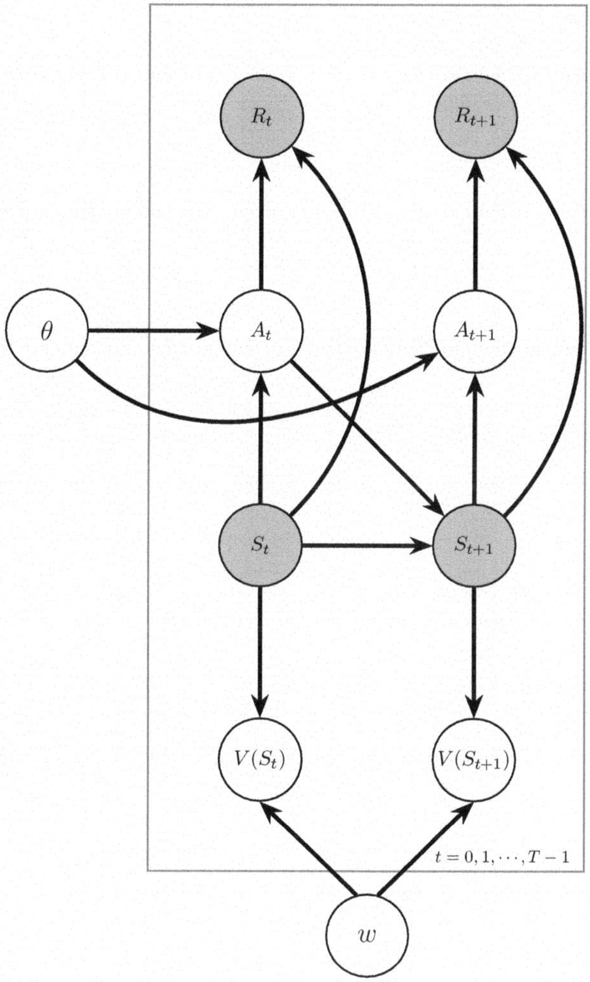

Fig. 2.15 Graphical model of MDP with parameterized policy and parameterized value functions

Recap of RL Skeleton

The **On-policy Value Function**, $v_\pi(s)$, which gives the expected return if you start in state s and always act according to policy π:

$$v_\pi(s) = \mathbb{E}_{\tau \sim \pi}[R(\tau)|S_0 = s] \quad (2.79)$$

Recall that the reinforcement learning optimization problem can be expressed as:

$$\pi_* = \arg\max_\pi J(\pi) \quad (2.80)$$

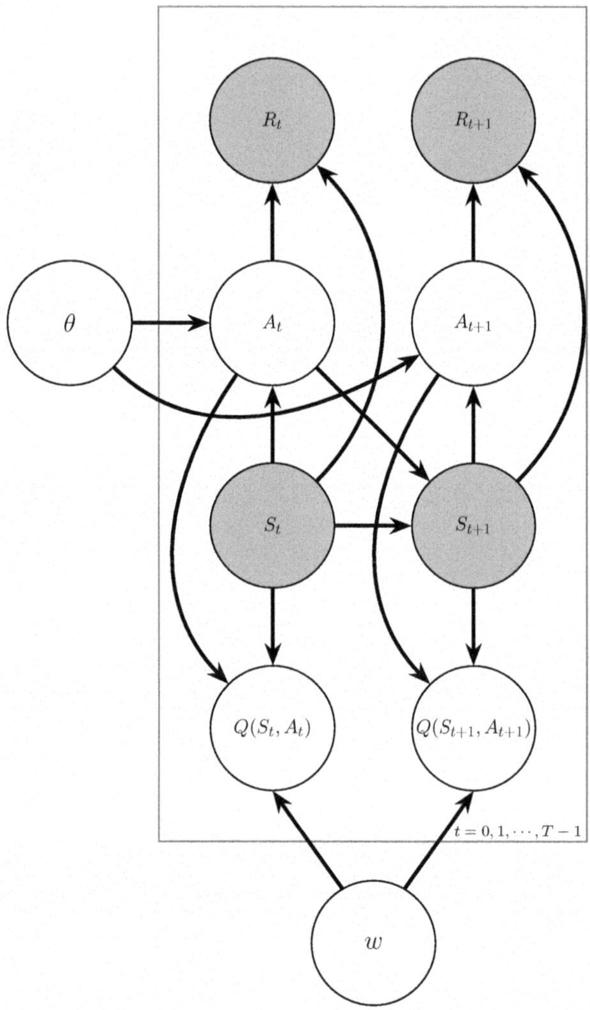

Fig. 2.16 Graphical model of MDP with parameterized policy and parameterized Q-value functions

The **Optimal Value Function**, $V^*(s)$, which gives the expected return if we start in state s and always act according to the optimal policy in the environment:

$$v_*(s) = \max_\pi v_\pi(s) \tag{2.81}$$

$$v_*(s) = \max_\pi \mathbb{E}_{\tau\sim\pi}[R(\tau)|S_0 = s] \tag{2.82}$$

The **On-Policy Action-Value Function**, $q_\pi(s, a)$, which gives the expected return if we start in state s, take an arbitrary action a (which may not come from the policy), and then forever after act according to policy π:

$$q_\pi(s, a) = \mathbb{E}_{\tau\sim\pi}[R(\tau)|S_0 = s, A_0 = a] \tag{2.83}$$

The **Optimal Action-Value Function**, $q_*(s, a)$, which gives the expected return if you start in state s, take an arbitrary action a, and then forever after act according to the optimal policy in the environment:

$$q_*(s, a) = \max_{\pi} q_\pi(s, a) \tag{2.84}$$

$$q_*(s, a) = \max_{\pi} \mathbb{E}_{\tau \sim \pi}[R(\tau)|S_0 = s, A_0 = a] \tag{2.85}$$

Value Function and Action-Value Function

$$v_\pi(s) = \mathbb{E}_{a \sim \pi}[q_\pi(s, a)] \tag{2.86}$$

$$v_*(s) = \max_{a} q_*(s, a) \tag{2.87}$$

Optimal Action

$$a_*(s) = \arg\max_{a} q_*(s, a) \tag{2.88}$$

Bellman Equations
Bellman equations for state value and action value are:

$$v_\pi(s) = \mathbb{E}_{a \sim \pi(\cdot|s), s' \sim p(\cdot|s,a)}[R(s, a) + \gamma v_\pi(s')] \tag{2.89}$$

$$q_\pi(s, a) = \mathbb{E}_{s' \sim p(\cdot|s,a)}[R(s, a) + \gamma \mathbb{E}_{a' \sim \pi(\cdot|s')}[q_\pi(s', a')]] \tag{2.90}$$

Bellman Optimality Equations
Bellman optimality equations for state value and action value are:

$$v_*(s) = \max_{a} \mathbb{E}_{s' \sim p(\cdot|s,a)}[R(s, a) + \gamma v_*(s')] \tag{2.91}$$

$$q_*(s, a) = \mathbb{E}_{s' \sim p(\cdot|s,a)}[R(s, a) + \gamma \max_{a'} q_*(s', a')] \tag{2.92}$$

2.7.2 *Value-Based Optimization*

A **value-based optimization** method always needs to alternate between value function estimation under the current policy and policy improvement with the estimated value function. However, the estimation of a complex value function may not be a trivial problem (Fig. 2.17).

From the previous sections, we see that the Q-learning can be used for solving some simple tasks in reinforcement learning. However, the real-world applications

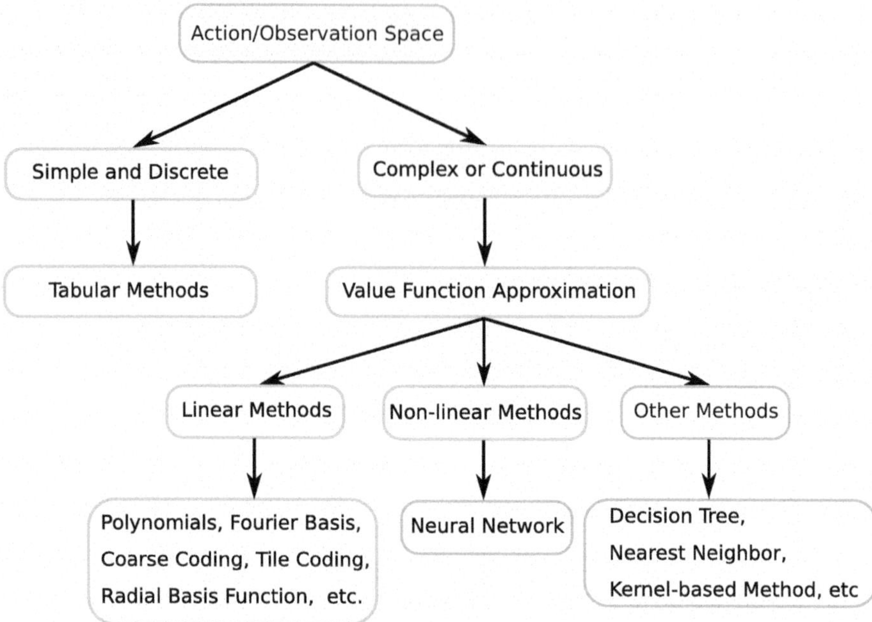

Fig. 2.17 An overview of methods for solving the value function

or even the quasi-real-world applications may have much larger and complicated state and action spaces, and the action is usually continuous in practice. For example, the Go game has 10^{170} states. In these cases, the traditional lookup table method in Q-learning cannot work well with the limitation of its scalability, because each state will have an entry $V(s)$ and each state-action pair will need an entry $Q(s, a)$. The values in the table are updated one-by-one in practice. Therefore the requirement of the memory and computational resources will be huge with tabular-based Q-learning. Moreover, state representations usually need to be manually specified with aligned data structures in practice.

Value Function Approximation

In order to apply the value-based reinforcement learning in relatively large-scale tasks, function approximators are applied to handle the above limitations (Fig. 2.17). Different types of value function approximation are summarized as follows and shown in Fig. 2.18:

- **Linear methods**: the approximated function is a linear combination of weights $\boldsymbol{\theta}$ and real-valued vector of features $\boldsymbol{\phi}(s) = (\phi_1(s), \phi_2(s)), \ldots, \phi_n(s))^T$, where s is the state. It is denoted as $v(s, \boldsymbol{\theta}) = \boldsymbol{\theta}^T \boldsymbol{\phi}(s)$. The TD($\lambda$) method is proven to be convergent with linear function approximators under certain conditions

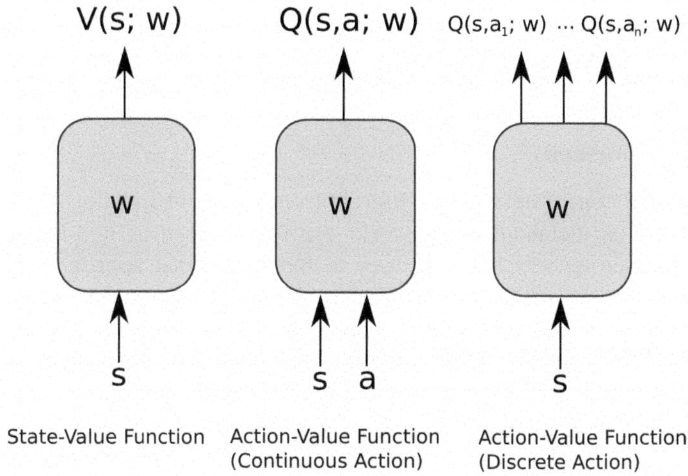

Fig. 2.18 Different value function approximation frameworks. The gray boxes with parameters w are the function approximators

as shown in Tsitsiklis and Roy (1997). Although the convergence guarantee of linear methods are attractive, the feature selection or feature representation $\boldsymbol{\phi}(s)$ can be critical in practice when applying linear representations. Different ways of constructing the features for linear methods are as follows:

- **Polynomials**: basic polynomial families can be used as feature vectors for function approximation. Assuming that every state $s = (S_1, S_2, \ldots, S_d)^T$ is a d-dimensional vector, then we have a d-dimensional polynomial basis as $\phi_i(s) = \prod_{j=1}^{d} S_j^{c_{i,j}}$, where each $c_{i,j}$ is an integer in set $\{0, 1, \ldots, N\}$. This forms order N polynomial basis, with $(N+1)^d$ different functions.
- **Fourier basis**: the Fourier transformation is usually used to represent sequential signals in the time/frequency domain. The one-dimensional order-N Fourier cosine basis with $N+1$ functions is: $\phi_i(s) = \cos(i\pi s)$ for $s \in [0, 1]$ and $i = 0, \ldots, N$.
- **Coarse coding**: the state space can be reduced from high-dimensional to low-dimensional, like binary representation through a region covering the determination process, which is called coarse coding.
- **Tile coding**: in the category of coarse coding, tile coding is an efficient approach for feature representation on multi-dimensional continuous spaces. The receptive field of features in tile coding are grouped into partitions of the input space. Each such partition is called a tilling, and each element of the partition is called a tile. Multiple tillings are usually applied in combination with overlapping receptive fields to give the feature vectors in practice.
- **Radial basis functions**: the radial basis functions (RBF) naturally generalize the coarse coding, which is binary-valued, to be continuous-valued features in

[0, 1]. The typical RBF is in Gaussian format $\phi_i(s) = \exp(-\frac{||s-c_i||^2}{2\sigma_i^2})$, where s is the state, c_i is the feature's prototypical or center state, and σ_i is the feature width.

- **Non-linear methods:**
  - **Artificial neural networks**: different from the above function approximation methods, artificial neural networks are widely used as non-linear function approximators, which are proven to have universal approximation ability under certain conditions (Leshno et al. 1993). Based on deep learning techniques, artificial neural networks form the main body of modern DRL methods with function approximation. Details of deep learning are introduced in Chap. 1. A typical example of it is the DQN algorithm, deploying an artificial neural network for Q-value approximation.
- **Other methods:**
  - **Decision trees** (Pyeatt et al. 2001): the decision trees can be used to represent the state space by dividing it with decision nodes, which forms a considerable method for state feature representation.
  - **Nearest neighbor method**: it measures the difference of current state and previous state in memory, and applies the value of the most similar state in memory to approximate the value of the current state.

The benefits of using value function approximation include not only the scalability to large-scale tasks, but also the ease to generalize to unseen states from the seen states given continuous state spaces. Moreover, ANN-based function approximation also reduces or eliminates the need for manually designing features to represent the states. For model-free methods, the parameters w of the approximators can be updated with Monte Carlo (MC) or TD learning. The updating of parameters can be conducted with a batch of samples instead of updating each value in a tabular-based method one-by-one. This makes it computational efficient when handling large-scale problems. For model-based methods, the parameters can be updated with dynamic programming. Details about MC, TD, and DP are introduced in previous sections.

Potential function approximators include a linear combination of features, neural networks, decision trees, the nearest neighbor method, etc. The most practical approximation method for present DRL algorithms is using the neural network, for its great scalability and generalization for various specific functions. A neural network is a differential method with gradient-based optimization, which has a guarantee of convergence to optimum within convex cases and can achieve near-optimal solutions for some non-convex functions approximation. However, it may require a large amount of data for training in practice and may cause other difficulties.

Extending deep learning problems to those of reinforcement learning comes with additional challenges including non-independently and identically distributed data (i.e. non-i.i.d.). Most supervised learning methods are constructed with the assumption that training data is from an i.i.d. and stationary distribution (Schmidhuber 2015). However, the training data in reinforcement learning usually consists of highly correlated samples from sequential agent–environment interactions, which violates the independence condition in supervised learning. Even worse, the training data in reinforcement learning is usually non-stationary as the value function is estimated with current policy, or at least the state-visit-frequency determined by current policy, and the policy is updated all the time during training. The agent learns through exploring different partitions of the state space. All these cases violate the condition of sampled data being identically distributed.

There are some practical requirements for the representations when using value function approximation in reinforcement learning, which may lead to divergence if not considered properly (Achiam et al. 2019). Specifically, the danger of instability and divergence arises whenever the three conditions are combined: (1) training on a distribution of transitions other than those naturally generated by the process whose expectation is being estimated (e.g., off-policy learning); (2) scalable function approximations (e.g., linear semi-gradient); (3) bootstrapping (e.g., DP, TD learning). These three main properties can lead to learning divergence only when they are combined, which is known as **the deadly triad** (Van Hasselt et al. 2018). Value-based methods using function approximation can also have an over-/under-estimation problem, if the way of leveraging function approximation is not fair enough. For example, original DQN has the problem of overestimating the Q-value (Van Hasselt et al. 2016), which decreases the learning performances in practice, and the double/dueling DQN techniques are proposed to alleviate the problem. Generally, policy-based methods with policy gradients have stronger convergence guarantee compared with value-based methods.

Gradient-Based Value Function Approximation

Considering the value function is parameterized as $V^{\pi}(s) = V^{\pi}(s; w)$ or $Q^{\pi}(s, a) = Q^{\pi}(s, a; w)$, we can derive the udpate rules with different methods of estimation. The optimization objective is set to be the mean-squared error (MSE) between the approximate function $V^{\pi}(s; w)$ (or $Q^{\pi}(s, a; w)$) and the true value function $v_{\pi}(s)$ (or $q_{\pi}(s, a)$):

$$J(w) = \mathbb{E}_{\pi}[(V^{\pi}(s; w) - v_{\pi}(s))^2] \tag{2.93}$$

or,

$$J(w) = \mathbb{E}_{\pi}[(Q^{\pi}(s, a; w) - q_{\pi}(s, a))^2] \tag{2.94}$$

Therefore the gradients with stochastic gradient descent are:

$$\Delta w = \alpha(V^{\pi}(s; w) - v_{\pi}(s))\nabla_w V^{\pi}(s; w) \tag{2.95}$$

or,

$$\Delta w = \alpha(Q^{\pi}(s, a; w) - q_{\pi}(s, a))\nabla_{w} Q^{\pi}(s, a; w) \tag{2.96}$$

where the gradients are estimated with each sample in the batch and the weights are updated in a stochastic manner. The target (true) value functions v_{π} or q_{π} in above equations are usually estimated, sometimes with a target network (DQN) or a max operator (Q-learning), etc. We show some basic estimations of the value functions here.

For **MC** estimation, the target value is estimated with the sampled return G_t. Therefore, the update gradients of value-function parameters are:

$$\Delta w_t = \alpha(V^{\pi}(S_t; w_t) - G_t)\nabla_{w_t} V^{\pi}(S_t; w_t) \tag{2.97}$$

or,

$$\Delta w_t = \alpha(Q^{\pi}(S_t, A_t; w_t) - G_{t+1})\nabla_{w_t} Q^{\pi}(S_t, A_t; w_t) \tag{2.98}$$

For **TD(0)**, the target is the TD target $R_t + \gamma V_{\pi}(S_{t+1}; w_t)$ according to the Bellman Optimality Equation as Eq. (2.92), therefore:

$$\Delta w_t = \alpha(V^{\pi}(S_t; w_t) - (R_t + \gamma V_{\pi}(S_{t+1}; w_t)))\nabla_{w_t} V^{\pi}(S_t; w_t) \tag{2.99}$$

or,

$$\Delta w_t = \alpha(Q^{\pi}(S_t, A_t; w_t) - (R_{t+1} + \gamma Q_{\pi}(S_{t+1}, A_{t+1}; w_t))\nabla_{w_t} Q^{\pi}(S_t, A_t; w_t)) \tag{2.100}$$

For **TD(λ)**, the target is the λ-return G_t^{λ}, so the update rule is:

$$\Delta w_t = \alpha(V^{\pi}(S_t; w_t) - G_t^{\lambda})\nabla_{w_t} V^{\pi}(S_t; w_t) \tag{2.101}$$

or,

$$\Delta w_t = \alpha(Q^{\pi}(S_t, A_t; w_t) - G_t^{\lambda})\nabla_{w_t} Q^{\pi}(S_t, A_t; w_t) \tag{2.102}$$

Different estimations have different preferences in bias and variances, which has already been discussed in previous sections about different estimation methods like MC and TD.

Example: Deep Q-Network

Deep Q-Network (DQN) is one of the most typical examples for value-based optimization. It uses a deep neural network for Q-value function approximation in

Q-Learning, and maintains an experience replay buffer to store transition samples during the agent–environment interactions. DQN also applies a target network Q^T, which is parameterized by a copy of the original network Q parameter and updated in a delayed manner, to stabilize the learning process, i.e. to alleviate the non-stationary data distribution problem in deep learning. It uses the MSE loss following the above Eq. (2.96), with the true value function q_π replaced by the approximation function $r + \gamma \max_{a'} Q^T(s', a')$ in a greedy manner.

The experience replay buffer provides stability for learning as random batches are sampled from the buffer to help to alleviate the problems of non-i.i.d. data. It makes the policy update to be an off-policy manner due to the mismatch between buffer content from the earlier policy and from the current policy. More details about the DQN algorithm are introduced in Chap. 4.

2.7.3 Policy-Based Optimization

Before we talk about policy-based optimization, we first introduce common policies in reinforcement learning. As introduced in previous sections, policies in reinforcement learning can be divided into deterministic and stochastic policies. In deep reinforcement learning, we use neural networks to represent the policies of both categories, which are called **parameterized policies**. Specifically, the parameterization here indicates the abstract policy is parameterized with the neural network (including single layer perceptrons), rather than other parametric representations. With the network parameters θ, the deterministic and stochastic policy can be written as $A_t = \mu_\theta(S_t)$ and $A_t \sim \pi_\theta(\cdot|S_t)$, respectively.

In deep reinforcement learning domain, there are several commonly seen specific distributions for representing the action distribution of a stochastic policy: the Bernoulli distribution, categorical distribution, and diagonal Gaussian distribution. The Bernoulli and categorical distributions work for the discrete action spaces, either binary or multi-category, while the diagonal Gaussian distributions work for the continuous action spaces.

The Bernoulli distribution of a single variable $x \in 0, 1$ with parameter θ is: $P(s; \theta) = \theta^x(1 - \theta)^{(1-x)}$. Therefore it can be used to represent the actions with binary value, for either single or multiple dimensions (with a vector of variables), which works for the so-called **binary-action policies**.

A **categorical policy** with categorical distribution as its output can be used in discrete and finite action spaces, it considers the policy as a classifier, which outputs the probabilities of each action in the finite action space conditioned on a state e.g., $\pi(a|s) = \boldsymbol{P}[A_t = a|S_t = s]$. The sum of all probabilities is equal to one, therefore the softmax activation function is usually applied in the last output layer when the categorical policy is parameterized. Instead of using probability function $p(\cdot|\cdot)$, here we use $\boldsymbol{P}[\cdot|\cdot]$ specifically for representing the cases with finite action space in a matrix. The agent can choose one action by sampling according to the categorical distribution. In practice, the action in this case is usually encoded as a one-hot vector with the same dimension as the action space as $a_i = (0, 0, \ldots, 1, \ldots, 0)$, so that

$a_i \odot \boldsymbol{p}(\cdot|s)$ gives $p(a_i|s)$, where $\odot$ is the element-wise product operator and $\boldsymbol{p}(\cdot|s)$ is the vector of matrix with fixed state s, usually also as the normalized output layer of the categorical policy. ***Gumbel-Softmax*** **trick** can be applied in practice to keep the sampling process of categorical distribution differentiable if the categorical policy is parameterized. Without specific tricks applied, the stochastic node with a sampling process and operations like arg max are usually non-differentiable, which is problematic when employed in parameterized policies depending on gradient-based optimization (introduced in later sections).

Gumbel-Softmax trick (Jang et al. 2016): first, the *Gumbel-Max* trick allows us to draw samples from categorical distribution π:

$$z = \text{one\_hot}[\arg\max_i(z_i + \log \pi_i)] \tag{2.103}$$

where "one_hot" is an operation transferring a scalar into a one-hot vector. However, as mentioned above, the arg max operation is generally non-differentiable. Therefore, in *Gumbel-Softmax* trick, a *Softmax* operation is applied to approximate the arg max continuously in *Gumbel-Max* trick:

$$a_i = \frac{\exp((\log \pi_i + g_i)/\tau)}{\sum_j \exp((\log \pi_j + g_j)/\tau}, \forall i = 0, \ldots, k \tag{2.104}$$

where k is the dimension of the desired variable $\boldsymbol{a}$ (the action for reinforcement learning policy) and g_i is the Gumbel variable sampled from the Gumbel distribution. The Gumbel (0,1) distribution can be sampled using inverse transform sampling by drawing $u \sim \text{Uniform}(0, 1)$ and computing $g = \log(\log(u))$ in practice.

The **diagonal Gaussian policy** outputting the means and variances of a diagonal Gaussian distribution can be used in continuous action spaces. A normal multi-variate Gaussian distribution contains a mean vector $\boldsymbol{\mu}$ and a covariance matrix $\boldsymbol{\Sigma}$, while the diagonal Gaussian distribution is a special case where only the diagonal of covariance matrix is non-zero, so we can use a vector $\boldsymbol{\sigma}$ to represent it. When applying the diagonal Gaussian distribution to represent the probabilistic actions, it removes the covariance relationships among different dimensions of the actions. When the policy is parameterized, the **reparametrization trick** as below (similar as in variational autoencoder by Kingma and Welling (2014)) can be applied to sample actions from the mean and variance vectors, as well as keeping the operations differentiable.

Reparameterization trick: sampling the action a from a diagonal Gaussian distribution $a \sim \mathcal{N}(\boldsymbol{\mu}_\theta, \boldsymbol{\sigma}_\theta)$ with the mean and variance vectors $\boldsymbol{\mu}_\theta$ and $\boldsymbol{\sigma}_\theta$ (parameterized) can be alternatively achieved with sampling a hidden vector z from a normal Gaussian $z \sim \mathcal{N}(0, \mathbf{1})$ and derive the action as:

$$a = \boldsymbol{\mu}_\theta + \boldsymbol{\sigma}_\theta \odot z \tag{2.105}$$

where $\odot$ is the elementwise product for two vectors of the same shape.

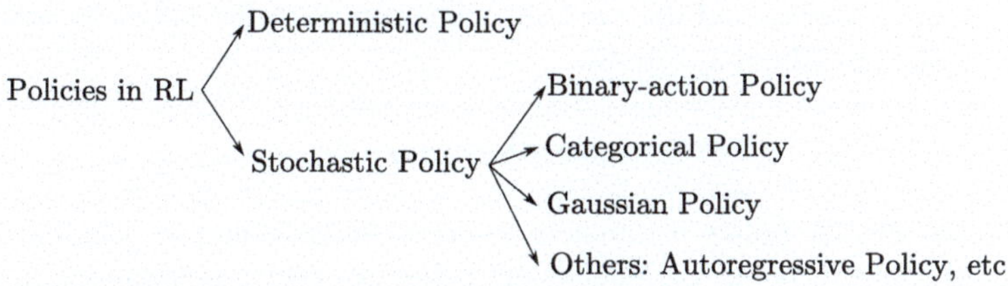

Fig. 2.19 Different policies in deep reinforcement learning

An overview of common policies in deep reinforcement learning is displayed in Fig. 2.19, for providing the readers a better understanding.

Policy-Based Optimization methods directly optimize the policy of the agent in reinforcement learning scenarios without estimating or learning an action-value function. The sampled reward values are usually used in the optimization process for improving action preferences. Either gradient-based or gradient-free methods are applied in the optimization process. Gradient-based methods always apply the policy gradient, which perhaps represents the most popular class of algorithms used in continuous-action reinforcement learning, benefiting from scalability to high-dimensional cases. The typical methods in gradient-based optimization include REINFORCE, etc. On the other hand, gradient-free algorithms usually have a faster learning process for relatively simple cases in policy searching, free from the computationally expensive process of calculating derivatives. The typical methods in gradient-free category include cross-entropy (CE) method and so on.

Recall that the goal of the agent in reinforcement learning is to maximize the cumulative discounted reward from the start state, in an expected or estimated view, which can be denoted as:

$$J(\pi) = \mathbb{E}_{\tau \sim \pi}[R(\tau)] \tag{2.106}$$

where $R(\tau) = \sum_{t=0}^{T} \gamma^t R_t$ as a discounted expected reward with finite steps (fits most scenarios), and τ are sampled trajectories.

The policy-based optimization will optimize the policy π with respect to the above goal $J(\pi)$, through gradient-based or gradient-free methods. We will first introduce **gradient-based** methods and give an example of REINFORCE algorithm, then introduce a **gradient-free** (non-gradient-based) algorithms and show the example CE method.

Gradient-Based Optimization

Gradient-based optimization uses an estimator for the gradients on the expected return (total reward) obtained from sample trajectories to improve the policy with gradient descent/ascent, and the gradient with respect to the policy parameter is called the **policy gradient** as follows:

$$\Delta\theta = \alpha\nabla_\theta J(\pi_\theta) \tag{2.107}$$

where θ indicates the policy parameters and α is the learning rate. Methods based on these gradients of policy parameters are called the policy gradient method. The **policy gradient theorem** proposed by Sutton et al. (2000) and Silver et al. (2014) is shown as follows and will be proved in the following sections.

Note: the representation θ of parameters in Eq. (2.107) is actually improper, which is supposed to be $\boldsymbol{\theta}$ for representing the vector as a default format of the book (see the chapter of math notation). However, here we apply the vanilla format θ as an interchangeable way of $\boldsymbol{\theta}$ whenever representing the model parameters. This follows the common format in literature and is also simple. One way to consider the rationality of this representation is: the gradients of parameters can be taken for each parameter individually, which is denoted by θ, while the equations are the same for all parameters. Therefore it also works for applying θ to represent all parameters. The rest of the book follows the above statements.

Theorem 2.2 (Policy Gradient Theorem)

$$\nabla_\theta J(\pi_\theta) = \mathbb{E}_{\tau\sim\pi_\theta}\left[\sum_{t=0}^{T}\nabla_\theta(\log\pi_\theta(A_t|S_t))Q^{\pi_\theta}(S_t, A_t)\right] \tag{2.108}$$

$$= \mathbb{E}_{S_t\sim\rho^\pi, A_t\sim\pi_\theta}[\nabla_\theta(\log\pi_\theta(A_t|S_t))Q^{\pi_\theta}(S_t, A_t)] \tag{2.109}$$

where the second form is derived through defining the discounted state distribution as in Silver et al. (2014) by $\rho^\pi(s') := \int_S\sum_{t=0}^{T}\gamma^{t-1}\rho_0(s)p(s'|s, t, \pi)ds$ *and* $p(s'|s, t, \pi)$ *is the transition probability of* s *to* s' *under policy* π *at time step* t.

The policy gradient theorem works for both stochastic policies and deterministic policies. It was originally proposed by Sutton et al. (2000) for stochastic policies, but extended to deterministic policies by Silver et al. (2014). For the deterministic cases, although the deterministic policy gradient theorem (introduced later) does not look like the above policy gradient theorem, it is proved that the deterministic policy gradient (DPG) is just a special (limiting) case of the stochastic policy gradient (SPG), if we parameterize the stochastic policy $\pi_{\mu_\theta,\sigma}$ by a deterministic policy $\mu_\theta : \mathcal{S} \to \mathcal{A}$ and a variance parameter σ, such that for $\sigma = 0$ the stochastic policy is equivalent to the deterministic policy, $\pi_{\mu_\theta,0} \equiv \mu$. A detailed proof will be provided in the section of the deterministic policy gradient.

1. Stochastic Policy Gradient

Now we first prove the policy gradient theorem for the stochastic policy, which is called the stochastic policy gradient method. For simplicity, we assume an episodic setting in finite MDP with the length of each trajectory fixed as $T+1$ in this section. Considering a parameterized stochastic policy $\pi_\theta(a|s)$, we then have the probability of trajectory $p(\tau|\pi) = \rho_0(S_0)\prod_{t=0}^{T} p(S_{t+1}|S_t, A_t)\pi(A_t|S_t)$ for MDP process with $\rho_0(S_0)$ as initial state distribution, we can get the logarithm of the probability of trajectory with parameterized policy π_θ as:

$$\log p(\tau|\theta) = \log \rho_0(S_0) + \sum_{t=0}^{T}\Bigg(\log p(S_{t+1}|S_t, A_t) + \log \pi_\theta(A_t|S_t)\Bigg). \tag{2.110}$$

We also need the **Log-Derivative Trick**: $\nabla_\theta p(\tau|\theta) = p(\tau|\theta)\nabla_\theta \log p(\tau|\theta)$

Therefore we can get the derivative of the log-probability of a trajectory as:

$$\nabla_\theta \log p(\tau|\theta) = \nabla_\theta \log \rho_0(S_0) + \sum_{t=0}^{T}\Bigg(\nabla_\theta \log p(S_{t+1}|S_t, A_t) + \nabla_\theta \log \pi_\theta(A_t|S_t)\Bigg) \tag{2.111}$$

$$= \sum_{t=0}^{T} \nabla_\theta \log \pi_\theta(A_t|S_t). \tag{2.112}$$

where the terms containing $\rho_0(S_0)$ and $p(S_{t+1}|S_t, A_t)$ are removed because they do not depend on parameters θ, although unknown.

Recall that the learning objective is to maximize the expected cumulative reward:

$$J(\pi_\theta) = \mathbb{E}_{\tau\sim\pi_\theta}[R(\tau)] = \mathbb{E}_{\tau\sim\pi_\theta}\left[\sum_{t=0}^{T} R_t\right] = \sum_{t=0}^{T}\mathbb{E}_{\tau\sim\pi_\theta}[R_t], \tag{2.113}$$

where $\tau = (S_0, A_0, R_0, \ldots, S_T, A_T, R_T, S_{T+1})$ and $R(\tau) = \sum_{t=0}^{T} R_t$. We can directly perform gradient ascent on the parameters of the policy θ to gradually improve the performance of the policy π_θ.

Note that R_t only depends on τ_t, where $\tau_t = (S_0, A_0, R_0, \ldots, S_t, A_t, R_t, S_{t+1})$.

$$\nabla_\theta \mathbb{E}_{\tau\sim\pi_\theta}[R_t] = \nabla_\theta \int_{\tau_t} R_t\, p(\tau_t|\theta)d\tau_t \qquad \text{Expand expectation} \tag{2.114}$$

$$= \int_{\tau_t} R_t \nabla_\theta p(\tau_t|\theta)d\tau_t \qquad \text{Exchange gradient and integral} \tag{2.115}$$

$$= \int_{\tau_t} R_t\, p(\tau_t|\theta)\nabla_\theta \log p(\tau_t|\theta)d\tau_t \qquad \text{Log-derivative trick} \tag{2.116}$$

$$= \mathbb{E}_{\tau\sim\pi_\theta}\big[R_t \nabla_\theta \log p(\tau_t|\theta)\big] \qquad \text{Return to expectation form} \tag{2.117}$$

The third equality above is due to the log-derivative trick introduced before.

Plug the above formula back to $J(\pi_\theta)$,

$$\nabla_\theta J(\pi_\theta) = \mathbb{E}_{\tau\sim\pi_\theta}\left[\sum_{t=0}^{T} R_t \nabla_\theta \log p(\tau_t|\theta)\right].$$

Now we need to compute $\nabla_\theta \log p_\theta(\tau_t)$, where $p_\theta(\tau_t)$ depends on both the policy π_θ and the ground truth of the model $p(R_t, S_{t+1}|S_t, A_t)$ which is not available to the agent. Luckily, to apply the policy gradient method, we only need the gradient of $\log p_\theta(\tau_t)$ instead of its original value, which can be derived easily by replacing the $\tau = \tau_{0:T}$ in Eq. (2.112) to be $\tau_t = \tau_{0:t}$, which gives:

$$\nabla_\theta \log p(\tau_t|\theta) = \sum_{t'=0}^{t} \nabla_\theta \log \pi_\theta(A_{t'}|S_{t'}). \tag{2.118}$$

Therefore,

$$\begin{aligned}\nabla_\theta J(\pi_\theta) &= \mathbb{E}_{\tau\sim\pi_\theta}\left[\sum_{t=0}^{T} R_t \nabla_\theta \sum_{t'=0}^{t} \log \pi_\theta(A_{t'}|\, S_{t'})\right]\\ &= \mathbb{E}_{\tau\sim\pi_\theta}\left[\sum_{t'=0}^{T} \nabla_\theta \log \pi_\theta(A_{t'}|\, S_{t'}) \sum_{t=t'}^{T} R_t\right].\end{aligned} \tag{2.119}$$

Here the last equality is simply by rearranging the summation.

Notice that in the above derivation process we use both the exchanging between sum and expectation and the exchanging between expectation and sum and derivative (both valid):

$$\nabla_\theta J(\pi_\theta) = \nabla_\theta \mathbb{E}_{\tau\sim\pi_\theta}[R(\tau)] = \nabla_\theta \mathbb{E}_{\tau\sim\pi_\theta}\left[\sum_{t=0}^{T} R_t\right] = \sum_{t=0}^{T} \nabla_\theta \mathbb{E}_{\tau\sim\pi_\theta}[R_t] \tag{2.120}$$

which ends up to take the integral in Eq. (2.114) over the partial trajectory τ_t of length $t + 1$. However, there is also other way of taking the expectation of the cumulative reward along the whole trajectory:

$$\nabla_\theta J(\pi_\theta) = \nabla_\theta \mathbb{E}_{\tau\sim\pi_\theta} R(\tau) \tag{2.121}$$

$$= \nabla_\theta \int_\tau p(\tau|\theta) R(\tau) \quad \text{Expand expectation} \tag{2.122}$$

$$= \int_{\tau} \nabla_{\theta} p(\tau|\theta) R(\tau) \quad \text{Exchange gradient and integral} \tag{2.123}$$

$$= \int_{\tau} p(\tau|\theta) \nabla_{\theta} \log p(\tau|\theta) R(\tau) \quad \text{Log-derivative trick} \tag{2.124}$$

$$= \mathbb{E}_{\tau \sim \pi_{\theta}} [\nabla_{\theta} \log p(\tau|\theta) R(\tau)] \quad \text{Return to expectation form} \tag{2.125}$$

$$\Rightarrow \nabla_{\theta} J(\pi_{\theta}) = \mathbb{E}_{\tau \sim \pi_{\theta}} \left[\sum_{t=0}^{T} \nabla_{\theta} \log \pi_{\theta}(A_t|S_t) R(\tau) \right] \tag{2.126}$$

$$= \mathbb{E}_{\tau \sim \pi_{\theta}} \left[\sum_{t=0}^{T} \nabla_{\theta} \log \pi_{\theta}(A_t|S_t) \sum_{t'=0}^{T} R_{t'} \right] \tag{2.127}$$

A careful reader may notice that the second result in Eq. (2.127) is slightly different from the first result as in Eq. (2.119). Specifically, the time scales of the cumulative reward are different. The first result uses only the cumulative future rewards $\sum_{t=t'}^{T} R_t$ after action A_t to evaluate the action, while the second result uses the cumulative rewards on the whole trajectory $\sum_{t=0}^{T} R_t$ to evaluate each action A_t on that trajectory, including the rewards before choosing that action. Intuitively, the action should not be evaluated by the rewards happened before that action is conducted, which is also reinforced by mathematical proof that the rewards obtained before the action have zero effects on the final expected gradients. Those past rewards can, therefore, be simply dropped in the derived policy gradient to have Eq. (2.119), which is called the "reward-to-go" policy gradient. A strict proof of the equivalence of the two policy gradient formulas is not provided here but can be referred to here.[1] The two derivations here can also be regarded as a proof of the equivalence of two results.

The ∇ in the above formulas called "nabla" is a specific computational operator with three basic meanings (gradient, divergence, and curl) in the physics and mathematics domains, depending on its operational objectives. But in the computer science domain, the "nabla" operator ∇ is usually used as partial derivative, which derives the derivative on the following objective explicitly containing the variable in the footnote position. As the $R(\tau)$ in above formulas does not explicitly contain θ, the ∇_{θ} does not operate on $R(\tau)$, although the τ implicitly depends on θ (according to the graphical model of MDP). We also notice that the expectation in Eq. (2.127) can be estimated with the sample mean. If we collect a set of trajectories $\mathcal{D} = \{\tau_i\}_{i=1,...,N}$ where each trajectory is obtained by letting the agent act in the

[1]Proof of equivalence of two versions of stochastic policy gradient: https://spinningup.openai.com/en/latest/spinningup/extra_pg_proof1.html.

environment using the policy π_θ, the policy gradient can be estimated with

$$\hat{g} = \frac{1}{|\mathcal{D}|} \sum_{\tau \in \mathcal{D}} \sum_{t=0}^{T} \nabla_\theta \log \pi_\theta(A_t|S_t) R(\tau), \tag{2.128}$$

The **Expected Grad-Log-Prob (EGLP)** lemma[2] is commonly used in policy gradient optimization, so we introduce it here.

Lemma 2.2 (EGLP Lemma) *Suppose that p_θ is a parameterized probability distribution over a random variable, x. Then:*

$$\mathbb{E}_{x \sim p_\theta}[\nabla_\theta \log P_\theta(x)] = 0. \tag{2.129}$$

Proof Recall that all probability distributions are normalized:

$$\int_x p_\theta(x) = 1. \tag{2.130}$$

Take the gradient of both sides of the normalization condition:

$$\nabla_\theta \int_x p_\theta(x) = \nabla_\theta 1 = 0. \tag{2.131}$$

Use the log derivative trick to get:

$$0 = \nabla_\theta \textstyle\int_x p_\theta(x) \tag{2.132}$$

$$= \textstyle\int_x \nabla_\theta p_\theta(x) \tag{2.133}$$

$$= \textstyle\int_x p_\theta(x) \nabla_\theta \log p_\theta(x) \tag{2.134}$$

$$\therefore 0 = \mathbb{E}_{x \sim p_\theta}[\nabla_\theta \log p_\theta(x)]. \tag{2.135}$$

From the EGLP lemma we can directly derive that:

$$\mathbb{E}_{A_t \sim \pi_\theta}[\nabla_\theta \log \pi_\theta(A_t|S_t) b(S_t)] = 0. \tag{2.136}$$

where $b(S_t)$ is called a baseline and is independent of the future trajectory the expectation is taken over. The baseline is any function dependent only on the currents state, without affecting the overall expected value in the optimization formula.

[2]Referred to OpenAI Spinning Up: https://spinningup.openai.com/en/latest/.

In the above formulas the optimization goal is finally:

$$\nabla_\theta J(\pi_\theta) = \mathbb{E}_{\tau\sim\pi_\theta}\left[\sum_{t=0}^{T} \nabla_\theta \log \pi_\theta(A_t|S_t) R(\tau)\right] \tag{2.137}$$

We can also modify the reward for total trajectory $R(\tau)$ to be reward-to-go G_t following time step t:

$$\nabla_\theta J(\pi_\theta) = \mathbb{E}_{\tau\sim\pi_\theta}\left[\sum_{t=0}^{T} \nabla_\theta \log \pi_\theta(A_t|S_t) G_t\right] \tag{2.138}$$

With the above EGLP lemma, the expected return can be generalized to be:

$$\nabla_\theta J(\pi_\theta) = \mathbb{E}_{\tau\sim\pi_\theta}\left[\sum_{t=0}^{T} \nabla_\theta \log \pi_\theta(A_t|S_t) \Phi_t\right] \tag{2.139}$$

where $\Phi_t = \sum_{t'=t}^{T}(R(S_{t'}, a_{t'}, S_{t'+1}) - b(S_t))$.

Actually Φ_t could be the following formats for more practical usage:

$$\Phi_t = Q^{\pi_\theta}(S_t, A_t) \tag{2.140}$$

or,

$$\Phi_t = A^{\pi_\theta}(S_t, A_t) = Q^{\pi_\theta}(S_t, A_t) - V^{\pi_\theta}(S_t) \tag{2.141}$$

which are both proven to be identical to the original format in the expected value, just with different variances in practice. The proof of these requires law of iterated expectations: $\mathbb{E}[X] = \mathbb{E}[\mathbb{E}[X|Y]]$ for two random variables (discrete or continuous). And this is easy to prove. The rest of the proof is given below:

$$\nabla_\theta J(\pi_\theta) = \mathbb{E}_{\tau\sim\pi_\theta}\left[\sum_{t=0}^{T} \nabla_\theta \log \pi_\theta(A_t|S_t) R(\tau)\right] \tag{2.142}$$

$$= \sum_{t=0}^{T} \mathbb{E}_{\tau\sim\pi_\theta}[\nabla_\theta \log \pi_\theta(A_t|S_t) R(\tau)] \tag{2.143}$$

$$= \sum_{t=0}^{T} \mathbb{E}_{\tau_{:t}\sim\pi_\theta}[\mathbb{E}_{\tau_{t:}\sim\pi_\theta}[\nabla_\theta \log \pi_\theta(A_t|S_t) R(\tau)|\tau_{:t}]] \tag{2.144}$$

$$= \sum_{t=0}^{T} \mathbb{E}_{\tau_{:t}\sim\pi_\theta}[\nabla_\theta \log \pi_\theta(A_t|S_t) \mathbb{E}_{\tau_{t:}\sim\pi_\theta}[R(\tau)|\tau_{:t}]] \tag{2.145}$$

$$= \sum_{t=0}^{T} \mathbb{E}_{\tau_{:t} \sim \pi_\theta}[\nabla_\theta \log \pi_\theta(A_t|S_t) \mathbb{E}_{\tau_{t:} \sim \pi_\theta}[R(\tau)|S_t, A_t]] \tag{2.146}$$

$$= \sum_{t=0}^{T} \mathbb{E}_{\tau_{:t} \sim \pi_\theta}[\nabla_\theta (\log \pi_\theta(A_t|S_t)) Q^{\pi_\theta}(S_t, A_t)] \tag{2.147}$$

in which $\mathbb{E}_\tau[\cdot] = \mathbb{E}_{\tau_{:t}}[\mathbb{E}_{\tau_{t:}}[\cdot|\tau_{:t}]]$ and $\tau_{:t} = (S_0, A_0, \ldots, S_t, A_t)$, and $Q^{\pi_\theta}(S_t, A_t) = \mathbb{E}_{\tau_{t:} \sim \pi_\theta}[R(\tau)|S_t, A_t]$.

Therefore, it's common to see

$$\nabla_\theta J(\pi_\theta) = \mathbb{E}_{\tau \sim \pi_\theta}\left[\sum_{t=0}^{T} \nabla_\theta (\log \pi_\theta(A_t|S_t)) Q^{\pi_\theta}(S_t, A_t)\right] \tag{2.148}$$

or,

$$\nabla_\theta J(\pi_\theta) = \mathbb{E}_{\tau \sim \pi_\theta}\left[\sum_{t=0}^{T} \nabla_\theta (\log \pi_\theta(A_t|S_t)) A^{\pi_\theta}(S_t, A_t)\right] \tag{2.149}$$

in the literature. In other words, it is equivalent to changing the optimization objective to be $J(\pi_\theta) = \mathbb{E}_{\tau \sim \pi}[Q^{\pi_\theta}(S_t, A_t)]$ or $J(\pi_\theta) = \mathbb{E}_{\tau \sim \pi}[A^{\pi_\theta}(S_t, A_t)]$ instead of original $\mathbb{E}_{\tau \sim \pi}[R(\tau)]$, in the sense of optimal policy. The $A^{\pi_\theta}(S_t, A_t)$ are usually estimated with TD-error in practice.

According to whether the environment model is used or not, reinforcement learning algorithms can be classified into model-free and model-based categories. For model-free reinforcement learning, pure gradient-based optimization algorithms are originated from the REINFORCE algorithm, or called the policy gradient method. For the model-based reinforcement learning category, there are also policy-based algorithms, like the method applying backpropagation through time (BPTT) for updating the policy using sampled rewards within episodes. No more details about model-based methods will be discussed here, and we instead direct the readers to Chap. 9.

Example: REINFORCE Algorithm

REINFORCE is an algorithm using stochastic policy gradient method as in Eq. (2.139), where $\Phi_t = Q^\pi(S_t, A_t)$ and it is estimated with sampled rewards along the trajectory $G_t = \sum_{t'=t}^{\infty} R_{t'}$ (or discounted version $G_t = \sum_{t'=t}^{\infty} \gamma^{t'-t} R_{t'}$) in REINFORCE. The gradients for updating the policy are:

$$g = \mathbb{E}\left[\sum_{t=0}^{\infty} \sum_{t'=t}^{\infty} R_{t'} \nabla_\theta \log \pi_\theta(A_t|S_t)\right] \tag{2.150}$$

Details of REINFORCE algorithm are introduced in Chap. 5.

2. Deterministic Policy Gradient

What has been described above belongs to stochastic policy gradient (SPG), and it works for optimizing the stochastic policy $\pi(a|s)$, which represents the action as a probabilistic distribution based on the current state. The contrary case to the stochastic policy is the deterministic policy, where $a = \pi(s)$ is a deterministic action instead of probability. We can derive the deterministic policy gradient (DPG) similarly as in SPG, and it also follows the policy gradient theorem numerically (as a limit case), although they have different explicit expressions.

Note: in the following part of this section, we use $\mu(s)$ instead of $\pi(s)$ as previously defined to represent the deterministic policy, for removing ambiguity in the distinction with stochastic policy $\pi(a|s)$.

For a more rigorous and general definition of DPG we refer to the deterministic policy gradient theorem proposed by Silver et al. (2014) in Eq. (2.159). We will introduce the deterministic policy gradient theorem and prove it, in an on-policy manner first and off-policy later, as well as discussing the relationship of DPG with SPG in detail.

First of all, we define the performance objective for the deterministic policy following the same expected discounted reward definition in stochastic policy gradient:

$$J(\mu) = \mathbb{E}_{S_t \sim \rho^\mu, A_t=\mu(S_t)}[\sum_{t=1}^{\infty} \gamma^{t-1} R(S_t, A_t)] \tag{2.151}$$

$$= \int_\mathcal{S} \int_\mathcal{S} \sum_{t=1}^{\infty} \gamma^{t-1} \rho_0(s) p(s'|s, t, \mu) R(s', \mu(s'))] \mathrm{d}s \mathrm{d}s' \tag{2.152}$$

$$= \int_\mathcal{S} \rho^\mu(s) R(s, \mu(s)) \mathrm{d}s \tag{2.153}$$

where $p(s'|s, t, \mu) = p(S_{t+1}|S_t, A_t) p^\mu(A_t|S_t)$, the first probability is the transition probability and the second is the probability of the action choice. Since it is deterministic policy, we have $p^\mu(A_t|S_t) = 1$ and therefore $p(s'|s, t, \mu) = p(S_{t+1}|S_t, \mu(S_t))$. Also, the state distribution in above formula is $\rho^\mu(s') := \int_\mathcal{S} \sum_{t=1}^{\infty} \gamma^{t-1} \rho_0(s) p(s'|s, t, \mu) \mathrm{d}s$.

As $V^\mu(s) = \mathbb{E}[\sum_{t=1}^{\infty} \gamma^{t-1} R(S_t, A_t)|S_1 = s; \mu] = \int_\mathcal{S} \sum_{t=1}^{\infty} \gamma^{t-1} p(s'|s, t, \mu) R(s', \mu(s'))] \mathrm{d}s'$ following the same definition in stochastic policy gradient except for applying the deterministic policy, we can also derive that

$$J(\mu) = \int_\mathcal{S} \rho_0(s) V^\mu(s) \mathrm{d}s \tag{2.154}$$

$$= \int_\mathcal{S} \int_\mathcal{S} \sum_{t=1}^{\infty} \gamma^{t-1} \rho_0(s) p(s'|s, t, \mu) R(s', \mu(s'))] \mathrm{d}s \mathrm{d}s' \tag{2.155}$$

which is the same as the above representation directly using discounted rewards. The relationships here also hold for stochastic policy gradient, just with the deterministic policy $\mu(s)$ replaced by the stochastic policy $\pi(a|s)$. For deterministic policy, we have $V^{\mu}(s) = Q^{\mu}(s, \mu(s))$ as the Q-value is an expectation over the action distribution for stochastic policy, but there is no action distribution but a single value for the deterministic policy. Therefore we also have the following representation for deterministic policy,

$$J(\mu) = \int_{\mathcal{S}} \rho_0(s) V^{\mu}(s) \mathrm{d}s \tag{2.156}$$

$$= \int_{\mathcal{S}} \rho_0(s) Q^{\mu}(s, \mu(s)) \mathrm{d}s \tag{2.157}$$

The different formats of performance objective will be used in the proof of DPG theorem, as well as several conditions. We list the conditions here without a detailed derivation process, which can be checked in the original paper by Silver et al. (2014):

- ***C.1 The Existence of Continuous Derivatives:*** $p(s'|s, a)$, $\nabla_a p(s'|s, a)$, $\mu_\theta(s)$, $\nabla_\theta \mu_\theta(s)$, $R(s, a)$, $\nabla_a R(s, a)$, $\rho_0(s)$ *are continuous in all parameters and variables* s, a, s', and x.
- ***C.2 The Boundedness Condition:*** *there exist* a, b, and L *such that* $\sup_s \rho_0(s) < b$, $\sup_{a,s,s'} p(s'|s, a) < b$, $\sup_{a,s} R(s, a) < b$, $\sup_{a,s,s'} ||\nabla_a p(s'|s, a)|| < L$, $\sup_{a,s} ||\nabla_a R(s, a)|| < L$.

Theorem 2.3 (Deterministic Policy Gradient Theorem) *suppose that the MDP satisfies conditions C.1 for the existence of* $\nabla_\theta \mu_\theta(s)$, $\nabla_a Q^{\mu}(s, a)$ *and the deterministic policy gradient, then,*

$$\nabla_\theta J(\mu_\theta) = \int_{\mathcal{S}} \rho^{\mu}(s) \nabla_\theta \mu_\theta(s) \nabla_a Q^{\mu}(s, a)|_{a=\mu_\theta(s)} ds \tag{2.158}$$

$$= \mathbb{E}_{s \sim \rho^{\mu}}[\nabla_\theta \mu_\theta(s) \nabla_a Q^{\mu}(s, a)|_{a=\mu_\theta(s)}] \tag{2.159}$$

Proof The proof of deterministic policy gradient theorem generally follows the same lines of the standard stochastic policy gradient theorem by Sutton et al. (2000). First of all, in order to exchange derivatives and integrals, and the order of integration whenever needed in the following proof, we need to use two lemmas, which are basic mathematical rules in calculus as follows:

- **Lemma 2.3 (Leibniz Integral Rule)** *let* $f(x, t)$ *be a function such that both* $f(x, t)$ *and its partial derivative* $f'_x(x, t)$ *are continuous in* t *and* x *in some region of the* (x, t)*-plane, including* $a(x) \leq t \leq b(x)$, $x_0 \leq x \leq x_1$*. Also suppose that the functions* $a(x)$ *and* $b(x)$ *are both continuous and both have continuous* *derivatives for $x_0 \leq x \leq x_1$. Then, for $x_0 \leq x \leq x_1$,*

$$\frac{d}{dx}\int_{a(x)}^{b(x)} f(x,t)dt = f(x,b(x))\cdot\frac{d}{dx}b(x) - f(x,a(x))\cdot\frac{d}{dx}a(x) + \int_{a(x)}^{b(x)}\frac{\partial}{\partial x}f(x,t)dt \tag{2.160}$$

- **Lemma 2.4 (Fubini's Theorem)** *Suppose $\mathcal{X}$ and $\mathcal{Y}$ are σ-finite measure spaces, and suppose that $\mathcal{X}\times\mathcal{Y}$ is given the product measure (which is unique as $\mathcal{X}$ and $\mathcal{Y}$ are σ-finite). Fubini's theorem states that if f is $\mathcal{X}\times\mathcal{Y}$ integrable, meaning that f is a measurable function and*

$$\int_{\mathcal{X}\times\mathcal{Y}} |f(x,y)|d(x,y) < \infty \tag{2.161}$$

then,

$$\int_{\mathcal{X}}\left(\int_{\mathcal{Y}} f(x,y)dy\right)dx = \int_{\mathcal{Y}}\left(\int_{\mathcal{X}} f(x,y)dx\right)dy = \int_{\mathcal{X}\times\mathcal{Y}} f(x,y)d(x,y) \tag{2.162}$$

To satisfy these two lemmas, we require the necessary conditions provided in C.1 as the Leibniz integral rule requires, which imply that $V^{\mu_\theta}(s)$ and $\nabla_\theta V^{\mu_\theta}(s)$ are continuous functions of θ and s. We also follow the assumption of the compactness of the state space $\mathcal{S}$, which is in C.2 required by Fubini's theorem and implies that for any θ, $||\nabla_\theta V^{\mu_\theta}(s)||$, $||\nabla_a Q^{\mu_\theta}(s,a)|_{a=\mu_\theta(s)}||$ and $||\nabla_\theta \mu_\theta(s)||$ are bounded functions of s. With above conditions, we have the following derivations:

$$\nabla_\theta V^{\mu_\theta}(s) = \nabla_\theta Q^{\mu_\theta}(s,\mu_\theta(s)) \tag{2.163}$$

$$= \nabla_\theta\Big(R(s,\mu_\theta(s)) + \int_{\mathcal{S}}\gamma p(s'|s,\mu_\theta(s))V^{\mu_\theta}(s')\mathrm{d}s'\Big) \tag{2.164}$$

$$= \nabla_\theta\mu_\theta(s)\nabla_a R(s,a)|_{a=\mu_\theta(s)} + \nabla_\theta\int_{\mathcal{S}}\gamma p(s'|s,\mu_\theta(s))V^{\mu_\theta}(s')\mathrm{d}s' \tag{2.165}$$

$$= \nabla_\theta\mu_\theta(s)\nabla_a R(s,a)|_{a=\mu_\theta(s)} + \int_{\mathcal{S}}\gamma\big(p(s'|s,\mu_\theta(s))\nabla_\theta V^{\mu_\theta}(s') + \nabla_\theta\mu_\theta(s)\nabla_a p(s'|s,a)V^{\mu_\theta}(s')\big)\mathrm{d}s' \tag{2.166}$$

$$= \nabla_\theta \mu_\theta(s) \nabla_a (R(s,a) + \int_{\mathcal{S}} \gamma p(s'|s,a) V^{\mu_\theta}(s') \mathrm{d}s')|_{a=\mu_\theta(s)})$$

$$+ \int_{\mathcal{S}} \gamma p(s'|s, \mu_\theta(s)) \nabla_\theta V^{\mu_\theta}(s') \mathrm{d}s' \tag{2.167}$$

$$= \nabla_\theta \mu_\theta(s) \nabla_a Q^{\mu_\theta}(s,a)|_{a=\mu_\theta(s)} + \int_{\mathcal{S}} \gamma p(s'|s, \mu_\theta(s)) \nabla_\theta V^{\mu_\theta}(s') \mathrm{d}s' \tag{2.168}$$

In the above derivations, the Leibniz integral rule is used to exchange the order of derivative and integration, requiring the continuity conditions of $p(s'|s,a)$, $\mu_\theta(s)$, $V^{\mu_\theta}(s)$ and their derivatives with respect to θ. Now we iterate the above formula with $\nabla_\theta V^{\mu_\theta}(s)$ to have:

$$\nabla_\theta V^{\mu_\theta}(s) = \nabla_\theta \mu_\theta(s) \nabla_a Q^{\mu_\theta}(s,a)|_{a=\mu_\theta(s)} \tag{2.169}$$

$$+ \int_{\mathcal{S}} \gamma p(s'|s, \mu_\theta(s)) \nabla_\theta \mu_\theta(s') \nabla_a Q^{\mu_\theta}(s',a)|_{a=\mu_\theta(s')} \mathrm{d}s' \tag{2.170}$$

$$+ \int_{\mathcal{S}} \gamma p(s'|s, \mu_\theta(s)) \int_{\mathcal{S}} \gamma p(s''|s', \mu_\theta(s')) \nabla_\theta V^{\mu_\theta}(s'') \mathrm{d}s'' \mathrm{d}s' \tag{2.171}$$

$$= \nabla_\theta \mu_\theta(s) \nabla_a Q^{\mu_\theta}(s,a)|_{a=\mu_\theta(s)} \tag{2.172}$$

$$+ \int_{\mathcal{S}} \gamma p(s \to s', 1, \mu_\theta(s)) \nabla_\theta \mu_\theta(s') \nabla_a Q^{\mu_\theta}(s',a)|_{a=\mu_\theta(s')} \mathrm{d}s' \tag{2.173}$$

$$+ \int_{\mathcal{S}} \gamma^2 p(s \to s', 2, \mu_\theta(s)) \nabla_\theta \mu_\theta(s') \nabla_a Q^{\mu_\theta}(s',a)|_{a=\mu_\theta(s')} \mathrm{d}s' \tag{2.174}$$

$$+ \dots \tag{2.175}$$

$$= \int_{\mathcal{S}} \sum_{t=0}^{\infty} \gamma^t p(s \to s', t, \mu_\theta(s)) \nabla_\theta \mu_\theta(s') \nabla_a Q^{\mu_\theta}(s',a)|_{a=\mu_\theta(s')} \mathrm{d}s' \tag{2.176}$$

where we use Fubini's theorem for changing the order of integration, which requires the condition that $||\nabla_\theta V^{\mu_\theta}(s)||$ is bound. The above integration contains a special case with $p(s \to s', 0, \mu_\theta(s)) = 1$ for $s' = s$ and is 0 for other s'. Now we take derivative on the modified performance objective, which is the expected value function,

$$\nabla_\theta J(\mu_\theta) = \nabla_\theta \int_{\mathcal{S}} \rho_0(s) V^{\mu_\theta}(s)\mathrm{d}s \tag{2.177}$$

$$= \int_{\mathcal{S}} \rho_0(s)\nabla_\theta V^{\mu_\theta}(s)\mathrm{d}s \tag{2.178}$$

$$= \int_{\mathcal{S}}\int_{\mathcal{S}} \sum_{t=0}^{\infty} \gamma^t \rho_0(s) p(s \to s', t, \mu_\theta(s))\nabla_\theta \mu_\theta(s')$$

$$\times \nabla_a Q^{\mu_\theta}(s', a)|_{a=\mu_\theta(s')}\mathrm{d}s'\mathrm{d}s \tag{2.179}$$

$$= \int_{\mathcal{S}} \rho^{\mu_\theta}(s)\nabla_\theta \mu_\theta(s)\nabla_a Q^{\mu_\theta}(s,a)|_{a=\mu_\theta(s)}\mathrm{d}s \tag{2.180}$$

where we use the Leibniz integral rule for exchanging the derivative and integral, requiring the conditions that $\rho_0(s)$ and $V^{\mu_\theta}(s)$ and their derivatives with respect to θ are continuous, and also the Fubini's theorem to exchange the order of integration with the boundedness conditions of integrand. Proof is completed.

Off-Policy Deterministic Policy Gradient

Apart from the on-policy version of DPG derived above, we can also derive the deterministic policy gradient in an off-policy manner, using the DPG theorem above and γ-discounted state distribution $\rho^\mu(s') := \int_{\mathcal{S}} \sum_{t=1}^{\infty} \gamma^{t-1} p(s) p(s'|s, t, \mu)\mathrm{d}s$. Off-policy deterministic policy gradient estimates current policy with samples from the behavior policy (e.g. previous policies if using replay buffer), which is different from current policy. In the off-policy settings, the gradients are estimated using trajectories sampled from a distinct behavior policy $\beta(s) \neq \mu_\theta(s)$, and the corresponding state distribution is $\rho^\beta(s)$, which is not dependent on the policy parameter θ. And in off-policy case, the performance objective is modified to be the value function of target policy averaged over the state distribution of the behavior policy $J_\beta(\mu_\theta) = \int_{\mathcal{S}} \rho^\beta(s) V^\mu(s)\mathrm{d}s = \int_{\mathcal{S}} \rho^\beta(s) Q^\mu(s, \mu_\theta(s))\mathrm{d}s$, while original objective follows Eq. (2.157) as $J(\mu_\theta) = \int_{\mathcal{S}} \rho_0(s) V^\mu(s)\mathrm{d}s$. Note that it is the first approximation we take in deriving the off-policy deterministic policy gradient, as $J(\mu_\theta) \approx J_\beta(u_\theta)$, and we will have another approximation in the following. We can directly apply the differential operator on the modified objective as follows:

$$\nabla_\theta J_\beta(\mu_\theta) = \int_{\mathcal{S}} \rho^\beta(s)(\nabla_\theta \mu_\theta(s)\nabla_a Q^{\mu_\theta}(s,a) + \nabla_\theta Q^{\mu_\theta}(s,a))|_{a=\mu(s)}\mathrm{d}s \tag{2.181}$$

$$\approx \int_{\mathcal{S}} \rho^\beta(s)\nabla_\theta \mu_\theta(s)\nabla_a Q^{\mu_\theta}(s,a)\mathrm{d}s \tag{2.182}$$

$$= \mathbb{E}_{s\sim\rho^\beta}[\nabla_\theta \mu_\theta(s)\nabla_a Q^{\mu_\theta}(s,a)|_{a=\mu(s)}] \tag{2.183}$$

The approximately equivalent symbol in above formulas indicates the difference between the on-policy DPG and off-policy DPG. The dependency relationships in above formula need to be carefully considered. The derivative of θ goes into the integration because $\rho^\beta(s)$ is independent on θ, therefore no term with derivative on $\rho^{\beta}(s)$. As the $Q^{\mu_\theta}(s, \mu_\theta(a))$ actually depends on θ in two ways (two μ_θ in the expression): (1) it depends on the action a determined by the deterministic policy μ_θ with current state s, and (2) the on-policy estimation of Q value also depends on the policy μ_θ for choosing actions for future states, as in $Q^{\mu_\theta}(s, a) = R(s, a) + \int_{\mathcal{S}} \gamma p(s'|s, a) V^{\mu_\theta}(s')\mathrm{d}s'$. So the derivative needs to be conducted separately. However, the second term $\nabla_\theta Q^{\mu_\theta}(s, a)|_{a=\mu(s)}$ in the first formula is dropped in the approximation due to the difficulty in estimation in practice, which has a similar corresponding operation in off-policy stochastic policy gradients (Degris et al. 2012).[3,4]

Relationship of Stochastic Policy Gradient and Deterministic Policy Gradient As shown in Eq. (2.148), the stochastic policy gradient has the same format as in policy gradient theorem in the previous paragraph, while the deterministic policy gradient in Eq. (2.159) seems to have an inconsistent format at first look. However, it can be proved that for a wide range of stochastic policies, the DPG is a special (limit) case of the SPG. In this sense, the DPG also satisfies the policy gradient theorem under certain conditions. In order to achieve that, we parameterize the stochastic policy $\pi_{\mu_\theta,\sigma}$ by a deterministic policy $\mu_\theta : \mathcal{S} \to \mathcal{A}$ and a variance parameter σ, such that for $\sigma = 0$ the stochastic policy is equivalent to the deterministic policy, $\pi_{\mu_\theta,0} \equiv \mu$. An additional condition is needed for defining the relationship between SPG and DPG, which is a composite condition to define the regular delta-approximation:

- ***C.3 Regular Delta-Approximation:*** *functions v_σ parameterized by σ are said to be a regular delta-approximation on $\mathcal{R} \in \mathcal{A}$ if they satisfy the conditions: (1) the distribution v_σ converges to a delta distribution $\lim_{\sigma\downarrow 0} \int_{\mathcal{A}} v_\sigma(a', a) f(a)\mathrm{d}a = f(a')$ for $a' \subseteq \mathcal{R}$ and suitably smooth f; (2) $v_\sigma(a', \cdot)$ is supported on compact $\mathcal{C}'_a \subseteq \mathcal{A}$ with Lipschitz boundary, vanishes on the boundary and is continuously differentiable on $\mathcal{C}_{a'}$; (3) the gradient $\nabla_{a'} v_\sigma(a', a)$ always exists; (4) translation invariance: $v(a', a) = v(a'+\delta, a+\delta)$ for any $a \in \mathcal{A}$, $a' \in \mathcal{R}$, $a+\delta \in \mathcal{A}$, $a'+\delta \in \mathcal{A}$.*

Theorem 2.4 (Deterministic Policy Gradient as Limit of Stochastic Policy Gradient) *Consider a stochastic policy $\pi_{\mu_\theta,\sigma}$ such that $\pi_{\mu_\theta,\sigma}(a|s) = v_\sigma(\mu_\theta(s), a)$, where σ is a parameter controlling the variance and $v_\sigma(\mu_\theta(s), a)$ satisfy C.3 and the MDP satisfies C.1 and C.2, then,*

$$\lim_{\sigma\downarrow 0} \nabla_\theta J(\pi_{\mu_\theta,\sigma}) = \nabla_\theta J(\mu_\theta) \tag{2.184}$$

[3] Details and arguments for this operation can be referred to the original paper.

[4] The paper of *Silver D, Lever G, Heess N, et al. Deterministic policy gradient algorithms[C]. 2014.* drops the ∇_a on the Q term after approximation in Eq. (15) of the paper, and here we modified this typo.

which indicates the gradient of the DPG (r.h.s) is the limit case of standard SPG (l.h.s).

The proof of the above relationship is an outline of this book and we will not discuss it here. Details refer to the original paper (Silver et al. 2014).

Applications and Variants of Deterministic Policy Gradient

One of the most famous algorithms of DPG is the deep deterministic policy gradient (DDPG), which is a deep variant of DPG. DDPG combines DQN and actor-critic algorithms to use deterministic policy gradients for updating the policy, via a deep-learning approach. It has the target networks for both the actor and the critic, and provides for sample-efficient learning but is notoriously challenging to use due to its extreme brittleness and hyperparameter sensitivity in practice (Duan et al. 2016). The details and implementation of DDPG will be introduced in later chapters.

From the above we can see that the policy gradient can be estimated in at least two approaches: SPG and DPG, depending on the type of policy. Actually, they use two kinds of different estimators, the score function estimator for SPG and the pathwise derivative estimators for DPG, in the terminology of variational inference (VI).

A reparameterization trick makes it possible to apply policy gradients derived from the value function for stochastic policy, which is called the **stochastic value gradients (SVG)** (Heess et al. 2015). In SVG algorithms, a value λ is usually used as SVG(λ) to indicate how many steps are expanded in Bellman recursion. For example, the SVG(0) and SVG(1) indicate the Bellman recursion expanded with 0 and 1 step, respectively, and SVG(∞) indicates the Bellman recursion is expanded along the whole episodic trajectory in a finite horizon. SVG(0) is a model-free method with the action-value estimated with current policy, and therefore the value gradients are back-propagated to the policy; while SVG(1) is a model-based method using a learned transition model to evaluate the value of next state, as in original paper (Heess et al. 2015).

A very simple but useful example of reparameterization trick is to write a conditional Gaussian density $p(y|x) = \mathcal{N}(\mu(x), \sigma^2(x))$ as the function $y(x) = \mu(x) + \sigma(x)\epsilon, \epsilon \sim \mathcal{N}(0, 1)$. So one can produce samples procedurally by first sampling ϵ then deterministically constructing y, which makes the sampling process from the stochastic policy trackable for gradients. And the same procedure for back-propagating the gradients from the action-value function to the policy is made feasible in practice. In order to get the gradients for the stochastic policy through the value function as in DPG, SVG applies the reparameterization trick and takes extra expectation on stochastic noise. Soft actor-critic (SAC) and the original SVG (Heess et al. 2015) algorithm both follow this routine to use stochastic policy for continuous control.

For example, in SAC, the stochastic policy is reparameterized with a mean and a variance, together with a noise term sampled from a normal distribution. The optimization objective in SAC with an additional entropy term is:

$$\pi^* = \arg\max_{\pi} \mathbb{E}_{\tau\sim\pi}\left[\sum_{t=0}^{\infty} \gamma^t (R(S_t, A_t, S_{t+1}) + \alpha H(\pi(\cdot|S_t)))\right] \tag{2.185}$$

and therefore the relationship of value function and Q-value function becomes:

$$V^{\pi}(s) = \mathbb{E}_{a\sim\pi}[Q^{\pi}(s,a)] + \alpha H(\pi(\cdot|s)) \tag{2.186}$$

$$= \mathbb{E}_{a\sim\pi}[Q^{\pi}(s,a) - \alpha \log \pi(a|s)] \tag{2.187}$$

The policy used in SAC is a normalized Gaussian distribution, which is different from traditional settings. The action in SAC can be represented as below via reparameterization trick:

$$a_{\theta}(s,\epsilon) = \tanh(\mu_{\theta}(s) + \sigma_{\theta}(s)\cdot\epsilon), \epsilon \sim \mathcal{N}(0, I) \tag{2.188}$$

Due to the stochasticity of the policy in SAC, the policy gradients are derived with the reparametrization trick through maximizing the expected value function, which is:

$$\max_{\theta} \mathbb{E}_{a\sim\pi_{\theta}}[Q^{\pi_{\theta}}(s,a) - \alpha \log \pi_{\theta}(a|s)] \tag{2.189}$$

$$= \max_{\theta} \mathbb{E}_{\epsilon\sim\mathcal{N}}[Q^{\pi_{\theta}}(s, a(s,\epsilon)) - \alpha \log \pi_{\theta}(a(s,\epsilon)|s)] \tag{2.190}$$

and the gradients can therefore go back through the Q-networks to the policy network, similar as in DPG, which are:

$$\nabla_{\theta} \frac{1}{|\mathcal{B}|} \sum_{S_t\in\mathcal{B}} (Q^{\pi_{\theta}}(S_t, a(S_t,\epsilon)) - \alpha \log \pi_{\theta}(a(S_t,\epsilon)|S_t)) \tag{2.191}$$

with a sampled batch $\mathcal{B}$ for updating the policy and $a(S_t,\epsilon)$ sampled from the stochastic policy with reparameterization trick. In this sense, the reparameterization trick makes it possible for the stochastic policy to be updated in a similar manner as DPG, and the resulting SVG is kind of an intermediate between DPG and SPG. DPG can also be regarded as a deterministic limit of SVG(0).

Gradient-Free Optimization

Apart from gradient-based optimization for policy-based learning, there are also non-gradient-based (also called gradient-free) optimization methods, which include cross-entropy (CE) method, covariance matrix adaptation (CMA) (Hansen and Ostermeier 1996), hill climbing, simplex/amoeba/Nelder-Mead algorithm (Nelder and Mead 1965), etc.

Example: Cross-Entropy (CE) Method

Instead of using gradient-based optimization for policies, CE method is usually faster for policy search in reinforcement learning as a non-gradient-based optimization method. In a CE method, the policy is updated iteratively and the optimization objective for parameters θ of the parameterized policy π_θ is:

$$\theta^* = \arg\max S(\theta) \tag{2.192}$$

where the $S(\theta)$ is the general objective function, and for our reinforcement learning cases, it could be the discounted expected return: $S(\theta) = R(\tau) = \sum_{t=0}^{T} \gamma^t R_t$.

The policy in the CE method can be parameterized as a multi-variate linear independent Gaussian distribution, and the distribution of the parameter vector at iteration t is $f_t \sim N(\mu_t, \sigma_t^2)$. After drawing n sample vectors $\theta_1, \ldots, \theta_n$ and evaluating their value $S(\theta_1), \ldots, S(\theta_i)$, we sort them and select the best $\lfloor \rho \cdot n \rfloor$ samples, where $0 < \rho < 1$ is the selection ratio. Denoting the set of indices of the selected samples by $I \in 1, 2, \ldots, n$, the mean of the distribution is updated using:

$$\mu_{t+1} =: \frac{\sum_{i \in I} \theta_i}{|I|} \tag{2.193}$$

and the deviation update is:

$$\sigma_{t+1}^2 := \frac{\sum_{i \in I} (\theta_i - \mu_{t+1})^T (\theta_i - \mu_{t+1})}{|I|} \tag{2.194}$$

The cross-entropy method is an efficient and general optimization algorithm. However, preliminary investigations showed that the applicability of CE to reinforcement learning problems is restricted severely by the phenomenon that the distribution concentrates to a single point too fast. Therefore, its applicability in reinforcement learning seems to be limited though fast, because it often converges to suboptimal policies. A standard technique for preventing early convergence is to introduce noise. General methods include adding a constant or an adaptive value on the standard deviation term during the iterative process for the Gaussian distribution like:

$$\sigma_{t+1}^2 := \frac{\sum_{i \in I} (\theta_i - \mu_{t+1})^T (\theta_i - \mu_{t+1})}{|I|} + Z_{t+1} \tag{2.195}$$

where $Z_t = \max(5 - \frac{t}{10}, 0)$ in work of Szita, et al.

2.7.4 Combination of Policy-Based and Value-Based Methods

With the above vanilla policy gradient method, some simple reinforcement learning tasks can be solved. However, there is usually a large variance in the evaluated updates if we choose to use Monte Carlo or TD(λ) estimation. We can use a critic as described in the value-based optimization to estimate the action-value function. So there will be two sets of parameters if we use parameterized value function as an approximation for policy optimization: the actor parameters and the critic parameters. This actually forms a very important algorithm architecture called actor-critic (AC), and typical algorithms include Q-value actor-critic, deep deterministic policy gradient (DDPG), etc.

Recall the policy gradient theorem introduced in previous sections, the derivatives of performance objective J with respect to the policy parameters θ are:

$$\nabla_\theta J(\pi_\theta) = \mathbb{E}_{\tau\sim\pi_\theta} \sum_{t=0}^{T} \nabla_\theta \log \pi_\theta(A_t|S_t) Q^\pi(S_t, A_t) \tag{2.196}$$

in which the $Q^\pi(S_t, A_t)$ are the true action-value function, and the simplest way of estimating $Q^\pi(S_t, A_t)$ is to use a sampled cumulative return $G_t = \sum_{t=0}^{\infty} \gamma^{t-1} R(S_t, A_t)$. In AC, we apply a critic to estimate the action-value function: $Q^w(S_t, A_t) \approx Q^\pi(S_t, A_t)$. And the update rule of the policy in AC is therefore:

$$\nabla_\theta J(\pi_\theta) = \mathbb{E}_{\tau\sim\pi_\theta} \sum_{t=0}^{T} \nabla_\theta \log \pi_\theta(A_t|S_t) Q^w(S_t, A_t) \tag{2.197}$$

where w are parameters of the critic for value function approximation. The critic can be evaluated with an appropriate policy evaluation algorithms such as temporal difference (TD) learning, like $\Delta w = \alpha(Q^\pi(S_t, A_t; w) - R_{t+1} + \gamma v_\pi(S_{t+1}, w))\nabla_w Q^\pi(S_t, A_t; w)$ for TD(0) estimation as in Eq. (2.100). More details about AC algorithm and implementation will be discussed in Chaps. 5 and 6.

Although the AC framework helps alleviate the variances in policy updates, it can introduce bias and potential instability due to replacing the true action-value function with the estimated one, which requires **compatible function approximation** to ensure its unbiased estimation as proposed by Sutton et al. (2000).

Compatible Function Approximation

The compatible function approximation conditions hold for both SPG and DPG. We will show them individually. The "compatible" here indicates that the approximate action-value function $Q^w(s, a)$ is compatible with the corresponding policy.

For SPG Specifically, the compatible function approximation proposes two conditions to ensure the unbiased estimation using the approximated action-value function $Q^\pi(s,a)$: (1) $Q^w(s,a) = \nabla_\theta \log \pi_\theta(a|s)^T w$ and (2) the parameters w are chosen to minimize the mean-squared error $\text{MSE}(w) = \mathbb{E}_{s\sim\rho^\pi, a\sim\pi_\theta}[(Q^w(s,a) - Q^\pi(s,a))^2]$. More intuitively, condition (1) says that compatible function approximators are linear in "features" of the stochastic policy, $\nabla_\theta \log \pi_\theta(a|s)$, and condition (2) requires that the parameters w are the solution to the linear regression problem that estimates $Q^\pi(s,a)$ from these features. In practice, condition (2) is usually relaxed in favor of policy evaluation algorithms that estimate the value function more efficiently by temporal difference learning.

If both conditions are satisfied, then the overall algorithm of AC is equivalent to not using the critic for approximation, like in the REINFORCE algorithm. This can be proved simply by setting the MSE in the condition (2) equivalent to 0 and taking gradients, then substituting the condition (1) into it:

$$\nabla_w \text{MSE}(w) = \mathbb{E}[2(Q^w(s,a) - Q^\pi(s,a))\nabla_w Q^w(s,a)] \tag{2.198}$$

$$= \mathbb{E}[2(Q^w(s,a) - Q^\pi(s,a))\nabla_\theta \log \pi_\theta(a|s)] \tag{2.199}$$

$$= 0 \tag{2.200}$$

$$\Rightarrow \mathbb{E}[Q^w(s,a)\nabla_\theta \log \pi_\theta(a|s)] = \mathbb{E}[Q^\pi(s,a)\nabla_\theta \log \pi_\theta(a|s)] \tag{2.201}$$

For DPG The two conditions in compatible function approximation are modified accordingly with respect to the deterministic policy $\mu_\theta(s)$: (1) $\nabla_a Q^w(s,a)|_{a=\mu_\theta(s)} = \nabla_\theta \mu_\theta(s)^T w$ and (2) w minimizes the mean-squared error, $\text{MSE}(\theta, w) = \mathbb{E}[\epsilon(s;\theta,w)^T \epsilon(s;\theta,w)]$ where $\epsilon(s;\theta,w) = \nabla_a Q^w(s,a)|_{a=\mu_\theta(s)} - \nabla_a Q^w(s,a)|_{a=\mu_\theta(s)}$. It can also be proved that these conditions ensure the unbiased estimation through transferring the approximation back to no-critic cases:

$$\nabla_w \text{MSE}(\theta, w) = 0 \tag{2.202}$$

$$\Rightarrow \mathbb{E}[\nabla_\theta \mu_\theta(s) \epsilon(s;\theta,w)] = 0 \tag{2.203}$$

$$\Rightarrow \mathbb{E}[\nabla_\theta \mu_\theta(s)\nabla_a Q^w(s,a)|_{a=\mu_\theta(s)}] = \mathbb{E}[\nabla_\theta \mu_\theta(s)\nabla_a Q^\mu(s,a)|_{a=\mu_\theta(s)}] \tag{2.204}$$

And it applies to both on-policy $\mathbb{E}_{s\sim\rho^\mu}[\cdot]$ and off-policy $\mathbb{E}_{s\sim\rho^\beta}[\cdot]$ cases.

Other Methods

If we replace the action-value function $Q^\pi(s,a)$ with advantage function $A^\pi(s,a)$ in Eq. (2.196) (as subtracting the baseline does not affect the gradients):

$$A^{\pi_\theta}(s,a) = Q^{\pi_\theta}(s,a) - V^{\pi_\theta}(s) \tag{2.205}$$

Then we actually get a more advanced algorithm called advantage actor-critic (A2C), which can use the TD-error to estimate the advantage function. This will not affect above theorems but changes the variances of gradient estimators.

Recently, people have proposed actor-free methods, like the QT-Opt algorithm (Kalashnikov et al. 2018) and the Q2-Opt algorithm (Bodnar et al. 2019) based on that. These methods are combinations of policy-based and value-based optimization as well, gradient-free CE method and DQN specifically. Instead of using the sampled discounted return as an estimation of actions sampled from Gaussian distributions, they apply action value approximation for learning $Q^{\pi_\theta}(s, a)$ instead, which are proved to be efficient and effective for robot learning in reality especially when there are demonstration datasets.

References

Achiam J, Knight E, Abbeel P (2019) Towards characterizing divergence in deep Q-learning. Preprint. arXiv:190308894

Auer P, Cesa-Bianchi N, Freund Y, Schapire RE (1995) Gambling in a rigged casino: the adversarial multi-armed bandit problem. In: Proceedings of IEEE 36th annual foundations of computer science. IEEE, Piscataway, pp 322–331

Bellman R et al (1954) The theory of dynamic programming. Bull Am Math Soc 60(6):503–515

Bodnar C, Li A, Hausman K, Pastor P, Kalakrishnan M (2019) Quantile QT-Opt for risk-aware vision-based robotic grasping. Preprint. arXiv:191002787

Bubeck S, Cesa-Bianchi N et al (2012) Regret analysis of stochastic and nonstochastic multi-armed bandit problems. Found Trends® Mach Learn 5(1):1–122

Degris T, White M, Sutton RS (2012) Linear off-policy actor-critic. In: In international conference on machine learning. Citeseer

Duan Y, Chen X, Houthooft R, Schulman J, Abbeel P (2016) Benchmarking deep reinforcement learning for continuous control. In: International conference on machine learning, pp 1329–1338

Fu J, Singh A, Ghosh D, Yang L, Levine S (2018) Variational inverse control with events: a general framework for data-driven reward definition. In: Advances in neural information processing systems, pp 8538–8547

Hansen N, Ostermeier A (1996) Adapting arbitrary normal mutation distributions in evolution strategies: The covariance matrix adaptation. In: Proceedings of IEEE international conference on evolutionary computation. IEEE, Piscataway, pp 312–317

Heess N, Wayne G, Silver D, Lillicrap T, Erez T, Tassa Y (2015) Learning continuous control policies by stochastic value gradients. In: Advances in neural information processing systems, pp 2944–2952

Jang E, Gu S, Poole B (2016) Categorical reparameterization with gumbel-softmax. Preprint. arXiv:161101144

Kalashnikov D, Irpan A, Pastor P, Ibarz J, Herzog A, Jang E, Quillen D, Holly E, Kalakrishnan M, Vanhoucke V et al (2018) Qt-opt: scalable deep reinforcement learning for vision-based robotic manipulation. Preprint. arXiv:180610293

Kingma DP, Welling M (2014) Auto-encoding variational Bayes. In: Proceedings of the international conference on learning representations (ICLR)

Leshno M, Lin VY, Pinkus A, Schocken S (1993) Multilayer feedforward networks with a nonpolynomial activation function can approximate any function. Neural Netw 6(6):861–867

Levine S (2018) Reinforcement learning and control as probabilistic inference: tutorial and review. Preprint. arXiv:180500909

Nelder JA, Mead R (1965) A simplex method for function minimization. Comput J 7(4):308–313
Peters J, Schaal S (2008) Natural actor-critic. Neurocomputing 71(7–9):1180–1190
Pyeatt LD, Howe AE et al (2001) Decision tree function approximation in reinforcement learning. In: Proceedings of the third international symposium on adaptive systems: evolutionary computation and probabilistic graphical models, Cuba, vol 2, pp 70–77
Schmidhuber J (2015) Deep learning in neural networks: an overview. Neural Netw 61:85–117
Silver D, Lever G, Heess N, Degris T, Wierstra D, Riedmiller M (2014) Deterministic policy gradient algorithms. In: Proceedings of the 31 st international conference on machine learning, Beijing
Singh S, Jaakkola T, Littman ML, Szepesvári C (2000) Convergence results for single-step on-policy reinforcement-learning algorithms. Mach Learn 38(3):287–308
Sutton RS, McAllester DA, Singh SP, Mansour Y (2000) Policy gradient methods for reinforcement learning with function approximation. In: Advances in neural information processing systems, pp 1057–1063
Szepesvári C (1998) The asymptotic convergence-rate of Q-learning. In: Advances in neural information processing systems, pp 1064–1070
Tsitsiklis JN, Roy BV (1997) An analysis of temporal-difference learning with function approximation. Technical Report. IEEE transactions on automatic control
Van Hasselt H, Guez A, Silver D (2016) Deep reinforcement learning with double Q-learning. In: Thirtieth AAAI conference on artificial intelligence
Van Hasselt H, Doron Y, Strub F, Hessel M, Sonnerat N, Modayil J (2018) Deep reinforcement learning and the deadly triad. Preprint. arXiv:181202648
Watkins CJ, Dayan P (1992) Q-learning. Mach Learn 8(3–4):279–292
Williams RJ, Baird III LC (1993) Analysis of some incremental variants of policy iteration: first steps toward understanding actor-critic learning systems. Technical Report. NU-CCS-93-11. Northeastern University, College of Computer Science

Chapter 3
Taxonomy of Reinforcement Learning Algorithms

Hongming Zhang and Tianyang Yu

Abstract In this chapter, we introduce and summarize the taxonomy and categories for reinforcement learning (RL) algorithms. Figure 3.1 presents an overview of the typical and popular algorithms in a structural way. We classify reinforcement learning algorithms from different perspectives, including model-based and model-free methods, value-based and policy-based methods (or combination of the two), Monte Carlo methods and temporal-difference methods, on-policy and off-policy methods. Most reinforcement learning algorithms can be classified under different categories according to the above criteria, hope this helps to provide the readers some overviews of the full picture before introducing the algorithms in detail in later chapters.

Keywords Model-based · Model-free · Value-based · Policy-based · Monte Carlo (MC) methods · Temporal-difference (TD) methods · On-policy · Off-policy

We introduce the classification of reinforcement learning from the following perspectives: model-based methods and model-free methods, value-based methods and policy-based methods, Monte Carlo methods and temporal-difference methods, on-policy methods and off-policy methods. Hopefully, the introduction of these new terms can make an easy start for a newcomer in the field of reinforcement learning. The algorithms introduced in Chaps. 4, 5, and 6 follow the taxonomy of value-based methods, policy-based methods, and the combination of the two methods, respectively.

H. Zhang (✉)
Peking University, Beijing, China
e-mail: zhanghongming@pku.edu.cn

T. Yu
Nanchang University, Nanchang, China

H. Dong et al. (eds.), *Deep Reinforcement Learning*,
https://doi.org/10.1007/978-981-15-4095-0_3

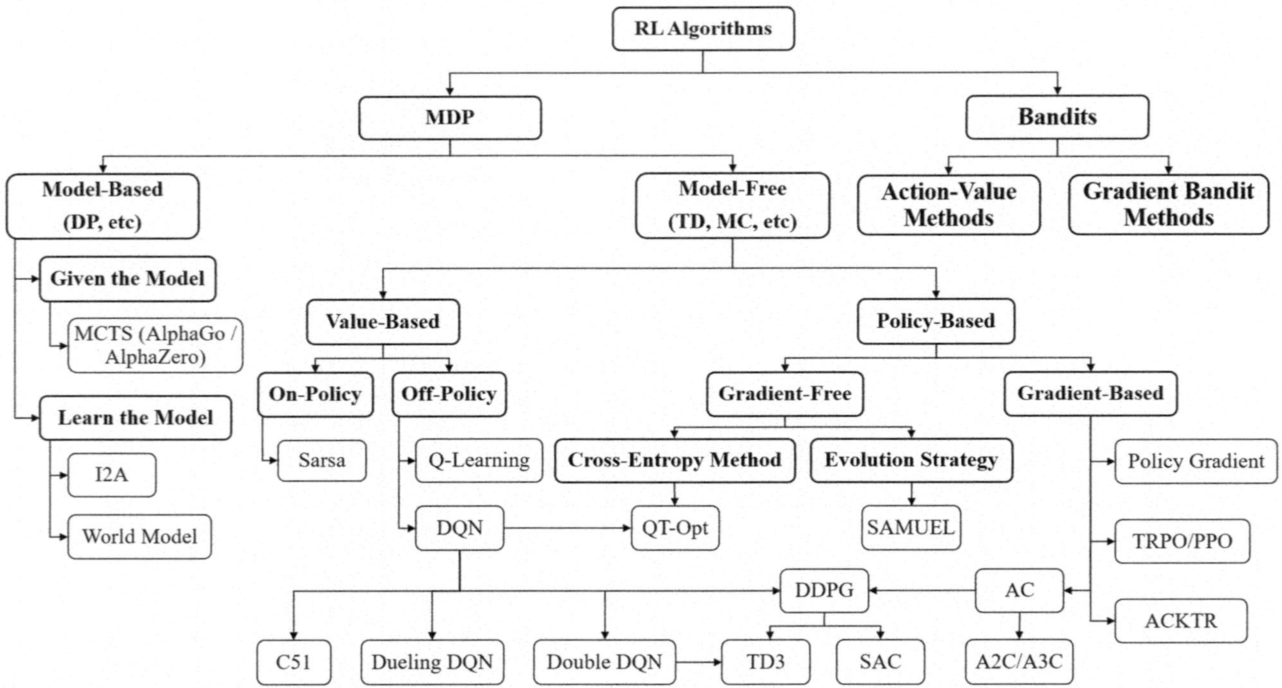

Fig. 3.1 Map of reinforcement learning algorithms. Boxes with thick lines denote different categories, others denote specific algorithms

3.1 Model-Based and Model-Free

We begin with the model-based methods and model-free methods for the discussion of the taxonomy in reinforcement learning. In deep learning, a model means a specific function with initialized parameters (pre-trained model) or learned parameters (well-trained model) such as a deep neural network. However, in model-based reinforcement learning, a "model" means an ensemble of acquired environmental knowledge. Recall that in Markov decision process (MDP), there are five elements denoted as: $\mathcal{S}, \mathcal{A}, \boldsymbol{P}, R, \gamma$. $\mathcal{S}$ and $\mathcal{A}$ denote the state space and the action space of the environment; $R(s, a)$ denotes a reward function that returns a reward when an agent takes action a at state s; $p(s'|s, a)$ denotes a transition function, it gives the transition probability of the environment transferring from s to s' when taking action a; γ denotes the discount factor. If all these elements are known, some planning methods introduced in Chap. 2 can be used directly without interaction with the environment, such as value iteration, policy iteration, etc. However, the reward function R and the transition function $p(s'|s, a)$ are usually unknown to the agent. The agent needs to go through lots of trials and errors and learns by observing the environment to leverage the reward feedback.

There are two ways for achieving the above learning process shown in Fig. 3.2. One way is to predict the elements of the environment. Even though the functions R and P are unknown, the agent can get some samples by taking actions in the environment. If the samples (s, a, s', r) are sufficient, the values of $p(s'|s, a)$ and r can be predicted by supervised learning. After that, all the elements are known and planning methods can be used directly. This way is called model-based. Another way is not to model the environment but to look for the optimal policy directly. For example, the Q-learning algorithm chooses each action with the highest Q-value and converges to an optimal Q-value function; the policy gradient algorithm searches the optimal policy in policy space directly. These two algorithms do not focus on the model and search for the highest reward directly. This approach is

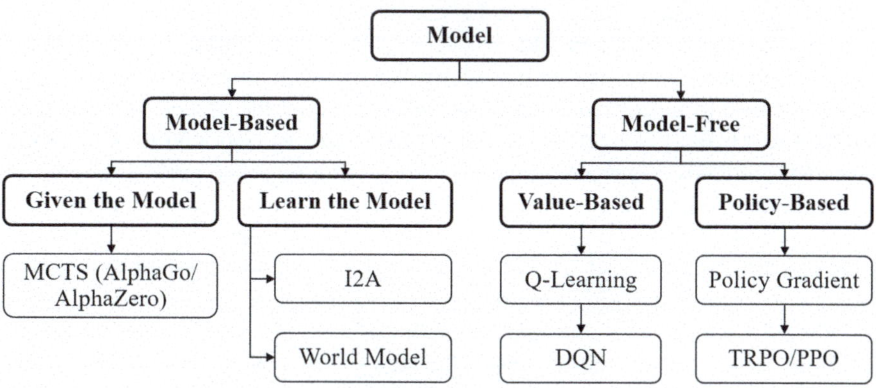

Fig. 3.2 Model-based methods and model-free methods

called model-free. The difference between model-based and model-free is whether the agent will get or learn the model (or dynamics) of the environment, such as the transition function and the reward function.

The model-based methods can be split into two categories: the methods that work with a given model and the methods that learn the model. For the methods that work with a given model, the models for the reward function and the transition process can be accessed directly by the agent. For example, in the AlphaGo algorithm (Silver et al. 2016), the rules of the Go game are specified, which can be described with the computer language easily. The transition function and reward function in Go are all known for the agent to evaluate and improve its policy. For the other category, the methods that learn the model cannot acquire the model directly due to the complexity or opaqueness of the environment, but the agent can learn a model from interactions with the environment first and then apply the model in policy improvement. Typical examples for the second category include the World Models algorithm (Ha and Schmidhuber 2018), the I2A algorithm (Racanière et al. 2017), etc. Like in the World Models algorithm, the agent collects some data (S_t, A_t, S_{t+1}) from a random policy and encodes it into a low dimensional latent vector z_t using a variational auto-encoder (VAE) (Baldi 2012). Then these data (Z_t, A_t, Z_{t+1}) are used to learn a predictive model of the future latent vector z. After that, the agent can improve its policy via the learned model.

The key advantage of model-based methods is that the future states and rewards can be anticipated in advance via the environment model, which helps the agent to make better planning. Some typical methods include pure planning and expert iteration (Sutton and Barto 2018). For example, the MBMF algorithm (Nagabandi et al. 2018) adopts pure planning techniques; the AlphaGo algorithm (Silver et al. 2016) adopts expert iteration. The disadvantage of model-based methods lies in the fact that the model is usually not available, and the dynamics of the environment can be complex, which may not even be represented explicitly. Moreover, the learned models are usually inaccurate in practice, which induce bias for estimation. The policy estimated and improved based on a biased model usually collapses when applying in the true environment.

Model-free methods do not try to build a model of the environment. The agent interacts with the environment directly and improves its performance based on the explored samples. Compared with model-based methods, model-free methods are straightforward to implement for they do not care about the model, which can be hard to learn if not given. However, model-free methods also tend to suffer from their own problems. Sometimes the cost of exploring in a real-world environment can be extremely high in terms of time consumption, tear and wear of equipment, and safety risks. For example, in the case of an automatic pilot, we cannot train an agent to explore in the real world with a model-free method without any further precautions because any traffic accident will be too much of a cost to bear.

Most of the algorithms introduced in Chaps. 4, 5, and 6 are model-free algorithms, including the deep Q-networks (DQN) algorithm (Mnih et al. 2015), policy gradient (PG) methods (Sutton et al. 2000), the deep deterministic policy gradient (DDPG) algorithm (Lillicrap et al. 2015), etc. But model-based methods are playing a more and more important role, due to the low sample efficiency of model-free methods (details in Chap. 7). For example, AlphaGo (Silver et al. 2016) and AlphaZero (Silver et al. 2018, 2017) in Chap. 15 belong to model-based algorithms.

3.2 Value-Based and Policy-Based

Recall that in Chap. 2, there are two main categories for policy optimization in deep reinforcement learning, the value-based methods and policy-based methods. A combination of the two engenders the actor-critic class of algorithms and other algorithms like QT-Opt (Kalashnikov et al. 2018), which leverage the value function for updating the policy. The relationship is shown in Fig. 3.3. The value-based methods usually imply the optimization of the action-value function $Q^{\pi}(s,a)$. The optimal value function after optimization is $Q^{\pi^*}(s,a) = \max_a Q^{\pi}(s,a)$, therefore the optimal policy can be derived by $\pi^* \approx \arg\max_\pi Q^{\pi}$ ("$\approx$" due to the approximation error). The advantages of value-based method lie in the sample efficiency is high, the variance of value function estimation is small, and it is not easy to fall into local optimum. The disadvantages are that it usually cannot handle the continuous action space problem, and the ϵ-greedy strategy and the max operator such as in DQN can easily result in overestimation.

Common value-based algorithms include Q-learning (Watkins and Dayan 1992), DQN (Mnih et al. 2015) and its variants: (1) Prioritized Experience Replay (Schaul et al. 2015) weights the data based on TD error to improve learning efficiency; (2) Dueling DQN (Wang et al. 2016) improves the network structure. It decomposes the

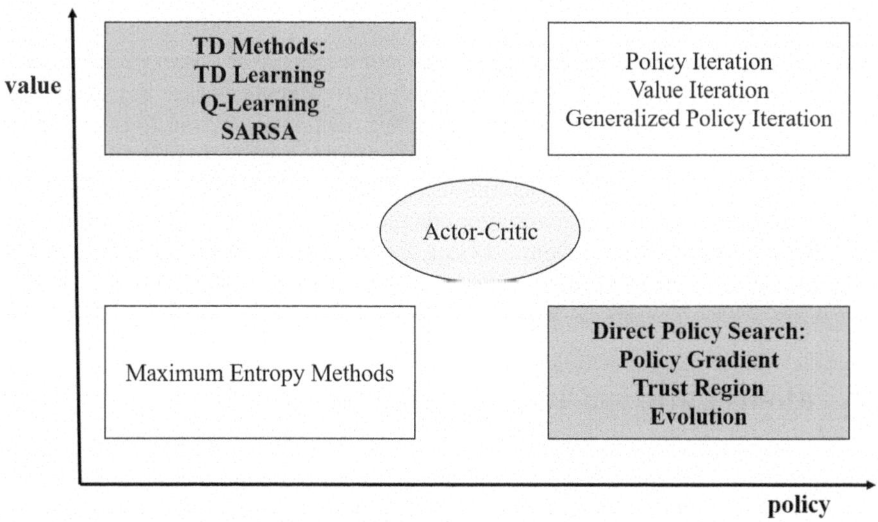

Fig. 3.3 Value-based and policy-based methods. Figure comes from Li (2017)

action-value function Q into the state-value function V and the advantage function A to improve the approximation capacity; (3) Double DQN (Van Hasselt et al. 2016) chooses and evaluates actions with different parameters to solve the overestimation problem; (4) Retrace (Munos et al. 2016) revises the calculation method of Q value and reduces the variance of value estimation; (5) Noisy DQN (Fortunato et al. 2017) adds noise to network parameters to increase exploration; (6) Distributed DQN (Bellemare et al. 2017) refines the estimation of Q value into the estimation of distribution.

The policy-based method optimizes the policy directly; it updates the policy iteratively until the accumulative return is maximized. Compared with the value-based method, a policy-based method has the advantages of simpler policy parameterization, better convergence, and is suitable for continuous or high dimensional action space. Some common policy-based algorithms include PG (Sutton et al. 2000), TRPO (Schulman et al. 2015), PPO (Schulman et al. 2017; Heess et al. 2017), etc. TRPO and PPO restrict the update step based on PG to prevent policy collapse and make the algorithm more stable.

In addition to the simple value-based methods and policy-based methods, the more popular methods are the combination of the two, which gives an actor-critic framework. The actor-critic method combines the merits of value-based method and policy-based method, using the value-based methods to learn a Q function or value function to improve sample efficiency and using the policy-based methods to learn the policy function, which is suitable for discrete or continuous action space. This kind of method can be regarded as an extension of the value-based methods in continuous action space, or as an improvement of the policy-based method for reducing sampling variance. Although this method absorbs the advantages of the two methods, it also inherits the corresponding disadvantages. For example, the critic also has the problem of overestimation, and the actor has the problem of insufficient exploration. Some common actor-critic deep reinforcement learning algorithms include the actor-critic (AC) algorithm (Sutton and Barto 2018) and a series of improvements: (1) A3C (Mnih et al. 2016) extends AC to asynchronous and parallel learning, disturbs the correlation between data, and improves the speed of data collection and training; (2) DDPG (Lillicrap et al. 2015) inherits DQN's target network, and the actor is a deterministic policy; (3) TD3 (Fujimoto et al. 2018) introduces clipped double Q-learning mode and delayed policy update strategy to solve the overestimation problem; (4) Entropy regularization is introduced in Q-value estimation by SAC (Haarnoja et al. 2018) to enhance exploration.

3.3 Monte Carlo and Temporal Difference

The differences between the Monte Carlo (MC) and temporal-difference (TD) methods have already discussed in Chap. 2 and some algorithms are shown in Fig. 3.4. Here we summarize their differences again for the completion of this chapter. TD is an intermediate form between dynamic programming (DP) and MC

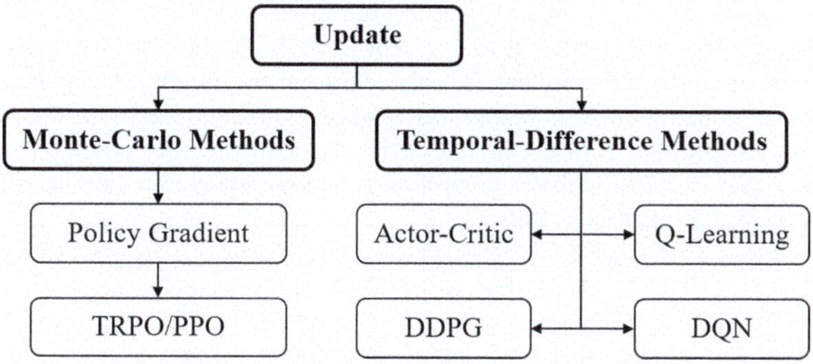

Fig. 3.4 Monte Carlo methods and temporal-difference methods

methods. Both TD and DP use bootstrapping for estimation and both TD and MC do not require the full knowledge of the environment. What makes MC differ from TD the most is how the learning update is done. MC has to wait until an episode is finished to update, whereas DP can do an update at each time step. This difference will let TD methods have larger biases, whereas MC methods have larger variances.

3.4 On-Policy and Off-Policy

The difference between on-policy and off-policy is from the perspective of policy (Fig. 3.5). On-policy methods attempt to evaluate or improve the policy that is used to make decisions, whereas off-policy methods evaluate or improve a policy different from that used to generate the data. The on-policy method requires the agent itself to interact with the environment; that is to say, the policy that interacts

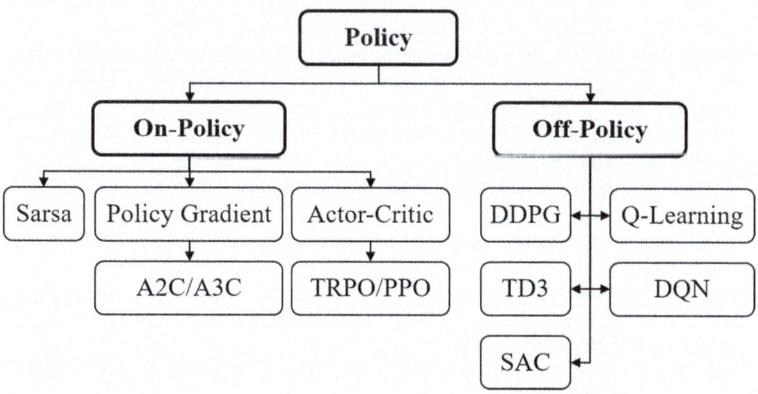

Fig. 3.5 On-policy and off-policy methods

with the environment and the policy to be improved must be the same one. The off-policy method does not need to conform to it, the experience of other agents interacting with the environment can also be used to improve the policy. The common on-policy method is Sarsa, which selects an action based on the current policy and executes the action, then it uses the data to update the current policy. So, the policy that interacts with the environment and the updated policy is the same one. It updates Q function as follows:

$$Q(S_t, A_t) \leftarrow Q(S_t, A_t) + \alpha[R_t + \gamma Q(S_{t+1}, A_{t+1}) - Q(S_t, A_t)]. \tag{3.1}$$

Q-learning is a typical off-policy method. It adopts the max operation and an ϵ-greedy policy when choosing actions, which makes the policy that interacts with the environment and the updated policy not the same policy. It updates Q function as follows:

$$Q(S_t, A_t) \leftarrow Q(S_t, A_t) + \alpha[R_t + \gamma \max_a Q(S_{t+1}, A_{t+1}) - Q(S_t, A_t)]. \tag{3.2}$$

References

Baldi P (2012) Autoencoders, unsupervised learning, and deep architectures. In: Proceedings of the international conference on machine learning (ICML), pp 37–50
Bellemare MG, Dabney W, Munos R (2017) A distributional perspective on reinforcement learning. In: Proceedings of the 34th international conference on machine learning, vol 70, pp 449–458. JMLR.org
Fortunato M, Azar MG, Piot B, Menick J, Osband I, Graves A, Mnih V, Munos R, Hassabis D, Pietquin O, et al (2017) Noisy networks for exploration. arXiv:170610295
Fujimoto S, van Hoof H, Meger D (2018) Addressing function approximation error in actor-critic methods. arXiv:180209477
Ha D, Schmidhuber J (2018) Recurrent world models facilitate policy evolution. In: Advances in neural information processing systems, pp 2450–2462
Haarnoja T, Zhou A, Abbeel P, Levine S (2018) Soft actor-critic: off-policy maximum entropy deep reinforcement learning with a stochastic actor. arXiv:180101290
Heess N, Sriram S, Lemmon J, Merel J, Wayne G, Tassa Y, Erez T, Wang Z, Eslami S, Riedmiller M, et al (2017) Emergence of locomotion behaviours in rich environments. arXiv:170702286
Kalashnikov D, Irpan A, Pastor P, Ibarz J, Herzog A, Jang E, Quillen D, Holly E, Kalakrishnan M, Vanhoucke V, et al (2018) QT-Opt: Scalable deep reinforcement learning for vision-based robotic manipulation. arXiv:180610293
Li Y (2017) Deep reinforcement learning: an overview. arXiv:170107274
Lillicrap TP, Hunt JJ, Pritzel A, Heess N, Erez T, Tassa Y, Silver D, Wierstra D (2015) Continuous control with deep reinforcement learning. arXiv:150902971
Mnih V, Kavukcuoglu K, Silver D, Rusu AA, Veness J, Bellemare MG, Graves A, Riedmiller M, Fidjeland AK, Ostrovski G, Petersen S, Beattie C, Sadik A, Antonoglou I, King H, Kumaran D, Wierstra D, Legg S, Hassabis D (2015) Human-level control through deep reinforcement learning. Nature 518(7540):529–533
Mnih V, Badia AP, Mirza M, Graves A, Lillicrap T, Harley T, Silver D, Kavukcuoglu K (2016) Asynchronous methods for deep reinforcement learning. In: International Conference on Machine Learning (ICML), pp 1928–1937

Munos R, Stepleton T, Harutyunyan A, Bellemare M (2016) Safe and efficient off-policy reinforcement learning. In: Advances in Neural Information Processing Systems, pp 1054–1062

Nagabandi A, Kahn G, Fearing RS, Levine S (2018) Neural network dynamics for model-based deep reinforcement learning with model-free fine-tuning. In: 2018 IEEE international conference on robotics and automation (ICRA), IEEE, Piscataway, pp 7559–7566

Racanière S, Weber T, Reichert D, Buesing L, Guez A, Rezende DJ, Badia AP, Vinyals O, Heess N, Li Y, et al (2017) Imagination-augmented agents for deep reinforcement learning. In: Advances in neural information processing systems, pp 5690–5701

Schaul T, Quan J, Antonoglou I, Silver D (2015) Prioritized experience replay. Preprint, arXiv:1511.05952

Schulman J, Levine S, Abbeel P, Jordan M, Moritz P (2015) Trust region policy optimization. In: International Conference on Machine Learning (ICML), pp 1889–1897

Schulman J, Wolski F, Dhariwal P, Radford A, Klimov O (2017) Proximal policy optimization algorithms. arXiv:170706347

Silver D, Huang A, Maddison CJ, Guez A, Sifre L, Van Den Driessche G, Schrittwieser J, Antonoglou I, Panneershelvam V, Lanctot M, et al (2016) Mastering the game of go with deep neural networks and tree search. Nature 529(7587):484

Silver D, Hubert T, Schrittwieser J, Antonoglou I, Lai M, Guez A, Lanctot M, Sifre L, Kumaran D, Graepel T, et al (2017) Mastering chess and shogi by self-play with a general reinforcement learning algorithm. arXiv:171201815

Silver D, Hubert T, Schrittwieser J, Antonoglou I, Lai M, Guez A, Lanctot M, Sifre L, Kumaran D, Graepel T, et al (2018) A general reinforcement learning algorithm that masters chess, shogi, and Go through self-play. Science 362(6419):1140–1144

Sutton RS, Barto AG (2018) Reinforcement learning: an introduction. MIT Press, Cambridge

Sutton RS, McAllester DA, Singh SP, Mansour Y (2000) Policy gradient methods for reinforcement learning with function approximation. In: Advances in neural information processing systems, pp 1057–1063

Van Hasselt H, Guez A, Silver D (2016) Deep reinforcement learning with double Q-learning. In: Thirtieth AAAI conference on artificial intelligence

Wang Z, Schaul T, Hessel M, Hasselt H, Lanctot M, Freitas N (2016) Dueling network architectures for deep reinforcement learning. In: International conference on machine learning, pp 1995–2003

Watkins CJ, Dayan P (1992) Q-learning. Mach. Learn. 8(3–4):279–292

Chapter 4
Deep Q-Networks

Yanhua Huang

Abstract This chapter aims to introduce one of the most important deep reinforcement learning algorithms, called deep Q-networks. We will start with the Q-learning algorithm via temporal difference learning, and introduce the deep Q-networks algorithm and its variants. We will end this chapter with code examples and experimental comparison of deep Q-networks and its variants in practice.

Keywords Temporal difference learning · DQN · Double DQN · Dueling DQN · Prioritized experience replay · Distributional reinforcement learning

4.1 Introduction

One of the most significant breakthroughs in reinforcement learning was the development of an off-policy temporal difference (TD) control algorithm, known as Q-learning, which is introduced in Chap. 2. Q-Learning has been proven to converge towards the optimal solution in a tabular case or using linear function approximation. However, it is known that Q-learning is unstable or even to diverge when using a non-linear function approximator such as a neural network to represent the Q-value function (Tsitsiklis and Van Roy 1996). With the advances in training deep neural networks, deep Q-networks (**DQN**) (Mnih et al. 2015) addressed this issue and ignited the research of deep reinforcement learning. In this chapter, we first review the background of Q-learning. Then we introduce DQN and its variants with detailed theories and explanations. Finally, in Sect. 4.10, we demonstrate their implementation details and empirical performance on the Atari games with code examples, for providing the readers a quick hands-on learning process. The

Y. Huang (✉)
Xiaohongshu Technology Co., Ltd., Shanghai, China

H. Dong et al. (eds.), *Deep Reinforcement Learning*,
https://doi.org/10.1007/978-981-15-4095-0_4

complete implementation of each algorithm is available in the repository provided together with the book.[1]

4.2 Background

Model-free methods provide a general way to tackle MDP-based decision-making problems, where "model" means an explicit model for the transition probability distribution and the reward function associated with the MDP. TD learning is a class of model-free methods. Recall that in Sect. 2.4 we discuss that if a perfect model of the MDP is available, one can get the optimal plan with dynamic programming by reusing the optimal solution of sub-problems recursively. TD learning follows such an idea that we can estimate the value of sub-problems with bootstrapping even though the estimation is not optimal all the time.

Sub-problems are represented by states in MDP. The value $v_\pi(s)$ of a state s with a policy π is defined by the expected return starting from s and acting with π:

$$v_\pi(s) = \mathbb{E}_\pi[R_t + \gamma v_\pi(S_{t+1})|S_t = s], \tag{4.1}$$

where $\gamma \in [0, 1]$ is the discount rate. TD learning decomposes the estimation above with bootstrapping. Given a value function $V : \mathcal{S} \to \mathbb{R}$, the simplest version, TD(0), is the following one-step bootstrapping:

$$V(S_t) \leftarrow V(S_t) + \alpha[R_t + \gamma V(S_{t+1}) - V(S_t)], \tag{4.2}$$

where $R_t + \gamma V(S_{t+1})$ and $R_t + \gamma V(S_{t+1}) - V(S_t)$ are known as the TD target and TD error, respectively.

The value of a policy provides a way to estimate the acting performance. To further know how to select the action in a particular state, we would like to calculate the quality of the state-action combinations. Q-value allows such estimation:

$$q_\pi(s, a) = \mathbb{E}_\pi[R_{t+1} + \gamma v_\pi(S_{t+1})|S_t = s, A_t = a]. \tag{4.3}$$

The simplest way to perform a better policy is acting greedily $\pi'(s, a) = \arg\max_{a'} q^\pi(s, a')$, where the improvement can be ensured with $q_{\pi'}(s, a) = \max_{a'} q_\pi(s, a') \geq q_\pi(s, a)$. An alternative for considering exploration is to act greedily most of the time, but with a small probability ϵ instead to select randomly from all actions with equal probability regardless of their Q-values. This method is

[1]Codes are available at: https://github.com/deep-reinforcement-learning-book/Chapter4-DQN.

called ϵ-greedy. We can calculate the Q-value of ϵ-greedy policy π' by

$$q_\pi(s, \pi'(s)) = (1-\epsilon) \max_{a\in\mathcal{A}} q_\pi(s, a) + \frac{\epsilon}{|\mathcal{A}|}\sum_{a\in\mathcal{A}} q_\pi(s, a). \tag{4.4}$$

Note that the sum of $\frac{\pi(s,a)-\epsilon/|\mathcal{A}|}{1-\epsilon}$ over $a \in \mathcal{A}$ is equal to 1. With the truth that the maximum is not less than the weighted average, we can get

$$\begin{aligned} q_\pi(s, \pi'(s)) &= (1-\epsilon) \max_{a\in\mathcal{A}} q_\pi(s, a) \sum_{a\in\mathcal{A}} \frac{\pi(s,a)-\epsilon/|\mathcal{A}|}{1-\epsilon} + \frac{\epsilon}{|\mathcal{A}|}\sum_{a\in\mathcal{A}} q_\pi(s, a) \\ &\geq (1-\epsilon)\sum_{a\in\mathcal{A}} \frac{\pi(s,a)-\epsilon/|\mathcal{A}|}{1-\epsilon} q_\pi(s,a) \\ &\quad + \frac{\epsilon}{|\mathcal{A}|}\sum_{a\in\mathcal{A}} q_\pi(s, a) = q_\pi(s, \pi(s)), \end{aligned} \tag{4.5}$$

which tells us that the Q-value of acting with the ϵ-greedy policy π' is not less than the origin policy π, i.e., ϵ-greedy method ensures policy improvement. We will discuss policy improvement with Q-value function in the next section.

4.3 Sarsa and Q-Learning

Similar to the update rule for the value function in TD(0), it is straightforward to update the Q-value function after every transition from a non-terminal state S_t:

$$Q(S_t, A_t) \leftarrow Q(S_t, A_t) + \alpha[R_t + \gamma Q(S_{t+1}, A_{t+1}) - Q(S_t, A_t)], \tag{4.6}$$

where both A_t and A_{t+1} are selected ϵ-greedily with respect to Q. If S_{t+1} is a terminal state, then $Q(S_{t+1}, A_{t+1})$ is defined as zero. We can continually estimate Q for the behavior policy π, and change π toward greediness with respect to Q at the same time. This algorithm is known as Sarsa. Note that π plays two roles in Sarsa—for experience generation and policy improvement. Basically, the policy used to generate behavior is called the behavior policy, and the policy that is evaluated and improved is called the target policy. The algorithm where the behavior policy and the target policy are the same, such as Sarsa, is known as the on-policy method.

On-policy methods are kinds of trial-and-error processes but only the experiences generated by the current policy are used for improvement. Off-policy methods address this issue with introspection, where the experience generated by the behavior policy is "off" (not following) the target policy. The off-policy technique allows reusing past experience. Q-learning is an off-policy method. Its simplest form, one-step Q-learning, follows the update rule below:

$$Q(S_t, A_t) \leftarrow Q(S_t, A_t) + \alpha[R_t + \gamma \max_{A_{t+1}} Q(S_{t+1}, A_{t+1}) - Q(S_t, A_t)], \tag{4.7}$$

where A_t is sampled by ϵ-greedy with respect to Q. Note that A_{t+1} is selected greedily, i.e., contrast to Sarsa, in Q-learning the behavior policy is also ϵ-greedy but the target policy is greedy. One-step Q-learning only takes in current transition. An alternative way to get more accurate Q-values in approximation case is to use multi-steps rewards, i.e., multi-steps Q-learning. Note that multi-steps Q-learning needs to consider the mismatches in subsequent rewards to keep the Q-value function approximating the expected return under the target policy as in Eq. (4.3). We will discuss multi-steps Q-learning in Sect. 4.9.

4.4 Why Deep Learning: Value Function Approximation

In tabular settings, the action-value functions can be represented by a big two-dimensional table, i.e., one entry for each discrete state and action. However, it is inefficient to deal with large-scale space such as raw pixels input, and let alone continuous control tasks. Fortunately, generalization from different inputs by function approximation has been widely studied, and we can utilize this technique in value-based reinforcement learning.

Let us consider the function approximation in Q-learning with some parameter θ. The approximator can be linear models, decision trees, or neural networks. Then the update rule in Eq. (4.7) is rewritten as

$$\theta_t \leftarrow \arg\min_{\theta} \mathcal{L}(Q(S_t, A_t; \theta), R_t + \gamma Q(S_{t+1}, A_{t+1}; \theta)), \tag{4.8}$$

where $\mathcal{L}$ represents the loss function, e.g., mean squared error. While one can solve the optimization problem above by collecting batch samples, which constructs the fitted Q iteration (Riedmiller 2005) shown as Algorithm 1, where S_i' is the successor state of S_i. An online stochastic variant with gradient is the online Q iteration algorithm presented in Algorithm 2.

Algorithm 1 Fitted Q iteration

```
for iteration i = 1, T do
    Collect D samples {(S_i, A_i, R_i, S_i')}_{i=1}^D
    for t = 1, K do
        Set Y_i ← R_i + γ max_a Q(S_i', a; θ)
        Set θ ← arg min_θ' 1/2 Σ_{i=1}^D (Q(S_i, A_i; θ') − Y_i)^2
    end for
end for
```

Algorithm 2 Online Q iteration

for iteration $= 1, T$ **do**
  Select action a and observe (s, a, r, s')
  Set $y \leftarrow r + \gamma \max_{a'} Q(s', a'; \theta)$
  Set $\theta \leftarrow \theta - \alpha(Q(s, a; \theta) - y)\frac{\mathrm{d}Q(s,a;\theta)}{\mathrm{d}\theta}$
end for

Note that both the fitted Q iteration and online Q iteration are off-policy algorithms so that the past experience can be reused many times. We will discuss this topic further in the next section.

Recall that we introduce the convergence of value iteration in Sect. 2.4.2 with the Bellman optimality backup operator $\mathcal{T}^*$. We define a new operator $\mathcal{B}$ with function approximation by $\mathcal{B}V = \arg\min_{V' \in \Omega} \mathcal{L}(V', V)$, where Ω is the set of all possible value functions that can be approximated. Note that the arg min in $\mathcal{B}$ can be viewed as a projection from $\mathcal{T}^* V$ to Ω. So the backup operator with function approximation can be represented by $\mathcal{B}\mathcal{T}^*$. While $\mathcal{T}^*$ is contracted with ∞-norm and $\mathcal{B}$ is contracted with L2-norm for MSE loss. However, $\mathcal{B}\mathcal{T}^*$ is not contracted of any kind. So value iteration is unstable and might even diverge when a non-linear function approximator such as a neural network is used to represent the value function (Tsitsiklis and Van Roy 1997). We leave the discussion about the stability of training with deep neural networks in the next section.

4.5 DQN

In the last section, we introduce the method to learn action-value functions with approximation and its instability of convergence. To achieve the end-to-end decision-making in complex problems with raw pixel input, DQN combines Q-learning with deep learning with two key ideas to address the instability issue and achieves significant progress on Atari games.

The first one is known as **replay buffer**, which is a biologically inspired mechanism termed experience replay (McClelland et al. 1995; O'Neill et al. 2010; Lin 1993). At each time step t, DQN stores the experience of the agent (S_t, A_t, R_t, S_{t+1}) into replay buffer, and then draws a mini-batch of samples from this buffer uniformly to apply the Q-learning update. Replay buffer has several advantages over the fitted Q iteration. First, the experience in each step can be reused to learn the Q-function, which allows for greater data efficiency. Second, if there is no replay buffer, as in the fitted Q iteration, mini-batch samples are collected consecutively, i.e., they are highly correlated, which increases the variance of the updates. Third, experience replay avoids the situation that the samples used to train are determined by the previous parameters, which smooths out learning and reduces oscillations or divergence in the parameters. In practice, only the last N experience tuples are stored in the replay buffer to save the memory.

Table 4.1 The effects of replay and separating the target Q-network

Game	With replay and target Q	With replay but without target Q	With target Q but without replay	Without replay and target Q
Breakout	316.8	240.7	10.2	3.2
Enduro	1006.3	831.4	141.9	29.1
River raid	7446.6	4102.8	2867.7	1453.0
Seaquest	2894.4	822.6	1003.0	275.8
Space invaders	1088.9	826.3	373.2	302.0

Data comes from Mnih et al. (2015)

The second idea aims to further improve the stability with neural networks. Instead of the desired Q-network, a separate network, known as **target network**, is used to generate the Q-learning targets. Furthermore, at every C steps, the target network will be synchronized with the primary Q-network by copying directly (hard update) or exponentially decaying average (soft update). The target network makes the generation of the Q-learning target delay with old parameters, which reduces the divergence and oscillations much more. For example, the update making Q-value increase on action (S_t, A_t) may increase $Q(S_{t+1}, a)$ for all action a because of the similarity between S_t and S_{t+1}, where the training target constructed by Q-network will be overestimated.

The effect of two enhancements above on five Atari games is shown in Table 4.1. Agents were trained for 1e7 frames with the hyperparameters search. Each agent was evaluated every 250,000 training frames for 135,000 validation frames, and the highest average episode score is reported.

Since it is challenging to feed histories of arbitrary length as inputs to a neural network, DQN instead works on the fixed-length representation of histories produced by a function ϕ. More precisely, ϕ concentrates on the current and the previous three frame, which is useful for tracking temporal information, e.g., object moving. The full algorithm is presented in Algorithm 3. The raw frames are resized to 84×84 and converted to gray-scale. The function ϕ stacks the 4 most recent frames as the input to the neural network. In addition, the architecture of the neural network consists of three convolutional layers and two fully connected layers with a single output for each valid action. We will discuss more training details in Sect. 4.10.2.

4.6 Double DQN

Double DQN is an enhancement of DQN for reducing overestimation (Van Hasselt et al. 2016). Before taking a closer look, let us first illustrate the overestimation problem in classic DQN. The Q-learning target $R_t + \gamma \max_a Q(S_{t+1}, a)$ contains a max operator. Q is noisy, which may be caused by environment, non-stationarity, function approximation or any other reasons. Note that the expectation of maximum

Algorithm 3 DQN

Hyperparameters: replay buffer capacity N, reward discount factor γ, delayed steps C for target action-value function update, ϵ-greedy factor ϵ
Input: empty replay buffer $\mathcal{D}$, initial parameters θ of action-value function Q
Initialize target action-value function $\hat{Q}$ with parameter $\hat{\theta} \leftarrow \theta$
for episode $= 0, 1, 2, \ldots$ **do**
    Initialize environment and get observation O_0
    Initialize sequence $S_0 = \{O_0\}$ and preprocess sequence $\phi_0 = \phi(S_0)$
    for t $= 0, 1, 2, \ldots$ **do**
        With probability ϵ select a random action A_t, otherwise select $A_t = \arg\max_a Q(\phi(S_t), a; \theta)$
        Execute action A_t and observe O_{t+1} and reward R_t
        If the episode has ended, set $D_t = 1$. Otherwise, set $D_t = 0$
        Set $S_{t+1} = \{S_t, A_t, O_{t+1}\}$ and preprocess $\phi_{t+1} = \phi(S_{t+1})$
        Store transition $(\phi_t, A_t, R_t, D_t, \phi_{t+1})$ in $\mathcal{D}$
        Sample random minibatch of transitions $(\phi_i, A_i, R_i, D_i, \phi_i')$ from $\mathcal{D}$
        If $D_i = 0$, set $Y_i = R_i + \gamma \max_{a'} \hat{Q}(\phi_i', a'; \hat{\theta})$. Otherwise, set $Y_i = R_i$
        Perform a gradient descent step on $(Y_i - Q(\phi_i, A_i; \theta))^2$ with respect to θ
        Synchronize the target $\hat{Q}$ every C steps
        If the episode has ended, break the loop
    end for
end for

noise is not less than the maximum expectation of noises, i.e., $\mathbb{E}[\max(\epsilon_1, \ldots, \epsilon_n)] \geq (\max(\mathbb{E}[\epsilon_1], \ldots, \mathbb{E}[\epsilon_n]))$. So the next Q-values are always overestimated. Thrun and Schwartz (1993) provides further theoretical analysis and experimental results.

Note that the training target in standard DQN can be rewritten by

$$R_t + \gamma \hat{Q}(S_{t+1}, \arg\max_a \hat{Q}(S_{t+1}, a; \hat{\theta}); \hat{\theta}), \tag{4.9}$$

where $\hat{\theta}$ is used in both action selection and value evaluation. The central idea of double DQN is to decorrelate the noises in selection and evaluation by using two different networks in these two stages. The Q-network in the DQN architecture provides a natural candidate for the extra network. Recall that it is the evaluation role of the target network that improves the stability more. As a consequence, the Q-learning target used in double DQN is

$$R_t + \gamma \hat{Q}(S_{t+1}, \arg\max_a Q(S_{t+1}, a; \theta); \hat{\theta}). \tag{4.10}$$

Following Wang et al. (2016), we measure improvement in percentage (positive or negative) in score over the better of human and baseline agent scores:

$$\frac{\text{Score}_{\text{Agent}} - \text{Score}_{\text{Baseline}}}{\max(\text{Score}_{\text{Baseline}}, \text{Score}_{\text{Human}}) - \text{Score}_{\text{Random}}}. \tag{4.11}$$

The improvement over DQN are available in Fig. 4.1.

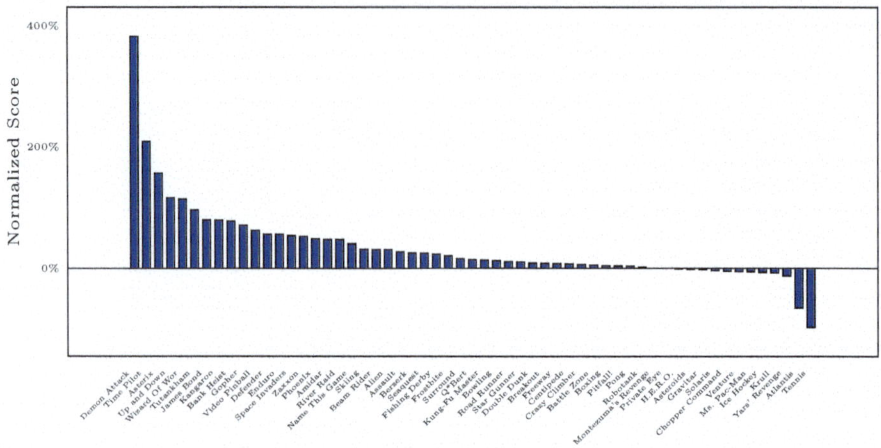

Fig. 4.1 Improvements of double DQN (Van Hasselt et al. 2016) over DQN (Mnih et al. 2015) in Atari benchmark, using the metric described in Eq. (4.11). All scores come from Wang et al. (2016) (Table 2)

4.7 Dueling DQN

For some states, different actions are not relevant to the expected value, and we do not need to learn the effect of each action for such states. For example, imagine standing on the mountain and watching the sunrise. The pleasant view comforts you a lot, which provides a high reward. You can stay here, and the Q-values of different actions do not matter. So decoupling the action-independent value of state and Q-value may lead to more robust learning.

Dueling DQN proposes a new network architecture to achieve this idea (Wang et al. 2016). More precisely, the Q-value can be split into state value part and action advantage part as following:

$$Q^{\pi}(s, a) = V^{\pi}(s) + A^{\pi}(s, a) \tag{4.12}$$

and dueling DQN separates the representations of these two parts by

$$Q(s, a; \theta, \theta_v, \theta_a) = V(s; \theta, \theta_v) + (A(s, a; \theta, \theta_a) - \max_{a'} A(s, a'; \theta, \theta_a)), \tag{4.13}$$

where θ_v and θ_a are parameters of the two streams of fully connected layers, θ represent the parameters in convolutional layers. Note that the max operator in Eq. (4.13) ensures identifiability that the Q-value recovers state value and action advantage uniquely. Otherwise, the training may ignore the state value term and make the advantage function converge to Q-value only. Moreover, Wang et al. (2016) also proposed to replace max with average as the following for better stability:

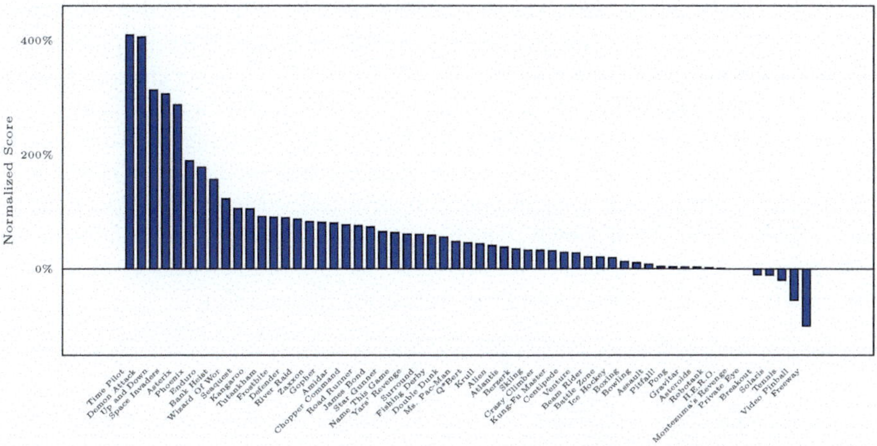

Fig. 4.2 Improvements of dueling DQN (Wang et al. 2016) over DQN (Mnih et al. 2015) in Atari benchmark, using the metric described in Eq. (4.11). All scores come from Wang et al. (2016) (Table 2)

$$Q(s, a; \theta, \theta_v, \theta_a) = V(s; \theta, \theta_v) + (A(s, a; \theta, \theta_a) - \frac{1}{|\mathcal{A}|} \sum_{a'} A(s, a'; \theta, \theta_a)) \tag{4.14}$$

by which the advantage function only need to adapt to the direct of mean advantage instead of pursuing the optimal advantage.

Training of the dueling architectures, as with standard DQN, requires only more layers. The experiments show that dueling architectures lead to better policy evaluation in the presence of many similar-valued actions. The improvement over DQN is available in Fig. 4.2.

4.8 Prioritized Experience Replay

One remaining area of improvement in standard DQN is a better sampling strategy for experience replay. **Prioritized experience replay (PER)** is a technique for prioritizing experience, so as to replay important transitions more frequently (Schaul et al. 2015). The central idea of PER is to consider the importance of transitions with

TD error δ, which can be viewed as a surprising measure. Why this can be of help is that some of the experience might contain more information to learn as compared to the others. Giving those more information-rich experience a greater chance of being replayed will help make the whole learning process faster and more efficient.

The most direct idea is using TD error for prioritization directly. However, it has several issues. First, sweeping over whole memory is inefficient. In addition, it is sensitive to noises such as approximation error and stochastic rewards. Finally, greedy makes error shrink slowly, which may cause the beginning transitions with high error replayed frequently. To overcome these issues, Schaul et al. (2015) proposed to use the following sampling probability for transition i:

$$P(i) = \frac{p_i^\alpha}{\sum_k p_k^\alpha}, \tag{4.15}$$

where $p_i > 0$, known as the priority of transition i, α is an exponent hyper-parameter with $\alpha = 0$ corresponding to the uniform case, and k is enumerated on sampled transitions. There are two variants of p_i. The first one is proportional prioritization $p_i = |\delta_i| + \epsilon$, where δ_i is the TD error of transition i and ϵ is a small positive value for numerical stability. The second one is rank-based prioritization $p_i = \frac{1}{\text{rank}(i)}$, where rank$(i)$ is the rank of transition i according to $|\delta_i|$.

Remind that it is the random sampling from a large replay buffer that helps to decorrelate the samples. But the purely random sampling is abandoned when adding priority sampling. Decreasing the training weight for high priority transitions may make sense. PER uses the importance-sampling weights to correct this bias for transition i:

$$w_i = (NP(i))^{-\beta}, \tag{4.16}$$

where N is the size of replay buffer, P is the probability defined in Eq. (4.15), and β is a hyper-parameter annealed up to 1 during training because the unbiased nature of the updates will nearly converge at the end of the training. This weight is usually folded into the loss function to construct weighted learning.

For efficient implementation, the cumulative density function of sampling probability is approximated by a piece-wise linear function with k segments. More precisely, the priorities are stored in a query-efficient data structure called the segment tree. During run-time, PER first samples a segment, and then sample uniformly among the transitions within it. The improvement over DQN is available in Fig. 4.3.

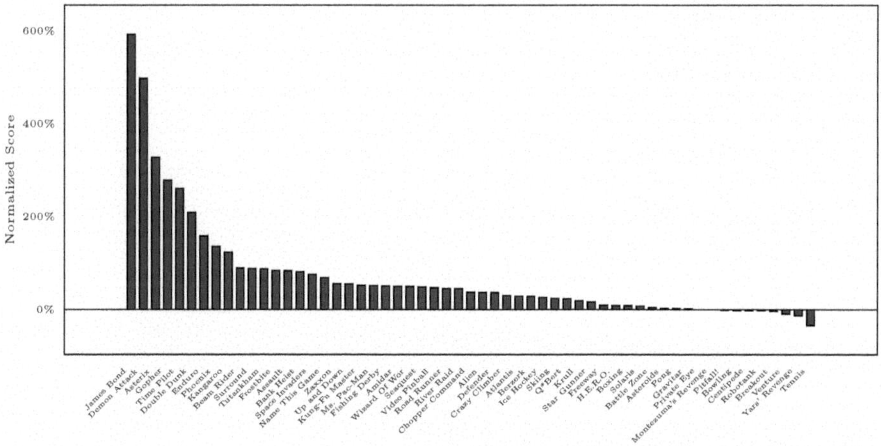

Fig. 4.3 Improvements of prioritized experience replay (Schaul et al. 2015) with rank-based prioritization over DQN (Mnih et al. 2015) in Atari benchmark, using the metric described in Eq. (4.11). All scores come from Wang et al. (2016) (Table 2)

4.9 Other Improvements: Multi-Step Learning, Noisy Nets, and Distributional Reinforcement Learning

Including double Q-learning, dueling architecture, and PER, **Rainbow** combines three more extensions to DQN and achieves significant results on the Atari domain (Hessel et al. 2018). We discuss them and their expansions in this section. The first one is multi-step learning. n-step return allows for accurate estimation and was proven to lead faster learning with suitably tuned n (Sutton and Barto 2018). However, there may exist a mismatch in the action selection between the target and behavior policy within the multi-steps during off-policy learning. One can find a systematic study about correcting this mismatch in Hernandez-Garcia and Sutton (2019). Rainbow uses the truncated n-step return $R_t^{(k)}$ from a given state S_t directly (Hessel et al. 2018; Castro et al. 2018), where $R_t^{(k)}$ is defined by

$$R_t^{(k)} = \sum_{k=0}^{n-1} \gamma^k R_{t+k}. \tag{4.17}$$

The Q-learning target in multi-step variant of Q-learning is then defined by

$$R_t^{(k)} + \gamma^k \max_a Q(S_{t+k}, a). \tag{4.18}$$

The second one is noisy nets (Fortunato et al. 2017). It is an alternative exploration algorithm for ϵ-greedy, especially for games requiring huge exploration such as Montezuma's Revenge. The noise is added into linear layer $\boldsymbol{y} = (\boldsymbol{W}\boldsymbol{x} + \boldsymbol{b})$

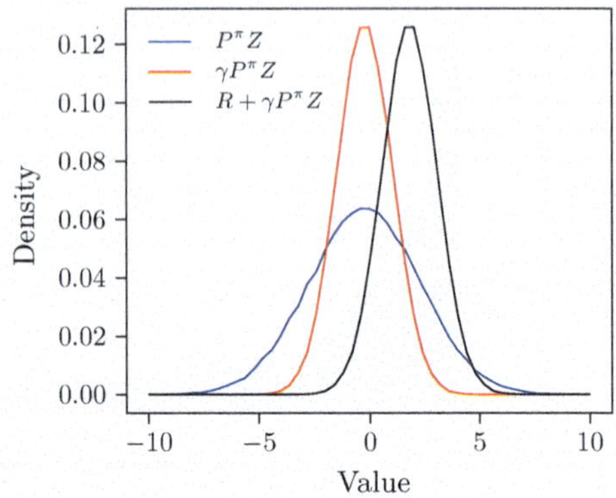

Fig. 4.4 A distributional Bellman operator in continuous case. Given the return distribution of the next state under policy π (blue curve), it is first discounted by the reward discounter γ (red curve), and then be shifted by the reward in current time step (black curve)

by an extra noisy stream

$$\boldsymbol{y} = (\boldsymbol{W}\boldsymbol{x} + \boldsymbol{b}) + ((\boldsymbol{W_{noisy}} \odot \epsilon_w)\boldsymbol{x} + \boldsymbol{b_{noisy}} \odot \epsilon_b), \tag{4.19}$$

where $\odot$ refers to the element-wise product, both $\boldsymbol{W_{noisy}}$ and $\boldsymbol{b_{noisy}}$ are trainable parameters whereas ϵ_w and ϵ_b are random scales annealed down to zero. The experiment shows that noisy nets yield substantially higher scores for a wide range of Atari games over several baselines.

The last one is distributional reinforcement learning (Bellemare et al. 2017), which gives a new perspective on value estimation. Instead of considering the expectation of returns represented by random variable Z^π, Bellemare et al. (2017) proposed to estimate the distribution of Z^π with the distributional Bellman operator $\mathcal{T}^\pi$:

$$\mathcal{T}^\pi Z = R + \gamma P^\pi Z. \tag{4.20}$$

Figure 4.4 shows a continuous case of $\mathcal{T}^\pi$.

The distributional variant of DQN used in Rainbow, known as categorical DQN (Bellemare et al. 2017), models the action-value distribution by a discrete distribution parameterized by a vector z with N elements (also known as atoms) $z_i = V_{min} + (i-1)\Delta z$, where $[V_{min}, V_{max}]$ is the action-value range and $\Delta z = \frac{V_{max}-V_{min}}{N-1}$. In practice, N is usually specified to 51 so sometimes this algorithm is also called **C51**. The parametric model θ of C51 outputs the probabilities $p_i(s,a) = e^{\theta_i(s,a)}/\sum_j e^{\theta_j(s,a)}$ on each atom as distribution Z_θ. Note that discrete approximation causes disjoint supports of the Bellman update $\mathcal{T}^\pi Z$ and the parametrization Z_θ. C51 addresses this issue by projecting the target distribution $\mathcal{T}^\pi Z_{\hat{\theta}}$ onto the support Z_θ. More precisely, given a transition (S_t, A_t, R_t, S_{t+1}), the i-th component of projected target $\Phi\mathcal{T}^\pi Z_{\hat{\theta}}(S_t, A_t)$ with double Q-learning is calculated by

$$\sum_{j=1}^{N} p_j\left(S_{t+1}, \arg\max_a z^\top p(S_{t+1}, a; \theta); \hat{\theta}\right)[1 - \frac{||[R_t + \gamma z_j]_{V_{min}}^{V_{max}} - z_i|}{\Delta z}]_0^1, \tag{4.21}$$

where $[\cdot]_a^b$ bounds its argument in the range $[a, b]$. TD error cannot measure the difference between value distributions. As a result, C51 proposes to use the following Kullback–Leibler divergence as training loss:

$$D_{\mathrm{KL}}(\Phi \mathcal{T}^\pi Z_{\hat{\theta}}(S_t, A_t)||Z_\theta(S_t, A_t)). \tag{4.22}$$

In addition, the priority is also replaced by KL-Divergence for prioritized experience replay. For dueling architecture, the output distribution is also split into value stream and advantage stream, and the aggregated distribution is estimated by

$$p_i(s, a) = \frac{exp(V_i(s) + A_i(s, a) - \bar{A}_i(s, a))}{\sum_j exp(V_j(s) + A_j(s, a) - \bar{A}_j(s, a))}, \tag{4.23}$$

where $\bar{A}_j(s, a)$ is defined by $\frac{1}{|\mathcal{A}|}\sum_{a'} A_j(s, a')$.

The main drawback of C51 to achieve distributional reinforcement learning is that it can only estimate values on a fixed discrete set. Dabney et al. (2018b) proposed **quantile regression DQN (QR-DQN)** to address this issue by estimating the quantiles of the full distribution with quantile regression. Before introducing QR-DQN, we first review the quantile regression. Recall that empirical risk minimization with absolute loss function makes the prediction fit the medium value (50% quantile). More precisely, given random variable x and its label y, for estimation function f, the empirical mean absolute error is $\mathcal{L}_{\mathrm{mae}} = \mathbb{E}[|f(x) - y|]$. Then with the following partial difference:

$$\begin{aligned}\frac{\partial \mathcal{L}_{\mathrm{mae}}}{\partial f(x)} &= \frac{\partial}{\partial f(x)}(P(f(x) > y)(f(x) - y) + P(f(x) \le y)(y - f(x))) \\ &= P(f(x) > y) - P(f(x) \le y) = 0,\end{aligned} \tag{4.24}$$

we can get $F(x) = 0.5$, where F is the primitive function of f. Generally, for quantile τ, the quantile loss is defined as $\mathcal{L}_{\mathrm{quantile}}(\tau) = \mathbb{E}[\rho_\tau(f(x) - y)]$ with

$$\rho_\tau(\alpha) = \begin{cases}\tau\alpha, & \text{if } \alpha > 0 \\ (\tau - 1)\alpha, & \text{otherwise.}\end{cases} \tag{4.25}$$

Similarly, by $\frac{\partial \mathcal{L}_{\mathrm{quantile}}}{\partial f(x)}$ we can get $F(x) = 1 - \tau$, i.e., $f(x)$ is the τ quantile value of random variable y.

Specifically, QR-DQN considers N uniform quantiles $q_i = \frac{1}{N}$ for the value distribution. For a QR-DQN model $\theta : \mathcal{S} \rightarrow \mathbb{R}^{N\times|\mathcal{A}|}$, during sampling, the Q-value

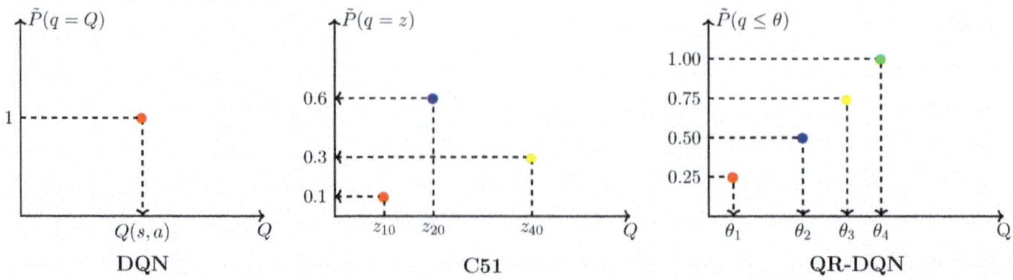

Fig. 4.5 Comparison of DQN, C51, and QR-DQN for state s and action a, where arrows point out the estimation and the number of quantiles in QR-DQN is specified to 4. The architecture of DQN only outputs the approximation of the actual Q-value. For distributional reinforcement learning, C51 estimates probabilities on several Q-value supports while QR-DQN provides quantiles of Q-value

of the state s and action a is the mean of N estimations: $Q(s,a) = \sum_{i=1}^{N} q_i\theta_i(s,a)$. During training, the greedy policy with respect to the Q-values in the next state provides $a^* = \arg\max_{a'} Q(s',a')$, and the distributional Bellman target is $\mathcal{T}\theta_j = r + \gamma\theta_j(s',a^*)$ according to Eq. (4.20). The Lemma 2 in Dabney et al. (2018b) points out that the following sum minimizes the 1-Wasserstein distance between the approximate value distribution and the ground truth:

$$\sum_{i=1}^{N} \mathbb{E}_j \left[\rho_{\hat{\tau}_i}(\mathcal{T}\theta_j - \theta_i(s,a))\right], \tag{4.26}$$

where $\hat{\tau}_i = \frac{i}{N} - \frac{1}{2N}$.

Figure 4.5 shows a comparison of DQN, C51, and QR-DQN. There are further works in the flexibility or robustness of parameterized distribution for distributional reinforcement learning. Readers with more interest in this topic can find related resources from Dabney et al. (2018a), Mavrin et al (2019), Yang et al. (2019).

4.10 DQN Examples

In this section, we discuss more training details in DQN and its variants. Before that, we first demonstrate the process of setting up Atari environments and how to implement some useful wrappers that make training easy and stable.

4.10.1 Related Gym Environment

OpenAI Gym is an open-source toolkit for developing and comparing reinforcement learning algorithms. It contains a collection of environments, as shown in Fig. 4.6. One can install it with Atari extension directly from PyPI

```
pip install gym[atari]
```

or from source

```
git clone https://github.com/openai/gym.git
cd gym
pip install -e .
```

An environment object `env` can be created by

```
import gym
env = gym.make(env_id)
```

where `env_id` is a string that represents an environment. All possible `env_ids` are available at https://github.com/openai/gym/wiki/Table-of-environments.

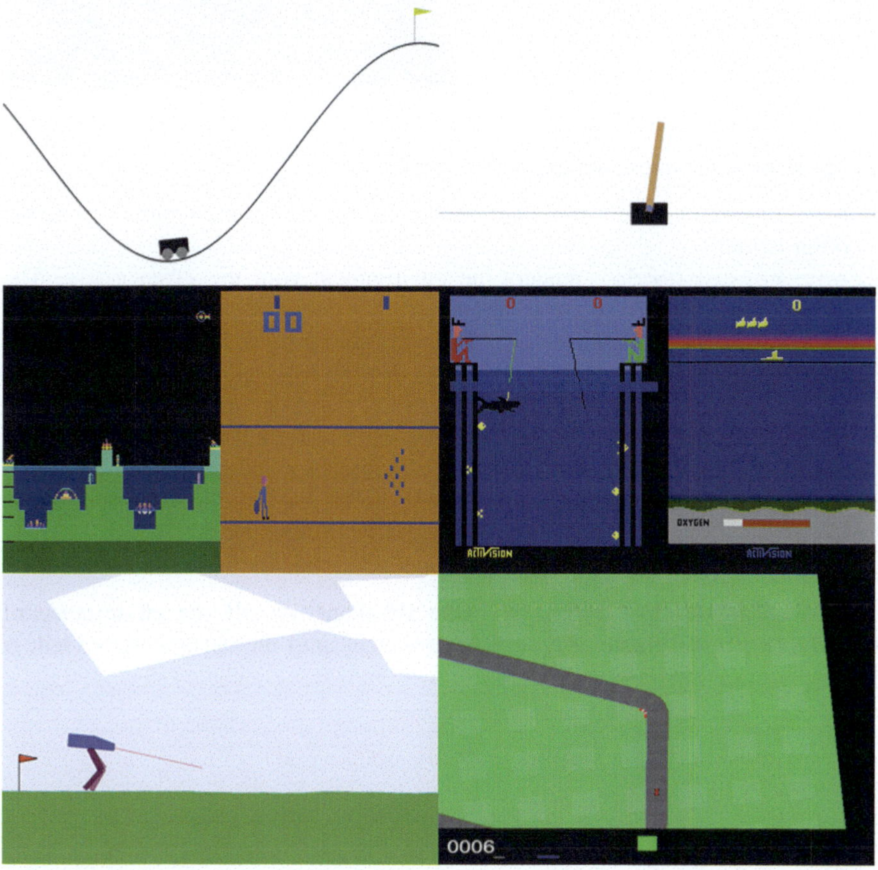

Fig. 4.6 Sample frames of some environments in OpenAI Gym

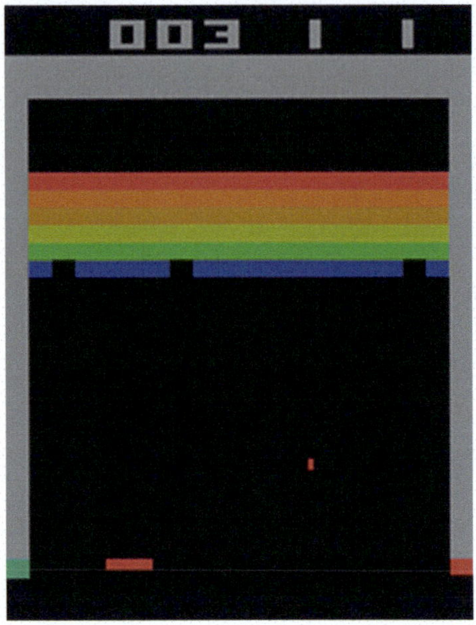

Fig. 4.7 An example frame of Breakout. There are several rows to destroy at the top of the screen. The agent can control the bar at the bottom of the screen to angle your shots at the images you want to smash with the ball. The observations are an RGB images of the screen with shape (210, 160, 3)

There are some important methods of `env`:

1. `env.reset()` resets the state of the environment and returns the initial observation.
2. `env.render(mode)` renders the environment with the given `mode`. The default mode is `human` which renders to the current display or terminal and returns nothing. You can specify `rgb_array` mode to make `env.render` return `numpy.ndarray` objects, which is suitable for generating videos.
3. `env.step(action)` runs one time step of the environment's dynamics with the given `action`, and then returns a tuple `(observation, reward, done, info)` where `observation` is the observation of the current environment, `reward` is the transition reward, `done` points out whether the episode has ended, and `info` contains some auxiliary information.
4. `env.seed(seed)` sets the seed manually, which is useful for reproduction.

Here is an example of classic game Breakout. It will run an instance of the `BreakoutNoFrameskip-v4` environment until the episode has ended. A sample frame shows in Fig. 4.7.

```
import gym
env = gym.make('BreakoutNoFrameskip-v4')
o = env.reset()
while True:
    env.render()
    # take a random action
    a = env.action_space.sample()
```

```
    o, r, done, _ = env.step(a)
    if done:
        break
env.close() # close and clean up
```

Note that `NoFrameskip` means no frame-skip and no action repeat, and `v4` means the 4th version which is the newest now. We will use this environment in following experiments.

Another useful feature in OpenAI Gym is the environment wrapper. It can wrap the environment object and make the training code more concise. Here is a time limit wrapper which limits the maximum length of each episode and is a default wrapper for Atari games.

```
class TimeLimit(gym.Wrapper):
    def __init__(self, env, max_episode_steps=None):
        super(TimeLimit, self).__init__(env)
        self._max_episode_steps = max_episode_steps
        self._elapsed_steps = 0

    def step(self, ac):
        o, r, done, info = self.env.step(ac)
        self._elapsed_steps += 1
        if self._elapsed_steps >= self._max_episode_steps:
            done = True
            info['TimeLimit.truncated'] = True
        return o, r, done, info

    def reset(self, **kwargs):
        self._elapsed_steps = 0
        return self.env.reset(**kwargs)
```

For efficient training, `gym.vector.AsyncVectorEnv` provides an implementation of vectorized wrapper that runs n environments in parallel. All interfaces receive and return n variables together. Furthermore, it is also possible to implement a vectorized wrapper with buffer whose interfaces also receive and return n variables but maintains $m > n$ workers in the background. It is efficient for environments where some transitions spend more time.

Included some classic control problems, Gym also provides standard interfaces of a collection of Atari 2600 games with RAM or screen images as input, using the Arcade Learning Environment (Bellemare et al. 2013). There are at most 18 different buttons in Atari 2600 games as follows:

1. Moving buttons: NOOP, UP, RIGHT, LEFT, DOWN, UPRIGHT, UPLEFT, DOWNRIGHT, DOWNLEFT
2. Fire buttons: FIRE, UPFIRE, RIGHTFIRE, LEFTFIRE, DOWNFIRE, UPRIGHTFIRE, UPLEFTFIRE, DOWNRIGHTFIRE, DOWNLEFTFIRE

where NOOP means do-nothing, and FIRE may also be used to start the game. For convenience, we will refer to buttons' names as actions later.

4.10.2 DQN

There are three more training tricks in DQN. First, the following wrappers are used in order for stable and efficient training:

1. `NoopResetEnv` takes random number of NOOPs in reset stage to ensure random initial states. The default maximum no-ops number is 30. This wrapper helps agent collect more beginning situations, which provides robust learning.
2. `MaxAndSkipEnv` repeats each action 4 times for efficient training. To further denoising observation, the returned frame is the max pooling result over pixels across the last two frames.
3. `Monitor` records the raw reward. We can also implement some useful functions such as speed tracer in this wrapper.
4. `EpisodicLifeEnv` makes the end of life equal to the end of episode, which helps value estimation (Roderick et al. 2017).
5. `FireResetEnv` takes action `FIRE` on reset for environments that need action `FIRE` to start the game. This is a prior knowledge for quick start of games.
6. `WarpFrame` converts the observations to 84×84 gray-scale images.
7. `ClipRewardEnv` wraps the rewards by their sign, which further improves the stability by not allowing any single mini-batch update to change the parameters drastically.
8. `FrameStack` stacks the last 4 frames. Recall that for capturing moving information, DQN preprocesses observations by concentrating the current frame and the previous three, represented by function ϕ. `FrameStack` and `WarpFrame` implement the ϕ. Note that we can optimize memory usage by storing common frames between the observations only once, which is also called lazy-frame trick.

Second, to avoid gradient explosion, DQN (Mnih et al. 2015; DeepMind 2015) uses gradient clipping of the squared error, which is equivalent to replace MSE by the Huber loss (Huber 1992) with $\delta = 1$. The Huber loss is given by

$$L_\delta(x) = \begin{cases} \frac{1}{2}x^2 & |x| \le \delta \\ \delta(|x| - \frac{1}{2}\delta) & \text{otherwise.} \end{cases} \tag{4.27}$$

Finally, replay buffer samples batch of experiences with replacement, and there are some warm start steps before updating for a stable beginning.

Note that all three tricks above are applied to all experiments in this section. Now we show how to build an agent to play Breakout. First of all, for reproducibility, we set random seeds in related libraries manually:

```
random.seed(seed)
np.random.seed(seed)
tf.random.set_seed(seed)
```

Then we build a Q-network with `tf.keras.Model`:

```
class QFunc(tf.keras.Model):
    def __init__(self, name):
        super(QFunc, self).__init__(name=name)
        self.conv1 = tf.keras.layers.Conv2D(
            32, kernel_size=(8, 8), strides=(4, 4),
            padding='valid', activation='relu')
        self.conv2 = tf.keras.layers.Conv2D(
            64, kernel_size=(4, 4), strides=(2, 2),
            padding='valid', activation='relu')
        self.conv3 = tf.keras.layers.Conv2D(
            64, kernel_size=(3, 3), strides=(1, 1),
            padding='valid', activation='relu')
        self.flat = tf.keras.layers.Flatten()
        self.fc1 = tf.keras.layers.Dense(512, activation='relu')
        self.fc2 = tf.keras.layers.Dense(action_dim,
            activation='linear')

    def call(self, pixels, **kwargs):
        # scale observation
        pixels = tf.divide(tf.cast(pixels, tf.float32),
            tf.constant(255.0))
        # extract features by convolutional layers
        feature =
            self.flat(self.conv3(self.conv2(self.conv1(pixels))))
        # calculate q-value
        qvalue = self.fc2(self.fc1(feature))

        return qvalue
```

The definition of a DQN object consists of attributes Q-network, target Q-network, number of time steps and optimizer, and synchronize Q-network and target Q-network as following:

```
class DQN(object):
    def __init__(self):
        self.qnet = QFunc('q')
        self.targetqnet = QFunc('targetq')
        sync(self.qnet, self.targetqnet)
        self.niter = 0
        self.optimizer = tf.optimizers.Adam(lr, epsilon=1e-5,
            clipnorm=clipnorm)
```

Declare an internal method to wrap the Q-network and then add a `get_action` method to DQN object for ϵ-greedy behavior:

```
@tf.function
def _qvalues_func(self, obv):
    return self.qnet(obv)

def get_action(self, obv):
```

```
    eps = epsilon(self.niter)
    if random.random() < eps:
        return int(random.random() * action_dim)
    else:
        obv = np.expand_dims(obv, 0).astype('float32')
        return self._qvalues_func(obv).numpy().argmax(1)[0]
```

where `epsilon` is a function that anneals ϵ linearly from 1.0 to 0.01 over the first 10% training time steps. For training, we use three common interfaces `train`, `_train_func`, `_tderror_func` for DQN and its variants in the following sections:

```
def train(self, b_o, b_a, b_r, b_o_, b_d):
    self._train_func(b_o, b_a, b_r, b_o_, b_d)

    self.niter += 1
    if self.niter % target_q_update_freq == 0:
        sync(self.qnet, self.targetqnet)

@tf.function
def _train_func(self, b_o, b_a, b_r, b_o_, b_d):
    with tf.GradientTape() as tape:
        td_errors = self._tderror_func(b_o, b_a, b_r, b_o_, b_d)
        loss = tf.reduce_mean(huber_loss(td_errors))

    grad = tape.gradient(loss, self.qnet.trainable_weights)
    self.optimizer.apply_gradients(zip(grad,
        self.qnet.trainable_weights))

    return td_errors

@tf.function
def _tderror_func(self, b_o, b_a, b_r, b_o_, b_d):
    b_q_ = (1 - b_d) * tf.reduce_max(self.targetqnet(b_o_), 1)
    b_q = tf.reduce_sum(self.qnet(b_o) * tf.one_hot(b_a,
        action_dim), 1)

    return b_q - (b_r + reward_gamma * b_q_)
```

where `train` calls `_train_func` and synchronizes the Q-network and target Q-network every `target_q_update_freq` time steps.

Finally, we build the main training procedure:

```
dqn = DQN()
buffer = ReplayBuffer(buffer_size)

o = env.reset()
nepisode = 0
t = time.time()
for i in range(1, number_time steps + 1):
    a = dqn.get_action(o)
```

```
    # execute action and feed to replay buffer
    # note that '_' tail in var name means next
    o_, r, done, info = env.step(a)
    buffer.add(o, a, r, o_, done)

    if i >= warm_start and i % train_freq == 0:
        transitions = buffer.sample(batch_size)
        dqn.train(*transitions)

    if done:
        o = env.reset()
    else:
        o = o_

    # episode in info is real (unwrapped) message
    if info.get('episode'):
        nepisode += 1
        reward, length = info['episode']['r'], info['episode']['l']
        print(
            'Time steps so far: {}, episode so far: {}, '
            'episode reward: {:.4f}, episode length: {}'
                .format(i, nepisode, reward, length)
        )
```

We run 10^7 time steps (4×10^7 frames) over three random seeds on Breakout. For better visualization, we smooth the episode rewards during training. Then we plot the mean and the standard deviation by following codes:

```
from matplotlib import pyplot as plt
plt.plot(xs, mean, color=color)
plt.fill_between(xs, mean - std, mean + std, color=color,
    alpha=.4)
```

The performance is shown in Fig. 4.8 with red area.

4.10.3 Double DQN

Double DQN can be implemented easily by using the following double Q estimation in `_tderror_func` of the agent:

```
# double Q estimation
b_a_ = tf.one_hot(tf.argmax(qnet(b_o_), 1), out_dim)
b_q_ = (1 - b_d) * tf.reduce_sum(targetqnet(b_o_) * b_a_, 1)
```

We also run 10^7 time steps over three random seeds on Breakout. The performance is shown in Fig. 4.8 with green area.

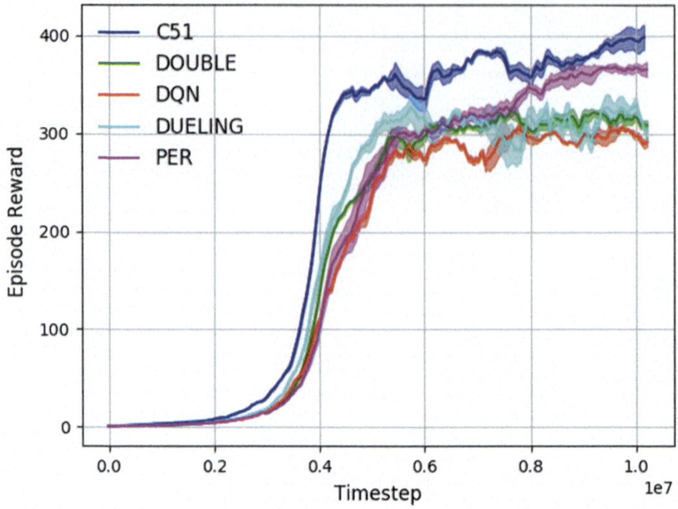

Fig. 4.8 Performances of DQN and its variants on breakout

4.10.4 Dueling DQN

The dueling architecture only changes the Q-network, which can be implemented by

```
class QFunc(tf.keras.Model):
    def __init__(self, name):
        super(QFunc, self).__init__(name=name)
        self.conv1 = tf.keras.layers.Conv2D(
            32, kernel_size=(8, 8), strides=(4, 4),
            padding='valid', activation='relu')
        self.conv2 = tf.keras.layers.Conv2D(
            64, kernel_size=(4, 4), strides=(2, 2),
            padding='valid', activation='relu')
        self.conv3 = tf.keras.layers.Conv2D(
            64, kernel_size=(3, 3), strides=(1, 1),
            padding='valid', activation='relu')
        self.flat = tf.keras.layers.Flatten()
        self.fc1q = tf.keras.layers.Dense(512, activation='relu')
        self.fc2q = tf.keras.layers.Dense(action_dim,
             activation='linear')
        self.fc1v = tf.keras.layers.Dense(512, activation='relu')
        self.fc2v = tf.keras.layers.Dense(1, activation='linear')

    def call(self, pixels, **kwargs):
        # scale observation
        pixels = tf.divide(tf.cast(pixels, tf.float32),
            tf.constant(255.0))
        # extract features by convolutional layers
```

```
        feature =
            self.flat(self.conv3(self.conv2(self.conv1(pixels))))
        # calculate q-value
        qvalue = self.fc2q(self.fc1q(feature))
        svalue = self.fc2v(self.fc1v(feature))

        return svalue + qvalue - tf.reduce_mean(qvalue, 1,
            keepdims=True)
```

We also run 10^7 time steps over three random seeds on Breakout. The performance is shown in Fig. 4.8 with cyan area.

4.10.5 Prioritized Experience Replay

There are three main changes in PER from standard DQN. First, the replay buffer maintains two segment trees with min operator and add operator to calculate the minimum priority and sum of priorities efficiently. More precisely, attribute `_it_sum` is the segment tree object with operation add with two interfaces, `sum` for getting the sum of elements in the given range and `find_prefixsum_idx` for finding the highest index i such that the sum of the smallest i elements is less than the input value.

Second, instead of uniform sampling, the sampling strategy of the proportional information is shown as follows:

```
res = []
p_total = self._it_sum.sum(0, len(self._storage) - 1)
every_range_len = p_total / batch_size
for i in range(batch_size):
    mass = random.random() * every_range_len + i * every_range_len
    idx = self._it_sum.find_prefixsum_idx(mass)
    res.append(idx)
return res
```

Third, apart from standard replay buffer, PER must return indexes and normalized weights of sampled experiences. Weights are used for weighting Huber loss, and indexes are used to update priorities. The sampling step is modified to

```
*transitions, idxs = buffer.sample(batch_size)
priorities = dqn.train(*transitions)
priorities = np.clip(np.abs(priorities), 1e-6, None)
buffer.update_priorities(idxs, priorities)
```

and the `_train_func` is modified to

```
@tf.function
def _train_func(self, b_o, b_a, b_r, b_o_, b_d, b_w):
    with tf.GradientTape() as tape:
```

```
        td_errors = self._tderror_func(b_o, b_a, b_r, b_o_, b_d)
        loss = tf.reduce_mean(huber_loss(td_errors) * b_w)

    grad = tape.gradient(loss, self.qnet.trainable_weights)
    self.optimizer.apply_gradients(zip(grad,
        self.qnet.trainable_weights))

    return td_errors
```

We also run 10^7 time steps over three random seeds on Breakout. The performance is shown in Fig. 4.8 with magenta area.

4.10.6 Distributed DQN

Distributional reinforcement learning estimates the distribution of the Q-value. In this section, we show how to implement one of these techniques, C51, to achieve distributed DQN. In game Breakout, the rewards are all positive. So we replace the value range $[-10, 10]$ used in Bellemare et al. (2017) by $[-1, 19]$, where -1 allows some approximation error. To implement C51, first of all, the Q-Network outputs 51 estimations for each action, which can be implemented by adding more output units in the last fully connection layer. Then instead of the TD error, KL-divergence between target Q distribution and the estimated distribution is used as loss:

```
@tf.function
def _kl_divergence_func(self, b_o, b_a, b_r, b_o_, b_d):
    b_r = tf.tile(
        tf.reshape(b_r, [-1, 1]),
        tf.constant([1, atom_num])
    ) # batch_size * atom_num
    b_d = tf.tile(
        tf.reshape(b_d, [-1, 1]),
        tf.constant([1, atom_num])
    )

    z = b_r + (1 - b_d) * reward_gamma * vrange # shift value
        distribution
    z = tf.clip_by_value(z, min_value, max_value) # clip the
        shifted distribution
    b = (z - min_value) / deltaz
    index_help = tf.expand_dims(tf.tile(
        tf.reshape(tf.range(batch_size), [batch_size, 1]),
        tf.constant([1, atom_num])
    ), -1)

    b_u = tf.cast(tf.math.ceil(b), tf.int32) # upper
    b_uid = tf.concat([index_help, tf.expand_dims(b_u, -1)], 2) #
        indexes
    b_l = tf.cast(tf.math.floor(b), tf.int32)
```

```
b_lid = tf.concat([index_help, tf.expand_dims(b_l, -1)], 2) #
    indexes

b_dist_ = self.targetqnet(b_o_) # whole distribution
b_q_ = tf.reduce_sum(b_dist_ * vrange_broadcast, axis=2)
b_a_ = tf.cast(tf.argmax(b_q_, 1), tf.int32)
b_adist_ = tf.gather_nd( # distribution of b_a_
   b_dist_,
   tf.concat([tf.reshape(tf.range(batch_size), [-1, 1]),
           tf.reshape(b_a_, [-1, 1])], axis=1)
)
b_adist = tf.gather_nd( # distribution of b_a
   self.qnet(b_o),
   tf.concat([tf.reshape(tf.range(batch_size), [-1, 1]),
           tf.reshape(b_a, [-1, 1])], axis=1)
) + 1e-8

b_l = tf.cast(b_l, tf.float32)
mu = b_adist_ * (b - b_l) * tf.math.log(tf.gather_nd(b_adist,
    b_uid))
b_u = tf.cast(b_u, tf.float32)
ml = b_adist_ * (b_u - b) * tf.math.log(tf.gather_nd(b_adist,
    b_lid))
kl_divergence = tf.negative(tf.reduce_sum(mu + ml, axis=1))

return kl_divergence
```

We also run 10^7 time steps over three random seeds on Breakout. The performance is shown in Fig. 4.8 with blue area.

References

Bellemare MG, Naddaf Y, Veness J, Bowling M (2013) The arcade learning environment: an evaluation platform for general agents. J Artif Intell Res 47:253–279

Bellemare MG, Dabney W, Munos R (2017) A distributional perspective on reinforcement learning. In: Proceedings of the 34th international conference on machine learning, vol 70, pp 449–458. JMLR.org

Castro PS, Moitra S, Gelada C, Kumar S, Bellemare MG (2018) Dopamine: a research framework for deep reinforcement learning. http://arxiv.org/abs/1812.06110

Dabney W, Ostrovski G, Silver D, Munos R (2018a) Implicit quantile networks for distributional reinforcement learning. In: International conference on machine learning, pp 1104–1113

Dabney W, Rowland M, Bellemare MG, Munos R (2018b) Distributional reinforcement learning with quantile regression. In: Thirty-second AAAI conference on artificial intelligence

DeepMind (2015) Lua/Torch implementation of DQN. https://github.com/deepmind/dqn

Fortunato M, Azar MG, Piot B, Menick J, Osband I, Graves A, Mnih V, Munos R, Hassabis D, Pietquin O, et al (2017) Noisy networks for exploration. arXiv:170610295

Hernandez-Garcia JF, Sutton RS (2019) Understanding multi-step deep reinforcement learning: a systematic study of the DQN target. In: Proceedings of the neural information processing systems (advances in neural information processing systems) workshop

Hessel M, Modayil J, Van Hasselt H, Schaul T, Ostrovski G, Dabney W, Horgan D, Piot B, Azar M, Silver D (2018) Rainbow: combining improvements in deep reinforcement learning. In: Thirty-second AAAI conference on artificial intelligence

Huber PJ (1992) Robust estimation of a location parameter. In: Breakthroughs in statistics, Springer, Berlin, pp 492–518

Lin LJ (1993) Reinforcement learning for robots using neural networks. Tech. Rep., Carnegie-Mellon Univ Pittsburgh PA School of Computer Science

Mavrin B, Yao H, Kong L, Wu K, Yu Y (2019) Distributional reinforcement learning for efficient exploration. In: International conference on machine learning, pp 4424–4434

McClelland JL, McNaughton BL, O'Reilly RC (1995) Why there are complementary learning systems in the hippocampus and neocortex: insights from the successes and failures of connectionist models of learning and memory. Psychol Rev 102(3):419

Mnih V, Kavukcuoglu K, Silver D, Rusu AA, Veness J, Bellemare MG, Graves A, Riedmiller M, Fidjeland AK, Ostrovski G, Petersen S, Beattie C, Sadik A, Antonoglou I, King H, Kumaran D, Wierstra D, Legg S, Hassabis D (2015) Human-level control through deep reinforcement learning. Nature 518(7540):529–533

O'Neill J, Pleydell-Bouverie B, Dupret D, Csicsvari J (2010) Play it again: reactivation of waking experience and memory. Trends Neurosci 33(5):220–229

Riedmiller M (2005) Neural fitted Q iteration–first experiences with a data efficient neural reinforcement learning method. In: European conference on machine learning. Springer, Berlin, pp 317–328

Roderick M, MacGlashan J, Tellex S (2017) Implementing the deep Q-network. arXiv:171107478

Schaul T, Quan J, Antonoglou I, Silver D (2015) Prioritized experience replay. In: International conference on learning representations

Sutton RS, Barto AG (2018) Reinforcement learning: an introduction. MIT Press, Cambridge

Thrun S, Schwartz A (1993) Issues in using function approximation for reinforcement learning. In: Proceedings of the 1993 Connectionist Models Summer School Hillsdale. Lawrence Erlbaum, New Jersey

Tsitsiklis J, Van Roy B (1996) An analysis of temporal-difference learning with function approximation technical. Report LIDS-P-2322) Laboratory for Information and Decision Systems, Massachusetts Institute of Technology, Tech Rep

Tsitsiklis JN, Van Roy B (1997) Analysis of temporal-difference learning with function approximation. In: Advances in Neural Information Processing Systems, pp 1075–1081

Van Hasselt H, Guez A, Silver D (2016) Deep reinforcement learning with double Q-learning. In: Thirtieth AAAI conference on artificial intelligence

Wang Z, Schaul T, Hessel M, Hasselt H, Lanctot M, Freitas N (2016) Dueling network architectures for deep reinforcement learning. In: International conference on machine learning, pp 1995–2003

Yang D, Zhao L, Lin Z, Qin T, Bian J, Liu TY (2019) Fully parameterized quantile function for distributional reinforcement learning. In: Advances in neural information processing systems, pp 6190–6199

Chapter 5
Policy Gradient

Ruitong Huang, Tianyang Yu, Zihan Ding, and Shanghang Zhang

Abstract Policy gradient methods are a type of reinforcement learning techniques that rely upon optimizing parameterized policies with respect to the expected return (long-term cumulative reward) by gradient descent. They do not suffer from many of the problems that have been traditional reinforcement learning approaches such as the lack of guarantees of an accurate value function, the intractability problem resulting from the uncertain state information, and the complexity arising from continuous states and actions. In this chapter, we will introduce a list of popular policy gradient methods. Starting with the basic policy gradient method REINFORCE, we then introduce the actor-critic method, the distributed versions of actor-critic, and trust region policy optimization and its approximate versions, each one improving its precedent. All the methods introduced in this chapter will be accompanied with its pseudo-code and, at the end of this chapter, a concrete implementation example.

Keywords Policy optimization · Policy gradient · Actor-critic · Trust region policy optimization · Proximal policy optimization

R. Huang
Borealis AI, Toronto, ON, Canada

T. Yu
Nanchang University, Nanchang, China

Z. Ding
Imperial College London, London, UK
e-mail: zhding@mail.ustc.edu.cn

S. Zhang (✉)
University of California, Berkeley, CA, USA

H. Dong et al. (eds.), *Deep Reinforcement Learning*,
https://doi.org/10.1007/978-981-15-4095-0_5

5.1 Introduction

We introduce policy gradient methods in this chapter. Different from the deep Q-learning in the last chapter that focuses on learning the Q value function, policy gradient methods directly perform learning on the parameterized policy function π_θ. By doing so, policy gradient methods are more suitable for high-dimensional or continuous action spaces, as they need no discretization or solving another layer of value maximization problem on the action space. Another benefit of policy gradient methods, compared to value-based methods, is that policy gradient methods can naturally model stochastic policies.[1] Moreover, policy gradient methods use the gradient information to guide the optimization, which generally enjoy better convergence properties.[2]

Policy gradient methods directly optimize the agent's policy with gradient ascent on the network's parameters. In this chapter, we will start with the basic gradient ascent idea to derive the vanilla policy gradient method, called REINFORCE, in Sect. 5.2. Vanilla policy gradient suffers the high variance problem. As we will see later, one way to mitigate this problem is by the idea of actor-critic in Sect. 5.3. Interestingly, actor-critic shares a similar design to the GAN framework, which we will discuss in Sect. 5.4. A distributed version of actor-critic will also be discussed in Sects. 5.5 and 5.6. We further improve policy gradient methods by considering the gradient step in the policy space rather than the parameter space. One of the popular algorithms doing so is trust region policy optimization (TRPO), as presented in Sect. 5.7, followed by its improved version Proximal Policy Optimization (PPO) in Sect. 5.8 and actor-critic K-factor trust region (ACKTR) in Sect. 5.9.

Furthermore, we provide examples with code implementation in Sect. 5.10 for each of the algorithms we introduced, for providing the readers a quick hands-on learning process. The full implementation of each algorithm is available in the repository provided together with the book.[3]

5.2 REINFORCE: Vanilla Policy Gradient

The algorithm REINFORCE follows a straightforward idea of performing gradient ascent on the parameters of the policy θ to gradually improve the performance of

[1] In the value learning setting, the agent needs to explicitly construct its exploration like ϵ-greedy to model the stochastic policies.

[2] But only to local optima rather than global ones.

[3] Code link: https://github.com/tensorlayer/tensorlayer/tree/master/examples/reinforcement_learning.

the policy π_θ. Recall that by Eq. (2.119) we have

$$\nabla_\theta J(\pi_\theta) = \mathbb{E}_{\tau\sim\pi_\theta}\left[\sum_{t=0}^{T} R_t \nabla_\theta \sum_{t'=0}^{t} \log \pi_\theta(A_{t'}|\, S_{t'})\right]$$

$$= \mathbb{E}_{\tau\sim\pi_\theta}\left[\sum_{t'=0}^{T} \nabla_\theta \log \pi_\theta(A_{t'}|\, S_{t'}) \sum_{t=t'}^{T} R_t\right]. \tag{5.1}$$

Remark 1 $\sum_{t=i}^{T} R_t$ can be seen as an estimate of the cumulative reward from time i for taking A_i on state S_i and following the current policy. In fact, $\sum_{t=i}^{T} R_t$ can be interpreted as $Q_i(A_i, S_i)$, the Q value of A_i on the state S_i at time i. Thus, one interpretation of REINFORCE is to weight the gradient by the cumulative reward of each action, encouraging the agent to take the action A_i that has greater cumulative reward.

One can easily extend the policy gradient to the infinite horizon setting with the discount factor γ by replacing T in the above equation by ∞ and weighting each R_t by γ^t,

$$\nabla J(\theta) = \mathbb{E}_{\tau\sim\pi_\theta}\left[\sum_{t'=0}^{\infty} \nabla_\theta \log \pi_\theta(A_{t'}|\, S_{t'})\gamma^{t'} \sum_{t=t'}^{\infty} \gamma^{t-t'} R_t\right]. \tag{5.2}$$

In fact, using a discount factor also helps reduce the high variance of the gradient estimate, as it assigns lower weights to future rewards that have higher variance. In practice, $\gamma^{t'}$ is usually removed to avoid overemphasizing on early states.

Despite of its simplicity, naive REINFORCE has been observed to suffer a large variance when estimating the gradient. Indeed, the complexity of the model, and thus the randomness of R_t, grows exponentially with the trajectory length L. To alleviate the large variance problem, we further introduce a baseline $b(S_i)$, where $b(S_i)$ is a function only depending on S_i (or more importantly, not dependent to A_i).

With the baseline $b(S_t)$, the gradients of the reinforcement learning objective can be represented as:

$$\nabla J(\theta) = \mathbb{E}_{\tau\sim\pi_\theta}\left[\sum_{t'=0}^{\infty} \nabla_\theta \log \pi_\theta(A_{t'}|\, S_{t'}) \left(\sum_{t=t'}^{\infty} \gamma^{t-t'} R_t - b(S_{t'})\right)\right]. \tag{5.3}$$

The above equality is due to one key observation that

$$\mathbb{E}_{\tau,\theta}\left[\nabla_\theta \log \pi_\theta(A_{t'}|\, S_{t'}) b(S_{t'})\right] = \mathbb{E}_{\tau,\theta}\left[b(S_{t'})\mathbb{E}_\theta\left[\nabla \log \pi_\theta(A_{t'}|\, S_{t'})|\, S_{t'}\right]\right] = 0, \tag{5.4}$$

where the last equality is due to the EGLP lemma in Lemma 2.2. The final REINFORCE algorithm with baseline is summarized in Algorithm 1.

Algorithm 1 REINFORCE with baseline

Hyperparameters: step size η_θ, reward discount factor γ, number of time steps L, batch size B, baseline value b
Input: initial policy parameters θ_0
Initialize $\theta = \theta_0$
for $k = 1, 2, \ldots,$ **do**
  Run policy π_θ for B trajectories, each one with L time steps, and collect $\{S_{t,\ell}, A_{t,\ell}, R_{t,\ell}\}$
  $\hat{A}_{t,\ell} = \sum_{\ell'=\ell}^{L} \gamma^{\ell'-\ell} R_{t,\ell} - b(S_{t,\ell})$
  $J(\theta) = \frac{1}{B} \sum_{t=1}^{B} \sum_{\ell=0}^{L} \log \pi_\theta(A_{t,\ell}|S_{t,\ell})\hat{A}_{t,\ell}$
  $\theta = \theta + \eta_\theta \nabla J(\theta)$
  Update $b(S_{t,\ell})$ by $\{S_{t,\ell}, A_{t,\ell}, R_{t,\ell}\}$
end for
Return θ

The intuition behind subtracting a baseline is a variance reduction technique. Consider to estimate $\mathbb{E}[X]$ for some random variable X. For any Y such that $\mathbb{E}[Y] = 0$, the sampled mean of $X - Y$ is still an unbiased estimator. Furthermore, the variance of $X - Y$

$$\mathbb{V}(X - Y) = \mathbb{V}(X) + \mathbb{V}(Y) - 2\text{cov}(X, Y), \tag{5.5}$$

where $\text{cov}(X, Y)$ is the covariance between X and Y. Therefore, if one can find another random variable Y that has small variance and is strongly correlated with X, then the sampled mean of $X - Y$ is another unbiased estimator of $\mathbb{E}[X]$ with a smaller variance. One common choice of $b(S_i)$ is its (estimated) state value $V(S_i)$, leading to our vanilla actor-critic method. Recent works have also proposed other candidates for the baseline function. Interested readers may refer to Liu et al (2017), Wu et al. (2018), Li and Wang (2018) for more details.

5.3 Actor-Critic

The actor-critic (AC) method (Konda and Tsitsiklis 2000; Sutton et al. 2000) is situated in the intersection of policy-based methods and value-based methods. As mentioned in the last section, one can use the state value as a baseline to reduce the variance of the estimated gradient in the vanilla policy gradient method. The actor-critic method follows this idea that learns together an **actor**, the policy function $\pi(\cdot|s)$, and the **critic**, the value function $V^\pi(s)$. Moreover, actor-critic also uses the idea of bootstrapping to estimate the Q value, replacing $\sum_{t=i}^{\infty} \gamma^{t-i} R_t - b(S_i)$ by the TD error, i.e., $R_i + \gamma V^\pi(S_{i+1}) - V^\pi(S_i)$. Here we use the L-step TD error.

The critic, $V_\psi^{\pi_\theta}(s)$, is optimized to minimize the square of the L-step TD error.

$$\psi = \psi - \eta_\psi \nabla J_{V_\psi^{\pi_\theta}}(\psi), \tag{5.6}$$

where η_ψ is the step size and

$$J_{V_\psi^{\pi_\theta}}(\psi) = \frac{1}{2}\left(\sum_{t=i}^{i+L-1} \gamma^{t-i} R_t + \gamma^L V_\psi^{\pi_\theta}(S') - V_\psi^{\pi_\theta}(S_i)\right)^2. \tag{5.7}$$

Here S' is the state reached after L steps of rollout under policy π_θ, and thus

$$\nabla J_{V_\psi^{\pi_\theta}}(\psi) = \left(V_\psi^{\pi_\theta}(S_i) - \sum_{t=i}^{i+L-1} \gamma^{t-i} R_t - \gamma^L V_\psi^{\pi_\theta}(S')\right) \nabla V_\psi^{\pi_\theta}(S_i). \tag{5.8}$$

Similarly the **actor**, $\pi_\theta(\cdot|s)$, takes a state s as its input and outputs a policy or an action. We update the policy function similar as in the vanilla policy gradient method.

$$\theta = \theta + \eta_\theta \nabla J_{\pi_\theta}(\theta), \tag{5.9}$$

where η_θ is the step size and

$$\nabla J(\theta) = \mathbb{E}_{\tau,\theta}\left[\sum_{i=0}^{\infty} \nabla \log \pi_\theta(A_i|\, S_i)\left(\sum_{t=i}^{i+L-1} \gamma^{t-i} R_t + \gamma^L V_\psi^{\pi_\theta}(S') - V_\psi^{\pi_\theta}(S_i)\right)\right]. \tag{5.10}$$

While θ and ψ are used to denote the parameters of π_θ and $V_\psi^{\pi_\theta}$ separately for the purpose of generality, they do not necessarily be different. In practice when using neural networks as the function approximators, the lower layers are usually shared between π_θ and $V_\psi^{\pi_\theta}$ as the common state representation. Moreover, in practice L is commonly set to be 1, which gives the TD(1) estimation. The AC algorithm is summarized in Algorithm 2.

It is worth mentioning that one can instead use the Q value function as the critic in the actor-critic method. In the update of the actor, the advantage function can now be estimated by

$$Q(s,a) - V(s) = Q(s,a) - \sum_a \pi(a|s) Q(s,a), \tag{5.11}$$

The loss function for the learning of the critic Q is now by

$$J_Q = \left(R_t + \gamma Q(S_{t+1}, A_{t+1}) - Q(S_t, A_t)\right)^2, \tag{5.12}$$

or

$$J_Q = \left(R_t + \gamma \sum_a \pi_\theta(a|S_{t+1}) Q(S_{t+1}, a) - Q(S_t, A_t)\right)^2, \tag{5.13}$$

where A_{t+1} is sampled from the current policy π_θ given S_{t+1}.

Algorithm 2 Actor-critic

Hyperparameters: step size η_θ and η_ψ, reward discount factor γ
Input: initial policy parameters θ_0, initial value function parameters ψ_0
Initialize $\theta = \theta_0$ and $\psi = \psi_0$
for $t = 0, 1, 2, \ldots$ **do**
  Run policy π_θ for one step, collection $\{S_t, A_t, R_t, S_{t+1}\}$
  Estimate advantages $\hat{A}_t = R_t + \gamma V_\psi^{\pi_\theta}(S_{t+1}) - V_\psi^{\pi_\theta}(S_t)$
  $J(\theta) = \sum_t \log \pi_\theta(A_t|S_t)\hat{A}_t$
  $J_{V_\psi^{\pi_\theta}}(\psi) = \sum_t \hat{A}_t^2$
  $\psi = \psi + \eta_\psi \nabla J_{V_\psi^{\pi_\theta}}(\psi), \theta = \theta + \eta_\theta \nabla J(\theta)$
end for
Return (θ, ψ)

5.4 Generative Adversarial Networks and Actor-Critic

On first look, generative adversarial networks (shortened as GAN, which is introduced in Chap. 1) (Goodfellow et al. 2014) and actor-critic are two different models proposed for different categories in machine learning. One is generative model and another is reinforcement learning algorithm. But they are actually very similar in their structures. For GAN, there are two parts: the generative network for generating objects with some inputs, and the discriminative network as a successive part after the generative network for judging how realistic the generated objects are. For the actor-critic framework, there are also two parts: the actor network who generates actions with respect to the state inputs, and a critic network as a successive part after the actor to evaluate the action with a value function (e.g., the value of next state or the Q-value).

So, GAN and actor-critic basically follow the same structure: there are two successive parts, with the first one for generating stuff and the second one for evaluating the generated stuff with a score (or value). An optimization method is chosen to increase the accuracy of the second evaluation part. It then back-propagates the gradients through the second part to the first part to let it generate desired objects, with the score or value function as a criteria.

The detailed comparison of GAN and actor-critic are discussed as follows, as well as depicted in Fig. 5.1:

- For the first generative part: the generator in GAN and the actor in AC are basically the same, for both the forward process of inference and the backward process with gradient-based optimization. The generator takes the random variables as inputs, and output generated objects for forward process; and for backward optimization process its goal is to maximize the discriminative score of the generated objects. The actor takes the states as inputs and outputs actions,

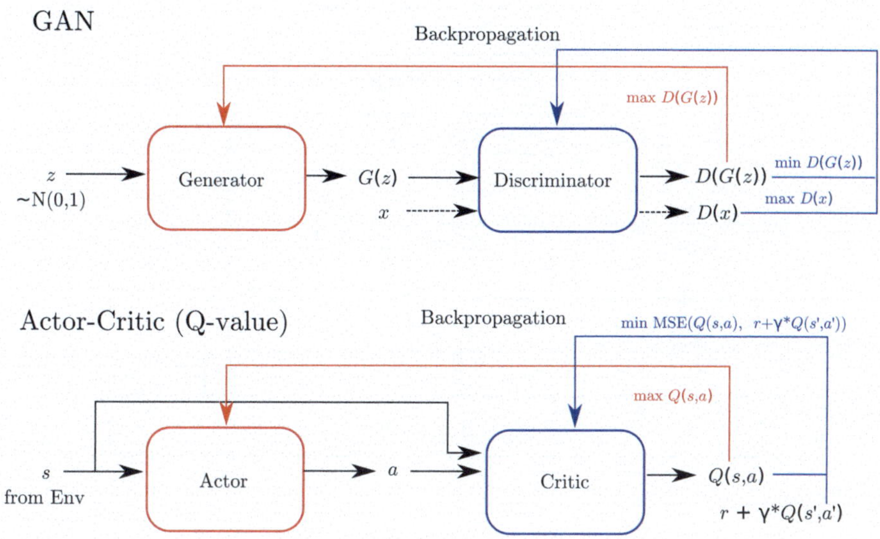

Fig. 5.1 The comparison of GAN and actor-critic structures. In GAN, the z is the input noise variable sampled from, for example, a normal distribution, and x is the data sample from the real objects. In actor-critic, the s and a are state and action, respectively

and for optimization its goal is to maximize the evaluated value of the action-state pairs.

- For the second evaluating part: the optimization formulas of the discriminator and the critic are different due to their different utilities, but following the similar objectives. There is an extra input from the real objects for the discriminator. Its optimization rule is to maximize the discriminative value of the real objects and minimize the discriminative value of the generated objects. This rule makes the discriminator more accurate, which satisfies the facts. For the critic, it uses the temporal difference (TD) error as a bootstrapping method according to optimal Bellman equation in reinforcement learning for optimizing the value function.

There are also other models which are very similar to each other. For example, the auto-encoder (AE) and GAN can simply be taken as a reverse structure of each other, etc. Noticing these similarities of different deep learning frameworks will help you to have a more comprehensive understanding of the commonality of present methods in different domains, which also helps to propose new frameworks for unsolved tasks.

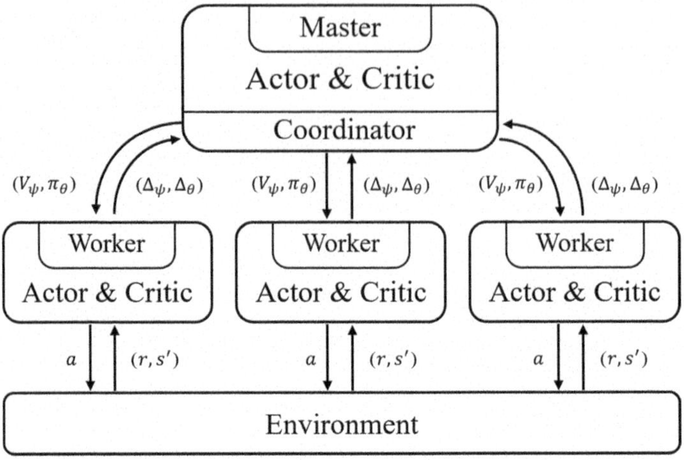

Fig. 5.2 The design of A2C

5.5 Synchronous Advantage Actor-Critic (A2C)

Synchronous advantage actor-critic (A2C) (Mnih et al. 2016) is similar to the actor-critic algorithm discussed in the last section, but focuses on parallel training.

As shown in Fig. 5.2, the global actor and the global critic are maintained and updated in the master node. The reinforcement learning agent in each worker talks to the master node via a coordinator. The coordinator is responsive for waiting until all the experience from each worker is collected and then performing a 1-step learning of the global actor and critic based on the collected trajectories. After the global actor is updated, each worker synchronizes with the master node, and continues to interact with the environment using the updated actor. In the master node, the way the actor and the critic are updated is the same as in the actor-critic algorithm, where a squared TD error is used as the loss for the learning on the critic, and policy gradient with TD error is used to update the actor.

Under such design, the workers are only responsible for interacting with the environment while all the updates happen in the master node. In practice, one may hope to alleviate the heavy computation in the master node as much as possible by delegating it to the worker nodes.[4] To do that, each worker may also keep a copy of the current global critic. After collecting a trajectory, the worker computes gradients for both the actor and critic. Instead of updating the local actor and critic, the gradients are sent back to the master. The task of the coordinator is then collecting and aggregating the gradients from all the workers and performing a simple update

[4]This usually depends on the computation resource of each worker, e.g., if GPU computing is available in the workers.

on the global models. Both the updated actor and critic are then synchronized again with all the workers. The outline of the algorithm is shown in Algorithm 3.

Algorithm 3 A2C

Master:
Hyperparameters: step size η_ψ and η_θ, set of workers $\mathcal{W}$
Input: initial policy parameters θ_0, initial value function parameters ψ_0
Initialize $\theta = \theta_0$ and $\psi = \psi_0$
for $k = 0, 1, 2, \ldots$ **do**
  $(g_\psi, g_\theta) = 0$
  for worker in $\mathcal{W}$ **do**
    $(g_\psi, g_\theta) = (g_\psi, g_\theta) +$ **worker**$(V_\psi^{\pi_\theta}, \pi_\theta)$
  end for
  $\psi = \psi - \eta_\psi g_\psi; \theta = \theta + \eta_\theta g_\theta.$
end for

Worker:
Hyperparameters: reward discount factor γ, the length of trajectory L
Input: value function $V_\psi^{\pi_\theta}$, policy π_θ
Run policy π_θ for L time steps, collection $\{S_t, A_t, R_t, S_{t+1}\}$
Estimate advantages $\hat{A}_t = R_t + \gamma V_\psi^{\pi_\theta}(S_{t+1}) - V_\psi^{\pi_\theta}(S_t)$
$J(\theta) = \sum_t \log \pi_\theta(A_t|S_t)\hat{A}_t$
$J_{V_\psi^{\pi_\theta}}(\psi) = \sum_t \hat{A}_t^2$
$(g_\psi, g_\theta) = (\nabla J_{V_\psi^{\pi_\theta}}(\psi), \nabla J(\theta))$
Return (g_ψ, g_θ)

5.6 Asynchronous Advantage Actor-Critic (A3C)

Asynchronous advantage actor-critic (A3C) (Mnih et al. 2016) is just an asynchronous version of A2C. In A3C, the coordinator is removed. Each worker talks directly to the global actor and critic. Instead of waiting until all the gradients from the workers are collected, the master updates the global actor-critic whenever there is a worker finishing the gradient computation, thus the computation efficiency is generally better compared to A2C. However, by doing so, each worker talks to the master independently, and thus it is also possible that the gradient computed is no longer for the current global actor-critic.

Remark 2 While it seems ad hoc to sacrifice the consistency among different workers for the learning efficiency, asynchronous update is actually very popular in parallel optimization for neural networks. It has been shown that, being asynchronous not only speeds up the learning but also automatically begets a momentum-like term to SGD (Mitliagkas et al. 2016).

Algorithm 4 A3C

Master:
Hyperparameters: step size η_ψ and η_θ, current policy π_θ, value function $V_\psi^{\pi_\theta}$
Input: gradients g_ψ, g_θ
$\psi = \psi - \eta_\psi g_\psi; \theta = \theta + \eta_\theta g_\theta.$
Return $(V_\psi^{\pi_\theta}, \pi_\theta)$

Worker:
Hyperparameters: reward discount factor γ, the length of trajectory L
Input: value function $V_\psi^{\pi_\theta}$, policy π_θ
$(g_\theta, g_\psi) = (0, 0)$
for $k = 1, 2, \ldots,$ **do**
  $(\theta, \psi) =$ **Master**(g_θ, g_ψ)
  Run policy π_θ for L time steps, collection $\{S_t, A_t, R_t, S_{t+1}\}$
  Estimate advantages $\hat{A}_t = R_t + \gamma V_\psi^{\pi_\theta}(S_{t+1}) - V_\psi^{\pi_\theta}(S_t)$
  $J(\theta) = \sum_t \log \pi_\theta(A_t|S_t)\hat{A}_t$
  $J_{V_\psi^{\pi_\theta}}(\psi) = \sum_t \hat{A}_t^2$
  $(g_\psi, g_\theta) = (\nabla J_{V_\psi^{\pi_\theta}}(\psi), \nabla J(\theta))$
end for

5.7 Trust Region Policy Optimization (TRPO)

In the previous sections, we have introduced the vanilla policy gradient method and its parallelized versions. Recall that in policy gradient with 1-step actor-critic, we update the policy by

$$\theta = \theta + \eta_\theta \nabla J(\theta), \tag{5.14}$$

where

$$\nabla J(\theta) = \mathbb{E}_{\tau,\theta}\left[\sum_{i=0}^{\infty} \nabla \log \pi_\theta(A_i|\, S_i) A^{\pi_\theta}(S_i, A_i)\right], \tag{5.15}$$

and the advantage function $A^{\pi_\theta}(s, a)$ is defined as

$$A^{\pi_\theta}(s, a) = Q^{\pi_\theta}(s, a) - V_\psi^{\pi_\theta}(s). \tag{5.16}$$

Similar to the standard gradient descent algorithm, vanilla policy gradient method also suffers the step size pitfall. In particular, the gradient $\nabla J(\theta)$ only provides the local first-order information at the current θ that completely ignores the curvature of the reward landscape. Therefore, if the step size η_θ is large in the highly curve area, the algorithm may suffer a performance collapse. On the other hand, if the step size is set too small, the learning would be too conservative to make progress. Even worse, $\nabla J(\theta)$ in policy gradient needs to be estimated by samples from the current policy π_θ, which makes the dependence of the learning performance on a proper step size more sensitive.

Another limitation of vanilla policy gradient methods is that the update happens in the parameter space rather than the policy space

$$\Pi = \{\pi \,|\, \pi \geq 0, \int \pi = 1\}. \tag{5.17}$$

This makes the step size η_θ more difficult to tune in practice. Why? Note that the same η_θ may have completely different update magnitudes in the policy space, depending on the current π_θ. For example, consider a policy $\pi = (\sigma(\theta), 1 - \sigma(\theta))$ in two cases where $\sigma(\theta)$ is the sigmoid function. Assume that in the first case θ is updated from $\theta = 6$ to $\theta = 3$, and in the second case θ is updated from $\theta = 1.5$ to $\theta = -1.5$. Both cases have the same update magnitude in the parameter space. However, in the first case, the update in the policy space is from $\pi \approx (1.00, 0.00)$ to $\pi \approx (0.95, 0.05)$, while in the second case, with the same update length in the parameter space, the update is from $\pi = (0.82, 0.18)$ to $\pi = (0.18, 0.82)$. The same update length of θ may result in completely different update scales in π_θ.

In this section we aim to develop an algorithm that can handle the step size more properly in policy gradient based on the idea of trust region, called trust region policy optimization (TRPO) (Schulman et al. 2015). Note that our goal is to find an updated policy π'_θ that improves the current policy π_θ. The following lemma provides an insightful connection between the performance of π_θ and π'_θ: the improvement from π_θ to π'_θ can be measured by the advantage function on π_θ, $A^{\pi_\theta}(s, a)$ (Kakade and Langford 2002). Denote the parameters of π'_θ by θ'.

Lemma 5.1

$$J(\theta') = J(\theta) + \mathbb{E}_{\tau \sim \pi'_\theta}\left[\sum_{t=0}^{\infty} \gamma^t A^{\pi_\theta}(S_t, A_t)\right], \tag{5.18}$$

where $J(\theta) = \mathbb{E}_{\tau \sim \pi_\theta}\left[\sum_{t=0}^{\infty} \gamma^t R(S_t, A_t)\right]$ *and* τ *is the state-action trajectory generated by* π'_θ.

Therefore, learning the optimal policy π_θ is equivalent to maximizing the bonus term

$$\mathbb{E}_{\tau \sim \pi'_\theta}\left[\sum_{t=0}^{\infty} \gamma^t A^{\pi_\theta}(S_t, A_t)\right]. \tag{5.19}$$

However, the above expectation is taken over π'_θ and thus difficult to optimize directly. Instead, TRPO optimizes an approximation of it, denoted by $\mathcal{L}_{\pi_\theta}(\pi'_\theta)$.

$$\mathbb{E}_{\tau\sim\pi'_\theta}\left[\sum_{t=0}^{\infty}\gamma^t A^{\pi_\theta}(S_t, A_t)\right] \tag{5.20}$$

$$= \mathbb{E}_{s\sim\rho_{\pi'_\theta}(s)}\left[\mathbb{E}_{a\sim\pi'_\theta(a|s)}\left[A^{\pi_\theta}(s,a)|\, s\right]\right] \tag{5.21}$$

$$\approx \mathbb{E}_{s\sim\rho_{\pi_\theta}(s)}\left[\mathbb{E}_{a\sim\pi'_\theta(a|s)}\left[A^{\pi_\theta}(s,a)|\, s\right]\right] \tag{5.22}$$

$$= \mathbb{E}_{s\sim\rho_{\pi_\theta}(s)}\left[\mathbb{E}_{a\sim\pi_\theta(a|s)}\left[\frac{\pi'_\theta(a|s)}{\pi_\theta(a|s)}A^{\pi_\theta}(s,a)|\, s\right]\right] \tag{5.23}$$

$$= \mathbb{E}_{\tau\sim\pi_\theta}\left[\sum_{t=0}^{\infty}\gamma^t \frac{\pi'_\theta(A_t|S_t)}{\pi_\theta(A_t|S_t)}A^{\pi_\theta}(S_t, A_t)\right] \tag{5.24}$$

$$\triangleq: \mathcal{L}_{\pi_\theta}(\pi'_\theta). \tag{5.25}$$

Although the above approximation seems coarse, it turns out its approximation error can be theoretically bounded, as shown in the next theorem.

Theorem 5.1 *Let* $D_{KL}^{max}(\pi_\theta\|\pi'_\theta) = \max_s D_{KL}(\pi_\theta\|\pi'_\theta)$, *then*

$$|J(\theta') - J(\theta) - \mathcal{L}_{\pi_\theta}(\pi'_\theta)| \leq C D_{KL}^{max}(\pi_\theta\|\pi'_\theta), \tag{5.26}$$

where C *is a constant independent to* π'_θ,

Therefore, if $D_{\mathrm{KL}}^{\max}(\pi_\theta\|\pi'_\theta)$ is small, it is reasonable to optimize $\mathcal{L}_{\pi_\theta}(\pi'_\theta)$, which is the idea of TRPO. In practice, TRPO tries to optimize $\mathcal{L}_{\pi_\theta}(\pi'_\theta)$ under the constraint that considers the average KL-divergence.

$$\begin{aligned} \max_{\pi'_\theta} \quad & \mathcal{L}_{\pi_\theta}(\pi'_\theta) \\ \text{s.t.} \quad & \mathbb{E}_{s\sim\rho_{\pi_\theta}}\left[D_{\mathrm{KL}}(\pi_\theta\|\pi'_\theta)\right] \leq \delta. \end{aligned} \tag{5.27}$$

5.7.1 *Natural Gradient*

It remains to solve the constraint optimization problem in TRPO. Here we use the first-order approximation for the objective function and the second-order approximation in the constraint. It turns out that the gradient of $\mathcal{L}_{\pi_\theta}(\pi'_\theta)$ at the policy

π_θ is the same as in actor-critic.

$$g = \nabla_\theta \mathcal{L}_{\pi_\theta}(\pi'_\theta)|_\theta = \mathbb{E}_{\tau\sim\pi_\theta}\left[\sum_{t=0}^{\infty} \gamma^t \frac{\nabla_\theta \pi'_\theta(A_t|\,S_t)}{\pi_\theta(A_t|\,S_t)} A^{\pi_\theta}(S_t, A_t)\right]\Bigg|_\theta \tag{5.28}$$

$$= \mathbb{E}_{\tau\sim\pi_\theta}\left[\sum_{t=0}^{\infty} \gamma^t \nabla_\theta \log \pi_\theta(A_t|\,S_t)\Big|_\theta A^{\pi_\theta}(S_t, A_t)\right]. \tag{5.29}$$

Moreover, let H denote the Hessian matrix of $\mathbb{E}_{s\sim\rho_{\pi_\theta}}\left[D_{\mathrm{KL}}(\pi_\theta \| \pi'_\theta)\right]$. The TRPO algorithm solves the optimization problem at the current π_θ,

$$\theta' = \arg\max_{\theta'} \quad g^\top(\theta' - \theta) \tag{5.30}$$

$$\text{s.t.} \quad (\theta' - \theta)^\top H(\theta' - \theta) \le \delta,$$

where the gradients are calculated in first order and the constraint is in second order. Solution to the above problem has an analytic form

$$\theta' = \theta + \sqrt{\frac{2\delta}{g^\top H^{-1} g}} H^{-1} g. \tag{5.31}$$

In practice, we use the conjugate gradient algorithm to approximate $H^{-1}g$.[5] We also pick a proper step size to enforce the sample KL-divergence constraint.

Lastly, the value function is learned by minimizing the MSE error. The complete TRPO algorithm is presented in Algorithm 5 which is based on paper (Schulman et al. 2015).

Remark 5.1 The negative Hessian matrix $-H$ is also called Fisher information matrix. In fact, using the Fisher information matrix in gradient descent has been well explored in the batch setting, called natural gradient descent. One nice property about natural gradient descent is it is invariant under reparameterization, i.e., the gradient remains the same regardless of the parametrization method used to compute it. We omit the details here. For more details, please refer to the paper (Amari 1998).

[5]In general, computing H^{-1} requires computation complexity $\Omega(N^3)$, which is too expensive in practice as here N is the number of the parameters.

Algorithm 5 TRPO

Hyperparameters: KL-divergence limit δ, backtracking coefficient α, maximum number of backtracking steps K

Input: empty replay buffer $\mathcal{D}_k$, initial policy parameters θ_0, initial value function parameters ϕ_0

for episode $= 0, 1, 2, \ldots$ **do**

Collect set of trajectories $\mathcal{D}_k = \{\tau_i\}$ by running policy $\pi_k = \pi(\theta_k)$ in the environment

Compute rewards-to-go $\hat{G}_t$

Compute advantage estimates, $\hat{A}_t$ (using any method of advantage estimation) based on the current value function V_{ϕ_k}

Estimate policy gradient as

$$\hat{g}_k = \frac{1}{|\mathcal{D}_k|} \sum_{\tau \in \mathcal{D}_k} \sum_{t=0}^{T} \nabla_\theta \log \pi_\theta(A_t|S_t)|_{\theta_k} \hat{A}_t \tag{5.32}$$

Use the conjugate gradient algorithm to compute

$$\hat{x}_k \approx \hat{H}_k^{-1} \hat{g}_k \tag{5.33}$$

where $\hat{H}_k$ is the Hessian of the sample average KL-divergence

Update the policy by backtracking line search with

$$\theta_{k+1} = \theta_k + \alpha^j \sqrt{\frac{2\delta}{\hat{x}_k^T \hat{H}_k \hat{x}_k}} \hat{x}_k \tag{5.34}$$

where $j \in \{0, 1, 2, \ldots K\}$ is the smallest value which improves the sample loss and satisfies the sample KL-divergence constraint

Fit value function by regression on mean-squared error:

$$\phi_{k+1} = \arg\min_{\phi} \frac{1}{|\mathcal{D}_k| T} \sum_{\tau \in \mathcal{D}_k} \sum_{t=0}^{T} \left(V_\phi(S_t) - \hat{G}_t \right)^2 \tag{5.35}$$

typically via some gradient descent algorithm

end for

5.8 Proximal Policy Optimization (PPO)

As one can easily see in the last section, TRPO is relatively complicated and suffers a computation burden in computing the natural gradient. Even with an approximation of $H^{-1}g$, it still requires many steps of conjugate gradient for every single parameter update. In this section, we introduce another policy gradient method, proximal policy optimization(PPO), that enforces the similarity between π_θ and π'_θ in a simpler and more efficient way (Schulman et al. 2017).

Recall that TRPO tends to optimize Eq (5.27).

$$\max_{\pi'_\theta} \quad \mathcal{L}_{\pi_\theta}(\pi'_\theta) \tag{5.36}$$

$$\text{s.t.} \quad \mathbb{E}_{s\sim\rho_{\pi_\theta}}\left[D_{\mathrm{KL}}(\pi_\theta\|\pi'_\theta)\right] \le \delta. \tag{5.37}$$

Instead of optimizing with a hard constraint, PPO tends to optimize its regularization version.

$$\max_{\pi'_\theta} \mathcal{L}_{\pi_\theta}(\pi'_\theta) - \lambda\mathbb{E}_{s\sim\rho_{\pi_\theta}}\left[D_{\mathrm{KL}}(\pi_\theta\|\pi'_\theta)\right]. \tag{5.38}$$

Here λ is the regularization coefficient. For every δ in Eq. (5.27), there existing a corresponding constant λ that recovers the same optimizer. However, such λ depends on π_θ. Thus it makes sense to have an adaptive λ for Eq. (5.38). Here we particularly check the KL-divergence constraint to decide if λ should be enlarged or reduced. We call such algorithm PPO-Penalty, as presented in Algorithm 6 which is based on these papers (Schulman et al. 2017; Heess et al. 2017).

Algorithm 6 PPO-penalty

Hyperparameters: reward discount factor γ, KL penalty coefficient λ, adaptive parameters $a = 1.5$, $b = 2$, the number of sub-iterations M, B
Input: initial policy parameters θ, initial value function parameters ϕ
for k = 0, 1, 2, ... **do**
  Run policy π_θ for T time steps, collection $\{S_t, A_t, R_t\}$
  Estimate advantages $\hat{A}_t = \sum_{t'>t}\gamma^{t'-t}R_{t'} - V_\phi(S_t)$
  $\pi_{\text{old}} \leftarrow \pi_\theta$
  for $m \in \{1, \dots, M\}$ **do**
    $J_{\text{PPO}}(\theta) = \sum_{t=1}^{T}\frac{\pi_\theta(A_t|S_t)}{\pi_{\text{old}}(A_t|S_t)}\hat{A}_t - \lambda\hat{\mathbb{E}}_t\left[D_{\mathrm{KL}}(\pi_{\text{old}}(\cdot|S_t)\|\pi_\theta(\cdot|S_t))\right]$
    Update θ by a gradient method w.r.t $J_{\text{PPO}}(\theta)$
  end for
  for $b \in \{1, \dots, B\}$ **do**
    $L_{BL}(\phi) = -\sum_{t=1}^{T}\left(\sum_{t'>t}\gamma^{t'-t}R_{t'} - V_\phi(S_t)\right)^2$
  end for
  Compute $d = \hat{\mathbb{E}}_t\left[D_{\mathrm{KL}}(\pi_{\text{old}}(\cdot|S_t)\|\pi_\theta(\cdot|S_t))\right]$
  if $d < d_{\text{target}}/a$ **then**
    $\lambda \leftarrow \lambda/b$
  else if $d > d_{\text{target}} \times a$ **then**
    $\lambda \leftarrow \lambda \times b$
  end if
end for

Another alternative approach to TRPO is to directly clipping the objective value for the policy gradient, thus resulting a more conservative update. Denote $\frac{\pi'_\theta(A_t|S_t)}{\pi_\theta(A_t|S_t)}$ by $\ell_t(\theta')$. It has been found that the following objective provides a simple and stable learning performance (empirically) for policy gradient method.

$$\mathcal{L}^{\text{PPO-Clip}}(\pi'_\theta) = \mathbb{E}_{\pi_\theta}\left[\min\left(\ell_t(\theta')A^{\pi_\theta}(S_t, A_t),\right.\right.$$
$$\left.\left.\text{clip}(\ell_t(\theta'), 1-\epsilon, 1+\epsilon)A^{\pi_\theta}(S_t, A_t)\right)\right], \tag{5.39}$$

where $\text{clip}(x, 1-\epsilon, 1+\epsilon)$ clips x within $[1-\epsilon, 1+\epsilon]$. The resulted algorithm is called PPO-Clip, as presented in Algorithm 7 which is based on paper (Schulman et al. 2017). In particular, PPO-Clip first clips $\ell_t(\theta')$ within $[1-\epsilon, 1+\epsilon]$ to ensure that π'_θ is close to π_θ. The final learning objective is then by taking the minimum between the clipped objected and the unclipped one. Thus PPO-Clip is maximizing a lower bound of the target objective, while maintaining the property that the update from π_θ to π'_θ is controllable.

Algorithm 7 PPO-clip

Hyperparameters: clip factor ϵ, the number of sub-iterations M, B
Input: initial policy parameters θ, initial value function parameters ϕ
for k = 0, 1, 2, ... **do**
  Collect set of trajectories $\mathcal{D}_k = \{\tau_i\}$ by running policy π_{θ_k} in the environment
  Compute rewards-to-go $\hat{G}_t$
  Compute advantage estimates, $\hat{A}_t$(using any method of advantage estimation) based on the current value function V_{ϕ_k}
  for $m \in \{1, \ldots, M\}$ **do**

$$\ell_t(\theta') = \frac{\pi_\theta(A_t|S_t)}{\pi_{\theta_{\text{old}}}(A_t|S_t)} \tag{5.40}$$

  Update the policy by maximizing the PPO-Clip objective:

$$\theta_{k+1} = \arg\max_\theta \frac{1}{|\mathcal{D}_k|T} \sum_{\tau \in \mathcal{D}_k} \sum_{t=0}^{T} \min(\ell_t(\theta')A^{\pi_{\theta_{\text{old}}}}(S_t, A_t), \tag{5.41}$$

$$\text{clip}(\ell_t(\theta'), 1-\epsilon, 1+\epsilon)A^{\pi_{\theta_{\text{old}}}}(S_t, A_t)) \tag{5.42}$$

  typically via stochastic gradient ascent with Adam
  end for
  for $b \in \{1, \ldots, B\}$ **do**
  Fit value function by regression on mean-squared error:

$$\phi_{k+1} = \arg\min_\phi \frac{1}{|\mathcal{D}_k|T} \sum_{\tau \in \mathcal{D}_k} \sum_{t=0}^{T} \left(V_\phi(S_t) - \hat{G}_t\right)^2$$

  typically via some gradient descent algorithm
  end for
end for

5.9 Actor Critic Using Kronecker-Factored Trust Region (ACKTR)

Actor critic using Kronecker-factored trust region (ACKTR) (Wu et al. 2017) is another alternative to alleviate the computation overhead in TRPO. The idea of ACKTR is to use the Kronecker-factored approximated curvature (K-FAC) (Martens and Grosse 2015; Grosse and Martens 2016) to compute the natural gradient. In this section, we present the idea of ACKTR for a MLP policy network.

Note that

$$\mathbb{E}\left[\frac{\partial^2}{\partial^2\theta} D_{\mathrm{KL}}(\pi_{\mathrm{old}}\|\pi_\theta)\right] \tag{5.43}$$

$$= -\mathbb{E}\left[\pi_{\mathrm{old}}\frac{\partial^2}{\partial^2\theta}\log\pi_\theta\right] \tag{5.44}$$

$$= -\mathbb{E}_{s\sim\rho_{\pi_{\mathrm{old}}}}\left[\mathbb{E}_{a\sim\pi_{\mathrm{old}}}\left[\frac{\partial^2}{\partial^2\theta}\log\pi_\theta(a|s)\right]\right] \tag{5.45}$$

$$= \mathbb{E}_{s\sim\rho_{\pi_{\mathrm{old}}}}\left[\mathbb{E}_{a\sim\pi_{\mathrm{old}}}\left[(\nabla_\theta\log\pi_\theta(a|s))\,(\nabla_\theta\log\pi_\theta(a|s))^\top\right]\right]. \tag{5.46}$$

Recall that TRPO needs to run multiple steps of conjugate gradient in order to approximate $H^{-1}g$. The idea of ACKTR is to approximate H^{-1} by a block diagonal matrix, where each block corresponds to the Fisher information matrix of each layer separately. Assume that for the ℓ layer, $x_{\mathrm{out}} = W_\ell x_{\mathrm{in}}$ and W_ℓ is a matrix in the dimension of $d_{\mathrm{in}} \times d_{\mathrm{in}}$. The idea of Kronecker factorization comes from the observation that the gradient $\nabla_{W_\ell} L$ is an outer product $(\nabla_{x_{\mathrm{out}}} L)x_{\mathrm{in}}^\top$, thus

$$(\nabla_\theta\log\pi_\theta(a|s))\,(\nabla_\theta\log\pi_\theta(a|s))^\top = x_{\mathrm{in}}x_{\mathrm{in}}^\top \otimes (\nabla_{x_{\mathrm{out}}}L)(\nabla_{x_{\mathrm{out}}}L)^\top, \tag{5.47}$$

where $\otimes$ denotes the Kronecker product. Furthermore,

$$\left((\nabla_\theta\log\pi_\theta(a|s))\,(\nabla_\theta\log\pi_\theta(a|s))^\top\right)^{-1} g \tag{5.48}$$

$$= \left(x_{\mathrm{in}}x_{\mathrm{in}}^\top \otimes (\nabla_{x_{\mathrm{out}}}L)(\nabla_{x_{\mathrm{out}}}L)^\top\right)^{-1} g \tag{5.49}$$

$$= \left[\left(x_{\mathrm{in}}x_{\mathrm{in}}^\top\right)^{-1} \otimes \left((\nabla_{x_{\mathrm{out}}}L)(\nabla_{x_{\mathrm{out}}}L)^\top\right)^{-1}\right] g. \tag{5.50}$$

Therefore, instead of inverting a $(d_{\mathrm{in}}d_{\mathrm{out}}) \times (d_{\mathrm{in}}d_{\mathrm{out}})$ matrix, which is $\Omega(d_{\mathrm{in}}^3 d_{\mathrm{out}}^3)$ naively, ACKTR only needs to invert two matrices in dimensions of $d_{\mathrm{in}} \times d_{\mathrm{in}}$ and $d_{\mathrm{out}} \times d_{\mathrm{out}}$, thus the computational complexity is $\Omega(d_{\mathrm{in}}^3 + d_{\mathrm{out}}^3)$.

The ACKTR algorithm is presented in Algorithm 8. The idea of ACKTR can also be applied to the learning of the value network. Interested readers can refer to Wu et al. (2017) for more details which we omit here.

Algorithm 8 ACKTR

1: **Hyperparameters**: learning rate η_{max}, trust region radius δ
2: **Input**: empty replay buffer $\mathcal{D}$, initial policy parameters θ_0, initial value function parameters ϕ_0
3: **for** k = 0, 1, 2, ... **do**
4: Collect set of trajectories $\mathcal{D}_k = \{\tau_i | i = 0, 1, \ldots\}$ by running policy $\pi_k = \pi(\theta_k)$ in the environment
5: Compute cumulative return G_t
6: Compute advantage estimates, $\hat{A}_t$ (using any method of advantage estimation) based on the current value function V_{ϕ_k}
7: Estimate policy gradient as

$$\hat{g}_k = \frac{1}{|\mathcal{D}_k|} \sum_{\tau \in \mathcal{D}_k} \sum_{t=0}^{T} \nabla_\theta \log \pi_\theta(A_t|S_t)|_{\theta_k} \hat{A}_t \tag{5.51}$$

8: **for** $l = 0, 1, 2, \ldots$ **do**

$$\text{vec}(\Delta\theta_k^l) = \text{vec}(A_l^{-1} \nabla_{\theta_k^l} \hat{g}_k S_l^{-1}) \tag{5.52}$$

where $A_l = \mathbb{E}[a_l a_l^T]$, $S_l = \mathbb{E}[(\nabla_{s_l} \hat{g}_k)(\nabla_{s_l} \hat{g}_k)^T]$ (A_l, S_l are calculated with running average among episodes), a_l is the input activation vector and $s_l = W_l a_l$ with the weight matrix W_l of lth layer, and vec($\cdot$) operation is used for transforming the matrix into one-dimensional vector
9: **end for**
10: Update the policy by K-FAC approximated natural gradient as

$$\theta_{k+1} = \theta_k + \eta_k \Delta\theta_k \tag{5.53}$$

where $\eta_k = \min(\eta_{max}, \sqrt{\frac{2\delta}{\theta_k^T \hat{H}_k \theta_k}})$ and $\hat{H}_k^l = A_l \otimes S_l$ for lth layer.
11: Fit value function by regression on mean-squared error:

$$\phi_{k+1} = \arg\min_\phi \frac{1}{|\mathcal{D}_k|T} \sum_{\tau \in \mathcal{D}_k} \sum_{t=0}^{T} (V_\phi(S_t) - G_t)^2 \tag{5.54}$$

via Gauss-Newton second-order gradient descent algorithm (also uses K-FAC approximation)
12: **end for**

5.10 Policy Gradient Examples

In the previous sections, we introduce several algorithms based on policy gradient in a theoretical perspective with pseudo codes, which involves REINFORCE (vanilla policy gradient), actor-citric (AC), synchronous advantage actor-critic (A2C), asynchronous advantage actor-critic (A3C), trust region policy optimization (TRPO), proximal policy optimization (PPO), and actor critic using Kronecker-factored trust region (ACKTR). In this section, we provide the Python code examples for above algorithms, and apply them on the OpenAI Gym environments. We will first briefly introduce the environments used in the examples, then provide detailed implementations for each algorithm. Although most of the algorithms introduced in this chapter are applicable to the environments of both continuous action space and discrete action space, the provided code examples may only work for specific environments of a single type of action space as a demonstration. Nevertheless, the readers can feel free to modify the algorithms slightly for adapting it to the other environments with different types of action space. The complete codes are provided at repository.[6]

5.10.1 Related Gym Environments

The following sections give some examples based on the OpenAI Gym environment. Examples can be divided into discrete and continuous action spaces.

```
import gym
env = gym.make('Pong-v0')
print(env.action_space)
```

The code above builds a 'Pong-v0' environment and shows the action space of it. Try to replace 'Pong-v0' to a different environment id like 'CartPole-v1' or 'Pendulum-v0' can build a different environment.

And the following sections also used some open source library. We can import them by using the following code:

```
import numpy as np
import tensorflow as tf
import tensorflow_probability as tfp
import tensorlayer as tl
...
```

[6]https://github.com/tensorlayer/tensorlayer/tree/master/examples/reinforcement_learning.

Discrete Action Space: Atari Pong Game and CartPole

Two games in OpenAI Gym with discrete action space are introduced here: Pong (Fig. 5.3) and CartPole (Fig. 5.4).

Pong

In the Pong game, the green tablet is controlled to move up and down to hit the ball. The example uses the 'Pong-v0' version. In this version, the state space is an RGB image tensor with shape (210, 160, 3). The input action is an integer in 0,1,2,3,4,5, which denotes the following meanings: $0 : NOOP, 1 : FIRE, 2 : RIGHT, 3 : LEFT, 4 : RIGHTFIRE, 5 : LEFTFIRE$.

CartPole

CartPole is a classical inverted pendulum environment. We will control the movement of the car to keep the pole erect. In the CartPole-v0 environment, the observation space is a four-dimensional vector representing the position and speed of the car, the angle of the pole, and the speed of the pole tip. The input action is 0 or 1 to control the car to move left or right.

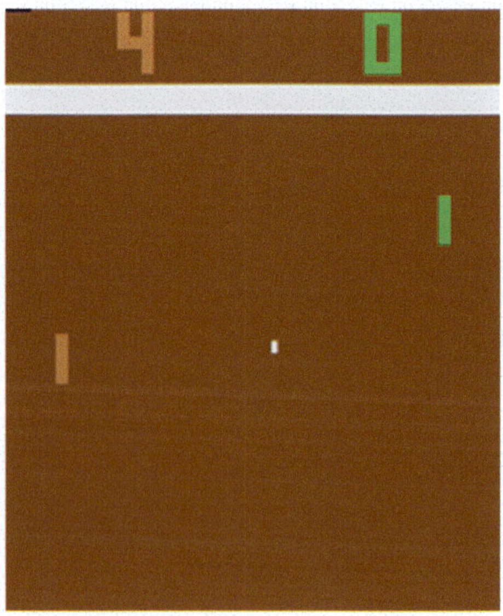

Fig. 5.3 Pong

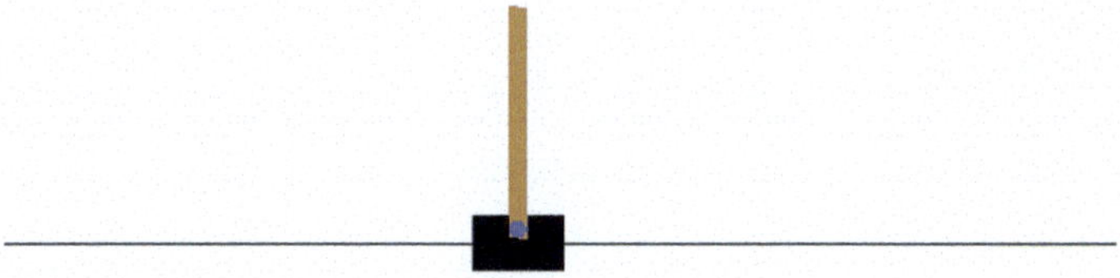

Fig. 5.4 CartPole

Fig. 5.5 BipedalWalker

Continuous Action Space: BipedalWalker-v2 and Pendulum-v0

In this part, we talk about environments with continuous action space: Bipedal-Walker (Fig. 5.5) and Pendulum (Fig. 5.6).

BipedalWalker-v2

BipedalWalker-v2 is a bipedal robot walking simulation environment. In the environment, we need to control the robot to walk on relatively flat ground and eventually reach its destination. The state space is a 24-dimensional vector, which denotes the angle, speed, and radars' information. The environmental action space is a four-dimensional continuous action space, which controls the rotation of four joints of the robot's feet.

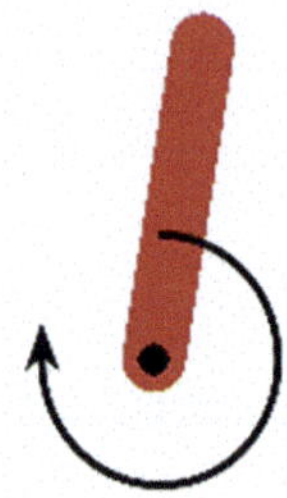

Fig. 5.6 Pendulum

Pendulum-v0

Pendulum-v0 is a classical inverted pendulum environment. In the environment, we need to control the rotation of the pole to make it erect. The state space is a three-dimensional vector which denotes $cos(\theta)$, $sin(\theta)$, $\Delta(\theta)$. Here, θ is the angle between the pole and the vertically upward direction. The action is only one dimension that controls the rotational moment.

It is worth noting that this environment has no terminal state, which means, the end of the game should be set artificially. By default, the environment has a maximum step size limit of 200. When running more than 200 steps, the *done* variable returned by `step()` function is True. Because of this limitation, when manually setting the maximum step loop size per round is greater than 200, it will still force the exit of the round in 200 steps because done is true. The following code can be used to lift the limit.

```
import gym
env = gym.make('Pendulum-v0')
env = env.unwrapped # lift the limit
```

5.10.2 REINFORCE: Atari Pong Game and CartPole-v0

Pong

Before we start, we need to prepare the environment, model, optimizer, and initialize some variables that we will use later.

```
env = gym.make("Pong-v0") # create environment
observation = env.reset() # reset environment
prev_x = None
running_reward = None
reward_sum = 0
episode_number = 0

# prepared to collect data
xs, ys, rs = [], [], []
```

```
epx, epy, epr = [], [], []

model = get_model([None, D]) # create model
train_weights = model.trainable_weights

optimizer = tf.optimizers.RMSprop(lr=learning_rate,
    decay=decay_rate) # create optimizer

model.train() # set model to train mode (in case you add dropout
    into the model)

start_time = time.time()
game_number = 0
```

After the preparation, the main loop can be started. At first we need to preprocess the observation. Then the preprocessed observation passed to variable x. After feeding x into the network, we will get the evaluation of the probability of each action by the network.

Here we only use three actions: $NOOP, UP, DOWN$. In REINFORCE algorithm, softmax function is used to output each action's probability and action is chosen according to probability finally.

```
while True:
   if render:
      env.render()

   cur_x = prepro(observation)
   x = cur_x - prev_x if prev_x is not None else np.zeros(D,
      dtype=np.float32)
   x = x.reshape(1, D)
   prev_x = cur_x

   _prob = model(x)
   prob = tf.nn.softmax(_prob)

   # action. 1: STOP 2: UP 3: DOWN
   action = tl.rein.choice_action_by_probs(prob[0].numpy(), [1,
      2, 3])
```

Now we have actions that we choose based on the current state. It is time to interact with the environment. The environment performs the next step according to the current action, and returns next observation, reward, done state, and additional information(the _ variable). We store these data for update.

```
   observation, reward, done, _ = env.step(action)

   reward_sum += reward
   xs.append(x) # all observations in an episode
   ys.append(action - 1) # all fake labels in an episode (action
      begins from 1, so minus 1)
   rs.append(reward) # all rewards in an episode
```

If the done state returned by `step()` function is true, it means that current episode is over. We can reset the environment and start a new episode. But before that, we need to preprocess the data we have just collected in this episode and store in the cross-episode data list.

```
if done:
    episode_number += 1
    game_number = 0

    epx.extend(xs)
    epy.extend(ys)
    disR = tl.rein.discount_episode_rewards(rs, gamma)
    disR -= np.mean(disR)
    disR /= np.std(disR)
    epr.extend(disR)
    xs, ys, rs = [], [], []
```

After the agent has played a lot of games and collected enough data, it can start to update. We use cross-entropy loss and gradient descent method to calculate the gradient of each parameter. Then apply the gradient to the corresponding parameter and an update is complete.

```
if episode_number % batch_size == 0:
    print('batch over...... updating parameters......')
    with tf.GradientTape() as tape:
        _prob = model(epx)
        _loss = tl.rein.cross_entropy_reward_loss(_prob,
            epy, disR)
    grad = tape.gradient(_loss, train_weights)
    optimizer.apply_gradients(zip(grad, train_weights))

    epx, epy, epr = [], [], []
```

The content above describes the main work, and the rest of the code is mainly used to display training related data for us to view the training trend. We use moving average to calculate running reward for each episode, just to reduce the degree of data jitter to facilitate observation of trends. Finally, do not forget to reset the environment, because at this point the current episode is over.

```
# if episode_number % (batch_size * 100) == 0:
#     tl.files.save_npz(network.all_params,
    name=model_file_name + '.npz')

running_reward = reward_sum if running_reward is None else
    running_reward * 0.99 + reward_sum * 0.01
print('resetting env. episode reward total was {}. running
    mean: {}'.format(reward_sum, running_reward))

reward_sum = 0
observation = env.reset() # reset env
```

```
        prev_x = None

    if reward != 0:
        print(
            ( 'episode %d: game %d took %.5fs, reward: %f' %
                (episode_number, game_number, time.time() -
                    start_time, reward)
            ), ('' if reward == -1 else ' !!!!!!!!')
        )
        start_time = time.time()
        game_number += 1
```

CartPole

The algorithm here is the same as Pong. We make the whole algorithm into a class, and write each code into corresponding functions. This can make the code more convenient to read and use. The structure of the policy gradient class is as follows:

```
class PolicyGradient:
    def __init__(self, state_dim, action_num, learning_rate=0.02,
        gamma=0.99): # Class initialization. Creates model,
        optimizer and required variables.
        ...
    def get_action(self, s, greedy=False): # Algorithm chooses
        actions based on action distribution.
        ...
    def store_transition(self, s, a, r): # Stores data generated
        by interaction with the environment.
        ...
    def learn(self): # Algorithm uses stored data to learn and
        update itself.
        ...
    def _discount_and_norm_rewards(self): # Calculate and
        normalize discounted rewards.
        ...
    def save(self): # save model
        ...
    def load(self): # load model
        ...
```

The initialization function firstly created some variables, then created the model, and finally created adam optimizer as policy optimizer. Through the code block, we can see the policy net we created here is just a network with just one hidden layer.

```
def __init__(self, state_dim, action_num, learning_rate=0.02,
    gamma=0.99):
    self.gamma = gamma

    self.state_buffer, self.action_buffer, self.reward_buffer
        = [], [], []

    input_layer = tl.layers.Input([None, state_dim],
        tf.float32)
    layer = tl.layers.Dense(
        n_units=30, act=tf.nn.tanh,
            W_init=tf.random_normal_initializer(mean=0,
            stddev=0.3),
        b_init=tf.constant_initializer(0.1)
    )(input_layer)
    all_act = tl.layers.Dense(
        n_units=action_num, act=None,
            W_init=tf.random_normal_initializer(mean=0,
            stddev=0.3),
        b_init=tf.constant_initializer(0.1)
    )(layer)

    self.model = tl.models.Model(inputs=input_layer,
        outputs=all_act)
    self.model.train()
    self.optimizer = tf.optimizers.Adam(learning_rate)
```

After we have got the policy net by initialization, we calculate the probability of each action with state by calling the `get_action()` function. By setting 'greedy=True' can simply get the action with the highest probability rather than get the action by sampling the action probability.

```
def get_action(self, s, greedy=False):
    _logits = self.model(np.array([s], np.float32))
    _probs = tf.nn.softmax(_logits).numpy()
    if greedy:
        return np.argmax(_probs.ravel())
    return tl.rein.choice_action_by_probs(_probs.ravel())
```

But the action we choose at this moment may not be correct. Only through continuous learning can the network make better and better decisions. This is what the `learn()` function does. This part of the function is basically the same as the code of pong example. We use the discounted and normalized rewards and cross-entropy loss to update the model. After each update, transition data will be discarded.

```
def learn(self):
    # discount and normalize episode reward
    discounted_ep_rs_norm = self._discount_and_norm_rewards()
    with tf.GradientTape() as tape:
        _logits = self.model(np.vstack(self.ep_obs))
```

```
        neg_log_prob =
            tf.nn.sparse_softmax_cross_entropy_with_logits
            (logits=_logits, labels=np.array(self.ep_as))
        loss = tf.reduce_mean(neg_log_prob *
            discounted_ep_rs_norm)

    grad = tape.gradient(loss, self.model.trainable_weights)
    self.optimizer.apply_gradients(zip(grad,
        self.model.trainable_weights))

    self.ep_obs, self.ep_as, self.ep_rs = [], [], [] # empty
        episode data
    return discounted_ep_rs_norm
```

The `learn()` function needs to use the stored data generated by the interaction between the agent and the environment. So we use the `store_transition()` function to store each state, action, and reward.

```
def store_transition(self, s, a, r):
    self.ep_obs.append(np.array([s], np.float32))
    self.ep_as.append(a)
    self.ep_rs.append(r)
```

Policy gradient algorithm uses Monte Carlo method. So we need to calculate the discounted rewards as follows. Normalize episode rewards is also helpful for learning.

```
def _discount_and_norm_rewards(self):
    # discount episode rewards
    discounted_ep_rs = np.zeros_like(self.ep_rs)
    running_add = 0
    for t in reversed(range(0, len(self.ep_rs))):
        running_add = running_add * self.gamma + self.ep_rs[t]
        discounted_ep_rs[t] = running_add

    # normalize episode rewards
    discounted_ep_rs -= np.mean(discounted_ep_rs)
    discounted_ep_rs /= np.std(discounted_ep_rs)
    return discounted_ep_rs
```

Like Pong code, let us prepare the environment and algorithm first. After creating the environment, we generate an instance of policy gradient class called agent.

```
env = gym.make(ENV_ID).unwrapped
# reproducible
np.random.seed(RANDOM_SEED)
tf.random.set_seed(RANDOM_SEED)
env.seed(RANDOM_SEED)
agent = PolicyGradient(
    action_num=env.action_space.n,
    state_dim=env.observation_space.shape[0],
)
t0 = time.time()
```

In training mode, we use the actions from the model to interact with the environment, store transition data, and update the policy in each episode. To simplify the code, agent is updated directly at the end of each round.

```
if args.train:
    all_episode_reward = []
    for episode in range(TRAIN_EPISODES):
        # reset the environment
        state = env.reset()
        episode_reward = 0

        for step in range(MAX_STEPS): # in one episode
            if RENDER:
                env.render()

            # choose action
            action = agent.get_action(state)

            # interact with the environment
            next_state, reward, done, info = env.step(action)

            # store transition
            agent.store_transition(state, action, reward)

            state = next_state
            episode_reward += reward
            # stop if environment returns done
            if done:
                break
        # update model at the end of each game
        agent.learn()
        print(
            'Training | Episode: {}/{} | Episode Reward: {:.0f}
                | Running Time: {:.4f}'.format(
                episode + 1, TRAIN_EPISODES, episode_reward,
                time.time() - t0))
```

We can add some code and extend some functions after each round to show the training process better. We print the total reward of each episode and calculated running reward by using moving average. After that plot the running reward for easy viewing of training trends. Finally, remember to save the model.

```
    agent.save()
    plt.plot(all_episode_reward)
    if not os.path.exists('image'):
        os.makedirs('image')
    plt.savefig(os.path.join('image', 'pg.png'))
```

If we use test mode, things are much easier. Just load the pre-trained model and use it to interact with the environment.

```
if args.test:
    # test
    agent.load()
    for episode in range(TEST_EPISODES):
        state = env.reset()
        episode_reward = 0
        for step in range(MAX_STEPS):
            env.render()
            state, reward, done, info =
                env.step(agent.get_action(state, True))
            episode_reward += reward
            if done:
                break
        print(
            'Testing | Episode: {}/{} | Episode Reward: {:.0f} |
             Running Time: {:.4f}'.format(
                episode + 1, TEST_EPISODES, episode_reward,
                time.time() - t0))
```

5.10.3 AC: CartPole-v0

Actor-critic computes the baseline (critic) using the TD method, it can update the policy at each step after interaction with the environment, which is quite different from the MC method.

In actor-critic algorithm, we made two classes, actor and critic. Their structures are as follows:

```
class Actor(object):
    def __init__(self, state_dim, action_num, lr=0.001): #Class
        initialization. Creates model, optimizer and required
        variables.
        ...
    def learn(self, state, action, td_error): # update model
        ...
    def get_action(self, state, greedy=False): # choose action by
        action probability distribution or greedy
        ...
    def save(self): # save trained weights
        ...
    def load(self): # load trained weights
        ...

class Critic(object):
```

```
    def __init__(self, state_dim, lr=0.01): #Class
        initialization. Creates model, optimizer and required
        parameters.
        ...
    def learn(self, state, reward, state_): # update model
        ...
    def save(self): # save trained weights
        ...
    def load(self): # load trained weights
        ...
```

The actor part is almost like the policy gradient algorithm. The only difference is that the `learn()` function uses the TD-error value as the estimated advantage value instead of the discounted reward value for updating.

```
    def learn(self, state, action, td_error):
        with tf.GradientTape() as tape:
            _logits = self.model(np.array([state]))
            _exp_v =
                tl.rein.cross_entropy_reward_loss(logits=_logits,
                actions=[action], rewards=td_error[0])
        grad = tape.gradient(_exp_v, self.model.trainable_weights)
        self.optimizer.apply_gradients(zip(grad,
            self.model.trainable_weights))
        return _exp_v
```

Unlike the PG algorithm, the AC algorithm has a critic with a value network that can estimate the value of each state. So the initialization function is quite clear, just create the network and the optimizer.

```
class Critic(object):
    def __init__(self, state_dim, lr=0.01):
        input_layer = tl.layers.Input([1, state_dim], name='state')
        layer = tl.layers.Dense(
            n_units=30, act=tf.nn.relu6,
                W_init=tf.random_uniform_initializer(0, 0.01),
                name='hidden'
        )(input_layer)
        layer = tl.layers.Dense(n_units=1, act=None,
            name='value')(layer)
        self.model = tl.models.Model(inputs=input_layer,
            outputs=layer, name="Critic")
        self.model.train()

        self.optimizer = tf.optimizers.Adam(lr)
```

After the initialization function, we have got a value network. The next step is the `learn()` function. The `learn()` function is very simple. Its main content is to calculate TD-error δ by using equation $\delta = R + \gamma V(s') - V(s)$. And use TD-error as an estimate of advantage to calculate the loss and return value.

```
def learn(self, state, reward, state_, done):
    d = 0 if done else 1
    v_ = self.model(np.array([state_]))
    with tf.GradientTape() as tape:
        v = self.model(np.array([state]))
        ## TD_error = r + d * lambda * V(newS) - V(S)
        td_error = reward + d * LAM * v_ - v
        loss = tf.square(td_error)
    grad = tape.gradient(loss, self.model.trainable_weights)
    self.optimizer.apply_gradients(zip(grad,
         self.model.trainable_weights))
```

The save and load functions are as usual. We can also save network parameters as `.npz` files.

```
def save(self): # save trained weights
    if not os.path.exists(os.path.join('model', 'ac')):
        os.makedirs(os.path.join('model', 'ac'))
    tl.files.save_npz(self.model.trainable_weights,
         name=os.path.join('model', 'ac', 'model_critic.npz'))

def load(self): # load trained weights
    tl.files.load_and_assign_npz(name=os.path.join('model',
         'ac', 'model_critic.npz'), network=self.model)
```

The code of the training main loop is very similar to the previous code. The only difference is the timing of the update. By using TD-error, we can update at each step.

```
if args.train:
    all_episode_reward = []
    for episode in range(TRAIN_EPISODES):
        # reset the environment
        state = env.reset().astype(np.float32)
        step = 0 # number of step in this episode
        episode_reward = 0 # rewards of all steps
        while True:
            if RENDER: env.render()
            # choose actions and interact with the environment
            action = actor.get_action(state)
            state_new, reward, done, info = env.step(action)
            state_new = state_new.astype(np.float32)

            if done: reward = -20 # reward shaping trick
            episode_reward += reward

            # update models at each step after interaction with
                the environment
            td_error = critic.learn(state, reward, state_new,
                done)
            actor.learn(state, action, td_error)
```

```
            state = state_new
            step += 1

            # run until the environment returns done or reach
                max step limit
            if done or step >= MAX_STEPS:
                break
```

The plot and test part is just the same as code in policy gradient code block.

5.10.4 A3C: BipedalWalker-v2

There are a global actor-critic and several workers in A3C. The global actor-critic's role is to update the network with the data collected by workers. Each worker has its own AC network, which interacts with the environment and pushes the collected data to Global AC, then pulls the latest network weights from Global AC to and substitutes its weights. The worker class structure is showed as follows:

```
class Worker(object):
    def __init__(self, name): # initialize worker
        ...
    def work(self, globalAC): # worker main function
        ...
```

As described above, each worker has its own actor and critic net. So in the initialization function, we created the model by instantiating the ACNet class.

```
class Worker(object):
    def __init__(self, name):
        self.env = gym.make(GAME)
        self.name = name
        self.AC = ACNet(name)
```

The work is the main function of the worker. It is similar to the main loop in the previous code, but it is different in the update part. As usual, the main content of the loop is to get actions from the agent and interact with the environment.

```
    def work(self, globalAC):
        global GLOBAL_RUNNING_R, GLOBAL_EP
        total_step = 1
        buffer_s, buffer_a, buffer_r = [], [], []

        while not COORD.should_stop() and GLOBAL_EP <
            MAX_GLOBAL_EP:

            # reset environment
```

```
        s = self.env.reset()
        ep_r = 0

        while True:
            # visualize Worker_0 during training
            if self.name == 'Worker_0' and total_step % 30 == 0:
                self.env.render()

            # choose actions and interact with the environment
            s = s.astype('float32') # double to float
            a = self.AC.choose_action(s)
            s_, r, done, _info = self.env.step(a)
            s_ = s_.astype('float32') # double to float

            # set robot falls reward to -2 instead of -100
            if r == -100: r = -2

            ep_r += r

            # store transitions
            buffer_s.append(s)
            buffer_a.append(a)
            buffer_r.append(r)
```

But when the agent gets enough data and starts to update, it updates the global networks. After that, parameters of local networks will be replaced by the latest updated global networks parameters.

```
            if total_step % UPDATE_GLOBAL_ITER == 0 or done: #
                update global and assign to local net
                if done:
                    v_s_ = 0 # terminal
                else:
                    v_s_ = self.AC.critic(s_[np.newaxis, :])[0,0]
                        # reduce dim from 2 to 0

                # discounted reward
                buffer_v_target = []
                for r in buffer_r[::-1]: # reverse buffer r
                    v_s_ = r + GAMMA * v_s_
                    buffer_v_target.append(v_s_)
                buffer_v_target.reverse()
                buffer_s =
                     tf.convert_to_tensor(np.vstack(buffer_s))
                buffer_a =
                     tf.convert_to_tensor(np.vstack(buffer_a))
                buffer_v_target = tf.convert_to_tensor
                     (np.vstack(buffer_v_target).astype('float32'))

                # update gradients on global network
                self.AC.update_global(buffer_s, buffer_a,
                     buffer_v_target.astype('float32'), globalAC)
                buffer_s, buffer_a, buffer_r = [], [], []
```

```
                # update local network from global network
                self.AC.pull_global(globalAC)

            s = s_
            total_step += 1
            if done:
                if len(GLOBAL_RUNNING_R) == 0: # record running
                     episode reward
                    GLOBAL_RUNNING_R.append(ep_r)
                else: # moving average
                    GLOBAL_RUNNING_R.append(0.95 *
                        GLOBAL_RUNNING_R[-1] + 0.05 * ep_r)

                print('Training | {}, Episode: {}/{} | Episode
                     Reward: {:.4f} | Running Time: {:.4f}'\
                .format(self.name, GLOBAL_EP, MAX_GLOBAL_EP,
                     ep_r, time.time()-T0 ))
                GLOBAL_EP += 1
                break
```

The ACNet class used in the above is a simple class that includes actors and critics. Its structure can be shown as follows:

```
class ACNet(object):
    def __init__(self, scope):
        ...
    def update_global(self, buffer_s, buffer_a, buffer_v_target,
         globalAC):
        ...
    def pull_global(self, globalAC): # run by a local, pull
         weights from the global nets
        ...
    def get_action(self, s, greedy=False): # run by a local
        ...
    def save(self): # save trained weights
        ...
    def load(self): # load trained weights
        ...
```

The `update_global()` function is the most important function. As we can see, the function uses sampled data to calculate the gradients but applied to the global networks. After that, update the data from global net and start the loop again. In this way, multiple workers are updated asynchronously.

```
    def update_global(
            self, buffer_s, buffer_a, buffer_v_target, globalAC
    ): # refer to the global Actor-Crtic network for updating it
         with samples
        ''' update the global critic '''
        with tf.GradientTape() as tape:
            self.v = self.critic(buffer_s)
```

```
        self.v_target = buffer_v_target
        td = tf.subtract(self.v_target, self.v, name='TD_error')
        self.c_loss = tf.reduce_mean(tf.square(td))
    self.c_grads = tape.gradient(self.c_loss,
        self.critic.trainable_weights)
    OPT_C.apply_gradients(zip(self.c_grads,
        globalAC.critic.trainable_weights)) # local grads
        applies to global net
    # del tape # Drop the reference to the tape
    ''' update the global actor '''
    with tf.GradientTape() as tape:
        self.mu, self.sigma = self.actor(buffer_s)
        self.test = self.sigma[0]
        self.mu, self.sigma = self.mu * A_BOUND[1], self.sigma
            + 1e-5

        normal_dist = tfd.Normal(self.mu, self.sigma) # no
            tf.contrib for tf2.0
        self.a_his = buffer_a # float32
        log_prob = normal_dist.log_prob(self.a_his)
        exp_v = log_prob * td # td is from the critic part, no
            gradients for it
        entropy = normal_dist.entropy() # encourage exploration
        self.exp_v = ENTROPY_BETA * entropy + exp_v
        self.a_loss = tf.reduce_mean(-self.exp_v)
    self.a_grads = tape.gradient(self.a_loss,
        self.actor.trainable_weights)
    OPT_A.apply_gradients(zip(self.a_grads,
        globalAC.actor.trainable_weights)) # local grads
        applies to global net
    return self.test # for test purpose
```

The function of updating the local network is very simple, just simply replace the parameters of local net with the parameters of global net.

```
def pull_global(self, globalAC): # run by a local, pull
    weights from the global nets
    for l_p, g_p in zip(self.actor.trainable_weights,
        globalAC.actor.trainable_weights):
        l_p.assign(g_p)
    for l_p, g_p in zip(self.critic.trainable_weights,
        globalAC.critic.trainable_weights):
        l_p.assign(g_p)
```

Finally, start the threads in turn in the main function.

```
env = gym.make(GAME)
N_S = env.observation_space.shape[0]
N_A = env.action_space.shape[0]

A_BOUND = [env.action_space.low, env.action_space.high]
A_BOUND[0] = A_BOUND[0].reshape(1, N_A)
A_BOUND[1] = A_BOUND[1].reshape(1, N_A)
```

```
with tf.device("/cpu:0"):
    GLOBAL_AC = ACNet(GLOBAL_NET_SCOPE) # we only need its
        params

T0 = time.time()
if args.train:
    with tf.device("/cpu:0"):
        OPT_A = tf.optimizers.RMSprop(LR_A, name='RMSPropA')
        OPT_C = tf.optimizers.RMSprop(LR_C, name='RMSPropC')

        workers = []
        # Create worker
        for i in range(N_WORKERS):
            i_name = 'Worker_%i' % i # worker name
            workers.append(Worker(i_name, GLOBAL_AC))

    COORD = tf.train.Coordinator()

    # start TF threading
    worker_threads = []
    for worker in workers:
        # t = threading.Thread(target=worker.work)
        job = lambda: worker.work(GLOBAL_AC)
        t = threading.Thread(target=job)
        t.start()
        worker_threads.append(t)
    COORD.join(worker_threads)

    GLOBAL_AC.save()
    plt.plot(GLOBAL_RUNNING_R)
    if not os.path.exists('image'):
        os.makedirs('image')
    plt.savefig(os.path.join('image', 'a3c.png'))
```

5.10.5 TRPO: Pendulum-v0

TRPO uses the trust region method to take the largest steps allowed under KL-divergence constraints to improve the strategy. The example uses generalized advantage estimator as well. Let us first see how `GAE_Buffer` is implemented.

```
class GAE_Buffer:
    def __init__(self, obs_dim, act_dim, size, gamma=0.99,
        lam=0.95): # initialize buffer
        ...
    def store(self, obs, act, rew, val, logp, mean, log_std): #
        store datas into the buffer
        ...
    def finish_path(self, last_val=0): # compute advantage
        estimates with GAE-Lambda
```

```
    ...
def _discount_cumsum(self, x, discount): # compute discounted
    cumulative sums
    ...
def is_full(self): # returns whether the buffer is full or not
    ...
def get(self): # get all data from the buffer
    ...
```

We create buffers and variables that will be used later in the initialization function.

```
class GAE_Buffer:
    def __init__(self, obs_dim, act_dim, size, gamma=0.99,
        lam=0.95):
        self.obs_buf = np.zeros((size, obs_dim), dtype=np.float32)
        self.act_buf = np.zeros((size, act_dim), dtype=np.float32)
        self.adv_buf = np.zeros(size, dtype=np.float32)
        self.rew_buf = np.zeros(size, dtype=np.float32)
        self.ret_buf = np.zeros(size, dtype=np.float32)
        self.val_buf = np.zeros(size, dtype=np.float32)
        self.logp_buf = np.zeros(size, dtype=np.float32)
        self.mean_buf = np.zeros(size, dtype=np.float32)
        self.log_std_buf = np.zeros(size, dtype=np.float32)
        self.gamma, self.lam = gamma, lam
        self.ptr, self.path_start_idx, self.max_size = 0, 0, size
```

In the `store()` function, we store the data in the corresponding buffer and then move the pointer.

```
    def store(self, obs, act, rew, val, logp, mean, log_std):
        assert self.ptr < self.max_size # buffer has to have room
            so you can store
        self.obs_buf[self.ptr] = obs
        self.act_buf[self.ptr] = act
        self.rew_buf[self.ptr] = rew
        self.val_buf[self.ptr] = val
        self.logp_buf[self.ptr] = logp
        self.mean_buf[self.ptr] = mean
        self.log_std_buf[self.ptr] = log_std
        self.ptr += 1
```

The `finish_path()` function is called at the end of a trajectory or when an epoch ends. It extracts the current trajectory to calculate GAE-Lambda advantage and rewards-to-go which will be targets for the value function.

```
    def finish_path(self, last_val=0):
        path_slice = slice(self.path_start_idx, self.ptr)
        rews = np.append(self.rew_buf[path_slice], last_val)
        vals = np.append(self.val_buf[path_slice], last_val)
        # the next two lines implement GAE-Lambda advantage
            calculation
```

```
    deltas = rews[:-1] + self.gamma * vals[1:] - vals[:-1]
    self.adv_buf[path_slice] = self._discount_cumsum(deltas,
        self.gamma * self.lam)

    # the next line computes rewards-to-go, to be targets for
        the value function
    self.ret_buf[path_slice] = self._discount_cumsum(rews,
        self.gamma)[:-1]

    self.path_start_idx = self.ptr
```

The `_discount_cumsum()` function we used in the previous function is shown below. We used SciPy(an open source library) build-in function to do this.

```
def _discount_cumsum(self, x, discount):
    return scipy.signal.lfilter([1], [1, float(-discount)],
        x[::-1], axis=0)[::-1]
```

The `is_full()` function simply checks whether the pointer reaches the end.

```
def is_full(self):
    return self.ptr == self.max_size
```

When the buffer is full, we fetch the data and reset the pointer. Here we use the advantage normalization trick.

```
def get(self):
    assert self.ptr == self.max_size # buffer has to be full
        before you can get
    self.ptr, self.path_start_idx = 0, 0

    # the next two lines implement the advantage normalization
        trick
    adv_mean, adv_std = np.mean(self.adv_buf),
        np.std(self.adv_buf)
    self.adv_buf = (self.adv_buf - adv_mean) / adv_std
    return [self.obs_buf, self.act_buf, self.adv_buf,
        self.ret_buf, self.logp_buf, self.mean_buf,
        self.log_std_buf]
```

Next we will describe the `TRPO` class. Its structure is shown below.

```
class TRPO:
    def __init__(self, state_dim, action_dim, action_bound): #
        create networks, optimizers and variables.
        ...
    def get_action(self, state, greedy=False): # get action and
        other values
        ...
    def pi_loss(self, states, actions, adv, old_log_prob): #
        calculate policy loss
        ...
```

```
    def gradient(self, states, actions, adv, old_log_prob): #
        calculate the gradient of the policy network
        ...
    def train_vf(self, states, rewards_to_go): # train value
        function
        ...
    def kl(self, states, old_mean, old_log_std): # calculate
        kl-divergence
        ...
    def _flat_concat(self, xs): # flatten variables
        ...
    def get_pi_params(self): # get parameters of the policy
        network
        ...
    def set_pi_params(self, flat_params): # set parameters of the
        policy network
        ...
    def save(self): # save parameters
        ...
    def load(self): # load parameters
        ...
    def cg(self, Ax, b): # conjugate gradient algorithm
        ...
    def hvp(self, states, old_mean, old_log_std, x): # calculate
        Hessian-vector product
        ...
    def update(self): # update all networks
        ...
    def finish_path(self, done, next_state): # finish a trajectory
        ...
```

As usual, we set up the network, optimizer, and other variables in the initialization function. Here the policy network only outputs the mean value of each action dimension. We use a single variable as the log std value to act on all action dimensions.

```
class TRPO:
    def __init__(self, state_dim, action_dim, action_bound):
        # critic
        with tf.name_scope('critic'):
            layer = input_layer = tl.layers.Input([None,
                state_dim], tf.float32)
            for d in HIDDEN_SIZES:
                layer = tl.layers.Dense(d, tf.nn.relu)(layer)
            v = tl.layers.Dense(1)(layer)
        self.critic = tl.models.Model(input_layer, v)
        self.critic.train()

        # actor
        with tf.name_scope('actor'):
            layer = input_layer = tl.layers.Input([None,
                state_dim], tf.float32)
            for d in HIDDEN_SIZES:
```

```
            layer = tl.layers.Dense(d, tf.nn.relu)(layer)
        mean = tl.layers.Dense(action_dim, tf.nn.tanh)(layer)
        mean = tl.layers.Lambda(lambda x: x *
            action_bound)(mean)
        log_std = tf.Variable(np.zeros(action_dim,
            dtype=np.float32))

    self.actor = tl.models.Model(input_layer, mean)
    self.actor.trainable_weights.append(log_std)
    self.actor.log_std = log_std
    self.actor.train()

    self.buf = GAE_Buffer(state_dim, action_dim, BATCH_SIZE,
        GAMMA, LAM)
    self.critic_optimizer =
        tf.optimizers.Adam(learning_rate=VF_LR)
    self.action_bound = action_bound
```

With the network, we can use the following function to get the action corresponding to the state. Besides, we need to calculate some additional data to store in the GAE buffer.

```
def get_action(self, state, greedy=False):
    state = np.array([state], np.float32)
    mean = self.actor(state)
    log_std = tf.convert_to_tensor(self.actor.log_std)
    std = tf.exp(log_std)
    std = tf.ones_like(mean) * std
    pi = tfp.distributions.Normal(mean, std)

    if greedy:
        action = mean
    else:
        action = pi.sample()
    action = np.clip(action, -self.action_bound,
        self.action_bound)
    logp_pi = pi.log_prob(action)

    value = self.critic(state)
    return action[0], value, logp_pi, mean, log_std
```

The following code shows how policy loss is calculated. We firstly calculate surrogate advantage, a measure of how current policy performs relative to the old policy using data from the old policy. And then use the negative surrogate advantage as policy loss.

```
def pi_loss(self, states, actions, adv, old_log_prob):
    mean = self.actor(states)
    pi = tfp.distributions.Normal(mean,
        tf.exp(self.actor.log_std))
    log_prob = pi.log_prob(actions)[:, 0]
    ratio = tf.exp(log_prob - old_log_prob)
```

```
        surr = tf.reduce_mean(ratio * adv)
        return -surr
```

By calling the previously defined `pi_loss()` function, we can easily calculate the gradient.

```
    def gradient(self, states, actions, adv, old_log_prob):
        pi_params = self.actor.trainable_weights
        with tf.GradientTape() as tape:
            loss = self.pi_loss(states, actions, adv, old_log_prob)
        grad = tape.gradient(loss, pi_params)
        gradient = self._flat_concat(grad)
        return gradient, loss
```

The way to train the value network is shown below. Just fit value function by regression on mean-squared error

```
    def train_vf(self, states, rewards_to_go):
        with tf.GradientTape() as tape:
            value = self.critic(states)
            loss = tf.reduce_mean((rewards_to_go - value[:, 0]) **
                2)
        grad = tape.gradient(loss, self.critic.trainable_weights)
        self.critic_optimizer.apply_gradients(zip(grad,
            self.critic.trainable_weights))
```

The way to calculate the KL-divergence is as follows. We first generate action distributions based on the mean and std values, and then calculate the KL-divergence.

```
    def kl(self, states, old_mean, old_log_std):
        old_mean = old_mean[:, np.newaxis]
        old_log_std = old_log_std[:, np.newaxis]
        old_std = tf.exp(old_log_std)
        old_pi = tfp.distributions.Normal(old_mean, old_std)

        mean = self.actor(states)
        std = tf.exp(self.actor.log_std)*tf.ones_like(mean)
        pi = tfp.distributions.Normal(mean, std)

        kl = tfp.distributions.kl_divergence(pi, old_pi)
        all_kls = tf.reduce_sum(kl, axis=1)
        return tf.reduce_mean(all_kls)
```

In this example code, many parameters are flattened by the `_flat_concat()` function. In this way we can simplify many calculations.

```
    def _flat_concat(self, xs):
        return tf.concat([tf.reshape(x, (-1,)) for x in xs],
            axis=0)
```

When using the `_flat_concat()` function, the process of get and set parameters needs some simple processing.

```
def get_pi_params(self):
    pi_params = self.actor.trainable_weights
    return self._flat_concat(pi_params)

def set_pi_params(self, flat_params):
    pi_params = self.actor.trainable_weights
    flat_size = lambda p: int(np.prod(p.shape.as_list())) #
        the 'int' is important for scalars
    splits = tf.split(flat_params, [flat_size(p) for p in
        pi_params])
    new_params = [tf.reshape(p_new, p.shape) for p, p_new in
        zip(pi_params, splits)]
    return tf.group([p.assign(p_new) for p, p_new in
        zip(pi_params, new_params)])
```

The save and load functions are the same as before.

```
def save(self):
    path = os.path.join('model', 'trpo')
    if not os.path.exists(path):
        os.makedirs(path)
    tl.files.save_weights_to_hdf5(os.path.join(path,
        'actor.hdf5'), self.actor)
    tl.files.save_weights_to_hdf5(os.path.join(path,
        'critic.hdf5'), self.critic)

def load(self):
    path = os.path.join('model', 'trpo')
    tl.files.load_hdf5_to_weights_in_order(os.path.join(path,
        'actor.hdf5'), self.actor)
    tl.files.load_hdf5_to_weights_in_order(os.path.join(path,
        'critic.hdf5'), self.critic)
```

The following code implements the conjugate gradient algorithm.[7] By using this function can compute the matrix-vector product instead of computing and storing the whole matrix directly.

```
def cg(self, Ax, b):
    x = np.zeros_like(b)
    r = copy.deepcopy(b) # Note: should be 'b - Ax(x)', but
        for x=0, Ax(x)=0. Change if doing warm start.
    p = copy.deepcopy(r)
    r_dot_old = np.dot(r, r)
    for _ in range(CG_ITERS):
        z = Ax(p)
        alpha = r_dot_old / (np.dot(p, z) + EPS)
```

[7] https://en.wikipedia.org/wiki/Conjugate_gradient_method.

```
        x += alpha * p
        r -= alpha * z
        r_dot_new = np.dot(r, r)
        p = r + (r_dot_new / r_dot_old) * p
        r_dot_old = r_dot_new
    return x
```

The following code shows the process of calculating Hessian-vector product by using the $Hx = \nabla_\theta \left(\left(\nabla_\theta \bar{D}_{KL}(\theta || \theta_k) \right)^T x \right)$ formula. Damping coefficient is used here to change calculation $Hx \rightarrow (\alpha I + H)x$ for numerical stability.

```
def hvp(self, states, old_mean, old_log_std, x):
    pi_params = self.actor.trainable_weights
    with tf.GradientTape() as tape1:
        with tf.GradientTape() as tape0:
            d_kl = self.kl(states, old_mean, old_log_std)
        g = self._flat_concat(tape0.gradient(d_kl, pi_params))
        l = tf.reduce_sum(g * x)
    hvp = self._flat_concat(tape1.gradient(l, pi_params))

    if DAMPING_COEFF > 0:
        hvp += DAMPING_COEFF * x
    return hvp
```

With the above preparations, we can finally start to update. Firstly, we sample data from the GAE buffer and calculate the gradient and loss. Next, we use the conjugate gradient algorithm to calculate variable x which corresponds to $\hat{x}_k$ in the formula $\hat{x}_k \approx \hat{H}_k^{-1} \hat{g}_k$. Then, we calculate the $\sqrt{\frac{2\delta}{\hat{x}_k^T \hat{H}_k \hat{x}_k}}$ part in the formula $\theta_{k+1} = \theta_k + \alpha^j \sqrt{\frac{2\delta}{\hat{x}_k^T \hat{H}_k \hat{x}_k}} \hat{x}_k$. After that, use the backtracking line search to update the policy net. And finally, update value function by MSE loss.

```
def update(self):
    states, actions, adv, rewards_to_go, logp_old_ph, old_mu,
        old_log_std = self.buf.get()
    g, pi_l_old = self.gradient(states, actions, adv,
        logp_old_ph)

    Hx = lambda x: self.hvp(states, old_mu, old_log_std, x)
    x = self.cg(Hx, g)

    alpha = np.sqrt(2 * DELTA / (np.dot(x, Hx(x)) + EPS))
    old_params = self.get_pi_params()

    def set_and_eval(step):
        params = old_params - alpha * x * step
        self.set_pi_params(params)
        d_kl = self.kl(states, old_mu, old_log_std)
        loss = self.pi_loss(states, actions, adv, logp_old_ph)
```

```
        return [d_kl, loss]

    # trpo with backtracking line search, hard kl
    for j in range(BACKTRACK_ITERS):
        kl, pi_l_new = set_and_eval(step=BACKTRACK_COEFF ** j)
        if kl <= DELTA and pi_l_new <= pi_l_old:
            # Accepting new params at step of line search
            break
    else:
        # Line search failed! Keeping old params.
        set_and_eval(step=0.)

    # Value function updates
    for _ in range(TRAIN_V_ITERS):
        self.train_vf(states, rewards_to_go)
```

A `finish_path()` function is also needed when cutting off a trajectory or when an epoch is over. If the trajectory finished because the agent reached a terminal state, the last value should be 0.

```
def finish_path(self, done, next_state):
    if not done:
        next_state = np.array([next_state], np.float32)
        last_val = self.critic(next_state)
    else:
        last_val = 0
    self.buf.finish_path(last_val)
```

The main loop of the code is shown below. We create the environment, agent, and some useful variable at first.

```
env = gym.make(ENV_ID).unwrapped

# reproducible
np.random.seed(RANDOM_SEED)
tf.random.set_seed(RANDOM_SEED)
env.seed(RANDOM_SEED)

state_dim = env.observation_space.shape[0]
action_dim = env.action_space.shape[0]
action_bound = env.action_space.high

agent = TRPO(state_dim, action_dim, action_bound)
t0 = time.time()
```

In the training mode, we store the data generated by the interaction between the agent and the environment into the buffer, and train it once the buffer is full.

```
if args.train: # train
    all_episode_reward = []
    for episode in range(TRAIN_EPISODES):
        state = env.reset()
```

```
        state = np.array(state, np.float32)
        episode_reward = 0
        for step in range(MAX_STEPS):
            if RENDER:
                env.render()
            action, value, logp, mean, log_std =
                agent.get_action(state)
            next_state, reward, done, _ = env.step(action)
            next_state = np.array(next_state, np.float32)
            agent.buf.store(state, action, reward, value, logp,
                mean, log_std)
            episode_reward += reward
            state = next_state
            if agent.buf.is_full():
                agent.finish_path(done, next_state)
                agent.update()
            if done:
                break
        agent.finish_path(done, next_state)
        if episode == 0:
            all_episode_reward.append(episode_reward)
        else:
            all_episode_reward.append(all_episode_reward[-1] *
                0.9 + episode_reward * 0.1)
        print(
            'Training | Episode: {}/{} | Episode Reward: {:.4f}
                | Running Time: {:.4f}'.format(
                episode+1, TRAIN_EPISODES, episode_reward,
                time.time() - t0
            )
        )
        if episode % SAVE_FREQ == 0:
            agent.save()
    agent.save()
```

Then add plot code to facilitate the observation of the training process

```
    plt.plot(all_episode_reward)
    if not os.path.exists('image'):
        os.makedirs('image')
    plt.savefig(os.path.join('image', 'trpo.png'))
```

After training is complete, we can start testing the model

```
if args.test:
    # test
    agent.load()
    for episode in range(TEST_EPISODES):
        state = env.reset()
        episode_reward = 0
        for step in range(MAX_STEPS):
            env.render()
            action, *_ = agent.get_action(state, greedy=True)
```

```
            state, reward, done, info = env.step(action)
            episode_reward += reward
            if done:
                break
        print(
            'Testing | Episode: {}/{} | Episode Reward: {:.4f} |
             Running Time: {:.4f}'.format(
                episode + 1, TEST_EPISODES, episode_reward,
                time.time() - t0))
```

5.10.6 PPO: Pendulum-v0

PPO is a family of first-order methods, which is different from TRPO's second-order method.

In PPO-Penalty, the objective function has a KL-divergence penalty term to solve a KL-constrained update like TRPO. The structure of the PPO class is shown as follows:

```
class PPO(object):
    def __init__(self, state_dim, action_dim, action_bound,
        method='clip'): # initialization function, create
        actor_old, actor and critic.
       ...
    def train_actor(self, state, action, adv, old_pi): # actor
        training function
       ...
    def train_critic(self, reward, state): # critic training
        function
       ...
    def update(self): # the main training function
       ...
    def get_action(self, s, greedy=False): # choose action
       ...
    def save(self): # save networks
       ...
    def load(self): # load networks
       ...
    def store_transition(self, state, action, reward): # Store
        state, action, reward at each step
       ...
    def finish_path(self, next_state): # Calculate cumulative
        reward
       ...
```

In the PPO algorithm, we build both actor and critic in the initialization function. PPO has two method: PPO-Penalty and PPO-Clip. We can set the corresponding parameters according to different methods. Since the environment is continuous motion control, we use the random stochastic policy network to output mean and log standard deviation to represent the action distribution. In addition, we used a lambda layer here to multiply mean value by 2 because the action range of the 'Pendulum-v0' environment is $[-2, 2]$.

```
class PPO(object):
  def __init__(self, state_dim, action_dim, action_bound,
      method='clip'):
    # critic
    with tf.name_scope('critic'):
      inputs = tl.layers.Input([None, state_dim], tf.float32,
          'state')
      layer = tl.layers.Dense(64, tf.nn.relu)(inputs)
      layer = tl.layers.Dense(64, tf.nn.relu)(layer)
      v = tl.layers.Dense(1)(layer)
    self.critic = tl.models.Model(inputs, v)
    self.critic.train()

    # actor
    with tf.name_scope('actor'):
      inputs = tl.layers.Input([None, state_dim], tf.float32,
          'state')
      layer = tl.layers.Dense(64, tf.nn.relu)(inputs)
      layer = tl.layers.Dense(64, tf.nn.relu)(layer)
      a = tl.layers.Dense(action_dim, tf.nn.tanh)(layer)
      mean = tl.layers.Lambda(lambda x: x * action_bound,
          name='lambda')(a)
      logstd = tf.Variable(np.zeros(action_dim,
          dtype=np.float32))
    self.actor = tl.models.Model(inputs, mean)
    self.actor.trainable_weights.append(logstd)
    self.actor.logstd = logstd
    self.actor.train()
    self.actor_opt = tf.optimizers.Adam(LR_A)
    self.critic_opt = tf.optimizers.Adam(LR_C)

    self.method = method
    if method == 'penalty':
      self.kl_target = KL_TARGET
      self.lam = LAM
    elif method == 'clip':
      self.epsilon = EPSILON

    self.state_buffer, self.action_buffer = [], []
    self.reward_buffer, self.cumulative_reward_buffer = [], []
    self.action_bound = action_bound
```

The actor training function updates actor by using PPO method. PPO uses specialized objective function to prevent the new strategy to get far from the old one.

```
def train_actor(self, state, action, adv, old_pi):
    with tf.GradientTape() as tape:
        mean, std = self.actor(state), tf.exp(self.actor.logstd)
        pi = tfp.distributions.Normal(mean, std)

        ratio = tf.exp(pi.log_prob(action) -
            old_pi.log_prob(action))
        surr = ratio * adv
        if self.method == 'penalty': # ppo penalty
            kl = tfp.distributions.kl_divergence(old_pi, pi)
            kl_mean = tf.reduce_mean(kl)
            aloss = -(tf.reduce_mean(surr - self.lam * kl))
        else: # ppo clip
            aloss = -tf.reduce_mean(
                tf.minimum(surr,
                        tf.clip_by_value(ratio, 1. -
                            self.epsilon, 1. + self.epsilon) *
                            adv)
            )
    a_gard = tape.gradient(aloss, self.actor.trainable_weights)
    self.actor_opt.apply_gradients(zip(a_gard,
         self.actor.trainable_weights))

    if self.method == 'kl_pen':
        return kl_mean
```

The critic training function update critic is shown as follows. Just calculate the advantage and minimize the loss function $\sum_t \hat{A}_t^2$.

```
def train_critic(self, reward, state):
    reward = np.array(reward, dtype=np.float32)
    with tf.GradientTape() as tape:
        advantage = reward - self.critic(state)
        loss = tf.reduce_mean(tf.square(advantage))
    grad = tape.gradient(loss, self.critic.trainable_weights)
    self.critic_opt.apply_gradients(zip(grad,
         self.critic.trainable_weights))
```

In the `update()` function, we first calculate the old pi distribution and then perform the update step. If we use the PPO-Penalty method, we also need to update the lambda value according to the KL-divergence after updating the actor.

```
def update(self):
    s = np.array(self.state_buffer, np.float32)
    a = np.array(self.action_buffer, np.float32)
    r = np.array(self.cumulative_reward_buffer, np.float32)
    mean, std = self.actor(s), tf.exp(self.actor.logstd)
    pi = tfp.distributions.Normal(mean, std)
    adv = r - self.critic(s)

    # update actor
```

```
        if self.method == 'kl_pen':
            for _ in range(A_UPDATE_STEPS):
                kl = self.a_train(s, a, adv, pi)
            if kl < self.kl_target / 1.5:
                self.lam /= 2
            elif kl > self.kl_target * 1.5:
                self.lam *= 2
        else:
            for _ in range(A_UPDATE_STEPS):
                self.a_train(s, a, adv, pi)

        # update critic
        for _ in range(C_UPDATE_STEPS):
            self.c_train(r, s)

        self.state_buffer.clear()
        self.action_buffer.clear()
        self.cumulative_reward_buffer.clear()
        self.reward_buffer.clear()
```

The `get_action()` function feeds the current state into the network to obtain the mean and standard deviation of the current action distribution. With this action distribution, we can sample actions.

```
    def get_action(self, s, greedy=False):
        state = state[np.newaxis, :].astype(np.float32)
        mean, std = self.actor(state), tf.exp(self.actor.logstd)
        if greedy:
            action = mean[0]
        else:
            pi = tfp.distributions.Normal(mean, std)
            action = tf.squeeze(pi.sample(1), axis=0)[0] # choosing
                action
        return np.clip(action, -self.action_bound,
            self.action_bound)
```

The `save()`, `load()`, `store_transition()` functions are similar to the previous codes and will not be explained here. The `finish_path()` function is used to calculate cumulative reward when the game ends or a batch of data is collected.

```
    def finish_path(self, next_state, done):
        if done:
            v_s_ = 0
        else:
            v_s_ = self.critic(np.array([next_state],
                np.float32))[0, 0]
        discounted_r = []
        for r in self.reward_buffer[::-1]:
            v_s_ = r + GAMMA * v_s_
            discounted_r.append(v_s_)
        discounted_r.reverse()
```

```
        discounted_r = np.array(discounted_r)[:, np.newaxis]
        self.cumulative_reward_buffer.extend(discounted_r)
        self.reward_buffer.clear()
```

The main function is quite similar. Firstly, create environment and PPO agent.

```
    env = gym.make(ENV_ID).unwrapped

    # reproducible
    env.seed(RANDOM_SEED)
    np.random.seed(RANDOM_SEED)
    tf.random.set_seed(RANDOM_SEED)

    state_dim = env.observation_space.shape[0]
    action_dim = env.action_space.shape[0]
    action_bound = env.action_space.high

    agent = PPO(state_dim, action_dim, action_bound)
    t0 = time.time()
```

Then use the agent to interact with the environment and store data. After a game ends or collected enough data, run `finish_path()` function to calculate cumulative reward. Update the agent when a batch of data is collected. After many studies, the agent can play the game very well.

```
    if args.train:
        all_episode_reward = []
        for episode in range(TRAIN_EPISODES):
            state = env.reset()
            episode_reward = 0
            for step in range(MAX_STEPS): # in one episode
                if RENDER:
                    env.render()
                action = agent.get_action(state)
                state_, reward, done, info = env.step(action)
                agent.store_transition(state, action, reward)
                state = state_
                episode_reward += reward

                # update ppo
                if len(agent.state_buffer) >= BATCH_SIZE:
                    agent.finish_path(state_, done)
                    agent.update()
                if done:
                    break
            agent.finish_path(state_, done)
            print(
                'Training | Episode: {}/{} | Episode Reward: {:.4f}
                    | Running Time: {:.4f}'.format(
                    episode + 1, TRAIN_EPISODES, episode_reward,
                        time.time() - t0)
            )
```

```
        if episode == 0:
            all_episode_reward.append(episode_reward)
        else:
            all_episode_reward.append(all_episode_reward[-1] *
                0.9 + episode_reward * 0.1)

    agent.save()
    plt.plot(all_episode_reward)
    if not os.path.exists('image'):
        os.makedirs('image')
    plt.savefig(os.path.join('image', 'ppo.png'))
```

Finally, test the agent as usual.

```
if args.test:
    # test
    agent.load()
    for episode in range(TEST_EPISODES):
        state = env.reset()
        episode_reward = 0
        for step in range(MAX_STEPS):
            env.render()
            state, reward, done, info =
                env.step(agent.get_action(state, greedy=True))
            episode_reward += reward
            if done:
                break
        print(
            'Testing | Episode: {}/{} | Episode Reward: {:.4f} |
                Running Time: {:.4f}'.format(
                episode + 1, TEST_EPISODES, episode_reward,
                time.time() - t0))
```

References

Amari SI (1998) Natural gradient works efficiently in learning. Neural Comput 10(2):251–276

Goodfellow I, Pouget-Abadie J, Mirza M, Xu B, Warde-Farley D, Ozair S, Courville A, Bengio Y (2014) Generative adversarial nets. In: Proceedings of the neural information processing systems (Advances in neural information processing systems) conference

Grosse R, Martens J (2016) A Kronecker-factored approximate fisher matrix for convolution layers. In: International conference on machine learning (ICML), pp 573–582

Heess N, Sriram S, Lemmon J, Merel J, Wayne G, Tassa Y, Erez T, Wang Z, Eslami S, Riedmiller M, et al (2017) Emergence of locomotion behaviours in rich environments. arXiv:170702286

Kakade S, Langford J (2002) Approximately optimal approximate reinforcement learning. In: Proceedings of the international conference on machine learning (ICML), vol 2, pp 267–274

Konda VR, Tsitsiklis JN (2000) Actor-critic algorithms. In: Advances in neural information processing systems, pp 1008–1014

Li J, Wang B (2018) Policy optimization with second-order advantage information. arXiv:180503586

Liu H, Feng Y, Mao Y, Zhou D, Peng J, Liu Q (2017) Action-dependent control variates for policy optimization via stein's identity. arXiv:171011198

Martens J, Grosse R (2015) Optimizing neural networks with Kronecker-factored approximate curvature. In: International conference on machine learning (ICML), pp 2408–2417

Mitliagkas I, Zhang C, Hadjis S, Ré C (2016) Asynchrony begets momentum, with an application to deep learning. In: 2016 54th annual Allerton conference on communication, control, and computing (Allerton). IEEE, Piscataway, pp 997–1004

Mnih V, Badia AP, Mirza M, Graves A, Lillicrap T, Harley T, Silver D, Kavukcuoglu K (2016) Asynchronous methods for deep reinforcement learning. In: International Conference on Machine Learning (ICML), pp 1928–1937

Schulman J, Levine S, Abbeel P, Jordan M, Moritz P (2015) Trust region policy optimization. In: International conference on machine learning (ICML), pp 1889–1897

Schulman J, Wolski F, Dhariwal P, Radford A, Klimov O (2017) Proximal policy optimization algorithms. arXiv:170706347

Sutton RS, McAllester DA, Singh SP, Mansour Y (2000) Policy gradient methods for reinforcement learning with function approximation. In: Advances in neural information processing systems, pp 1057–1063

Wu Y, Mansimov E, Grosse RB, Liao S, Ba J (2017) Scalable trust-region method for deep reinforcement learning using Kronecker-factored approximation. In: Advances in neural information processing systems, pp 5279–5288

Wu C, Rajeswaran A, Duan Y, Kumar V, Bayen AM, Kakade S, Mordatch I, Abbeel P (2018) Variance reduction for policy gradient with action-dependent factorized baselines. arXiv:180307246

Chapter 6
Combine Deep Q-Networks with Actor-Critic

Hongming Zhang, Tianyang Yu, and Ruitong Huang

Abstract The deep Q-network algorithm is one of the most well-known deep reinforcement learning algorithms, which combines reinforcement learning with deep neural networks to approximate the optimal action-value functions. It receives only the pixels as inputs and achieves human-level performance on Atari games. Actor-critic methods transform the Monte Carlo update of the REINFORCE algorithm into the temporal-difference update for learning the policy parameters. Recently, some algorithms that combine deep Q-networks with actor-critic methods such as the deep deterministic policy gradient algorithm are very popular. These algorithms take advantages of both methods and perform well in most environments especially with continuous action spaces. In this chapter, we give a brief introduction of the advantages and disadvantages of each kind of method, then introduce some classical algorithms that combine deep Q-networks and actor-critic like the deep deterministic policy gradient algorithm, the twin delayed deep deterministic policy gradient algorithm, and the soft actor-critic algorithm.

Keywords Deep Q-network · Actor-critic · Deep deterministic policy gradient · Twin delayed deep deterministic policy gradient · Soft actor-critic

6.1 Introduction

The deep Q-network (DQN) (Mnih et al. 2015) algorithm is a classical off-policy method. It combines the Q-learning algorithm with a deep neural network to realize end-to-end learning from visual inputs to decision outputs. This algorithm has

H. Zhang (✉)
Peking University, Beijing, China
e-mail: zhanghongming@pku.edu.cn

T. Yu
Nanchang University, Nanchang, China

R. Huang
Borealis AI, Toronto, ON, Canada

H. Dong et al. (eds.), *Deep Reinforcement Learning*,
https://doi.org/10.1007/978-981-15-4095-0_6

Table 6.1 Characteristics of DQN and actor-critic

Algorithm	On-policy/off-policy	Sample efficiency	Action space
DQN	Off-policy	High	Discrete
Actor-critic	On-policy	Low	Continuous
DQN+Actor-critic	Off-policy	High	Discrete and continuous

achieved human-level performance on Atari games using raw pixel inputs. However, while the inputs can be unprocessed and high-dimensional observation spaces, DQN can only handle discrete and low-dimensional action spaces. For continuous and high-dimensional action spaces, DQN fails to calculate the Q value of each action.

The actor-critic (AC) (Sutton and Barto 2018) method is an extension of the REINFORCE (Sutton and Barto 2018) algorithm. By incorporating a critic, this method transforms policy gradient's Monte Carlo update into a temporal-difference update. Through this multi-step method, the degree of bootstrapping can be flexibly selected, so the update of policy does not need to wait until the end of the game. Though some bias will be introduced in temporal-difference update, it can reduce variances and accelerate the learning. However, the original actor-critic method is still an on-policy algorithm, and the sample efficiency of on-policy methods is much lower than off-policy methods.

Combining DQN with actor-critic can take advantages of both algorithms. Because of DQN, actor-critic methods are transformed into off-policy methods. Networks can be trained with samples from a replay buffer that improves sample efficiency. Sampling from a replay buffer can also minimize correlations between samples, which can learn value functions in a stable and robust way. Also, due to the actor-critic method, we can easily handle problems with continuous action spaces by using a network to learn a policy π (Table 6.1).

Next, we introduce some classical algorithms: the deep deterministic policy gradient (DDPG) algorithm (Lillicrap et al. 2015), and its improvements: the twin delayed deep deterministic policy gradient (TD3) algorithm (Fujimoto et al. 2018) and the soft actor-critic (SAC) algorithm (Haarnoja et al. 2018a).

6.2 Deep Deterministic Policy Gradient (DDPG)

The deep deterministic policy gradient (DDPG) algorithm can be regarded as a combination of the deterministic policy gradient (DPG) algorithm (Silver et al. 2014) and deep neural networks; The DDPG algorithm can also be viewed as an extension of the DQN algorithm in continuous action space. It wants to tackle the problem with continuous action spaces that DQN cannot be straightforwardly applied to. DDPG establishes a Q function (critic) and a policy function (actor) simultaneously. The Q function (critic) is the same with DQN, temporal-difference methods (TD methods) are used to update it. The policy gradient algorithm is used to update the policy function (actor) through the value from Q function (critic).

In DDPG, the actor is a deterministic policy function, denoted as $\pi(s)$, and the parameter is denoted as θ^{π}. The action of each step is calculated directly by $A_t = \pi(S_t|\theta_t^{\pi})$, which does not need to sample from a stochastic policy.

A critical problem here is how to balance exploration and exploitation with this deterministic policy. In DDPG, noises sampled from a noise process N are added to actions when training. The action is $A_t = \pi(S_t|\theta_t^{\pi})+N_t$. N can be chosen according to the specific task; the original paper uses an Ornstein–Uhlenbeck process (O–U process) (Uhlenbeck and Ornstein 1930).

The O–U process satisfies the following stochastic differential equation:

$$\mathrm{d}X_t = \theta(\pi - X_t)\mathrm{d}t + \sigma \mathrm{d}W_t, \tag{6.1}$$

where X_t is a random variable, $\theta > 0, x, \sigma > 0$ are parameters. W_t is a Wiener process or named Brownian Motion (It and McKean 1965), which has the following properties:

- W_t is a process with independent increments, which means for times $T_0 < T_1 < \ldots < T_n$, random variables $W_{T_0}, W_{T_1} - W_{T_0}, \ldots, W_{T_n} - W_{T_{n-1}}$ are independent.
- For any time t and Δt, $W(t + \Delta_t) - W(t) \sim N(0, \sigma_W^2 \Delta t)$.
- W_t is a continuous function about t.

We know the Markov Decision Process (MDP) is based on Markov Chains; MDP satisfies the property $p(X_{t+1}|X_t, \ldots, X_1) = p(X_{t+1}|X_t)$, where X_t is a random variable at time step t. This means the random variable X_t is conditioned on the last time step's random variable X_{t-1}, which is time-correlated. The O–U noise is also time-correlated, which conforms to the property of Markov Decision Process (MDP). However, more recent results suggest that time-uncorrelated, mean-zero Gaussian noise also works well.

Back to the algorithm, the action-value function $Q(s, a|\theta^Q)$ is learned using the Bellman equation as in DQN.

In the state S_t, the next state S_{t+1} and the return R_t are obtained by executing action $A_t = \pi(S_t|\theta_t^{\pi})$ through the policy π. We have

$$Q^{\pi}(S_t, A_t) = \mathbb{E}[r(S_t, A_t) + \gamma Q^{\pi}(S_{t+1}, \pi(S_{t+1}))]. \tag{6.2}$$

Then we can compute the Q value:

$$Y_i = R_i + \gamma Q^{\pi}(S_{t+1}, \pi(S_{t+1})). \tag{6.3}$$

Using gradient descent to minimize the loss function:

$$L = \frac{1}{N}\sum_i \left(Y_i - Q(S_i, A_i|\theta^Q)\right)^2. \tag{6.4}$$

The policy function π is updated by applying the chain rule to the expected return from the start distribution J. Here, $J = \mathbb{E}_{R_i, S_i \sim E, A_i \sim \pi}[R_1]$ (E denotes the environment) and $R_t = \sum_{i=t}^{T} \gamma^{(i-t)} r(S_i, A_i)$. We have

$$\begin{aligned} \nabla_{\theta^\pi} J \approx & \mathbb{E}_{S_t \sim \rho^\beta}\left[\nabla_{\theta^\pi} Q\left(s, a|\theta^Q\right)|_{s=S_t, a=\pi(S_t|\theta^\pi)}\right], \\ = & \mathbb{E}_{S_t \sim \rho^\beta}\left[\nabla_a Q\left(s, a|\theta^Q\right)|_{s=S_t, a=\pi(S_t)} \nabla_{\theta_\pi} \pi\left(s|\theta^\pi\right)|_{s=S_t}\right]. \end{aligned} \tag{6.5}$$

Learning in mini-batches:

$$\nabla_{\theta^\pi} J \approx \frac{1}{N} \sum_i \nabla_a Q\left(s, a|\theta^Q\right)|_{s=S_i, a=\pi(S_i)} \nabla_{\theta^\pi} \pi\left(s|\theta^\pi\right)|_{S_i}. \tag{6.6}$$

In addition, DDPG adopts a similar way of the target network like DQN, but it updates network parameters by exponential smoothing rather than directly copying the parameters:

$$\theta^{Q'} \leftarrow \rho\theta^Q + (1-\rho)\theta^{Q'}, \tag{6.7}$$

$$\theta^{\pi'} \leftarrow \rho\theta^\pi + (1-\rho)\theta^{\pi'}. \tag{6.8}$$

Since the hyperparameter $\rho \ll 1$ here, the target network changes very slowly and smoothly, which improves the stability of learning.

The whole pseudocode shows in Algorithm 1.

6.3 Twin Delayed Deep Deterministic Policy Gradient (TD3)

The twin delayed deep deterministic policy gradient (TD3) algorithm is an improvement of DDPG, where three critical techniques are used:

1. Clipped double Q-learning for actor-critic: learn two Q-value functions, which is similar to double Q-learning.
2. Target networks and delayed policy updates: update the policy (and the target network) less frequently than the Q-value function.
3. Target policy smoothing regularization: Add noise to the target action to smooth the Q-value function and avoid overfitting.

For the first technique, we know that in DQN, there is an overestimation problem due to the existence of the max operation, this problem also exists in DDPG, because $Q(s, a)$ is updated in the same way as DQN

$$Q(s, a) \leftarrow R_s^a + \gamma \max_{\hat{a}} Q\left(s', \hat{a}\right). \tag{6.9}$$

Algorithm 1 DDPG

Hyperparameters: soft update factor ρ, reward discount factor γ
Input empty replay buffer $\mathcal{D}$, initialize parameters θ^Q of critic network $Q(s, a|\theta^Q)$ and parameters θ^π of actor network $\pi(s|\theta^\pi)$, target network Q' and π'
Initialize target network Q' and π' with weights $\theta^{Q'} \leftarrow \theta^Q, \theta^{\pi'} \leftarrow \theta^\pi$
for episode = 1, M **do**
  Initialize a random process $\mathcal{N}$ for action exploration
  Receive initial observation state S_1
  for t = 1, T **do**
    Selection action $A_t = \pi(S_t|\theta^\pi) + \mathcal{N}_t$
    Execute action A_t and observe reward R_t, and observe new state S_{t+1}
    Store transion $(S_t, A_t, R_t, D_t, S_{t+1})$ in $\mathcal{D}$
    Set $Y_i = R_i + \gamma(1 - D_t)Q'(S_{t+1}, \pi'(S_{t+1}|\theta^{\pi'})|\theta^{Q'})$
    Update critic by minimizing the loss:

$$L = \frac{1}{N}\sum_i (Y_i - Q(S_i, A_i|\theta^Q))^2$$

    Update the actor policy using the sampled policy gradient:

$$\nabla_{\theta^\pi} J \approx \frac{1}{N}\sum_i \nabla_a Q(s, a|\theta^Q)|_{s=S_i, a=\pi(S_i)} \nabla_{\theta^\pi} \pi(s|\theta^\pi)|_{S_i}$$

    Update the target networks:
    $\theta^{Q'} \leftarrow \rho\theta^Q + (1 - \rho)\theta^{Q'}$
    $\theta^{\pi'} \leftarrow \rho\theta^\pi + (1 - \rho)\theta^{\pi'}$
  end for
end for

This is not a problem in the tabular case, because the Q-values are stored precisely. However, when we use a network as a function approximator in a more complex case, the estimation of Q-value will be noisy. That is to say:

$$Q^{approx}(s', \hat{a}) = Q^{target}\left(s', \hat{a}\right) + Y_{s'}^{\hat{a}}, \tag{6.10}$$

where $Y_{s'}^{\hat{a}}$ is a noise with zero mean. But with the max operator, the noise induces a difference between Q^{approx} and Q^{target}. Denote the difference as Z_s, we have

$$\begin{aligned} Z_s &\overset{\text{def}}{=} R_s^a + \gamma \max_{\hat{a}} Q^{approx}\left(s', \hat{a}\right) - \left(R_s^a + \gamma \max_{\hat{a}} Q^{target}(s', \hat{a})\right), \\ &= \gamma\left(\max_{\hat{a}} Q^{approx}(s', \hat{a}) - \max_{\hat{a}} Q^{target}(s', \hat{a})\right). \end{aligned} \tag{6.11}$$

Considering the noise $Y_{s'}^{\hat{a}}$, some of the Q-values might be too small, while others might be too large. The max operator always picks the largest value for each state, which will make it sensitive to overestimate the correct Q-values for some actions. In this case, this noise will lead to $\mathbb{E}[Z_s] > 0$ and cause the overestimation problem.

The TD3 algorithm incorporates the idea of double Q-learning in DDPG; it establishes two Q-value networks to compute the value of the next state:

$$Q_{\theta_1'}\left(s', a'\right) = Q_{\theta_1'}\left(s', \pi_{\phi_1}(s')\right), \tag{6.12}$$

$$Q_{\theta_2'}\left(s', a'\right) = Q_{\theta_2'}\left(s', \pi_{\phi_1}(s')\right). \tag{6.13}$$

Use the minimum of the two values (clipped) to compute the Bellman equation:

$$Y_1 = r + \gamma \min_{i=1,2} Q_{\theta_i'}\left(s', \pi_{\phi_1}(s')\right). \tag{6.14}$$

With clipped double Q-learning, the value target will not introduce additional overestimation over using the standard Q-learning target. While this update rule may induce an underestimation bias, this is preferable to an overestimation bias. Because unlike overestimated actions, the value of underestimated actions will not be explicitly propagated through the policy update (Fujimoto et al. 2018).

For the second technique, we know that the target network is a good tool to achieve stability in deep reinforcement learning. As deep function approximators require multiple gradient updates to converge, target networks provide a stable objective in the learning procedure and allow better coverage of the training data. So, if target networks can be used to reduce the error over multiple updates, and policy updates on high-error states cause divergent behavior, then the policy network should be updated at a lower frequency than the value network, to first minimize error before introducing a policy update. In this way, the TD3 algorithm reduces the update frequency of the policy function. It only updates the policy and target networks after a fixed number of updates d to the critic. The less frequent policy updates can make the update of Q-value function has a smaller variance, and thus a higher quality policy can be obtained.

For the third technique, a concern with deterministic policies is that they can overfit to narrow peaks in the value estimate. In TD3 paper, the author enforces the notion that similar actions should have similar value, so fitting the value of a small area around the target action makes sense:

$$y = r + \mathbb{E}_\epsilon\left[Q_{\theta'}\left(s', \pi_{\phi'}(s') + \epsilon\right)\right]. \tag{6.15}$$

By adding a truncated normal distribution noise to each action as a regularization, the computation of Q-values can be smoothed to avoid overfitting.

This makes a modified target update:

$$y = r + \gamma Q_{\theta'}\left(s', \pi_{\phi'}(s') + \epsilon\right), \epsilon \sim clip(N(0, \sigma), -c, c). \tag{6.16}$$

The whole pseudocode shows in Algorithm 2.

Algorithm 2 TD3

Hyperparameters: soft update factor ρ, reward discount factor γ, clip factor c
Input: empty replay buffer $\mathcal{D}$, initial parameters θ_1, θ_2 of critic networks $Q_{\theta_1}, Q_{\theta_2}$, initial parameters ϕ of actor network π_ϕ
Initialize target networks $\hat{\theta}_1 \leftarrow \theta_1, \hat{\theta}_2 \leftarrow \theta_2, \hat{\phi} \leftarrow \phi$
for $t = 1$ to T do **do**
    Select action with exploration noise $A_t \sim \pi_\phi(S_t) + \epsilon, \epsilon \sim \mathcal{N}(0, \sigma)$
    Observe reward R_t and new state S_{t+1}
    Store transition tuple $(S_t, A_t, R_t, D_t, S_{t+1})$ in $\mathcal{D}$
    Sample mini-batch of N transitions $(S_t, A_t, R_t, D_t, S_{t+1})$ from $\mathcal{D}$
    $\tilde{a}_{t+1} \leftarrow \pi_{\hat{\phi}'}(S_{t+1}) + \epsilon, \epsilon \sim clip(\mathcal{N}(0, \tilde{\sigma}, -c, c))$
    $y \leftarrow R_t + \gamma(1 - D_t) \min_{i=1,2} Q_{\theta_i'}(S_{t+1}, \tilde{a}_{t+1})$
    Update critics $\theta_i \leftarrow \arg\min_{\theta_i} N^{-1} \sum (y - Q_{\theta_i}(S_t, A_t))^2$
    if t mod d **then**
        Update ϕ by the deterministic policy gradient:
        $\nabla_\phi J(\phi) = N^{-1} \sum \nabla_a Q_{\theta_1}(S_t, A_t)|_{A_t = \pi_\phi(S_t)} \nabla_\phi \pi_\phi(S_t)$
        Update target networks:
        $\hat{\theta}_i \leftarrow \rho\theta_i + (1 - \rho)\hat{\theta}_i$
        $\hat{\phi} \leftarrow \rho\phi + (1 - \rho)\hat{\phi}$
    end if
end for

6.4 Soft Actor-Critic (SAC)

The soft actor-critic (SAC) algorithm follows the idea of maximum entropy reinforcement learning, where instead of maximizing the discounted cumulative reward, the optimal policy aims to maximize its entropy regularized reward, thus encouraging the exploration of the policy.

$$\max_{\pi_\theta} \mathbb{E}\left[\sum_t \gamma^t \left(r(S_t, A_t) + \alpha\mathcal{H}(\pi_\theta(\cdot|S_t))\right)\right], \tag{6.17}$$

where α is the regularization coefficient. Maximum entropy reinforcement learning has been well explored in the literature (Ziebart et al. 2008; Levine and Koltun 2013; Fox et al. 2016; Nachum et al. 2017; Haarnoja et al. 2017). Here we only introduce the idea of soft policy iteration which serves as the fundamental of the SAC method.

6.4.1 Soft Policy Iteration

Soft policy iteration is a general algorithm for learning the optimal maximum entropy policies with provable guarantees. Similar to policy iteration, soft policy iteration also has two steps: soft policy evaluation and soft policy improvement.

Let

$$V^{\pi}(s) = \mathbb{E}\left[\sum_{t} \gamma^{t} \left(r(S_t, A_t) + \alpha \mathcal{H}(\pi(\cdot|S_t))\right)\right], \tag{6.18}$$

where $S_0 = s$. Further let

$$Q(s, a) = r(s, a) + \gamma \mathbb{E}\left[V(s')\right], \tag{6.19}$$

where $s' \sim \Pr(\cdot|s, a)$ is the next state. It is straightforward to verify that

$$V^{\pi}(s) = \mathbb{E}_{a\sim\pi}\left[Q(s, a) - \alpha \log(a|s)\right]. \tag{6.20}$$

In the soft policy evaluation step, define the Bellman backup operator $\mathcal{T}$ by

$$\mathcal{T}^{\pi} Q(s, a) = r(s, a) + \gamma \mathbb{E}\left[V^{\pi}(s')\right]. \tag{6.21}$$

Similar to policy evaluation, one can prove that for any map $Q^0 : \mathcal{S} \times \mathcal{A} \to \mathbb{R}$, Q^k will converge to the soft Q-value of π, where $Q^k = \mathcal{T}^{\pi} Q^{k-1}$.

In the policy improvement step, we solve the entropy regularized reward maximization problem with the current Q values.

$$\pi(\cdot|s) = \arg\max_{\pi} \mathbb{E}_{a\sim\pi}\left[Q(s, a) + \alpha \mathcal{H}(\pi)\right]. \tag{6.22}$$

Solving the above optimization problem (Fox et al. 2016; Nachum et al. 2017), one can get that

$$\pi(\cdot|s) = \frac{\exp\left(\frac{1}{\alpha} Q(s, \cdot)\right)}{Z(s)}, \tag{6.23}$$

where $Z(s)$ is the normalizing factor, i.e. $Z(s) = \sum_a \exp\left(\frac{1}{\alpha} Q(s, a)\right)$. Given that the optimal π may not be representable in the policy model, we instead update the policy by

$$\pi(\cdot|s) = \arg\min_{\pi\in\Pi} D_{\mathrm{KL}}\left(\pi(\cdot|s) \| \frac{\exp\left(\frac{1}{\alpha} Q(s, \cdot)\right)}{Z(s)}\right). \tag{6.24}$$

Not surprising, one can also prove the policy monotonically improvement property for the above soft policy improvement step, even with the projection to Π using KL-divergence. The next theorem shows that soft policy iteration, similar to policy iteration, converges to the optimal solution.

Theorem 6.1 *Let $\pi_0 \in \Pi$ be any initialized policy. Assume that by performing soft policy iteration steps, π_0 converges to $\pi*$. Then $Q^{\pi*}(s,a) \geq Q^{\pi}(s,a)$ for any $(s,a) \in \mathcal{S} \times \mathcal{A}$ and any $\pi \in \Pi$.*

We omit all the proofs of this section of this book. Interested readers can refer to (Haarnoja et al. 2018b) for more details.

6.4.2 SAC

SAC extends soft policy iteration to the setting with function approximation which is more practical. Instead of estimating the true Q value of the policy π for policy improvement, SAC performs an alternative optimization on both the value function and the policy.

Consider a parametrized Q function $Q_\phi(s,a)$ and policy π_θ. Here we consider the continuous action setting, where the output of π_θ is a Gaussian mean and covariance. Similarly, the Q function can be learned by minimizing the soft Bellman residual,

$$J_Q(\phi) = \mathbb{E}\left[\left(Q(S_t, A_t) - r(S_t, A_t) - \gamma \mathbb{E}_{S_{t+1}}\left[V_{\tilde{\phi}}(S_{t+1})\right]\right)^2\right]. \tag{6.25}$$

where $V_{\tilde{\phi}}(s) = \mathbb{E}_{\pi_\theta}\left[Q_{\tilde{\phi}}(s,a) - \alpha \log \pi_\theta(a|s)\right]$, and $Q_{\tilde{\phi}}$ is a target Q network, whose parameter $\tilde{\phi}$ is obtained as an exponentially moving average of ϕ. Moreover, the policy π_θ can be learned by minimizing the expected KL-divergence.

$$J_\pi(\theta) = \mathbb{E}_{s\sim\mathcal{D}}\left[\mathbb{E}_{a\sim\pi_\theta}\left[\alpha \log \pi_\theta(a|s) - Q_\phi(s,a)\right]\right]. \tag{6.26}$$

In practice, SAC uses two Q-networks (as well as two target Q-networks) to mitigate the biased Q value problem, i.e. $Q_\phi(s,a) = \min\left(Q_{\phi_1}(s,a), Q_{\phi_2}(s,a)\right)$. Note that $J_\pi(\theta)$ has the expectation taken on π_θ. To optimize $J_\pi(\theta)$, one option is to use the idea of likelihood ratio gradient estimator (Williams 1992). However, in the continuous action setting, one can instead use the reparametrization trick for the policy network, which usually results in a lower variance estimator. To do that, we reparametrize π_θ as an action network taking both state s and a standard Gaussian noise ϵ as its input.

$$a = f_\theta(s, \epsilon). \tag{6.27}$$

Plugging into $J_\pi(\theta)$,

$$J_\pi(\theta) = \mathbb{E}_{s\sim\mathcal{D},\epsilon\sim\mathcal{N}}\left[\alpha \log \pi_\theta(f_\theta(s,\epsilon)|s) - Q_\phi(s, f_\theta(s,\epsilon))\right], \tag{6.28}$$

where $\mathcal{N}$ is the standard Gaussian distribution, and π_θ is now defined implicitly in terms of f_θ.

Finally, SAC also provides a way in automatically update the regularization coefficient α, by minimizing the following loss:

$$J(\alpha) = \mathbb{E}_{a\sim\pi_\theta}\left[-\alpha \log \pi_\theta(a|s) - \alpha\kappa\right], \tag{6.29}$$

where κ is a hyperparameter interpreted as the target entropy. Such updating schemes for α are also called automating entropy adjustments. The intuition behind $J(\alpha)$ is the dual form of the original policy optimization problem with the constraint that the average entropy at each time step should be at least κ. For more rigorous statement around automating entropy adjustment, please refer to the SAC paper (Haarnoja et al. 2018b). We summarize the SAC algorithm in Algorithm 3.

Algorithm 3 Soft actor-critic (SAC)

Hyperparameters: target entropy κ, step sizes λ_Q, λ_π, λ_α, exponentially moving average coefficient τ
Input: initial policy parameters θ, initial Q value function parameters ϕ_1 and ϕ_2
$\mathcal{D} = \emptyset$; $\tilde{\phi}_i = \phi_i$, for $i = 1, 2$
for $k = 0, 1, 2, \ldots$ **do**
  for $t = 0, 1, 2, \ldots$ **do**
    Sample A_t from $\pi_\theta(\cdot|S_t)$, collect (R_t, S_{t+1})
    $\mathcal{D} = \mathcal{D} \cup \{S_t, A_t, R_t, S_{t+1}\}$
  end for
  Perform multiple step of gradients:
    $\phi_i = \phi_i - \lambda_Q \nabla J_Q(\phi_i)$ for $i = 1, 2$
    $\theta = \theta - \lambda_\pi \nabla_\theta J_\pi(\theta)$
    $\alpha = \alpha - \lambda_\alpha \nabla J(\alpha)$
    $\tilde{\phi}_i = (1-\tau)\phi_i + \tau\tilde{\phi}_i$ for $i = 1, 2$
end for
Output θ, ϕ_1, ϕ_2

6.5 Examples

This section will share examples of DDPG, TD3, and SAC. They are all actor-critic methods and use a Q-network as a critic. Examples are based on the OpenAI Gym Environment. Since these algorithms are based on continuous action space, "Pendulum-v0" environment is used.

6.5.1 Related Gym Environment

As mentioned above, Pendulum-v0 is a classical inverted pendulum environment with three-dimensional observation space and one-continuous action space. At each step, the environment returns a reward affected by the current rotation angle, speed, and acceleration. The goal of this task is to turn the pendulum upside down to gain the maximum score.

6.5.2 DDPG: Pendulum-v0

DDPG uses off-policy data and TD methods. The structure of DDPG class can be shown as follows:

```
class DDPG(object):
    def __init__(self, action_dim, state_dim, action_range):
        ...
    def ema_update(self):
        ...
    def get_action(self, s, greedy=False):
        ...
    def learn(self):
        ...
    def store_transition(self, s, a, r, s_):
        ...
    def save(self):
        ...
    def load(self):
        ...
```

There are four networks created in the initialization function. They are the actor, critic, actor target, and critic target. The parameters of the target network will be replaced with the corresponding network parameters.

```
class DDPG(object):
    def __init__(self, action_dim, state_dim, action_range):
        self.memory = np.zeros((MEMORY_CAPACITY, state_dim * 2 +
            action_dim + 1), dtype=np.float32)
        self.pointer = 0
        self.action_dim, self.state_dim, self.action_range =
            action_dim, state_dim, action_range
        self.var = VAR

        W_init = tf.random_normal_initializer(mean=0, stddev=0.3)
        b_init = tf.constant_initializer(0.1)
```

```
def get_actor(input_state_shape, name=''):
    input_layer = tl.layers.Input(input_state_shape,
        name='A_input')
    layer = tl.layers.Dense(n_units=64, act=tf.nn.relu,
        W_init=W_init, b_init=b_init,
        name='A_l1')(input_layer)
    layer = tl.layers.Dense(n_units=64, act=tf.nn.relu,
        W_init=W_init, b_init=b_init, name='A_l2')(layer)
    layer = tl.layers.Dense(n_units=action_dim,
        act=tf.nn.tanh, W_init=W_init, b_init=b_init,
        name='A_a')(layer)
    layer = tl.layers.Lambda(lambda x: action_range *
        x)(layer)
    return tl.models.Model(inputs=input_layer,
        outputs=layer, name='Actor' + name)

def get_critic(input_state_shape, input_action_shape,
    name=''):
    state_input = tl.layers.Input(input_state_shape,
        name='C_s_input')
    action_input = tl.layers.Input(input_action_shape,
        name='C_a_input')
    layer = tl.layers.Concat(1)([state_input, action_input])
    layer = tl.layers.Dense(n_units=64, act=tf.nn.relu,
        W_init=W_init, b_init=b_init, name='C_l1')(layer)
    layer = tl.layers.Dense(n_units=64, act=tf.nn.relu,
        W_init=W_init, b_init=b_init, name='C_l2')(layer)
    layer = tl.layers.Dense(n_units=1, W_init=W_init,
        b_init=b_init, name='C_out')(layer)
    return tl.models.Model(inputs=[state_input,
        action_input], outputs=layer, name='Critic' + name)

self.actor = get_actor([None, state_dim])
self.critic = get_critic([None, state_dim], [None,
    action_dim])
self.actor.train()
self.critic.train()

def copy_para(from_model, to_model):
    for i, j in zip(from_model.trainable_weights,
        to_model.trainable_weights):
       j.assign(i)

self.actor_target = get_actor([None, state_dim],
    name='_target')
copy_para(self.actor, self.actor_target)
self.actor_target.eval()

self.critic_target = get_critic([None, state_dim], [None,
    action_dim], name='_target')
copy_para(self.critic, self.critic_target)
self.critic_target.eval()
```

```
    self.ema = tf.train.ExponentialMovingAverage(decay=1 -
        TAU) # soft replacement

    self.actor_opt = tf.optimizers.Adam(LR_A)
    self.critic_opt = tf.optimizers.Adam(LR_C)
```

During the training process, the parameters of the target network will be updated by a moving average.

```
def ema_update(self):
    paras = self.actor.trainable_weights +
        self.critic.trainable_weights
    self.ema.apply(paras)
    for i, j in zip(self.actor_target.trainable_weights +
        self.critic_target.trainable_weights, paras):
       i.assign(self.ema.average(j))
```

As the policy network is a deterministic policy network, we need to add some randomness if we do not choose actions by greedy methods. We used a normal distribution here and the value of variance decreases in the process of the update iteration. The randomness can also be changed to other methods, such as an O–U noise.

```
def get_action(self, state, greedy=False):
    a = self.actor(np.array([s], dtype=np.float32))[0]
    if greedy:
       return a

    # add randomness to action selection for exploration
    return np.clip(np.random.normal(a, self.var),
                   -self.action_range,
                   self.action_range)
```

In the `learn()` function, we sample the off-policy data from the replay buffer and use the Bellman equation to learn the Q-function. After that, the policy can be learned by maximizing the Q-value. Finally, target networks will be updated with Polyak averaging (Polyak 1964) by using formula $\theta^{Q'} \leftarrow \rho\theta^{Q} + (1-\rho)\theta^{Q'}$, $\theta^{\pi'} \leftarrow \rho\theta^{\pi} + (1-\rho)\theta^{\pi'}$.

```
def learn(self):
    self.var *= .9995
    indices = np.random.choice(MEMORY_CAPACITY,
        size=BATCH_SIZE)
    bt = self.memory[indices, :]
    bs = bt[:, :self.s_dim]
    ba = bt[:, self.s_dim:self.s_dim + self.a_dim]
    br = bt[:, -self.s_dim - 1:-self.s_dim]
    bs_ = bt[:, -self.s_dim:]

    with tf.GradientTape() as tape:
       a_ = self.actor_target(bs_)
```

```
        q_ = self.critic_target([bs_, a_])
        y = br + GAMMA * q_
        q = self.critic([bs, ba])
        td_error = tf.losses.mean_squared_error(y, q)
    c_grads = tape.gradient(td_error,
        self.critic.trainable_weights)
    self.critic_opt.apply_gradients(zip(c_grads,
        self.critic.trainable_weights))

    with tf.GradientTape() as tape:
        a = self.actor(bs)
        q = self.critic([bs, a])
        a_loss = -tf.reduce_mean(q) # maximize the q
    a_grads = tape.gradient(a_loss,
        self.actor.trainable_weights)
    self.actor_opt.apply_gradients(zip(a_grads,
        self.actor.trainable_weights))
    self.ema_update()
```

The `store_transition()` function uses a replay buffer to store the transition of each step.

```
def store_transition(self, s, a, r, s_):
    s = s.astype(np.float32)
    s_ = s_.astype(np.float32)
    transition = np.hstack((s, a, [r], s_))
    index = self.pointer % MEMORY_CAPACITY # replace the old
        memory with new memory
    self.memory[index, :] = transition
    self.pointer += 1
```

The main function is straightforward, the agent interacts with the environment at each step, stores data into the replay buffer, and uses sampled batch data from the replay buffer to update the networks.

```
env = gym.make(ENV_ID).unwrapped

# reproducible
env.seed(RANDOM_SEED)
np.random.seed(RANDOM_SEED)
tf.random.set_seed(RANDOM_SEED)

state_dim = env.observation_space.shape[0]
action_dim = env.action_space.shape[0]
action_range = env.action_space.high # scale action,
    [-action_range, action_range]

agent = DDPG(action_dim, state_dim, action_range)
t0 = time.time()

if args.train: # train
    all_episode_reward = []
```

```
    for episode in range(TRAIN_EPISODES):
        state = env.reset()
        episode_reward = 0
        for step in range(MAX_STEPS):
            if RENDER:
                env.render()
            # Add exploration noise
            action = agent.get_action(state)
            state_, reward, done, info = env.step(action)
            agent.store_transition(state, action, reward, state_)
            if agent.pointer > MEMORY_CAPACITY:
                agent.learn()
            state = state_
            episode_reward += reward
            if done:
                break

        if episode == 0:
            all_episode_reward.append(episode_reward)
        else:
            all_episode_reward.append(all_episode_reward[-1] *
                0.9 + episode_reward * 0.1)
        print(
            'Training | Episode: {}/{} | Episode Reward: {:.4f}
                | Running Time: {:.4f}'.format(
                episode+1, TRAIN_EPISODES, episode_reward,
                time.time() - t0
            )
        )

    agent.save()
    plt.plot(all_episode_reward)
    if not os.path.exists('image'):
        os.makedirs('image')
    plt.savefig(os.path.join('image', 'ddpg.png'))
```

The agent can be tested after training.

```
if args.test:
    # test
    agent.load()
    for episode in range(TEST_EPISODES):
        state = env.reset()
        episode_reward = 0
        for step in range(MAX_STEPS):
            env.render()
            state, reward, done, info =
                env.step(agent.get_action(state, greedy=True))
            episode_reward += reward
            if done:
                break
        print(
```

```
            'Testing | Episode: {}/{} | Episode Reward: {:.4f} |
                Running Time: {:.4f}'.format(
              episode + 1, TEST_EPISODES, episode_reward,
              time.time() - t0))
```

6.5.3 TD3: Pendulum-v0

TD3 code uses these classes: `ReplayBuffer` class, `QNetwork` class, `PolicyNetwork` class, and `TD3` class.

`ReplayBuffer` class is used to build a replay buffer, so it should have the function of `push()` and `sample()`.

```
class ReplayBuffer:
    def __init__(self, capacity):
        ...
    def push(self, state, action, reward, next_state, done):
        ...
    def sample(self, batch_size):
        ...
    def __len__(self):
        ...
```

The `push()` function appends data to the buffer and move the pointer.

```
    def push(self, state, action, reward, next_state, done):
        if len(self.buffer) < self.capacity:
            self.buffer.append(None)
        self.buffer[self.position] = (state, action, reward,
             next_state, done)
        self.position = int((self.position + 1) % self.capacity) #
             as a ring buffer
```

The `sample()` function simply samples data from the buffer and returns.

```
    def sample(self, batch_size):
        batch = random.sample(self.buffer, batch_size)
        state, action, reward, next_state, done = map(np.stack,
             zip(*batch)) # stack for each element
        return state, action, reward, next_state, done
```

By refactoring the `__len__()` function, the buffer size will be returned when called by the `len()` function.

```
    def __len__(self):
        return len(self.buffer)
```

The `QNetwork` class is used to build the *Q*-network for critic. It is another coding way for building networks.

```
class QNetwork(Model):
    def __init__(self, num_inputs, num_actions, hidden_dim,
        init_w=3e-3):
        super(QNetwork, self).__init__()
        input_dim = num_inputs + num_actions
        w_init = tf.random_uniform_initializer(-init_w, init_w)
        self.linear1 = Dense(n_units=hidden_dim, act=tf.nn.relu,
            W_init=w_init, in_channels=input_dim, name='q1')
        self.linear2 = Dense(n_units=hidden_dim, act=tf.nn.relu,
            W_init=w_init, in_channels=hidden_dim, name='q2')
        self.linear3 = Dense(n_units=1, W_init=w_init,
            in_channels=hidden_dim, name='q3')

    def forward(self, input):
        x = self.linear1(input)
        x = self.linear2(x)
        x = self.linear3(x)
        return x
```

The `PolicyNetwork` class is used to build the policy network for the actor. While building the network, it also adds `evaluate()`, `get_action()`, `sample_action()` functions.

```
class PolicyNetwork(Model):
    def __init__(self, num_inputs, num_actions, hidden_dim,
        action_range=1., init_w=3e-3):
        ...
    def forward(self, state):
        ...
    def evaluate(self, state, eval_noise_scale):
        ...
    def get_action(self, state, explore_noise_scale,
        greedy=False):
        ...
    def sample_action(self):
        ...
```

The following part of the code shows how to build the network:

```
class PolicyNetwork(Model):
    def __init__(self, num_inputs, num_actions, hidden_dim,
        action_range=1., init_w=3e-3):
        super(PolicyNetwork, self).__init__()
        w_init = tf.random_uniform_initializer(-init_w, init_w)
        self.linear1 = Dense(n_units=hidden_dim, act=tf.nn.relu,
            W_init=w_init, in_channels=num_inputs, name='policy1')
        self.linear2 = Dense(n_units=hidden_dim, act=tf.nn.relu,
            W_init=w_init, in_channels=hidden_dim, name='policy2')
```

```
        self.linear3 = Dense(n_units=hidden_dim, act=tf.nn.relu,
            W_init=w_init, in_channels=hidden_dim, name='policy3')
        self.output_linear = Dense(n_units=num_actions,
            W_init=w_init,
            b_init=tf.random_uniform_initializer(-init_w, init_w),
            in_channels=hidden_dim, name='policy_output')
        self.action_range = action_range
        self.num_actions = num_actions

    def forward(self, state):
        x = self.linear1(state)
        x = self.linear2(x)
        x = self.linear3(x)
        output = tf.nn.tanh(self.output_linear(x)) # unit range
            output [-1, 1]
        return output
```

The `evaluate()` function generates actions with the states for calculating gradients. It uses the trick of target policy smoothing for generating noisy actions.

```
    def evaluate(self, state, eval_noise_scale):
        state = state.astype(np.float32)
        action = self.forward(state)
        action = self.action_range * action
        # add noise
        normal = Normal(0, 1)
        eval_noise_clip = 2 * eval_noise_scale
        noise = normal.sample(action.shape) * eval_noise_scale
        noise = tf.clip_by_value(noise, -eval_noise_clip,
            eval_noise_clip)
        action = action + noise
        return action
```

The `get_action()` function generates actions with the states to interact with the environment.

```
    def get_action(self, state, explore_noise_scale,
        greedy=False):
        action = self.forward([state])
        action = self.action_range * action.numpy()[0]
        if greedy:
            return action
        # add noise
        normal = Normal(0, 1)
        noise = normal.sample(action.shape) * explore_noise_scale
        action += noise
        return action.numpy()
```

The `sample_action()` function is used to generate random actions at the beginning of training.

```
def sample_action(self, ):
    a = tf.random.uniform([self.num_actions], -1, 1)
    return self.action_range * a.numpy()
```

The TD3 class is the core content of the TD3 algorithm.

```
class TD3():
    def __init__(self, state_dim, action_dim, replay_buffer,
        hidden_dim, action_range,
        policy_target_update_interval=1, q_lr=3e-4,
        policy_lr=3e-4): # create replay buffer and networks
        ...
    def target_ini(self, net, target_net): # hard-copy update for
        initializing target networks
        ...
    def target_soft_update(self, net, target_net, soft_tau):
        #soft update the target net with Polyak averaging
        ...
    def update(self, batch_size, eval_noise_scale,
        reward_scale=10., gamma=0.9, soft_tau=1e-2): # update all
        networks in TD3
        ...
    def save(self): # save trained weights
        ...
    def load(self): # load trained weights
        ...
```

The initialization function creates twin q-networks, policy-networks, and their target networks. Six networks are established in total.

```
class TD3():
    def __init__(self, state_dim, action_dim, replay_buffer,
        hidden_dim, action_range,
        policy_target_update_interval=1, q_lr=3e-4,
        policy_lr=3e-4):
        self.replay_buffer = replay_buffer

        # initialize all networks
        self.q_net1 = QNetwork(state_dim, action_dim, hidden_dim)
        self.q_net2 = QNetwork(state_dim, action_dim, hidden_dim)
        self.target_q_net1 = QNetwork(state_dim, action_dim,
            hidden_dim)
        self.target_q_net2 = QNetwork(state_dim, action_dim,
            hidden_dim)
        self.policy_net = PolicyNetwork(state_dim, action_dim,
            hidden_dim, action_range)
        self.target_policy_net = PolicyNetwork(state_dim,
            action_dim, hidden_dim, action_range)
        print('Q Network (1,2): ', self.q_net1)
```

```
print('Policy Network: ', self.policy_net)

# initialize weights of target networks
self.target_q_net1 = self.target_ini(self.q_net1,
    self.target_q_net1)
self.target_q_net2 = self.target_ini(self.q_net2,
    self.target_q_net2)
self.target_policy_net = self.target_ini(self.policy_net,
    self.target_policy_net)

# set train mode
self.q_net1.train()
self.q_net2.train()
self.target_q_net1.train()
self.target_q_net2.train()
self.policy_net.train()
self.target_policy_net.train()

self.update_cnt = 0
self.policy_target_update_interval =
    policy_target_update_interval

self.q_optimizer1 = tf.optimizers.Adam(q_lr)
self.q_optimizer2 = tf.optimizers.Adam(q_lr)
self.policy_optimizer = tf.optimizers.Adam(policy_lr)
```

The `target_ini()` function and `target_soft_update()` function are used to update the parameters of target networks. The difference is that the former one is a hard copy replacement and the latter one updates the parameters by Polyak averaging.

```
def target_ini(self, net, target_net):=
    for target_param, param in
        zip(target_net.trainable_weights,
        net.trainable_weights):
      target_param.assign(param)
    return target_net

def target_soft_update(self, net, target_net, soft_tau):=
    for target_param, param in
        zip(target_net.trainable_weights,
        net.trainable_weights):
      target_param.assign(target_param * (1.0 - soft_tau) +
          param * soft_tau) # copy weight value into target
          parameters
    return target_net
```

Next is the most critical part: the `update()` function. This part fully reflects the three tricks of the TD3 algorithm.

At the beginning of the function, we first sample data from the replay buffer.

```
def update(self, batch_size, eval_noise_scale,
    reward_scale=10., gamma=0.9, soft_tau=1e-2):
    ''' update all networks in TD3 '''
    self.update_cnt += 1

    # sample batch data
    state, action, reward, next_state, done =
        self.replay_buffer.sample(batch_size)
    reward = reward[:, np.newaxis] # expand dim
    done = done[:, np.newaxis]
```

Next, we use a target policy smoothing trick by adding noise to the target action. This makes it harder for the policy to exploit Q-value function errors by smoothing out Q values along changes in action.

```
    # Trick Three: Target Policy Smoothing. Add noise to the
        target action
    new_next_action = self.target_policy_net.evaluate(
       next_state, eval_noise_scale=eval_noise_scale
    ) # clipped normal noise

    # normalize with batch mean and std
    reward = reward_scale * (reward - np.mean(reward, axis=0))
        / np.std(reward, axis=0)
```

The next trick is clipped double-Q learning. It learns two Q-value functions and uses the smaller Q-value to form the targets in the Bellman error loss function. In this way, the overestimation of the Q-value can be mitigated.

```
    # Training Q Function
    target_q_input = tf.concat([next_state, new_next_action],
        1) # the dim 0 is number of samples

    # Trick One: Clipped Double-Q Learning. Use the smaller
        Q-value.
    target_q_min =
        tf.minimum(self.target_q_net1(target_q_input),
        self.target_q_net2(target_q_input))

    target_q_value = reward + (1 - done) * gamma *
        target_q_min # if done==1, only reward
    q_input = tf.concat([state, action], 1) # input of q_net

    with tf.GradientTape() as q1_tape:
       predicted_q_value1 = self.q_net1(q_input)
       q_value_loss1 =
           tf.reduce_mean(tf.square(predicted_q_value1 -
           target_q_value))
    q1_grad = q1_tape.gradient(q_value_loss1,
        self.q_net1.trainable_weights)
```

```
    self.q_optimizer1.apply_gradients(zip(q1_grad,
        self.q_net1.trainable_weights))

    with tf.GradientTape() as q2_tape:
       predicted_q_value2 = self.q_net2(q_input)
       q_value_loss2 =
           tf.reduce_mean(tf.square(predicted_q_value2 -
           target_q_value))
    q2_grad = q2_tape.gradient(q_value_loss2,
        self.q_net2.trainable_weights)
    self.q_optimizer2.apply_gradients(zip(q2_grad,
        self.q_net2.trainable_weights))
```

The last trick is the "delayed" policy updates trick. The policy and its target networks are updated less frequently than the Q-value function. The paper recommends one policy update for every two Q-value function updates.

```
    # Training Policy Function
    # Trick Two: ''Delayed''İ Policy Updates. Update the
        policy less frequently
    if self.update_cnt % self.policy_target_update_interval ==
        0:
       with tf.GradientTape() as p_tape:
          new_action = self.policy_net.evaluate(
             state, eval_noise_scale=0.0
          ) # no noise, deterministic policy gradients
          new_q_input = tf.concat([state, new_action], 1)
          # ''' implementation 1 '''
          # predicted_new_q_value =
              tf.minimum(self.q_net1(new_q_input),self.q_net2
              (new_q_input))
          ''' implementation 2 '''
          predicted_new_q_value = self.q_net1(new_q_input)
          policy_loss = -tf.reduce_mean(predicted_new_q_value)
       p_grad = p_tape.gradient(policy_loss,
           self.policy_net.trainable_weights)
       self.policy_optimizer.apply_gradients(zip(p_grad,
           self.policy_net.trainable_weights))

       # Soft update the target nets
       self.target_q_net1 =
           self.target_soft_update(self.q_net1,
           self.target_q_net1, soft_tau)
       self.target_q_net2 =
           self.target_soft_update(self.q_net2,
           self.target_q_net2, soft_tau)
       self.target_policy_net =
           self.target_soft_update(self.policy_net,
           self.target_policy_net, soft_tau)
```

The following code is part of the training code. We first create the environment and the agent.

```
# initialization of env
env = gym.make(ENV_ID).unwrapped
state_dim = env.observation_space.shape[0]
action_dim = env.action_space.shape[0]
action_range = env.action_space.high # scale action,
    [-action_range, action_range]

# reproducible
env.seed(RANDOM_SEED)
random.seed(RANDOM_SEED)
np.random.seed(RANDOM_SEED)
tf.random.set_seed(RANDOM_SEED)

# initialization of buffer
replay_buffer = ReplayBuffer(REPLAY_BUFFER_SIZE)
# initialization of trainer
agent = TD3(state_dim, action_dim, action_range, HIDDEN_DIM,
    replay_buffer,
        POLICY_TARGET_UPDATE_INTERVAL, Q_LR, POLICY_LR)
t0 = time.time()
```

Before starting each episode, it is necessary to do some initialization. In this code, the training time is limited by the total number of steps instead of the max episode iterations. And because the network is built in different ways, you need an extra call of the function before using it.

```
# training loop
if args.train:
    frame_idx = 0
    all_episode_reward = []
    # need an extra call here to make inside functions be able
        to use model.forward
    state = env.reset().astype(np.float32)
    agent.policy_net([state])
    agent.target_policy_net([state])
```

At the very beginning, a random sample was used by the agent to provide enough data for the update. After that the agent was left to interact with the environment and update at every step as usual.

```
    for episode in range(TRAIN_EPISODES):
        state = env.reset().astype(np.float32)
        episode_reward = 0
        for step in range(MAX_STEPS):
            if RENDER:
                env.render()
            if frame_idx > EXPLORE_STEPS:
                action = agent.policy_net.get_action(state,
                    EXPLORE_NOISE_SCALE)
```

```
            else:
                action = agent.policy_net.sample_action()

            next_state, reward, done, _ = env.step(action)
            next_state = next_state.astype(np.float32)
            done = 1 if done is True else 0

            replay_buffer.push(state, action, reward,
                next_state, done)
            state = next_state
            episode_reward += reward
            frame_idx += 1

            if len(replay_buffer) > BATCH_SIZE:
                for i in range(UPDATE_ITR):
                    agent.update(BATCH_SIZE, EVAL_NOISE_SCALE,
                        REWARD_SCALE)
            if done:
                break
```

Finally, we provide the necessary functions to visualize the training process and save the agent.

```
        if episode == 0:
            all_episode_reward.append(episode_reward)
        else:
            all_episode_reward.append(all_episode_reward[-1] *
                0.9 + episode_reward * 0.1)
        print(
            'Training | Episode: {}/{} | Episode Reward: {:.4f}
                | Running Time: {:.4f}'.format(
                episode+1, TRAIN_EPISODES, episode_reward,
                time.time() - t0
            )
        )
    agent.save()
    plt.plot(all_episode_reward)
    if not os.path.exists('image'):
        os.makedirs('image')
    plt.savefig(os.path.join('image', 'td3.png'))
```

6.5.4 SAC: Pendulum-v0

SAC is an entropy method. The target q value uses the minimum of the two Q network and log probability of policy $\pi(\tilde{a}|s)$. The example code used `ReplayBuffer` class, `SoftQNetwork` class, `PolicyNetwork` class, and `SAC` class.

The `ReplayBuffer` class and the `SoftQNetwork` class are almost the same as `ReplayBuffer` class and `QNetwork` class in TD3 code, so we can skip them and see the code after.

```
class ReplayBuffer: # a ring buffer for storing transitions and
    sampling for training
   def __init__(self, capacity):
       ...
   def push(self, state, action, reward, next_state, done):
       ...
   def sample(self, batch_size):
       ...
   def __len__(self):
       ...

class SoftQNetwork(Model): # the network for evaluate values of
    state-action pairs: Q(s,a)
   def __init__(self, num_inputs, num_actions, hidden_dim,
       init_w=3e-3):
       ...
   def forward(self, input):
       ...
```

The `PolicyNetwork` class is also similar. The difference is that SAC uses a stochastic policy network instead of a deterministic policy network, which causes some differences.

```
class PolicyNetwork(Model):
   def __init__(self, num_inputs, num_actions, hidden_dim,
       action_range=1., init_w=3e-3, log_std_min=-20,
       log_std_max=2):
       ...
   def forward(self, state):
       ...
   def evaluate(self, state, epsilon=1e-6):
       ...
   def get_action(self, state, greedy=False):
       ...
   def sample_action(self):
       ...
```

The stochastic network outputs mean values and log standard deviations to depict an action distribution. Therefore, the network has two outputs.

```
class PolicyNetwork(Model):
   def __init__(self, num_inputs, num_actions, hidden_dim,
       action_range=1., init_w=3e-3, log_std_min=-20,
       log_std_max=2):
       super(PolicyNetwork, self).__init__()
       self.log_std_min = log_std_min
       self.log_std_max = log_std_max
       w_init = tf.keras.initializers.glorot_normal(seed=None)
```

```
    self.linear1 = Dense(n_units=hidden_dim, act=tf.nn.relu,
        W_init=w_init, in_channels=num_inputs, name='policy1')
    self.linear2 = Dense(n_units=hidden_dim, act=tf.nn.relu,
        W_init=w_init, in_channels=hidden_dim, name='policy2')
    self.linear3 = Dense(n_units=hidden_dim, act=tf.nn.relu,
        W_init=w_init, in_channels=hidden_dim, name='policy3')
    self.mean_linear = Dense(n_units=num_actions,
        W_init=w_init,
        b_init=tf.random_uniform_initializer(-init_w, init_w),
        in_channels=hidden_dim, name='policy_mean')
    self.log_std_linear = Dense(n_units=num_actions,
        W_init=w_init,
        b_init=tf.random_uniform_initializer(-init_w, init_w),
        in_channels=hidden_dim, name='policy_logstd')
    self.action_range = action_range
    self.num_actions = num_actions
```

A clip method on log standard deviations is used in the `forward()` function to prevent the standard deviation value from becoming too large.

```
def forward(self, state):
    x = self.linear1(state)
    x = self.linear2(x)
    x = self.linear3(x)
    mean = self.mean_linear(x)
    log_std = self.log_std_linear(x)
    log_std = tf.clip_by_value(log_std, self.log_std_min,
        self.log_std_max)
    return mean, log_std
```

The `evaluate()` function uses a reparameterization trick to sample actions from the action distribution so that the gradient here can be propagated back. It also calculates the log probability of the sampled actions on the original action distribution.

```
def evaluate(self, state, epsilon=1e-6):
    state = state.astype(np.float32)
    mean, log_std = self.forward(state)
    std = tf.math.exp(log_std) # no clip in evaluation, clip
        affects gradients flow
    normal = Normal(0, 1)
    z = normal.sample(mean.shape)
    action_0 = tf.math.tanh(mean + std * z) # TanhNormal
        distribution as actions; reparameterization trick
    action = self.action_range * action_0
    # according to original paper, with an extra last term for
        normalizing different action range
    log_prob = Normal(mean, std).log_prob(mean + std * z) -
        tf.math.log(1. - action_0 ** 2 + epsilon) -
        np.log(self.action_range)
    # both dims of normal.log_prob and -log(1-a**2) are
        (N,dim_of_action);
    # the Normal.log_prob outputs the same dim of input
        features instead of 1 dim probability,
```

```
    # needs sum up across the dim of actions to get 1 dim
        probability; or else use Multivariate Normal.
    log_prob = tf.reduce_sum(log_prob, axis=1)[:, np.newaxis]
        # expand dim as reduce_sum causes 1 dim reduced
    return action, log_prob, z, mean, log_std
```

The `get_action()` function is a simplified version of the previous function. It only samples actions from the action distributions.

```
def get_action(self, state, greedy=False):
    mean, log_std = self.forward([state])
    std = tf.math.exp(log_std)
    normal = Normal(0, 1)
    z = normal.sample(mean.shape)
    action = self.action_range * tf.math.tanh(
        mean + std * z
    ) # TanhNormal distribution as actions; reparameterization
        trick

    action = self.action_range * tf.math.tanh(mean) if greedy
        else action
    return action.numpy()[0]
```

The `sample_action()` function is much more simple. It is only used at the very beginning of training to sample data for the first update.

```
def sample_action(self, ):
    a = tf.random.uniform([self.num_actions], -1, 1)
    return self.action_range * a.numpy()
```

The structure of the SAC we will talk about next is as follows:

```
class SAC():
    def __init__(self, state_dim, action_dim, replay_buffer,
        hidden_dim, action_range, soft_q_lr=3e-4, policy_lr=3e-4,
        alpha_lr=3e-4): # create networks and variables
        ...
    def target_ini(self, net, target_net): # hard-copy update for
        initializing target networks
        ...
    def target_soft_update(self, net, target_net, soft_tau): #
        soft update the target net with Polyak averaging
        ...
    def update(self, batch_size, reward_scale=10.,
        auto_entropy=True, target_entropy=-2, gamma=0.99,
        soft_tau=1e-2): # update all networks in SAC
        ...
    def save(self): # save trained weights
        ...
    def load(self): # load trained weights
        ...
```

There are 5 networks in SAC algorithm. They are two soft Q-networks and their target networks, and a stochastic policy network. An alpha variable is also needed as a trade-off coefficient for the entropy regularization.

```
class SAC():
    def __init__(self, state_dim, action_dim, replay_buffer,
        hidden_dim, action_range, soft_q_lr=3e-4, policy_lr=3e-4,
        alpha_lr=3e-4):
        self.replay_buffer = replay_buffer

        # initialize all networks
        self.soft_q_net1 = SoftQNetwork(state_dim, action_dim,
            hidden_dim)
        self.soft_q_net2 = SoftQNetwork(state_dim, action_dim,
            hidden_dim)
        self.target_soft_q_net1 = SoftQNetwork(state_dim,
            action_dim, hidden_dim)
        self.target_soft_q_net2 = SoftQNetwork(state_dim,
            action_dim, hidden_dim)
        self.policy_net = PolicyNetwork(state_dim, action_dim,
            hidden_dim, action_range)
        self.log_alpha = tf.Variable(0, dtype=np.float32,
            name='log_alpha')
        self.alpha = tf.math.exp(self.log_alpha)
        print('Soft Q Network (1,2): ', self.soft_q_net1)
        print('Policy Network: ', self.policy_net)
        # set mode
        self.soft_q_net1.train()
        self.soft_q_net2.train()
        self.target_soft_q_net1.eval()
        self.target_soft_q_net2.eval()
        self.policy_net.train()

        # initialize weights of target networks
        self.target_soft_q_net1 =
            self.target_ini(self.soft_q_net1,
            self.target_soft_q_net1)
        self.target_soft_q_net2 =
            self.target_ini(self.soft_q_net2,
            self.target_soft_q_net2)

        self.soft_q_optimizer1 = tf.optimizers.Adam(soft_q_lr)
        self.soft_q_optimizer2 = tf.optimizers.Adam(soft_q_lr)
        self.policy_optimizer = tf.optimizers.Adam(policy_lr)
        self.alpha_optimizer = tf.optimizers.Adam(alpha_lr)
```

Let us introduce the `update()` function next. The other functions are the same as the previous code, so we can skip them. As usual, at the beginning of the `update()` function, we sample data from the replay buffer first. A normalization on reward values can improve the training effect.

```
    def update(self, batch_size, reward_scale=10.,
        auto_entropy=True, target_entropy=-2, gamma=0.99,
        soft_tau=1e-2):
        state, action, reward, next_state, done =
            self.replay_buffer.sample(batch_size)
        reward = reward[:, np.newaxis] # expand dim
        done = done[:, np.newaxis]
        reward = reward_scale * (reward - np.mean(reward, axis=0))
            / (
              np.std(reward, axis=0) + 1e-6
        ) # normalize with batch mean and std; plus a small number
            to prevent numerical problem
```

After that, we will calculate the target Q-value based on the next state. SAC used the minimum of the two target networks which is the same as TD3. But they differ in that SAC adds entropy regularization when calculating target q value. The `log_prob` part here is the entropy which is a measure of randomness in the policy.

```
        # Training Q Function
        new_next_action, next_log_prob, _, _, _ =
            self.policy_net.evaluate(next_state)
        target_q_input = tf.concat([next_state, new_next_action],
            1) # the dim 0 is number of samples
        target_q_min = tf.minimum(
           self.target_soft_q_net1(target_q_input),
               self.target_soft_q_net2(target_q_input)
        ) - self.alpha * next_log_prob
        target_q_value = reward + (1 - done) * gamma *
            target_q_min # if done==1, only reward
```

After calculating the target Q-value, training the Q function is simple.

```
        q_input = tf.concat([state, action], 1) # the dim 0 is
            number of samples
        with tf.GradientTape() as q1_tape:
           predicted_q_value1 = self.soft_q_net1(q_input)
           q_value_loss1 =
               tf.reduce_mean(tf.losses.mean_squared_error
               (predicted_q_value1, target_q_value))
        q1_grad = q1_tape.gradient(q_value_loss1,
            self.soft_q_net1.trainable_weights)
        self.soft_q_optimizer1.apply_gradients(zip(q1_grad,
            self.soft_q_net1.trainable_weights))
        with tf.GradientTape() as q2_tape:
           predicted_q_value2 = self.soft_q_net2(q_input)
           q_value_loss2 =
               tf.reduce_mean(tf.losses.mean_squared_error
               (predicted_q_value2, target_q_value))
```

```
q2_grad = q2_tape.gradient(q_value_loss2,
    self.soft_q_net2.trainable_weights)
self.soft_q_optimizer2.apply_gradients(zip(q2_grad,
    self.soft_q_net2.trainable_weights))
```

The policy loss is the expected future return plus expected future entropy. By maximizing the loss function, the policy will be trained to maximize a trade-off between expected return and entropy.

```
# Training Policy Function
with tf.GradientTape() as p_tape:
   new_action, log_prob, z, mean, log_std =
       self.policy_net.evaluate(state)
   new_q_input = tf.concat([state, new_action], 1) # the
       dim 0 is number of samples
   ''' implementation 1 '''
   predicted_new_q_value =
       tf.minimum(self.soft_q_net1(new_q_input),
       self.soft_q_net2(new_q_input))
   # ''' implementation 2 '''
   # predicted_new_q_value = self.soft_q_net1(new_q_input)
   policy_loss = tf.reduce_mean(self.alpha * log_prob -
       predicted_new_q_value)
p_grad = p_tape.gradient(policy_loss,
    self.policy_net.trainable_weights)
self.policy_optimizer.apply_gradients(zip(p_grad,
    self.policy_net.trainable_weights))
```

Finally, we update the entropy trade-off coefficient alpha and target networks.

```
# Updating alpha w.r.t entropy
# alpha: trade-off between exploration (max entropy) and
    exploitation (max Q)
if auto_entropy is True:
   with tf.GradientTape() as alpha_tape:
       alpha_loss = -tf.reduce_mean((self.log_alpha *
           (log_prob + target_entropy)))
   alpha_grad = alpha_tape.gradient(alpha_loss,
       [self.log_alpha])
   self.alpha_optimizer.apply_gradients(zip(alpha_grad,
       [self.log_alpha]))
   self.alpha = tf.math.exp(self.log_alpha)
else: # fixed alpha
   self.alpha = 1.
   alpha_loss = 0

# Soft update the target value nets
self.target_soft_q_net1 =
    self.target_soft_update(self.soft_q_net1,
    self.target_soft_q_net1, soft_tau)
```

```
    self.target_soft_q_net2 =
        self.target_soft_update(self.soft_q_net2,
        self.target_soft_q_net2, soft_tau)
```

The main loop process of training is the same as TD3. First, build the environment and the agent.

```
# initialization of env
env = gym.make(ENV_ID).unwrapped
state_dim = env.observation_space.shape[0]
action_dim = env.action_space.shape[0]
action_range = env.action_space.high # scale action,
    [-action_range, action_range]

# reproducible
env.seed(RANDOM_SEED)
random.seed(RANDOM_SEED)
np.random.seed(RANDOM_SEED)
tf.random.set_seed(RANDOM_SEED)

# initialization of buffer
replay_buffer = ReplayBuffer(REPLAY_BUFFER_SIZE)
# initialization of trainer
agent = SAC(state_dim, action_dim, action_range, HIDDEN_DIM,
            replay_buffer, SOFT_Q_LR, POLICY_LR, ALPHA_LR)
t0 = time.time()
```

Then, use the agent to interact with the environment and store sampled data for updates. Before the first update, a random action sample is used.

```
# training loop
if args.train:
    frame_idx = 0
    all_episode_reward = []

    # need an extra call here to make inside functions be able
        to use model.forward
    state = env.reset().astype(np.float32)
    agent.policy_net([state])

    for episode in range(TRAIN_EPISODES):
        state = env.reset().astype(np.float32)
        episode_reward = 0
        for step in range(MAX_STEPS):
            if RENDER:
                env.render()
            if frame_idx > EXPLORE_STEPS:
                action = agent.policy_net.get_action(state)
            else:
                action = agent.policy_net.sample_action()
            next_state, reward, done, _ = env.step(action)
            next_state = next_state.astype(np.float32)
```

```
            done = 1 if done is True else 0
            replay_buffer.push(state, action, reward,
                next_state, done)
            state = next_state
            episode_reward += reward
            frame_idx += 1
```

When enough data are collected, we can start to update at each step.

```
            if len(replay_buffer) > BATCH_SIZE:
                for i in range(UPDATE_ITR):
                    agent.update(
                        BATCH_SIZE, reward_scale=REWARD_SCALE,
                            auto_entropy=AUTO_ENTROPY,
                        target_entropy=-1. * action_dim
                    )
            if done:
                break
```

Through the above steps, the agent can become more and more powerful through updates. The next step is to better represent the training process.

```
        if episode == 0:
            all_episode_reward.append(episode_reward)
        else:
            all_episode_reward.append(all_episode_reward[-1] *
                0.9 + episode_reward * 0.1)
        print(
            'Training | Episode: {}/{} | Episode Reward: {:.4f}
                | Running Time: {:.4f}'.format(
                episode+1, TRAIN_EPISODES, episode_reward,
                time.time() - t0
            )
        )
```

Finally, save the agent and plot the figure.

```
    agent.save()
    plt.plot(all_episode_reward)
    if not os.path.exists('image'):
        os.makedirs('image')
    plt.savefig(os.path.join('image', 'sac.png'))
```

References

Fox R, Pakman A, Tishby N (2016) Taming the noise in reinforcement learning via soft updates. In: Proceedings of the thirty-second conference on uncertainty in artificial intelligence. AUAI Press, Corvallis, pp 202–211

Fujimoto S, van Hoof H, Meger D (2018) Addressing function approximation error in actor-critic methods. arXiv:180209477

Haarnoja T, Tang H, Abbeel P, Levine S (2017) Reinforcement learning with deep energy-based policies. In: Proceedings of the 34th international conference on machine learning, vol 70, pp 1352–1361. JMLR.org

Haarnoja T, Zhou A, Abbeel P, Levine S (2018a) Soft actor-critic: off-policy maximum entropy deep reinforcement learning with a stochastic actor. arXiv:180101290

Haarnoja T, Zhou A, Hartikainen K, Tucker G, Ha S, Tan J, Kumar V, Zhu H, Gupta A, Abbeel P, et al (2018b) Soft actor-critic algorithms and applications. arXiv:181205905

It K, McKean H (1965) Diffusion processes and their sample paths. Die Grundlehren der math Wissenschaften, vol 125. Springer, Berlin

Levine S, Koltun V (2013) Guided policy search. In: International conference on machine learning, pp 1–9

Lillicrap TP, Hunt JJ, Pritzel A, Heess N, Erez T, Tassa Y, Silver D, Wierstra D (2015) Continuous control with deep reinforcement learning. arXiv:150902971

Mnih V, Kavukcuoglu K, Silver D, Rusu AA, Veness J, Bellemare MG, Graves A, Riedmiller M, Fidjeland AK, Ostrovski G, Petersen S, Beattie C, Sadik A, Antonoglou I, King H, Kumaran D, Wierstra D, Legg S, Hassabis D (2015) Human-level control through deep reinforcement learning. Nature 518(7540):529–533

Nachum O, Norouzi M, Xu K, Schuurmans D (2017) Bridging the gap between value and policy based reinforcement learning. In: Advances in neural information processing systems, pp 2775–2785

Polyak BT (1964) Some methods of speeding up the convergence of iteration methods. USSR Comput Math Math Phys 4(5):1–17

Silver D, Lever G, Heess N, Degris T, Wierstra D, Riedmiller M (2014) Deterministic policy gradient algorithms. In: Proceedings of the 31st international conference on machine learning

Sutton RS, Barto AG (2018) Reinforcement learning: an introduction. MIT Press, Cambridge

Uhlenbeck GE, Ornstein LS (1930) On the theory of the Brownian motion. Phys. Rev. 36(5):823

Williams RJ (1992) Simple statistical gradient-following algorithms for connectionist reinforcement learning. Mach Learn 8(3–4):229–256

Ziebart BD, Maas AL, Bagnell JA, Dey AK (2008) Maximum entropy inverse reinforcement learning. In: Proceedings of the AAAI conference on artificial intelligence, Chicago, vol 8, pp 1433–1438

Part II
Research

Zihan Ding
e-mail: zhding@mail.ustc.edu.cn

The following chapters introduce selected research topics in deep reinforcement learning, which would be useful if the readers would like to have a deeper understanding or start the related research.

We first discuss several challenges of deep reinforcement learning in Chap. 7, including sample efficiency, learning stability, catastrophic interference, exploration, meta-learning and representation learning, multi-agent reinforcement learning, sim2real, and large-scale reinforcement learning. Then we compose six chapters to introduce different advanced research topics and algorithms in deep reinforcement learning, as well as indicating how they are related to and applied for solving those challenges. From a research perspective, lots of recent advances are introduced in this part of contents with seven chapters in total.

Chapter 8 introduces the imitation learning in relatively comprehensive perspectives. Combination of imitation learning and reinforcement learning helps to alleviate the challenge of low sample efficiency in deep reinforcement learning, through leveraging the expert demonstrations in the learning process.

Chapter 9 introduces model-based reinforcement learning, which also improves the learning efficiency in deep reinforcement learning, but through leveraging the models of environments. Model-based reinforcement learning is a worthwhile direction with fruitful advanced contents when facing real-world applications, as well as a frontier hotspot.

Chapter 10 describes hierarchical reinforcement learning, which helps with problems including catastrophic interference and hard exploration in deep reinforcement learning, as well as improving the overall learning efficiency. Options framework and feudal reinforcement learning are highlighted in the description.

Chapter 11 describes the concept of multi-agent reinforcement learning, as an extension of reinforcement learning to tasks with more than one agent. Competitive and collaborative relationships among agents, Nash equilibrium and some multi-agent reinforcement learning algorithms are detailed in this chapter.

Chapter 12 introduces parallel computing in deep reinforcement learning, for solving the scalability challenge and improving the learning speed in wall-clock time. Different parallel training frameworks are introduced in this chapter, which helps to employ deep reinforcement learning in large-scale real-world applications.

The related codes are released in the following link: https://github.com/deep-reinforcement-learning-book.

Chapter 7
Challenges of Reinforcement Learning

Zihan Ding and Hao Dong

Abstract This chapter introduces the existing challenges in deep reinforcement learning research and applications, including: (1) the sample efficiency problem; (2) stability of training; (3) the catastrophic interference problem; (4) the exploration problems; (5) meta-learning and representation learning for the generality of reinforcement learning methods across tasks; (6) multi-agent reinforcement learning with other agents as part of the environment; (7) sim-to-real transfer for bridging the gaps between simulated environments and the real world; (8) large-scale reinforcement learning with parallel training frameworks to shorten the wall-clock time for training, etc. This chapter proposes the above challenges with potential solutions and research directions, as the primers of the advanced topics in the second main part of the book, including Chaps. 8–12, to provide the readers a relatively comprehensive understanding about the deficiencies of present methods, recent development, and future directions in deep reinforcement learning.

Keywords Sample efficiency · Stability · Catastrophic interference · Exploration · Meta-learning · Representation learning · Generality · Multi-agent reinforcement learning · Sim2real · Scalability

7.1 Sample Efficiency

A **sample-efficient** (or **data-efficient**) algorithm in reinforcement learning means that the algorithm can make better use of the collected samples, so that it can learn to improve the policy faster. With the same number of training samples (e.g., the time steps in reinforcement learning), a sample-efficient method can provide a

Z. Ding (✉)
Imperial College London, London, UK
e-mail: zhding@mail.ustc.edu.cn

H. Dong
Peking University, Beijing, China
e-mail: hao.dong@pku.edu.cn

H. Dong et al. (eds.), *Deep Reinforcement Learning*,
https://doi.org/10.1007/978-981-15-4095-0_7

superior performance on the learning curve or final results, compared with other "sample-inefficient" methods. Take the game Pong as an example: a normal human only needs dozens of trials to basically master the game and achieve a relatively good score. However, for present reinforcement learning algorithms (especially with model-free methods), it may need at least tens of thousands of samples to gradually learn some useful policies. This forms a crucial problem in reinforcement learning: how can we design a more efficient reinforcement learning algorithm for an agent to learn faster with fewer examples?

The importance of this problem is mostly due to the cost of real-time or real-world interactions between the agent and the environment, or even the time and energy consumption of the interactions in simulated environments at present. Most of present reinforcement learning algorithms are of such a low learning efficiency on a large-scale or continuous-space problem that a typical training process even with fast simulation still requires unbearable waiting time with current computational power. It can only be worse for real-world interactions. Potential problems of time consumption, wear and tear of equipment, safety during the exploration of reinforcement learning and risks of failure cases all make stricter requirements on the learning efficiency of reinforcement learning methods in practice.

Improving data efficiency requires either having informative prior knowledge or extracting information more efficiently from available data. Starting from these two points, there are several approaches in present literature for solving the problem of learning efficiency:

- Learning from expert demonstrations. The idea of learning from demonstrations requires an expert to provide samples with high reward values. It actually falls into an category called "imitation learning," which tries not only to mimic the expert actions but also to learn a generalized policy for handling unseen cases as well. The combination of imitation learning and reinforcement learning is actually a very promising area which has been heavily investigated in recent years for applications like the game of Go, robotic learning, etc., to alleviate the problem of low learning efficiency of reinforcement learning.

  The key of learning from expert demonstrations is to extract the underlying principles for generating good actions from the available demonstration dataset, and apply it to more general cases. More contents about learning from demonstrations are discussed in Chap. 8.
- Model-based reinforcement learning rather than model-free reinforcement learning. As introduced in the previous chapter, a model-based reinforcement learning method usually indicates that the agent not only learns a policy for predicting its action, but also learns a model of the environment for assisting the planning, thereby accelerating the learning process of the policy. The model of the environment basically contains two models: a transition model, which gives the state change after the agent makes an action, and a reward model, which determines how much reward the agent will get from the environment as a feedback of its action.

Learning an accurate model of the environment provides additional information for better evaluation of the agent's current policy, which could potentially make the learning process more efficient. However, the model-based methods have their own drawbacks. For instance, model-based methods always suffer from **"model bias"** problem in practice, i.e. the model-based methods usually inherently assume that the learned dynamic model of the environment sufficiently and accurately resembles the real one. But this may not always be true if there are only a few samples for the model to learn from, leading to an inaccurate model. It could be problematic when the policy learned together with the inaccurate or biased model is employed in the true environment.

One of the efficient model-based reinforcement learning algorithms is called PILCO (Deisenroth and Rasmussen 2011), which applies non-parametric probabilistic model Gaussian Processes (GPs) to resemble the dynamic model of the environment. It leverages the straightforward solving process of GP methods for efficient model learning, instead of using the neural network approximation. The policy evaluation and improvement are performed based on the learned probabilistic model. For a cart-double-pendulum swing up task in the real world, the PILCO method only takes about 20–30 trials to learn an effective policy for controlling, while the other methods like multi-layered perceptrons for learning a dynamic model will finally take at least hundreds of trials. However, the PILCO method has its own problems as well in that it cannot guarantee to search an optimal control as the non-convex optimization problem for learning the policy parameters, and the solving process of a GP is not scalable to high-dimensional parameter space for complicated models. Other model-based methods together with a general overview of model-based reinforcement learning are introduced in Chap. 9.

- Design more efficient learning algorithms through solving existing defects. The above two methods are trying to solve the learning efficiency problem through leveraging additional external information. If no extra information can be leveraged or the dynamic model of environments is hard to learn accurately, we should improve the efficiency of algorithms without extra information. There are usually two categories of reinforcement learning algorithms according to their updating manner: on-policy and off-policy, as described in previous chapters. The on-policy methods can evaluate the policy with less bias but larger variances, while the off-policy can leverage a large batch of randomly sampled data to achieve lower variances.

Many advanced and efficient algorithms have been proposed in recent years. Most of them are targeted at some specific defects of conventional algorithms. For reducing the variance of policy gradients, the critic network is introduced to evaluate the action-value function in actor-critic; for scaling reinforcement learning tasks from small scale to large scale, deep neural networks are employed in DQN to improve the tabular-based Q-learning algorithm. To address the overestimation problem of the max operator in DQN updating rules, the double DQN method is proposed with an additional Q-network. For boosting exploration, a noisy DQN is proposed with parameter noise, and soft actor-critic (abbreviated as SAC, introduced in Chap. 6) is created with adaptive entropy for the probabilistic distribution given by the policy. To extend the DQN methods from solving only the discrete tasks to continuous cases, deep deterministic policy gradient (abbreviated as DDPG, introduced in Chap. 6) algorithm is proposed. In order to stabilize the learning process of DDPG, twin-delayed DDPG (abbreviated as TD3, introduced in Chap. 6) is proposed with additional networks and delayed update schedule. To ensure a safe update in on-policy reinforcement learning policy optimization, trust-region-based algorithms like trust region policy optimization (abbreviated as TRPO, introduced in Chap. 5) are proposed. To reduce the computational time with second-order optimization in TRPO, the PPO (abbreviated as PPO, introduced in Chap. 5) algorithm is proposed with first-order approximation. For accelerating the second-order natural gradient descent methods, the algorithm actor-critic using Kronecker-factored trust region (abbreviated as ACKTR, introduced in Chap. 5) is proposed to use the Kronecker-factored method for approximating the inverse fisher information matrix in second-order optimization process. Maximum a posteriori policy optimization (MPO) (Abdolmaleki et al. 2018) and its on-policy variant V-MPO (Song et al. 2019) relate to the policy optimization in a perspective of "reinforcement learning as inference." The MPO employs probabilistic inference tools like expectation maximization (EM) for optimizing a maximum entropy reinforcement learning objective. The above algorithms are just a small proportion of the overall development in the field of reinforcement learning algorithms. We direct the readers to the literature for more algorithms for improving the efficiency and other drawbacks of reinforcement learning. At the same time, the structures of proposed reinforcement learning algorithms are becoming more and more complicated, with more flexible parameters either being learned adaptively or being manually chosen, which requires more delicate considerations in the research of reinforcement learning. Sometimes those additional hyper-parameters improve the learning performances greatly, but sometimes they could also make the learning process more sensitive, of which you should take care case by case.

- In the above cases we assume the data samples are information-rich, but the learning efficiency of reinforcement learning algorithms is low. In practice, it is usually common to see the samples' lack of useful information, especially for the sparse-reward tasks. For example, for a single binary valued success label of task completion, the intermediate samples may all have an immediate zero reward without any discrimination. The information contained in those samples are naturally scarce. In cases like this, the way to effectively explore the learning space without the reward instructions can be crucial. Techniques like hindsight experience replay (Andrychowicz et al. 2017), hierarchical learning structure (Kulkarni et al. 2016), intrinsic reward (Sukhbaatar et al. 2018), curiosity-driven exploration (Pathak et al. 2017), and other effective exploration strategies (Houthooft et al. 2016) are applied in some works. The learning efficiency in reinforcement learning is significantly affected by the exploration process due to the intrinsic properties of reinforcement learning, and effective exploration can improve the efficiency of learning from samples through gathering more informative samples. As exploration is another big challenge in reinforcement learning, it will be discussed individually in one of the following sections.

7.2 Learning Stability

Deep reinforcement learning can be terribly unstable or stochastic. Here the term "unstable" indicates the differences of the learning performances during time for single run or for horizontal comparison across multiple runs. The unstable learning process during time shows up as the large local variances or non-monotonicity on the single learning curve, e.g. sometimes the learning performance even degrades for some reasons. And the unstable learning for different runs displays as a large difference in the performances across trials at each stage during training, resulting in large variances for horizontal comparisons.

The unstable and unpredictable properties of deep neural networks are further exacerbated in the deep reinforcement learning domain, due to shifting objective distribution, unsatisfied requirements of independent and identically distributed (i.i.d.) data, the unstable biased estimation of value function approximation, and so on. These factors lead to noise in the gradient estimators, which further causes the unstable learning performances. Different from learning a fixed training dataset in supervised learning (not considering the batch constrained reinforcement learning), reinforcement learning methods usually learn from the samples that are highly correlated. For example, the learning agent mostly takes samples explored with the policy, either by the current policy for on-policy learning or the previous policy for off-policy learning (sometimes even other policies). Samples generated during the sequential interactions between the agent and the environment can be highly correlated, which breaks the independent requirement for effective learning with neural networks. Since the value function is evaluated on the trajectories chosen by the current policy, there is a dependency relationship of value function on the policy for estimating it. Due to the policy changing over the training time, the optimization manifold of the parameterized value function changes over time as well. Considering the policy is usually stochastic for the benefits of exploration during training, the value function is even more untraceable. This ends up with the unsatisfied condition of identically distributed data for learning. The unstable learning is mostly caused by the variances in policy gradient or value function estimation. However, the biased estimation is another source of unstable performances in reinforcement learning, especially when the bias is unstable itself. For example, recall that in Chap. 2, the compatible function approximation condition needs to be satisfied so as to provide an unbiased estimation of action-value function $Q^{\pi}(s, a)$ with $Q^{w}(s, a)$. There are also several other conditions to ensure an unbiased estimation of value functions, as well as further requirements to ensure advanced reinforcement learning algorithms have accurate and correct gradients for improving the policy. However, in practice, those requirements or conditions are usually relaxed, which ends up with unstable biased estimation of value function, or large variances in the policy gradients. In

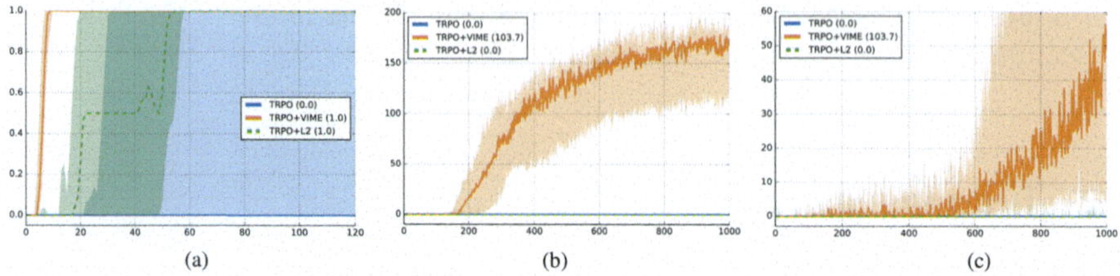

Fig. 7.1 Learning curves in experiments of VIME. Figure is adapted from Houthooft et al. (2016). (**a**) MountainCar. (**b**) CartPoleSwingup. (**c**) HalfCheetah

most cases people are talking about the bias and variance trade-off for estimation in reinforcement learning algorithms, but the unstable bias term itself can contribute in the unstable learning performances as well. There are also other factors contributing to the unstable learning performances, like randomness in the exploration strategy, randomness in the environment, random seeds for numerical calculations, etc.

Take the example of some experiments in the paper by Houthooft et al. (2016), which proposes the variational information maximizing exploration (VIME) as an exploration strategy to be applied on general reinforcement learning algorithms. Some learning performances are displayed in their comparison of algorithms, and the learning results using TRPO or TRPO+VIME for three different environments almost all show large variances in their learning curves, as shown in Fig. 7.1. For the environment MountainCar, the learning curves of TRPO algorithm could cover the whole range of reward value [0, 1], and it is almost a similar case with TRPO+VIME method for HalfCheetah environment. We need to note that TRPO is already a relatively stable reinforcement learning algorithm than other algorithms for most cases, with the second-order optimization in gradient descent and the trust-region constraint. Other algorithms like DDPG can be even more unstable during training, the noisy exploration can even deteriorate the learning performance after training for a long time (Fujimoto et al. 2018).

The randomness of the learning process with reinforcement learning makes it hard to evaluate the performance of algorithms accurately, which also addresses the importance of applying different random seeds to get averaged results.

Previous investigations (Henderson et al. 2018) about deep reinforcement learning give some conclusions about the instability and sensitivity of experiments in deep reinforcement learning:

- The policy network architecture can significantly impact results in both TRPO and DDPG.
- For hidden layers of policy network or value network, usually ReLU or leaky ReLU activations perform the best across environments and algorithms. The effects are not consistent across algorithms or environments.

- Reward rescaling can have a large effect, but results were inconsistent across environments and scaling values.
- Five random seeds (a common reporting metric) may not be enough to argue significant results, since with careful selection you can get non-overlapping confidence intervals for different random seeds even with exactly the same implementation.
- The stability of environment dynamics can severely affect the learning performance of reinforcement learning algorithms. For example, an unstable environment could diminish the effective learning performance of DDPG rapidly.

People have been working on solving the stability problem in reinforcement learning for a long time. To solve the variances in the cumulative reward function for the original REINFORCE algorithm, the value function approximation is introduced to estimate the reward value. Furthermore, the action-value function is also used for reward function approximation, which reduces the variances even if it is biased. Methods like this form a mainstream of deep reinforcement learning algorithms combining Q-learning with the policy gradient methods, as introduced in previous Chap. 6. In the original DQN (Mnih et al. 2013), the methods of using the target network with delayed update and the replay buffer help alleviate the problem of unstable learning. Usually a deep function approximator requires multiple gradient updates to converge instead of a single update, and the target network provides a stable objective during the learning process, which help with the convergence on the training data. To some extent, it satisfies the identically distributed requirement that is broken by reinforcement learning without the target networks. The replay buffer provides the DQN an off-policy learning manner, and randomly sampled data from the buffer for training is more close to the independent distributed data, which helps to stabilize the learning process as well. More details of the DQN are introduced in Chap. 4. Moreover, the TD3 algorithm (details in Chap. 6) applies the target policy smooth regularization on top of the stable techniques applied in DQN, with the smoothness assumption that similar actions should have similar value. Therefore the target value is estimated with noise on the action to reduce the variance. TD3 also employs a pair of critics instead of a single one in DDPG, the further stabilize the learning performances. On the other hand, for policy-gradient based methods, TRPO uses second-order optimization to provide more stable updates with more comprehensive information, as well as applying the constraints on updated policy to ensure conservative but steady improvements.

However, even with the above works, instability, randomness, and sensitivity to initialization and hyper-parameters make it difficult for reinforcement learning researchers to evaluate the algorithms across tasks and reproduce the results, which still forms a big challenge for the reinforcement learning community.

7.3 Catastrophic Interference

As reinforcement learning usually has a dynamics learning process instead of learning with a fixed dataset as in supervised learning, it can be regarded as a process of chasing a running goal with dataset being updated during the whole period. For example, in Chap. 2 we introduce the on-policy value function $V^{\pi}(s)$ and action-value function $Q^{\pi}(s, a)$, which are both estimated with the current policy π. But the policy is updated all the time during the learning process, which leads to a dynamic estimation of the value functions. Although applying the off-policy replay buffer helps to alleviate the problem with relatively stationary training dataset, the samples in the buffer still change along with the agent's exploration process. Therefore, a problem called **catastrophic interference** or **catastrophic forgetting** (Kirkpatrick et al. 2017) can happen during learning process especially when the policy or the value function is learnt based on the deep learning method with neural networks, and it describes the poor ability in handling this kind of incremental learning mentioned above. The new data usually makes the trained network change a lot to fit it, but forgets what it has learned in previous training process even if it is useful. This is a limitation of applying neural networks as approximators in reinforcement learning methods.

The natural and human-like learning process is actually on-policy learning, instead of the off-policy approach. Humans keep learning new things everyday in real time instead of learning from their memories all the time. However, the on-policy reinforcement learning still struggles to improve learning efficiency, and tries to prevent the catastrophic interference problem. Trust-region-based algorithms like TRPO and PPO make a constraint about the potential range of updated policy during learning, to ensure a steady but relatively slow improvement in learning performance. For on-policy learning, the data is usually collected as correlated data, which contributes to the catastrophic interference a lot. Therefore, off-policy learning methods apply an experience replay buffer for alleviating this problem, so that the old data will remain during learning to some extent. Techniques like prioritized experience replay and hindsight experience replay are proposed to leverage the data stored in the replay buffer according to their importance or goals in a sophisticated manner.

Catastrophic forgetting also happens when the learning process has multiple stages. For example, in the sim-to-real policy transfer process, the policy usually needs to be pre-trained in the simulated environment and then fine-tuned with the real-world data. However, the loss function for the two processes may be different in practice, and may not always be consistent with the overall reinforcement learning objective. Like in the work by Jeong et al. (2019a), the image observations are embedded in to latent representation as inputs of the policy, and the embedding network is fine-tuned for sim-to-real adaptation with a self-supervised loss instead of the original reinforcement learning loss in simulation training process. This kind of mismatch of the loss function in a multi-stage training process will cause catastrophic forgetting in practice, which means that the policy has chances to forget the skills obtained in pre-training. To solve that, freezing partial layers of the network and keep updating the network with previous loss function can help during the post-training process, which tries to maintain the pre-trained network to the best during post-training process. Another similar idea is the residual policy learning mentioned in Chap. 8 Sect. 8.6, which also freezes the weights of the pre-trained network but applies an additional network alongside to learn the corrections.

7.4 Exploration

Exploration is another main challenge in reinforcement learning, which greatly affects the learning efficiency as mentioned in previous section. Rather than discussing the exploration-exploitation trade-off, which is a classical and well-known problem in reinforcement learning mentioned in Chap. 2, we focus on the challenge of exploration itself here in this section. The hardness of exploration in reinforcement learning lies in sparse rewards, large action space, and non-stationary environments for exploration, as well as the safety problems in real-world exploration, etc. Exploration means finding more information about the environments through interactions, usually counter to exploitation, which denotes exploiting known information to maximize reward. The learning process of reinforcement learning is based on trial-and-error. An optimal policy cannot be learned unless those optimal trajectories have been explored before. For example, Atari games like Montezuma's Revenge, Pitfall in OpenAI Gym are hard to solve for general reinforcement learning algorithms due to the hardness of exploration, and the game scenes of them are shown in Fig. 7.2,[1] which usually contain a complicated maze to be solved with a long sequence of operations. They are like a maze solving problem but via more complicated structures and hierarchies. Montezuma's Revenge is a very typical example with sparse rewards in the task, which makes the exploration in reinforcement learning very hard to conduct. Within one game scene, the agent in Montezuma's Revenge has to finish dozens of subsequent actions to pass one room, while there are 23 rooms of different game scenes that the agent needs to navigate itself. A wrong action at each time step could potentially make the agent fail to pass. A similar case happens in the game Pitfall. These games are usually used as a benchmark for evaluating the exploration ability of reinforcement learning methods. OpenAI[2] and Deepmind (Aytar et al. 2018) have both claimed they have solved this game Montezuma's Revenge with efficient deep reinforcement learning methods. However, the results are actually not very satisfying. In both of their solutions, the expert demonstrations are leveraged to assist exploration. For example, in Deepmind's solution they let the agent watch the YouTube videos, while OpenAI uses human demonstrations for better initialization of agent's position.

[1] Figures source: https://gym.openai.com/envs/#atari.

[2] https://openai.com/blog/learning-montezumas-revenge-from-a-single-demonstration/.

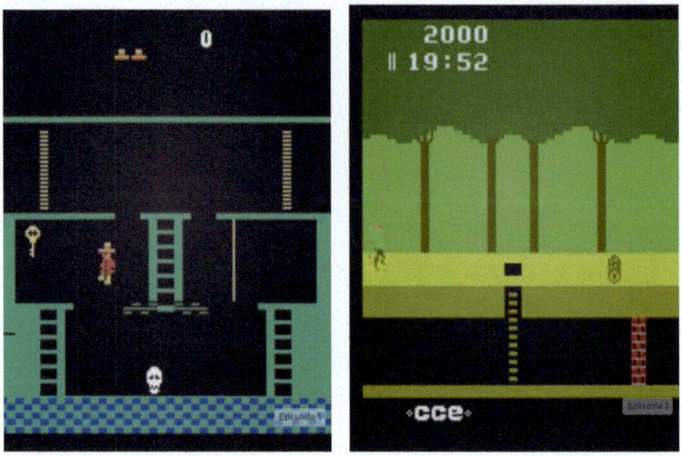

Fig. 7.2 Atari games that are hard to learn: Montezuma's revenge (left) and pitfall (right)

The bottleneck of this kind of sparse-reward task actually lies in exploration. Sparse rewards can make the value networks and policy networks optimized on hyper-surfaces that are not smooth and not convex, or even discontinuous at some stages of training. Therefore, the policy after one-step optimization may not help with exploring higher-reward regions. The agent would find it very hard to explore a high-reward trajectory during its exploration with traditional exploration strategy like random actions or ϵ-greedy policy. Even if they have sampled one near-optimal trajectory, the value-based or policy-based optimization methods may not pay enough attention to it, which could also end up with a failure or slow process of learning a good policy. The problems described above address the defects of current deep reinforcement learning methods.

Apart from the sparse rewards, large action space and non-stationary environments also raise the difficulty of exploration for reinforcement learning agents. A typical example is the StarCraft II game solved by Vinyals et al. (2019). Table 7.1[3] compares the Atari games, Go, and StarCraft in their information types, action space, moves in a game, and number of players. The large action space and length of game control sequences make it extremely hard for exploring a good policy in StarCraft. Moreover, the multi-player settings make opponents part of the game environment for the agent, which increases the hardness of exploration as well.

To solve the problem of exploration, researchers have been looking into concepts including imitation learning (as in Chap. 8), intrinsic reward/motivation, hierarchical learning (as in Chap. 10), etc. With imitation learning, the agent tries to mimic expert demonstrations from human or other sources to improve the efficiency of its learning with less difficulty in exploring near-optimal samples. Intrinsic motivation

[3]Data source: Oriol Vinyals, Deep Reinforcement Learning Workshop, NeurIPS 2019.

Table 7.1 Comparison of different games

	Atari	Go	StarCraft
Information type	Near-perfect	Perfect	Imperfect
Action space	17	361	10^{26}
Moves per game	100's	100's	1000's
Players	Single	Two	Multiple

is based on the notion that behavior is not just the result of external reward, but is also driven by internal desires, like acquiring more effective information about the unknown. For example, babies can learn about the world so fast with curiosity-driven exploration. Curiosity is one of the internal drives to improve the agent's learning towards the final goal. More internal drives are worth exploring in the research. Hierarchical learning decomposes the complicated and hard-to-explore tasks into smaller sub-tasks, which are easier to learn. For example, the feudal network (FuN) as a key method in feudal reinforcement learning applies a hierarchical structure with manager and worker to solve the Montezuma's Revenge via more effective exploration and learning (Vezhnevets et al. 2017).

In recent years, some new methods have been proposed to solve the exploration problem, one of them being Go-Explore, which is not a deep reinforcement learning solution. The main idea of Go-Explore is to first explore the game world using deterministic training without neural networks, i.e. not using deep reinforcement learning approaches, then to apply a deep neural network for imitation learning on the best trajectories, to make the policy robust to randomness of the environments. To solve the large-scale highly complicated game like StarCraft II, DeepMind's researchers (Vinyals et al. 2019) apply the population-based training framework to effectively explore the global optimal strategies, and the set of agents is called the league. Different agents are initialized around different clusters on the distribution, to ensure the diversity during exploration. The population-based training provides more thorough explorations than a single agent in the policy space.

Exploration in real-world tasks also corresponds with the safety problem. For example, when considering an autonomous driving car controlled by an agent, the failure cases with car accidents are what the agent is supposed to learn from. But an actual car cannot be used in reality to collect those failure cases for the agent to learn with a low and acceptable consumption. A real car cannot even take random actions for exploration, which could lead to disastrous results. The same problem happens in other real-world applications like robotic manipulations, robotic surgery, and so on. To solve this problem, sim-to-real transfer is developed for applying reinforcement learning in the real world, which achieves the training process in simulation and transfers the policy into reality.

7.5 Meta-Learning and Representation Learning

Apart from improving the learning efficiency on a specific task, researchers are seeking a way to improve the overall learning performance on different tasks, which relates to the **generality** and **versatility** of models. So how can we make the agent learn faster on a new task based on what it has learned from an old task? Several concepts can be introduced here, including meta-learning, representation learning, transfer learning, etc.

The problem of meta-learning can actually be traced back to 1980s–1990s (Bengio et al. 1990). Recent fast development deep learning and deep reinforcement learning bring this problem back into our sight. A lot of exciting new ideas are proposed, such as those based on model-agnostic meta-learning, and more powerful frameworks for learning across tasks are invented in recent years, which makes this area develop very fast. The original goal of meta-learning is to let the agent learn to solve different tasks or grasp different skills. However, we cannot suffer learning from scratch for each task, especially with deep learning methods for approximation. **Meta-learning**, also called **learning to learn**, is proposed to let the agent learn faster on a new task with previous experience, rather than regarding each task as an independent task. Usually a standard learner for learning a specific task is taken as an inner-loop learning process for meta-learning, while a meta-learner for learning to update the inner-loop learners is regarded as an outer-loop learning process. These two learning processes are optimized at the same time or in an iterative manner. Three main categories of meta-learning are: recurrent models, metric learning, and learning the optimizers. The combination of meta-learning and reinforcement learning gives the **meta-reinforcement learning** methods. An effective meta-reinforcement learning method like model-agnostic meta-learning (Finn et al. 2017) can solve a simple but new task with few-shot learning, or few steps for updating.

For a specific task domain, there may be some hidden correlations among different tasks. Can we enable the agent to master these underlying principles from some sampled tasks in this domain, and therefore generalize what have learned to other tasks, so as to learn them faster? Learning the underlying relationships or principles is related to a concept called **representation learning** (Bengio et al. 2013). Representation learning is originally proposed in machine learning, and is defined as learning the representations from the raw data and extract useful information or features for the classifiers or the predictors (like policies in reinforcement learning). Representation learning tries to learn some abstract and compact features to represent the raw materials, and with this kind of abstraction, the predictors or classifiers will not degrade their performances, but with a higher learning efficiency. Learning the hidden representation can be extremely useful for improving the learning efficiency of reinforcement learning, and transferring these general principles will benefit the learning process on different tasks. The representation learning is usually used for learning compact representation of complex states of reinforcement learning environments, which is called **state**

representation learning (SRL). The representation contains the invariance and distinction properties in a proper abstract space, which is distilled from variant domains. For example, in a sequence of frames of a video capturing the motion of objects, the set of the key points on the corners (or other specific points on the surfaces) of the object is an invariant and robust representation of the object motion, although the pixels in frames are always variant along with the objects' motion. Those key points are sometimes called the descriptors in computer vision terminology, within a descriptor space. Under this representation, the positions of those key points are changing during the object motion, and therefore can represent the motion of the object. Different objects will have different sets of key points, which can be used to distinguish them from each other. This area of representation learning for reinforcement learning is important when the reinforcement learning policy is transferred across domains, including different task domains, simulation-to-reality domain transfer, and so on. It is promising and still under exploration, which provides a direction for exploring how humans leverage the knowledge for planning.

7.6 Multi-Agent Reinforcement Learning

In the chapters we introduced above, there is only one agent trying to find its optimal policy in an environment, which belongs to the category of single-agent reinforcement learning. Apart from single-agent reinforcement learning, we can actually set several agents inside the same scene, to explore the policies for multi-agents at the same time in an alternating or simultaneous manner, which is called multi-agent reinforcement learning (MARL). MARL is promising and worth exploring as it provides a way to investigate the swarm intelligence, more dynamic environments for each agent, and innovations from the agents themselves, etc.

Modern learning algorithms are more so outstanding test-takers, but less so innovators. The ceiling of an agent's intelligence may be limited by the complexity of its environment. Thus, the emergence of innovation is becoming a hot topic for artificial intelligence (AI). One of the most promising paths towards such a vision is learning via social interaction with multi-agent learning. In multi-agent learning, how the agents beat the opponents or collaborate with each other is not defined by the builder of the environment. For example, the inventor of the ancient game of Go never defines what strategies are good enough to beat the opponent, but the opponent usually forms part of the dynamic environment. However, enormous and sophisticated strategies are invented while a population of human players/artificial agents evolve by improving themselves over the others, i.e. each agent is acting as an environment for the others and improving itself means proposing new problems for the others.

Combinations of traditional game theory and modern deep reinforcement learning are explored (Lanctot et al. 2017; Nowé et al. 2012) in recent years for MARL, as well as new ideas like self-play (Silver et al. 2018a; Heinrich and Silver 2016;

Shoham et al. 2003; Berner et al. 2019), prioritized fictitious self-play (Vinyals et al. 2019), population-based training (PBT) (Jaderberg et al. 2017; Vinyals et al. 2019), and independent reinforcement learning (InRL) (Tan 1993; Lanctot et al. 2017). MARL not only makes it possible to explore the distributional intelligence in a multi-agent environment, but can also help to learn the near-optimal or close-to-equilibrium agent policy in a complex large-scale environment, like in Deepmind's AlphaStar for mastering the game of StarCraft II shown in Fig. 7.3. The AlphaStar framework applies PBT, by employing a league of agents, each of which is a single colored block with index in Fig. 7.3, to ensure sufficient exploration in the policy space. The unit of policy optimization is no longer the single policy for each agent in PBT, but rather the league of agents. The overall strategy will not merely care about the improvement of a single policy, but more about the overall performances in the agent league. More contents about MARL are introduced in Chap. 11.

7.7 Sim to Real

Reinforcement learning methods can successfully solve a large variety of tasks in simulated environments, and can sometimes even beat the best human performance for specific areas as in the game of Go. However, the challenge of applying reinforcement learning methods for real-world tasks remains unsolved. Apart from playing Atari games, strategy computer games, or board games, potential applications of reinforcement learning in real world include robotics control, autonomous driving vehicles, autonomous drone control, etc. These tasks which involve real-world hardware usually have high requirements for safety and accuracy. For these cases, a single operation by mistake can even lead to disastrous results. This is a more considerable problem when the policy is learned with reinforcement learning methods, of which the exploration process makes great differences for the learning agent without even considering the sample complexity in real world. Modern machine control in industry still depends heavily on traditional control methods, instead of state-of-the-art machine learning or reinforcement learning solutions. However, it is still a wonderful dream of controlling those physical machines with a smart agent that plenty of researchers in corresponding areas are working towards.

Recent years have seen the application of deep reinforcement learning to a growing repertoire of control problems. But due to the high sample complexity of reinforcement learning algorithms and other physical limitations, many of the capabilities demonstrated in simulation have yet to be replicated in the physical world. We will demonstrate the ideas mainly with the robot learning example, which is a more and more active research direction attracting attentions from both the academia and the industry.

Guided policy search (GPS) (Levine and Koltun 2013) represents one of the few algorithms capable of training policies directly on a real robot within limited time. By leveraging trajectory optimization with learned linear dynamics models, the method is able to develop complex manipulation skills with relatively small numbers

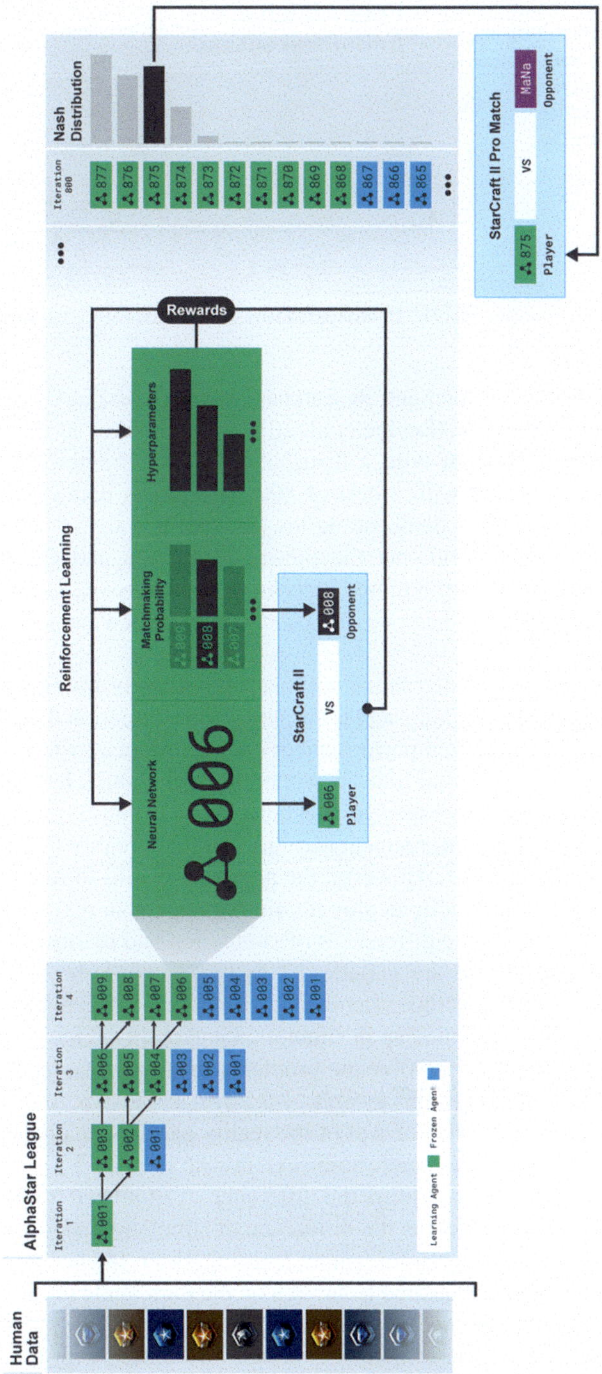

Fig. 7.3 Training scheme of AlphaStar. Each small block indicates an agent trained in the AlphaStar league

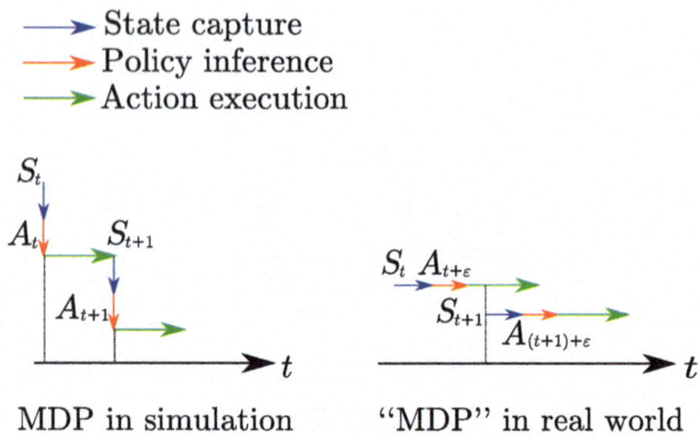

Fig. 7.4 The figure shows the difference of MDP in both simulation and in reality due to the time delays for both the state capture and the policy inference processes, which form one of the factors for reality gap

of interactions with the environment. Researchers have also explored parallelizing training across multiple robots (Levine et al. 2018). Kalashnikov et al. (2018) also propose the QT-Opt algorithm with a distributed training framework on 7 real robots at the same time, but with a cost of 800 robot hours data collection over the course of 4 months. They demonstrate the successful cases of robot learning directly deployed in real world, but the time consumption and requirement of resources are unbearable. Furthermore, successful examples of training policies directly on physical systems have so far been demonstrated only on relatively restrictive domains.

Sim-to-real transfer is an alternative approach for directly training deep reinforcement learning agents in reality, and is attracting more attention than before due to the development of simulation performances and other facts. Instead of directly training in real world, sim-to-real transfer works through a quick learning process in simulation. Recent years have seen great achievements in sim-to-real approaches for successfully deploying reinforcement learning agents in reality (Andrychowicz et al. 2018; Akkaya et al. 2019). However, the approach of sim-to-real has its own drawbacks compared with directly deploy training processes in real environments, which are mostly caused by the differences of simulation and reality environments called the **reality gap**. There are varieties of factors causing the reality gap in practice, depending on the specific systems. For example, the differences in system dynamics will cause the dynamic gap in simulation and reality (Fig. 7.4). Different approaches are also proposed to solve the problems in sim-to-real transfer, which will be discussed in later paragraphs as well.

We first try to understand the concept of the reality gap. The reality gap in real-world application can be understood to some extent with the Fig. 7.5 from the work of Jeong et al. (2019b), which displays the difference of simulated trajectories and real trajectories on robot as well as the difference of simulation and the reference. For robotic control tasks with reinforcement learning, the reference is the control signal sent by the agent or the desired behaviors on the joint angle of the robot arm. Due to the latency, inertance and other dynamic inaccuracy, both the trajectories in simulations and in reality have quite significant differences with the reference. Moreover, the trajectory in reality differs from the one in simulation as well, which

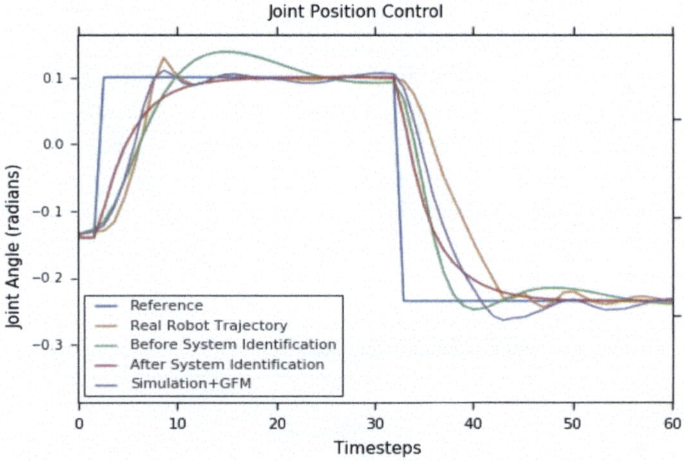

Fig. 7.5 Differences of robotic control among reference, simulation and reality, for a simple control process on the joint angle. Figure is adapted from the paper by Jeong et al. (2019b)

is the reality gap. The system identification in the graph is a method for configuring the values of dynamics parameters in the system, which can be applied by the policy or the simulator to mitigate the differences between the simulated dynamics and the real dynamics. The generalized force mode (GFM) is a newly proposed method in their paper (Jeong et al. 2019b) for calibrating the simulator with extra forces, which provides more similar trajectory in simulation as the real trajectory. However, the reality gap still exists even with the identification and calibration approaches, which will affect the transferred policy from simulation to reality.

Apart from the difference in the trajectories for simulation and reality at each time step due to different dynamic processes, there are also other sources of the reality gap. For example, the time delay of system response or system observation construction in continuous real-world control system, which may not exist in ideal simulation cases with discrete time steps. As shown in Fig. 7.4, in the MDP of simulated environments or conventional reinforcement learning settings, the state capture and policy inference process are assumed to have zero time consumption all the time, while in real-world cases, both of these two processes can take a considerable amount of time, which makes the agent always making action choices based on lagged observation from previous states during the previous action execution.

The above problem can also affect the trajectories to display different patterns for simulation and reality, as shown in Fig. 7.6. Considering an object manipulation task, even if we neglect the time consumption of the policy inference due to the fast forward process of a neural network, the position of the object in real world may need to be captured with a camera and tracked with some localization techniques, which may require some considerable time to process. This process will induce the time delay during the observation construction, and it displays the

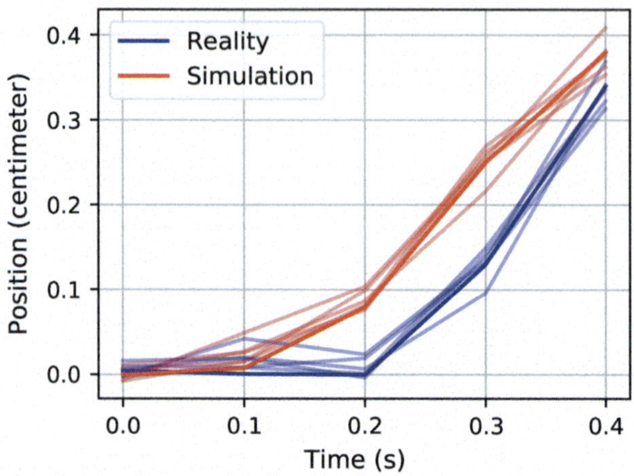

Fig. 7.6 The figure displays the time delay in the observed state (position) of the object under the same control signals. The real-world trajectory (below) is delayed compared with the simulated trajectory (above) due to the extra observation construction process in reality. Different lines show several trials and the bold ones are the means

time gap in the figure between the real-world trajectories and the simulated ones, even with the same control signals. These kind of **delayed observations** make the reinforcement learning agent in the real world only capable of receiving the previous observation O_{t-1} to make an action choice A_t for the current step instead of directly observing the current state S_t. So the policies may generally have the form $\pi(A_t|O_{t-\delta})$ according to the time delay δ in practice, which is different from the policy trained in simulation with real-time observation and therefore with a bad performance. One way of solving this is to modify the simulator to have the same time delay for the reinforcement learning agent to learn. However, this induces other problems like how to accurately represent and measure the time delay between simulation and real world, how to ensure the performance of the learned agent based on the delayed observation, etc. Recently, Ramstedt et. al. proposed real-time reinforcement learning (Ramstedt and Pal 2019) method, and Xiao et. al. proposed the method of "thinking while moving" with continuous-time MDP settings to mitigate the problem of real-time environments with delayed observations and concurrent action choices for reinforcement learning, which displays smoother trajectories for controlling in real world.

As shown above, the main problem for sim-to-real transfer in reinforcement learning perspective is: the policies trained in simulation cannot work all the time in the real world due to the reality gap, which describes the differences of simulation and reality. Due to this modeling error, policies that are successful in simulation may not be transferred well to their real-world counterparts. Generally, the methods for solving sim-to-real transfer can be divided into at least two main categories: the zero-shot methods and adaptive learning methods. The problem of transfer learning for control policies from simulation to the real world can be viewed as an instance of **domain adaptation**, where a model trained in a source domain is transferred to a new target domain. One of the key assumptions behind these methods is that the different domains share common characteristics such that representations and behaviors learned in one will prove useful for the other. Domain adaptation requires data in the new domain to adapt the pre-trained policies in the new domain. Due to the complexity or harness for acquiring data in the new domain, e.g. collecting samples in reality, the efficiency of this kind of adaptive learning needs to be high. Methods like meta-learning (Arndt et al. 2019; Nagabandi et al. 2018) and residual policy learning (Silver et al. 2018b; Johannink et al. 2019), progressive networks (Rusu et al. 2016a,b) are applied in these scenarios. **Zero-shot** transfer is a complementary class of techniques for domain adaptation that is particularly well suited for learning in simulation. This means no further learning process on real-world data is applied during the transfer process. **Domain randomization** is one typical type of method within the category of zero-shot transfer. With domain randomization, discrepancies between the source and target domains are modeled as variability in the source domain. Instead of overfitting to the characteristics of the specific simulator settings, more general policies can be learned through domain randomization. Randomization can be applied on different characteristics according to the specific application. For example, for robotic manipulation task, the amount of friction and mass, the errors in torques and velocities will all affect the control accuracy when applied in real robot. Those parameters can therefore be randomized in simulators for training a more robust policy with reinforcement learning (Peng et al. 2018), which is called dynamics randomization. Randomization in the visual domain has been used to directly transfer vision-based policies from simulation to the real world without requiring real images during training (Sadeghi and Levine 2016; Tobin et al. 2017). Potential components for visual feature randomization include texture, lighting, objects positions, etc.

The reality gap is usually task-dependent, and it can be caused by the differences in dynamic parameters or even the definitions of dynamic process. Apart from the dynamics randomization (Peng et al. 2018) or visual feature (observation) randomization, there are some other methods for bridging the reality gap. Learning a dynamics-aware policy with system identification (Yu et al. 2017; Zhou et al. 2019) is a promising direction, which tries to learn a policy conditioned on the system characteristics like dynamics parameters or embeddings of trajectories. There are also methods trying to minimize the discrepancies between sim and real, like the GFM method mentioned previously for force calibration, etc. Sim-to-real via sim-to-sim (James et al. 2019) is another approach for crossing the reality gap using Randomized-to-Canonical Adaptation Networks (RCANs). It transforms randomized or real-world images to their equivalent non-randomized canonical versions, which are similar to ones in simulation. The progressive nets (Rusu et al. 2016a) can be applied for sim-to-real transfer (Rusu et al. 2016b), which is a general framework that enables reuse of everything from low-level visual features to high-level policies for transfer to new tasks, enabling a compositional, yet simple, approach to building complex skills.

The computational framework nowadays deploys the discrete computation process based on binary operations, so we should always admit the difference of simulation and the real world to some extent. This is because the latter is continuous in space and time (in classical physics systems at least). As long as the learning algorithms are not efficient enough to be directly applied in real world like a human's mind (or even so), it is always useful to achieve some pre-trained model in simulation. And it can be better if the model has certain level of generalization ability in real-world cases, which is the significance of the algorithms for sim-to-real transfer. In other words, the sim-to-real methods provide the methodology of learning a model always with respect to the reality gap, no matter how accurate the simulators can be.

7.8 Large-Scale Reinforcement Learning

As discussed in previous sections, reinforcement learning applications in real world suffer from several problems at present, like delayed observations, domain shifts, etc., generally within the scope of reality gap. However, there are other factors that hinder the application of reinforcement learning, either in simulated cases and in real-world cases. One of the most challenging problems is the **scalability** of reinforcement learning, although deep reinforcement learning is leveraging the general representative ability of deep neural networks. This proposes the challenge of large-scale reinforcement learning.

We can take a look at some examples first. In the applications of mastering the large-scale real-time computer games like StarCraft II and Dota 2, teams of DeepMind and OpenAI propose methods AlphaStar (Vinyals et al. 2019) and OpenAI Five (Berner et al. 2019), respectively. In AlphaStar, both deep reinforcement learning methods and supervised learning (e.g., behavioral cloning in imitation learning) are applied in a population-based training (PBT) framework, as well as advanced network structures like scatter connections, transformer, and pointer networks, which make the deep reinforcement learning methods liable to only a small fraction of the overall strategy. The steps which become more critical for finally solving the task in AlphaStar are how to efficiently learn from existing demonstration data and apply the pre-trained policy as initialization of reinforcement learning agents, and how to effectively combine different sub-optimal policies explored by different agents in the league. In OpenAI Five, a self-play framework is applied instead of the PBT framework, but it also leverages the imitation learning from human demonstration. The above facts show that, present deep reinforcement learning algorithms themselves are still not effective and efficient enough to solve a large-scale task perfectly in an end-to-end manner for most cases. Some other techniques like imitation learning (in Chap. 8), hierarchical reinforcement learning strategies (in Chap. 10), and so on are generally required for solving the large-scale problems.

Moreover, a parallel learning framework is usually employed in the large-scale problems as well. For example, in the algorithm QT-Opt (Kalashnikov et al. 2018) for solving real-world robot learning tasks, to handle the paralleled robot sampling, a replay buffer containing both on-policy and off-policy data is applied, as well as distributed training workers to learn the policy efficiently with data from the buffer. A distributed or paralleled sampling and training framework is critical for solving the large-scale problems, especially for high-dimensional state and action spaces. Espeholt et al. (2018) proposed the method called importance weighted actor-learner architecture (IMPALA) and Espeholt et al. (2019) proposed SEED (Scalable, Efficient Deep-RL) for large-scale distributed reinforcement learning. Furthermore, the distributed framework for reinforcement learning usually concerns the balance between different computational devices (e.g., CPUs and GPUs), as discussed in Chap. 18. In terms of reinforcement learning algorithms, asynchronous advantage actor-critic (A3C) (Mnih et al. 2016), distributed proximal policy optimization (DPPO) (Heess et al. 2017), recurrent replay distributed DQN (R2D2) (Kapturowski et al. 2018) are proposed in recent years for supporting better parallel sampling and training in reinforcement learning. More contents about parallel computation for reinforcement learning are introduced in Chap. 12.

7.9 Others

Apart from the above-mentioned challenges in (deep) reinforcement learning, there are also other challenges like the explainability (Madumal et al. 2019) of deep reinforcement learning, the safety problem (Berkenkamp et al. 2017; Garcıa and Fernández 2015) in applications of reinforcement learning, hardness in theoretical proofs of complexity (Lattimore et al. 2013; Koenig and Simmons 1993), efficiency (Jin et al. 2018), and convergence property (Papavassiliou and Russell 1999) for reinforcement learning algorithms, and figuring out the role of reinforcement learning methods in general artificial intelligence, etc. These contents are beyond the scope of the book, readers with interests are encouraged to explore the frontiers of these domains.

At the end of this chapter, we quote some words by Richard Sutton (2019).[4] "*One thing that should be learned from the bitter lesson is the great power of general purpose methods, of methods that continue to scale with increased computation even as the available computation becomes very great. The two methods that seem to scale arbitrarily in this way are search and learning.* " This is based on the observations that the previous success in computer chess and computer Go, as well as in speech recognition and computer vision, the general statistical methods (e.g., neural networks) won over the human-knowledge-based methods. So the built-in knowledge in intelligent agents may satisfy the researchers within a short term,

[4]Richard S. Sutton. "The Bitter Lesson." March 13, 2019.

but may hinder the general progress of general artificial intelligence in a long run. "*The second general point to be learned from the bitter lesson is that the actual contents of minds are tremendously, irredeemably complex; we should stop trying to find simple ways to think about the contents of minds, such as simple ways to think about space, objects, multiple agents, or symmetries. All these are part of the arbitrary, intrinsically-complex, outside world. They are not what should be built in, as their complexity is endless; instead we should build in only the meta-methods that can find and capture this arbitrary complexity.* " This emphasizes the importance of proposing meta-methods that can handle the complexity of the world naturally, rather than applying the relatively simple cognitive structures and decision-making mechanisms that are manually built by humans for special functionalities.

References

Abdolmaleki A, Springenberg JT, Tassa Y, Munos R, Heess N, Riedmiller M (2018) Maximum a posteriori policy optimisation. arXiv:180606920

Akkaya I, Andrychowicz M, Chociej M, Litwin M, McGrew B, Petron A, Paino A, Plappert M, Powell G, Ribas R, et al (2019) Solving Rubik's cube with a robot hand. arXiv:191007113

Andrychowicz M, Wolski F, Ray A, Schneider J, Fong R, Welinder P, McGrew B, Tobin J, Abbeel OP, Zaremba W (2017) Hindsight experience replay. In: Advances in neural information processing systems, pp 5048–5058

Andrychowicz M, Baker B, Chociej M, Jozefowicz R, McGrew B, Pachocki J, Petron A, Plappert M, Powell G, Ray A, et al (2018) Learning dexterous in-hand manipulation. arXiv:180800177

Arndt K, Hazara M, Ghadirzadeh A, Kyrki V (2019) Meta reinforcement learning for sim-to-real domain adaptation. arXiv:190912906

Aytar Y, Pfaff T, Budden D, Paine T, Wang Z, de Freitas N (2018) Playing hard exploration games by watching YouTube. In: Advances in neural information processing systems, pp 2930–2941

Bengio Y, Bengio S, Cloutier J (1990) Learning a synaptic learning rule. Université de Montréal, Département d'informatique et de recherche opérationnelle

Bengio Y, Courville A, Vincent P (2013) Representation learning: a review and new perspectives. IEEE Trans Pattern Anal Mach Intell 35(8):1798–1828

Berkenkamp F, Turchetta M, Schoellig A, Krause A (2017) Safe model-based reinforcement learning with stability guarantees. In: Advances in neural information processing systems, pp 908–918

Berner C, Brockman G, Chan B, Cheung V, Dębiak P, Dennison C, Farhi D, Fischer Q, Hashme S, Hesse C, et al (2019) Dota 2 with large scale deep reinforcement learning. arXiv:191206680

Deisenroth M, Rasmussen CE (2011) PILCO: a model-based and data-efficient approach to policy search. In: Proceedings of the 28th international conference on machine learning (ICML-11), pp 465–472

Espeholt L, Soyer H, Munos R, Simonyan K, Mnih V, Ward T, Doron Y, Firoiu V, Harley T, Dunning I, et al (2018) IMPALA: scalable distributed deep-RL with importance weighted actor-learner architectures. arXiv:180201561

Espeholt L, Marinier R, Stanczyk P, Wang K, Michalski M (2019) Seed RL: Scalable and efficient deep-RL with accelerated central inference. arXiv:191006591

Finn C, Abbeel P, Levine S (2017) Model-agnostic meta-learning for fast adaptation of deep networks. In: Proceedings of the 34th international conference on machine learning, vol 70, pp 1126–1135. JMLR.org

Fujimoto S, van Hoof H, Meger D (2018) Addressing function approximation error in actor-critic methods. arXiv:180209477

Garcıa J, Fernández F (2015) A comprehensive survey on safe reinforcement learning. J Mach Learn Res 16(1):1437–1480
Heess N, Sriram S, Lemmon J, Merel J, Wayne G, Tassa Y, Erez T, Wang Z, Eslami S, Riedmiller M, et al (2017) Emergence of locomotion behaviours in rich environments. arXiv:170702286
Heinrich J, Silver D (2016) Deep reinforcement learning from self-play in imperfect-information games. arXiv:160301121
Henderson P, Islam R, Bachman P, Pineau J, Precup D, Meger D (2018) Deep reinforcement learning that matters. In: Thirty-second AAAI conference on artificial intelligence
Houthooft R, Chen X, Duan Y, Schulman J, Turck FD, Abbeel P (2016) VIME: variational information maximizing exploration. https://1605.09674
Jaderberg M, Dalibard V, Osindero S, Czarnecki WM, Donahue J, Razavi A, Vinyals O, Green T, Dunning I, Simonyan K, et al (2017) Population based training of neural networks. arXiv:171109846
James S, Wohlhart P, Kalakrishnan M, Kalashnikov D, Irpan A, Ibarz J, Levine S, Hadsell R, Bousmalis K (2019) Sim-to-real via sim-to-sim: data-efficient robotic grasping via randomized-to-canonical adaptation networks. In: Proceedings of the IEEE conference on computer vision and pattern recognition, pp 12627–12637
Jeong R, Aytar Y, Khosid D, Zhou Y, Kay J, Lampe T, Bousmalis K, Nori F (2019a) Self-supervised sim-to-real adaptation for visual robotic manipulation. arXiv:191009470
Jeong R, Kay J, Romano F, Lampe T, Rothorl T, Abdolmaleki A, Erez T, Tassa Y, Nori F (2019b) Modelling generalized forces with reinforcement learning for sim-to-real transfer. arXiv:191009471
Jin C, Allen-Zhu Z, Bubeck S, Jordan MI (2018) Is Q-learning provably efficient? In: Advances in neural information processing systems, pp 4863–4873
Johannink T, Bahl S, Nair A, Luo J, Kumar A, Loskyll M, Ojea JA, Solowjow E, Levine S (2019) Residual reinforcement learning for robot control. In: 2019 international conference on robotics and automation (ICRA). IEEE, Piscataway, pp 6023–6029
Kalashnikov D, Irpan A, Pastor P, Ibarz J, Herzog A, Jang E, Quillen D, Holly E, Kalakrishnan M, Vanhoucke V, et al (2018) QT-opt: scalable deep reinforcement learning for vision-based robotic manipulation. arXiv:180610293
Kapturowski S, Ostrovski G, Quan J, Munos R, Dabney W (2018) Recurrent experience replay in distributed reinforcement learning. In: International conference on learning representations. https://openreview.net/forum?id=r1lyTjAqYX
Kirkpatrick J, Pascanu R, Rabinowitz N, Veness J, Desjardins G, Rusu AA, Milan K, Quan J, Ramalho T, Grabska-Barwinska A, et al (2017) Overcoming catastrophic forgetting in neural networks. Proc Natl Acad Sci 114(13):3521–3526
Koenig S, Simmons RG (1993) Complexity analysis of real-time reinforcement learning. In: Proceedings of the AAAI conference on artificial intelligence, pp 99–107
Kulkarni TD, Narasimhan K, Saeedi A, Tenenbaum J (2016) Hierarchical deep reinforcement learning: integrating temporal abstraction and intrinsic motivation. In: Advances in neural information processing systems, pp 3675–3683
Lanctot M, Zambaldi V, Gruslys A, Lazaridou A, Tuyls K, Pérolat J, Silver D, Graepel T (2017) A unified game-theoretic approach to multiagent reinforcement learning. In: Advances in neural information processing systems, pp 4190–4203
Lattimore T, Hutter M, Sunehag P, et al (2013) The sample-complexity of general reinforcement learning. In: Proceedings of the 30th international conference on machine learning
Levine S, Koltun V (2013) Guided policy search. In: International conference on machine learning, pp 1–9
Levine S, Pastor P, Krizhevsky A, Ibarz J, Quillen D (2018) Learning hand-eye coordination for robotic grasping with deep learning and large-scale data collection. Int J Robot Res 37(4–5):421–436
Madumal P, Miller T, Sonenberg L, Vetere F (2019) Explainable reinforcement learning through a causal lens. arXiv:190510958

Mnih V, Kavukcuoglu K, Silver D, Graves A, Antonoglou I, Wierstra D, Riedmiller M (2013) Playing Atari with deep reinforcement learning. arXiv:13125602

Mnih V, Badia AP, Mirza M, Graves A, Lillicrap T, Harley T, Silver D, Kavukcuoglu K (2016) Asynchronous methods for deep reinforcement learning. In: International conference on machine learning (ICML), pp 1928–1937

Nagabandi A, Clavera I, Liu S, Fearing RS, Abbeel P, Levine S, Finn C (2018) Learning to adapt in dynamic, real-world environments through meta-reinforcement learning. arXiv:180311347

Nowé A, Vrancx P, De Hauwere YM (2012) Game theory and multi-agent reinforcement learning. In: Reinforcement learning. Springer, Berlin, pp 441–470

Papavassiliou VA, Russell S (1999) Convergence of reinforcement learning with general function approximators. In: International joint conference on artificial intelligence, vol 99, pp 748–755

Pathak D, Agrawal P, Efros AA, Darrell T (2017) Curiosity-driven exploration by self-supervised prediction. In: Proceedings of the international conference on machine learning (ICML)

Peng XB, Andrychowicz M, Zaremba W, Abbeel P (2018) Sim-to-real transfer of robotic control with dynamics randomization. In: 2018 IEEE international conference on robotics and automation (ICRA). IEEE, Piscataway, pp 1–8

Ramstedt S, Pal C (2019) Real-time reinforcement learning. In: Advances in neural information processing systems, pp 3067–3076

Rusu AA, Rabinowitz NC, Desjardins G, Soyer H, Kirkpatrick J, Kavukcuoglu K, Pascanu R, Hadsell R (2016a) Progressive neural networks. arXiv:160604671

Rusu AA, Vecerik M, Rothörl T, Heess N, Pascanu R, Hadsell R (2016b) Sim-to-real robot learning from pixels with progressive nets. arXiv:161004286

Sadeghi F, Levine S (2016) Cad2rl: Real single-image flight without a single real image. arXiv:161104201

Shoham Y, Powers R, Grenager T (2003) Multi-agent reinforcement learning: a critical survey. Web manuscript

Silver D, Hubert T, Schrittwieser J, Antonoglou I, Lai M, Guez A, Lanctot M, Sifre L, Kumaran D, Graepel T, et al (2018a) A general reinforcement learning algorithm that masters chess, shogi, and Go through self-play. Science 362(6419):1140–1144

Silver T, Allen K, Tenenbaum J, Kaelbling L (2018b) Residual policy learning. arXiv:181206298

Song HF, Abdolmaleki A, Springenberg JT, Clark A, Soyer H, Rae JW, Noury S, Ahuja A, Liu S, Tirumala D, et al (2019) V-MPO: On-policy maximum a posteriori policy optimization for discrete and continuous control. arXiv:190912238

Sukhbaatar S, Lin Z, Kostrikov I, Synnaeve G, Szlam A, Fergus R (2018) Intrinsic motivation and automatic curricula via asymmetric self-play. In: International conference on learning representations. https://openreview.net/forum?id=SkT5Yg-RZ

Tan M (1993) Multi-agent reinforcement learning: independent vs. cooperative agents. In: Proceedings of the international conference on machine learning (ICML)

Tobin J, Fong R, Ray A, Schneider J, Zaremba W, Abbeel P (2017) Domain randomization for transferring deep neural networks from simulation to the real world. In: International conference on intelligent robots and systems (IROS)

Vezhnevets AS, Osindero S, Schaul T, Heess N, Jaderberg M, Silver D, Kavukcuoglu K (2017) Feudal networks for hierarchical reinforcement learning. In: Proceedings of the 34th international conference on machine learning, vol 70, pp 3540–3549. JMLR.org

Vinyals O, Babuschkin I, Czarnecki WM, Mathieu M, Dudzik A, Chung J, Choi DH, Powell R, Ewalds T, Georgiev P, et al (2019) Grandmaster level in StarCraft II using multi-agent reinforcement learning. Nature 575(7782):350–354

Yu W, Tan J, Liu CK, Turk G (2017) Preparing for the unknown: learning a universal policy with online system identification. arXiv:170202453

Zhou W, Pinto L, Gupta A (2019) Environment probing interaction policies. arXiv:190711740

Chapter 8
Imitation Learning

Zihan Ding

Abstract To alleviate the low sample efficiency problem in deep reinforcement learning, imitation learning, or called apprenticeship learning, is one of the potential approaches, which leverages the expert demonstrations in sequential decision-making process. In order to provide the readers a comprehensive understanding about how to effectively extract information from the demonstration data, we introduce the most important categories in imitation learning, including behavioral cloning, inverse reinforcement learning, imitation learning from observations, probabilistic methods, and other methods. Imitation learning can either be regarded as an initialization or a guidance for training the agent in the scope of reinforcement learning. Combination of imitation learning and reinforcement learning is a promising direction for efficient learning and faster policy optimization in practice.

Keywords Imitation learning · Apprenticeship learning · Demonstration · Reinforcement learning · Behavioral cloning · Inverse reinforcement learning · Generative adversarial networks · Sample efficiency

8.1 Introduction

As we know, reinforcement learning (RL), especially model-free reinforcement learning, suffers from low sample efficiency as discussed in the chapter of present challenges in reinforcement learning (Chap. 7). Hundreds of thousands of examples are usually needed to solve an uncomplicated task with human-level performance. However, humans can learn to solve the tasks with significantly shorter periods of time and a much smaller number of samples. Apart from improving the efficiency of reinforcement learning algorithms themselves through more elaborate algorithm design with mathematical guarantees, we can actually let the agent leverage

---

Z. Ding (✉)
Imperial College London, London, UK
e-mail: zhding@mail.ustc.edu.cn

H. Dong et al. (eds.), *Deep Reinforcement Learning*,
https://doi.org/10.1007/978-981-15-4095-0_8

additional information resource, like expert demonstrations. The expert demonstrations contain biased choices for the policy with prior knowledge, which can be distilled and transferred into agent policies in reinforcement learning, through a proper learning process. The task of learning from an expert is called **imitation learning (IL)** (also known as apprenticeship learning). Humans and animals are born to learn from mimicking other individuals of the same kind, which inspires the method of imitation learning for an intelligent agent to learn from demonstrations by others. On the other hand, supervised learning is much more efficient in the aspect of data usage compared with reinforcement learning, due to the benefits of labeled data. Therefore, the method of supervised learning can be incorporated into the agent's learning process to improve its efficiency if demonstrations are provided in a labeled format.

In this chapter, we introduce different approaches for learning a policy with demonstrations. The overview of algorithms and methods in imitation learning categories is shown in Fig. 8.1. We will introduce detailed methods of imitation learning in the following sections, summarized in several main categories including (1) behavioral cloning (BC), (2) inverse reinforcement learning (IRL), (3) imitation learning from observations (IfO, or ILFO in some other literature (Sun et al. 2019)), (4) probabilistic methods, and (5) other approaches. BC is the most simple and straightforward way of using the demonstration data in a supervised learning manner, which is widely applied due to its simplicity and is usually regarded as a cornerstone to build more advanced methods on. IRL is useful in applications where it may be difficult to write down an explicit reward function specifying exactly how different desiderata should be traded off. For example, how much attention should be paid on taking care of different reflectors for automatic driving vehicles based on visual observations is hard to specify through reward engineering. IRL is an approach to recover the unknown reward function from the demonstration data and uses it for further reinforcement learning process. The IfO actually solves the drawback of imitation learning that it usually requires actions as labels for the state inputs, which often happens in human imitation learning process. The methods from a probabilistic inference view include using Gaussian mixture regression or Gaussian process regression to represent the demonstration data and therefore guide the action policy, which is a more efficient alternative for deep neural network methods in some cases. There are also other approaches like directly feeding demonstrations into a replay buffer for off-policy reinforcement learning, etc. After introducing basic categories of different imitation learning methods, we will discuss the relationship of imitation learning and reinforcement learning, like applying imitation learning as an initialization of reinforcement learning in order to improve the learning efficiency of reinforcement learning. Finally, we introduce some other specific methods in imitation learning with reinforcement learning, which are either combinations of previous conceptions or outliers of the summarized categories as Fig. 8.1.

The concept of imitation learning can be defined with the apprenticeship learning formalism (Abbeel and Ng 2004): the learner finds a policy π that performs no worse than expert π_E with respect to an unknown reward function $r(s, a)$. We

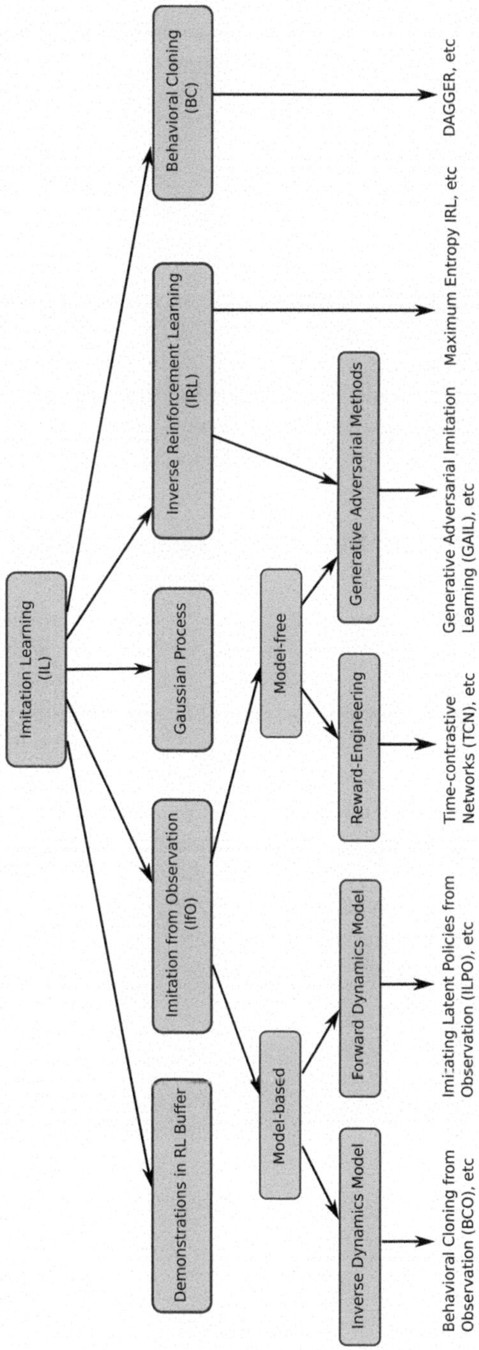

Fig. 8.1 Overview of imitation learning algorithms

define the occupancy measure $\rho_\pi \in \mathcal{D} : \mathcal{S} \times \mathcal{A} \rightarrow \mathbb{R}$ of a policy $\pi \in \Pi$ as: $\rho_\pi(s, a) = \pi(a|s) \sum_{t=0}^{\infty} \gamma^t p(S_t = s|\pi)$ (Puterman 2014), which is a joint distribution of state and action estimated with current policy. Owing to the one-to-one correspondence between Π and $\mathcal{D}$, an imitation learning problem is equivalent to a matching problem between $\rho_\pi(s, a)$ and $\rho_{\pi_E}(s, a)$. A general objective of imitation learning is to learn a policy:

$$\hat{\pi} = \arg\min_{\pi \in \Pi} \psi^*(\rho_\pi - \rho_{\pi_E}) - \lambda H(\pi) \tag{8.1}$$

where ψ^* is a distance measurement between ρ_π and ρ_{π_E}, and $H(\pi)$ is a normalization term with trade-off factor λ. For instance, the normalization term can be defined as the γ-discounted causal entropy of policy π: $H(\pi) \triangleq \mathbb{E}_\pi[-\log \pi(s, a)]$. The overall goal of imitation learning is to increase the similarity of the distribution of $\{(s, a)\}$ samples from current policy and the distribution of those in demonstration dataset, with respect to the some constraints on policy parameters.

8.2 Behavioral Cloning: Supervised Learning Approach

The imitation learning with demonstrations can be naturally regarded as a supervised learning task, if the demonstration data is provided with labels (e.g., a good action for the state can be regarded as a label). In reinforcement learning circumstances, the labeled demonstration data $\mathcal{D}$ usually contains the pairs of state and action as: $\mathcal{D} = \{(s_i, a_i)|i = 1, \ldots, N\}$, where N is the size of the demonstration dataset and index i indicates the s_i and a_i are at the same time step. The state-action pairs can be shuffled for training under the MDP assumption, i.e. the optimal action only depends on the current state. Considering an initial policy π_θ parameterized by θ with input state s and output deterministic action $\pi_\theta(s)$ in reinforcement learning settings, we have demonstrations dataset $\mathcal{D} = \{(s_i, a_i)|i = 1, \ldots, N\}$ generated from experts, which could be used to train the policy, with an objective as follows:

$$\min_\theta \sum_{(s_i, a_i) \sim \mathcal{D}} ||a_i - \pi_\theta(s_i)||_2^2 \tag{8.2}$$

The cases with stochastic policies $\pi_\theta(\tilde{a}|s)$ in some specific formats, e.g. Gaussian policy, etc., can be handled as well using the reparameterization trick:

$$\min_\theta \sum_{\tilde{a}_i \sim \pi(\cdot|s_i), (s_i, a_i) \sim \mathcal{D}} ||a_i - \tilde{a}_i||_2^2 \tag{8.3}$$

This supervised learning approach to directly imitate the expert demonstration is called the **behavioral cloning (BC)** in the literature.

8.2.1 *Challenges of BC*

- **Covariate Shift**: Although imitation learning can provide a relatively good performance for the cases similar to samples in the demonstration dataset (used for policy training), it could still suffer from bad generalization for samples it never meets during training, as the demonstration dataset only contains finite samples. For example, the new samples during testing can be around another cluster in distribution rather than the same one for training if it is a multimodal distribution, like applying the classifier for cats on distinguishing dogs in practice. As the BC approach boils down the decision-making problem to a supervised learning problem, this well-known problem of covariate shift (Ross and Bagnell 2010) in machine learning can potentially make the learned policy brittle, which is challenging in BC methods. Figure 8.2 further elaborates the covariate shift in BC.
- **Compounding Errors**: BC suffers greatly from the compounding error, a situation where minor errors are compounded over time and finally induce a dramatically different state distribution (Ross et al. 2011). The MDP property of reinforcement learning tasks is the key factor leading to the compounding error, i.e. the amplification effect of consecutive errors. The main reason of the error for each time step can actually be caused by the covariance shift described above, in BC methods. Figure 8.3 shows the compounding errors.

8.2.2 *Dataset Aggregation*

Dataset Aggregation (DAgger) (Ross et al. 2011) is a more advanced no-regret iterative algorithm for imitation learning from demonstrations following the approach of BC. It proactively chooses demonstration samples that the policy has larger chances to meet afterwards, according to the previous training iterations,

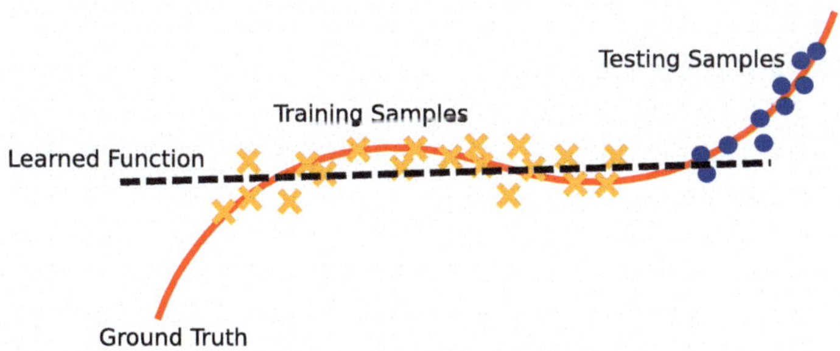

Fig. 8.2 The covariate shift: the learned function (black dashed line) fits well on the training samples (orange "cross"), but has large prediction bias on the testing samples (blue dot). The red line is the ground truth

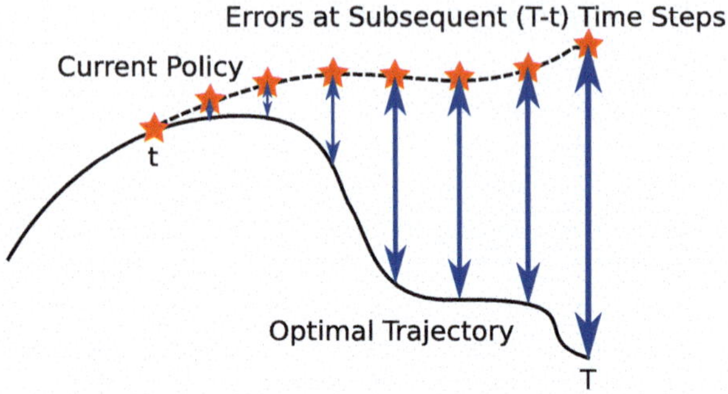

Fig. 8.3 The compounding errors increase along the trajectory chosen by current policy in a task with sequential decisions

which makes DAgger more effective and efficient for online imitation learning in sequential prediction problem like reinforcement learning. The demonstration dataset $\mathcal{D}$ is continuously aggregated with new dataset $\mathcal{D}_i$ for time step i containing expert actions and corresponding states visited by current policy during the whole imitation learning process. So, DAgger also has a drawback that it needs to iteratively interact with the expert, which is usually demanding in real-world applications. The algorithm of DAgger is shown in Algorithm 1, where π^* is the expert policy and β_i is the parameter for soft-updating the policy at iteration i.

Algorithm 1 DAgger

Initialize $\mathcal{D} \leftarrow \emptyset$.
Initialize the policy $\hat{\pi}_1$ to any policy in policy set Π.
for i $= 1, 2, \ldots, N$ **do**
    $\pi_i \leftarrow \beta_i \pi^* + (1 - \beta_i)\hat{\pi}_i$.
    Sample several T-step trajectories using π_i.
    Get dataset $\mathcal{D}_i = \{(s, \pi^*(s))\}$ of visited states by π_i and actions given by the expert.
    Aggregate datasets: $\mathcal{D} \leftarrow \mathcal{D} \cup \mathcal{D}_i$. Train current policy $\hat{\pi}_{i+1}$ on $\mathcal{D}$.
end for
Return policy $\hat{\pi}_{N+1}$.

8.2.3 *Variational Dropout*

A method for alleviating the generalization problem in imitation learning is pre-training with **variational dropout** (Blau et al. 2018), instead of fully cloning the behavior of expert demonstrations in BC methods. The weights pre-trained with the demonstration dataset are parameterized as Gaussian distributions with Gaussian dropout of a certain variance threshold value for initializing reinforcement learning policies. Variational dropout approach for imitation learning (Molchanov et al. 2017) can be taken as a more advanced method for generalization than noise injection in the weights of pre-trained neural networks, it reduces the sensitivity of choosing the magnitude of noise, which is a useful technique when applying imitation learning for initializing reinforcement learning.

8.2.4 *Other Methods in BC*

Some other concepts are involved in behavioral cloning as well. For example, some methods provide ways to generalize demonstrations to more general scenarios in a task using framework like **dynamic movement primitives (DMPs)** (Pastor et al. 2009), which apply a set of differential equations to represent any recorded movement. The differential equations in DMP usually contain adjustable weights, as well as non-linear functions to allow the generation of arbitrarily complex movements. Therefore DMP is more of an analytical-form solution compared with the "black-box" deep learning methods in behavioral cloning. Moreover, there exists a method in one-shot imitation learning (Duan et al. 2017) using **soft attention** on demonstrations to generalize model to unseen scenarios in training data. It is a meta-learning scheme to map one demonstration of one task to an effective policy for a variety of tasks. There are some other methods, which will not be discussed here.

8.3 Inverse Reinforcement Learning Approach

Another major category of imitation learning approaches is composed of techniques based on **inverse reinforcement learning (IRL)** (Ng et al. 2000; Russell 1998). The IRL problem is defined to be the problem of extracting a reward function given observed, optimal behavior, represented as expert policy π_E. Instead of directly learning a mapping from states to actions using the demonstration data, IRL-based methods iteratively alternate between using the demonstration to infer a hidden reward/cost function and using reinforcement learning with the inferred reward function to learn an imitating policy. IRL chooses the reward function R to make the policy optimal and moreover to favor solutions that make any single-step deviation from π_E as costly as possible. For all reward functions R satisfying $|R(s)| \leq R_{\max}, \forall s$, IRL method chooses the R^* following:

$$R^* = \arg\max_R \sum_{s\in\mathcal{S}} \left(Q^\pi(s, a_E) - \max_{a\in A\setminus a_E} Q^\pi(s, a) \right) \tag{8.4}$$

where $a_E = \pi_E(s)$ or $a_E \sim \pi(\cdot|s)$ is the expert (optimal) action. IRL-based techniques have been used for a variety of tasks such as maneuvering a helicopter (Abbeel and Ng 2004) and object manipulation (Finn et al. 2016b). IRL (Ng et al. 2000; Russell 1998) tries to extract a reward function from observed optimal behavior, like the expert demonstrations, but the reward function may not be unique (discussed later). A typical method in IRL is to use maximum causal entropy regularization, which is maximum entropy (MaxEnt) IRL (Ziebart et al. 2010). The MaxEnt IRL can be represented as the following two stages:

$$\text{IRL}(\pi_E) = \arg\max_R \mathbb{E}_{\pi_E}[R(s,a)] - \text{RL}(R) \tag{8.5}$$

$$\text{RL}(R) = \max_\pi H(\pi(\cdot|s)) + \mathbb{E}_\pi[R(s,a)] \tag{8.6}$$

which forms the RL $\circ$ IRL(π_E) policy learning framework. The first formula IRL(π_E) learns a reward function to maximize the difference of reward values between the expert policy and the reinforcement learning policy, and it can be replaced by Eq. (8.4) as the Q value is an estimation of rewards. The second formula RL(R) is the entropy-regularized (forward) reinforcement learning with the learned reward function R from the first formula. The entropy $H(\pi(\cdot|s))$ here is the entropy of the policy distribution given a specific state.

Shannon's information entropy of distribution P over random variable X measures the uncertainty of that probability distribution.

Definition 8.1 The **entropy** of a discrete random variable, X, distributed according to p is

$$H_p(X) = \mathbb{E}_{p(X)}[-\log p(X)] = -\sum_{X\in\mathcal{X}} p(X)\log p(X) \tag{8.7}$$

For the case of stochastic policies in reinforcement learning, the random variables are usually aligned in a vector with the same dimension as the action space. The commonly used distributions are diagonal Gaussian distributions and categorical distributions, the derivation of their entropy is trivial (referred to the chapters for algorithms implementation).

It is also common to see the cost function $c(s,a) = -R(s,a)$, which is minimized in the reinforcement learning process as follows:

$$\text{RL}(c) = \arg\min_\pi -H(\pi) + \mathbb{E}_\pi[c(s,a)] \tag{8.8}$$

where $H(\pi) = \mathbb{E}_\pi[-\log\pi(a|s)]$ is called entropy of policy π. The cost function $c(s,a)$ is usually a measurement of the similarity between distributions from current policy π and the demonstrations dataset. The entropy term $H(\pi)$ can be regarded as a normalization term for the uniqueness of optimality.

By substituting the above formula into the IRL formula Eq. (8.5), we can represent the objective of IRL in a max–min form, which tries to learn a cost function $c(s,a)$ of state s and action a with the objective of maximizing the entropy-regularized reward, as well as learning the policy π.

$$\max_c \left(\min_\pi -\mathbb{E}_\pi[-\log\pi(a|s)] + \mathbb{E}_\pi[c(s,a)]\right) - \mathbb{E}_{\pi_E}[c(s,a)] \tag{8.9}$$

where the π_E denotes the expert policy for generating expert demonstrations and π is the policy trained in reinforcement learning process. The learned cost function will assign high entropy to expert policy and low entropy to other policies.

8.3.1 Challenges of IRL

- **Non-uniqueness of Reward** (or **Reward Ambiguity**): The function search in IRL is ill-posed as the demonstrated behavior could be induced by multiple reward/cost functions. It originates from the concept of reward shaping (Ng et al. 1999), which describes a class of reward transformations that preserve the optimal policy. The main result is that under the following reward transformation:

$$\hat{r}(s, a, s') = r(s, a, s') + \gamma\phi(s') - \phi(s) \tag{8.10}$$

  the optimal policy remains unchanged for any function $\phi : \mathcal{S} \rightarrow \mathbb{R}$. The reward function learned with IRL methods from demonstration data only cannot disambiguate between reward functions within the class of above transformations.

  Constraints are thereby imposed on the rewards or the policy to ensure the optimality uniqueness of the demonstrated behavior. For example, the reward function is usually defined to be a linear (Ng et al. 2000; Abbeel and Ng 2004) or convex (Syed et al. 2008) combination of the state features. The learned policy is also assumed to have the maximum entropy (Ziebart et al. 2008) or the maximum causal entropy (Ziebart et al. 2010). These explicit constraints potentially limit the generability of the proposed methods (Ho and Ermon 2016).
- **Intensive Computational Cost**: The IRL could learn a better policy from demonstrations and interactions in the general reinforcement learning process. However, using reinforcement learning to optimize the policy given the inferred reward function requires the agent to interact with its environment, which can be costly from the perspectives of time and safety. Moreover, the IRL step typically requires the agent to solve an MDP in the inner loop of iterative reward optimization (Abbeel and Ng 2004; Ziebart et al. 2008), which can be extremely costly from a computational perspective. However, recently, a number of methods have been developed, which relieve this requirement (Finn et al. 2016b; Ho and Ermon 2016). One of these approaches is called generative adversarial imitation from observation (GAIL) (Ho and Ermon 2016), which will be described in Sect. 8.3.2.

8.3.2 Generative Adversarial Approach

The generative adversarial imitation learning (GAIL) (Ho and Ermon 2016) borrows the idea of generative adversarial training in generative adversarial networks

(GANs) (Goodfellow et al. 2014). The associated algorithm can be thought of as trying to induce an imitator state-action occupancy measure that is similar to that of the demonstrator. It applies a discriminator in GAN for providing the estimation of an action-value function based on demonstrations. In a general process of action-value based reinforcement learning with algorithms like TRPO, PPO, etc., the action value is provided from the demonstrations with a generative approach as:

$$Q(s,a) = \mathbb{E}_{\mathcal{T}_i}[\log(D_{\omega_{i+1}}(s,a))] \tag{8.11}$$

where $\mathcal{T}_i$ are samples from explorations for iteration i and $D_{\omega_{i+1}}(s,a)$ is the output value from the discriminator with parameters ω_{i+1}. The ω_{i+1} indicates the Q-value is estimated with one-step updated discriminator weights, therefore with iteration $i+1$. The loss function of the discriminator is defined in a general way:

$$\text{Loss} = \mathbb{E}_{\mathcal{T}_i}[\nabla_\omega \log(D_\omega(s,a))] + \mathbb{E}_{\mathcal{T}_E}[\nabla_\omega \log(1 - D_\omega(s,a))] \tag{8.12}$$

where the $\mathcal{T}_i$, $\mathcal{T}_E$ are samples from explorations and expert demonstrations, respectively. ω are parameters of the discriminator. Figure 8.4 shows the architecture of GAIL.

With the method of GAIL, the policy can be learned with samples generalized from demonstrations with lower computational cost, compared with methods via IRL. It does not need to interact with the expert during training, which happens in method like DAgger and is sometimes not accessible in practice.

This approach can be further generated to multimodal policy for learning across tasks. Multimodal imitation learning with GAN (Hausman et al. 2017) applies a more advanced objective function (additional latent indices for different tasks)

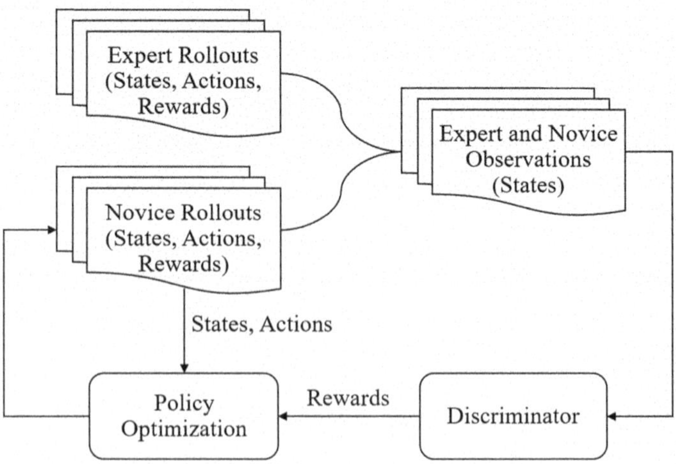

Fig. 8.4 The architecture of GAIL, adapted from Ho and Ermon (2016)

in a generative adversarial form, to automatically segment the demonstrations for different tasks and learn a multimodal policy in an imitation manner.

According to Goodfellow et al. (2014) (details of GANs are introduced in Chap. 1), with infinite data and infinite computation, at optimality, the distribution of generated state-action pairs should exactly match the distribution of demonstrated state-action pairs under the GAIL objective. The downside to this approach, however, is that we bypass the intermediate step of recovering rewards. Specifically, note that we cannot extract reward functions from the discriminator, as $D_\omega(s, a)$ will converge to 0.5 for all (s, a) pairs.

8.3.3 Generative Adversarial Network Guided Cost Learning (GAN-GCL)

As mentioned above, the GAIL method cannot recover the reward function from the demonstration data. A similar work named generative adversarial network guided cost learning (GAN-GCL) optimizes the guided cost learning (GCL) method based on GAN's structure, to extract an optimal reward function from the optimal discriminator trained with the demonstration data. We will describe this method in details.

The GAN-GCL method (specifically the GCL) is based on the maximum causal entropy IRL described above, which considers an entropy-regularized Markov decision process (MDP). The goal in entropy-regularized MDP for reinforcement learning is to maximize the expected entropy-regularized discounted reward:

$$\pi^* = \arg\max_{\pi} \mathbb{E}_{\tau\sim\pi}\left[\sum_{t=0}^{T} \gamma^t (r(S_t, A_t) + H(\pi(\cdot|S_t)))\right] \tag{8.13}$$

which is a more specific term used for learning the policy in practice originated from Eq. (8.5). It can be shown that the optimal policy $\pi^*(a|s)$ gives trajectory distribution satisfying $\pi^*(a|s) \propto \exp(Q^*_{soft}(s, a))$ (Ziebart et al. 2010), where $Q^*_{soft}(S_t, A_t) = r(S_t, A_t) + \mathbb{E}_{\tau\sim\pi}[\sum_{t'=t}^{T} \gamma^{t'-t}(r(s_{t'}, a_{t'}) + H(\pi(\cdot|s_{t'})))]$ denotes the soft Q-function (also used in soft actor-critic algorithm).

The IRL problem can be interpreted as solving the maximum likelihood problem:

$$\max_{\theta} \mathbb{E}_{\tau\sim\pi_E}[\log p_\theta(\tau)] \tag{8.14}$$

where π_E is the expert policy for providing demonstrations, and $p_\theta(\tau) \propto p(S_0)\prod_{t=0}^{T} p(S_{t+1}|S_t, A_t)e^{\gamma^t r_\theta(S_t, A_t)}$ parameterizes the reward function $r_\theta(s, a)$ but with the dynamics (or transition) and initial state distribution of the MDP. $p_\theta(\tau)$ is the trajectory-centric distribution of the demonstration data derived from state-centric π_E, $p_\theta(\tau) \sim \pi_E$. With deterministic transition $p(S_{t+1}|S_t, A_t) = 1$, this simplifies to an energy-based model $p_\theta(\tau) \propto e^{\sum_{t=0}^{T} \gamma^t r_\theta(S_t, A_t)}$ (Ziebart et al. 2008). The parameterized reward function can be learned through optimizing parameters θ w.r.t the above objective. Similar to processes before, we can introduce the cost function here as the negative discounted cumulative rewards $c_\theta = -\sum_{t=0}^{T} \gamma^t r_\theta(S_t, A_t)$, parameterized by θ. Then the MaxEnt IRL can be viewed as modeling the demonstrations using a Boltzmann distribution in a trajectory-centric formulation, where the energy is given by the cost function c_θ:

$$p_\theta(\tau) = \frac{1}{Z} \exp(-c_\theta(\tau)) \tag{8.15}$$

where τ is the state-action trajectory and $c_\theta(\tau) = \sum_t c_\theta(S_t, A_t)$, and the partition function Z is the integral of $\exp(-c_\theta(\tau))$ over all trajectories consistent with the environment dynamics, for normalizing the probability. Estimating the partition function Z is difficult for large-scale or continuous domains, as precise estimation with dynamic programming for Z can only work in small and discrete domains. Otherwise we need to use approximated estimation methods, like the sampling-based GCL method.

GCL uses importance sampling for estimating Z with a new distribution $q(\tau)$ (the original demonstration distribution is $p(\tau)$) in MaxEnt IRL formulation:

$$\theta^* = \arg\min_\theta \mathbb{E}_{\tau\sim p}[-\log p_\theta(\tau)] \tag{8.16}$$

$$= \arg\min_\theta \mathbb{E}_{\tau\sim p}[c_\theta(\tau)] + \log Z \tag{8.17}$$

$$= \arg\min_\theta \mathbb{E}_{\tau\sim p}[c_\theta(\tau)] + \log\left(\mathbb{E}_{\tau'\sim q}\left[\frac{\exp(-c_\theta(\tau'))}{q(\tau')}\right]\right) \tag{8.18}$$

where the τ' is sampled from distribution q and $q(\tau')$ gives its probability. Therefore q can be optimized through minimizing the KL-divergence between $q(\tau')$ and $\frac{1}{Z}\exp(-c_\theta(\tau'))$ for updating the $q(\tau')$ during learning θ or equivalently as following:

$$q^* = \min \mathbb{E}_{\tau\sim q}[c_\theta(\tau)] + \mathbb{E}_{\tau\sim q}[\log q(\tau)] \tag{8.19}$$

Finn et al. (2016a) proposed to use GAN's manner for the above optimization problem, which optimizes the GCL with GAN structure and is similar to the GAIL method but with different specific formulations.

Note that in GAN the discriminator also tries to approximate one distribution with the other as:

$$D^*(\tau) = \frac{p(\tau)}{p(\tau) + q(\tau)} \tag{8.20}$$

We can apply it here in the GCL of MaxEnt IRL formulation,

$$D_\theta(\tau) = \frac{\frac{1}{Z}\exp(-c_\theta(\tau))}{\frac{1}{Z}\exp(-c_\theta(\tau)) + q(\tau)} \tag{8.21}$$

which leads to the method GAN-GCL. The policy π is trained to maximize $R_\theta(\tau) = \log(1 - D_\theta(\tau)) - \log D_\theta(\tau)$, and the reward function is therefore learned through optimizing the discriminator. The policy is learned through updating the sampling distribution $q(\tau)$ used to estimate the partition function. If the optimality is reached, the optimal cost function $c^*_\theta = -R^*_\theta(\tau) = -\sum_{t=0}^{T} \gamma^t r^*_\theta(S_t, A_t)$ can be learned for evaluating the optimal reward function, and the optimal policy can be derived with $\pi^* = q^*$. GAN-GCL provides an alternative approach for optimizing the MaxEnt IRL problem instead of directly maximizing the likelihood.

8.3.4 Adversarial Inverse Reinforcement Learning (AIRL)

As the above GAN-GCL is trajectory-centric, which means the full trajectories are estimated, it has high variance in estimation compared with estimating the single state-action pair. The adversarial inverse reinforcement learning (AIRL) (Fu et al. 2017) proposes to directly estimate the single state and action:

$$D_\theta(s, a) = \frac{\exp(f_\theta(s, a))}{\exp(f_\theta(s, a)) + \pi(a|s)} \tag{8.22}$$

where the $\pi(a|s)$ is the sampling distribution to be updated and the $f_\theta(s, a)$ is the learned function. The partition function is ignored in the above formula and the normalization of probability values can be guaranteed with *SoftMax* operator or sigmoid output activation in practice. It is proven that at optimality, $f^*(s, a) = \log \pi^*(a|s) = A^*(s, a)$, which gives the advantage function of optimal policy. However, the advantage function is a heavily entangled reward function with a baseline value subtracted. Fu et al. (2017) argue that the reward function cannot be robustly recovered for the changes in environment dynamics. Therefore, they also propose to learn the disentangled reward with AIRL through decoupling the reward function from the advantage function:

$$D_{\theta,\phi}(s, a, s') = \frac{\exp(f_{\theta,\phi}(s, a, s'))}{\exp(f_{\theta,\phi}(s, a, s')) + \pi(a|s)} \tag{8.23}$$

where $f_{\theta,\phi}$ is restricted to a reward approximator g_θ and a shaping term h_ϕ as:

$$f_{\theta,\phi}(s, a, s') = g_\theta(s, a) + \gamma h_\phi(s') - h_\phi(s) \tag{8.24}$$

where the extra approximation of h_ϕ is required.

8.4 Imitation Learning from Observation (IfO)

First, imitation learning from observation (IfO) is imitation learning without fully observable actions. One typical example of IfO is learning from the videos, in which the ground-truth actions of objects are not available from the frames only, but humans can still learn from videos like mimicking the actions. Therefore the examples of learning from videos are common to see in the literature of IfO. IfO regards imitation learning from a different perspective, compared with other methods introduced above. Therefore, there are inevitable overlappings of some specific methods introduced in this section with some methods introduced above, but in the IfO category. When you read this section, you should keep in mind that the IfO methods are in most cases orthogonal to other categories of methods as it looks at the imitation learning in a different perspective and focuses on the problem of unobservable actions.

The aforementioned algorithms, however, can hardly handle the demonstrations with partial or unobservable actions. One idea to learning from these demonstrations is to first recover actions from states and then adopt standard imitation learning algorithms to learn a policy from the recovered state-action pairs. For example, Torabi et al. recover actions from states by learning a dynamic model of state transitions, and then use a BC algorithm to find the optimal policy (Torabi et al. 2018a). However, the performance of this method is highly dependent on the learned dynamic model and may fail when the states transit with noise. Instead, Merel et al. proposed to learn from only state (or state feature) trajectories. They extended the GAIL framework to learn a control policy from only states of motion capture demonstrations (Merel et al. 2017) and showed that partial state features without demonstrator actions suffice for adversarial imitation. Similarly, Eysenbach et al. pointed out that the policy should control which states the agent visits, and thus use the states to train a policy by maximizing mutual information between the policy and the state trajectories (Eysenbach et al. 2018). Other studies have also tried to learn from raw observations instead of states. For instance, Stadie et al. extracted features from observations by the domain adaptation method to ensure that experts and novices are in the same feature space (Stadie et al. 2017). However, only using demonstrated states or state features may require a huge number of environmental interactions during the training since any possible information from actions is ignored.

In order to provide more clear structure about advanced methods in IfO, we organize the methods of IfO in the literature into two general groups: (1) model-based algorithms, and (2) model-free algorithms, which also follow one of the main taxonomies in reinforcement learning (as shown in Chap. 3). Next, we discuss the features of each group and present relevant algorithms from the literature as examples.

8.4.1 Model-Based

Similar to model-based reinforcement learning (as in Chap. 9), if the model of the environment can be learned precisely with low consumption, it can benefit the learning process through efficient planning. As imitation learning is about mimicking a sequential of actions instead of a single one in the interactive process with the environment, it inevitably involves the dynamics of the environment, which can be learned with model-based approaches. According to different types of dynamics models, the model-based IfO methods can be categorized as: (1) inverse dynamics models or (2) forward dynamics models.

Inverse Dynamics Models An inverse dynamics model is a mapping from state transitions $\{(S_t, S_{t+1})\}$ to actions $\{A_t\}$ (Hanna and Stone 2017). One work by Nair et al. (2017) in this category learns to predict a sequence of actions for rope manipulation with the sequences of images of a human manipulating a rope from initial condition to a goal condition, which requires to learn a pixel-level inverse dynamics model as follows:

$$A_t = M_\theta(I_t, I_{t+1}) \tag{8.25}$$

where the A_t is the predicted action by the inverse dynamics model M with the input pair of images I_t, I_{t+1}, and the model is parameterized by θ. A convolutional neural network is used for learning the inverse dynamics model. The robot collects rope manipulation samples with an exploration policy automatically. The collected samples are used for learning the inverse dynamics model, after which the robot conducts planning with the learned model and desired states from the human demonstration. The learned inverse dynamics model M_θ^* can actually serve as the policy for choosing actions similar to the demonstration with respect to the desired frame I^e:

$$A_t = M_\theta^*\left(I_t, I_{t+1}^e\right) \tag{8.26}$$

Another work called reinforced inverse dynamics modeling (RIDM) (Pavse et al. 2019) applies a reinforced post-training for fine-tuning the learned inverse dynamics model after training on samples collected with a pre-defined exploration policy. The pre-trained inverse dynamics model, as said above, is regarded as the policy for the agent in a reinforcement learning setting and a sparse reward function R can be applied for reinforcement learning fine-tuning process:

$$\theta^* = \arg\max_\theta \sum_t R\left(S_t, M_\theta^{pre}\left(S_t, S_{t+1}^e\right)\right) \tag{8.27}$$

where M_θ^{pre} is the pre-trained model and fine-tuned here in a reinforcement learning manner.

The covariance matrix adaptation evolution strategy (CMA-ES) or Bayesian optimization (BO) methods can be applied for optimizing the model for a low-dimensional cases. However, the author assumes that each observation transition is reachable through the application of a single action. Targeting at removing this unnecessary assumption, Pathak et al. (2018) allow the agent to execute multiple actions until it gets close enough to the next demonstrated frame.

The algorithms introduced above try to recover the policy with the inverse dynamics model for each single demonstration state. The behavioral cloning from observation (BCO) algorithm proposed by Torabi et al. (2018a), on the other hand, tries to recover the demonstration dataset with full observation-action pairs using the learned inverse dynamics model, and then learn the policy with the augmented demonstration dataset in a regular imitation learning manner, as shown in Fig. 8.5.

Guo et al. (2019) propose to apply a tensor-based model to infer the unobserved actions of the expert state sequences (the IfO problem), which is shown in Fig. 8.6. The policy of the agent is then optimized via a hybrid objective combining reinforcement learning and imitation learning as:

$$\theta^* = \arg\min_{\theta} \mathrm{L}_{RL}(\pi(a|s;\theta)) - \mathbb{E}_{\left(S_t^e, S_{t+1}^e\right)\sim\mathcal{D}}\left[\log \pi_\theta\left(M\left(S_t^e, S_{t+1}^e\right)|S_t^e\right)\right] \quad (8.28)$$

where the L_{RL} is a regular reinforcement learning loss term with policy π parameterized by θ. $\mathcal{D}$ is the demonstration dataset, and the second term is the behavioral cloning loss for maximizing the likelihood of predicting "expert" actions given the expert states s^e and inverse dynamics model M. Guo's method is, in a way, a combination of RIDM and BCO methods. Instead of using a parameterized inverse dynamics model like in other methods above, the inverse dynamics model M here is a low-rank tensor model with advantages over deep neural networks. The reward signals are required for providing the reinforcement learning loss, which is similar to RIDM.

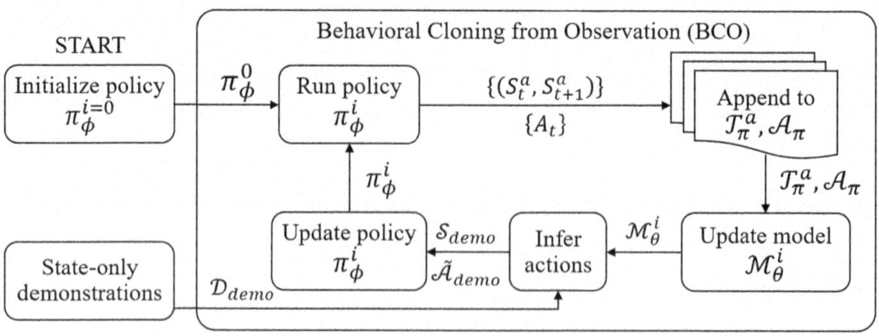

Fig. 8.5 The learning framework of Behavioral Cloning from Observation (BCO), adapted from Torabi et al. (2018a)

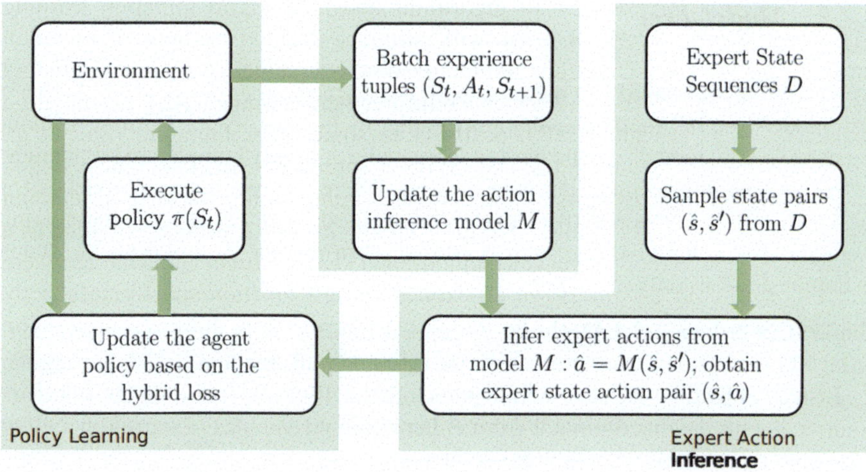

Fig. 8.6 The learning framework of hybrid reinforcement learning with expert state sequences framework, adapted from Guo et al. (2019)

Forward Dynamics Models A forward dynamics model is a mapping from state-action pairs, $\{(S_t, A_t)\}$, to the next states, $\{S_{t+1}\}$. One typical method leveraging the forward dynamics model in IfO is called imitating latent policies from observation (ILPO) (Edwards et al. 2018). ILPO applies two networks in its learning process: the latent policy network and the action remapping network. The latent policy network includes an action inference module which maps the state S_t to a latent action z, and a forward dynamics module which predicts the next state S_{t+1} given current state S_t and the latent action z. The update rules of these two modules are as follows:

$$\omega^* = \arg\min \mathbb{E}_{\left(S_t^e, S_{t+1}^e\right)\sim\mathcal{D}}\left[||G_\omega(S_t^e, z) - S_{t+1}^e||_2^2\right] \tag{8.29}$$

for the latent dynamics model G_ω and

$$\theta^* = \arg\max \mathbb{E}_{\left(S_t^e, S_{t+1}^e\right)\sim\mathcal{D}}\left[||\sum_z \pi_\theta(z|S_t^e)G_\omega(S_t^e, z) - S_{t+1}^e||_2^2\right] \tag{8.30}$$

for the latent policy $\pi_\theta(\cdot|z)$, where $\mathcal{D}$ is the expert demonstration dataset.

However, since the latent action produced in the latent policy network may not necessarily be the true action in real dynamics of the environment, the action remapping network is applied for associating the latent actions to the true actions. The usage of latent actions requires no interactions with the environment during learning the latent model and latent policy, while the remapping action network only needs limited number of interactions with the environment, which makes the algorithm efficient in the learning process.

8.4.2 Model-Free

Apart from model-based IfO methods with the learned dynamics models, there are also model-free IfO methods, which is another main category for learning without the models. The models can be hard to learn well for highly complicated dynamics, as in regular reinforcement learning settings. There are two main approaches for model-free IfO: (1) generative adversarial methods and (2) reward-engineering methods. The generative approach is similar as in regular IL, but with the states as demonstrations only.

Generative Adversarial Methods A general framework in the generative adversarial IfO is modified from previously introduced GAIL method in IRL for regular IL. Instead of feeding the state-action pairs into the discriminator, only the states are compared with the discriminator from either explored samples of current policy or the expert demonstration, which gives the loss:

$$\text{Loss} = \mathbb{E}_{\mathcal{D}}[\nabla_\omega \log(D_\omega(s))] + \mathbb{E}_{\mathcal{D}^e}[\nabla_\omega \log(1 - D_\omega(s))] \tag{8.31}$$

where $\mathcal{D}$ is the explored sample set with current policy and $\mathcal{D}^e$ is the demonstration dataset. Different specific algorithms will have different specific forms and modifications based on that.

For example, Merel et al. (2017) developed a variant of GAIL with only partially observed state features and without access to actions to provide human-like motions for humanoids via the GAN's structure. Similar as RIDM method and hybrid reinforcement learning method in model-based IfO, it also applies the reinforcement learning module together with an imitation learning module, but with a hierarchical structure. The reinforcement learning module is a high-level controller built on the low-level controller with the BC method for capturing the motion features of humanoid. Trajectories of states and actions are collected during the interaction process of a stochastic policy π and the environment, which corresponds to the generator in GAN framework. The state-action pairs are transformed into features, z, in which the actions may be excluded. The demonstration data are assumed to be in the same feature space according to the original paper. Either demonstration data or generated data are evaluated by the discriminator to yield a probability of the data being demonstration data. The output value of the discriminator is then used as a reward to update the imitation policy using reinforcement learning, similar to Eq. (8.12) in GAIL. An additional context variable is also applied for learning multi-behavior policies. The loss of the discriminator can be written as:

$$\text{Loss} = \mathbb{E}_{z\sim s, s\sim\mathcal{D}}[\nabla_\omega \log(D_\omega(z, c))] + \mathbb{E}_{z^e\sim s^e, s^e\sim\mathcal{D}^e}\left[\nabla_\omega \log(1 - D_\omega(z^e, c^e))\right] \tag{8.32}$$

where z, z^e are encoded features of s, s^e sampled from reinforcement learning explorations $\mathcal{D}$ and expert demonstrations $\mathcal{D}^e$, respectively, and c, c^e are the context variables indicating different behaviors.

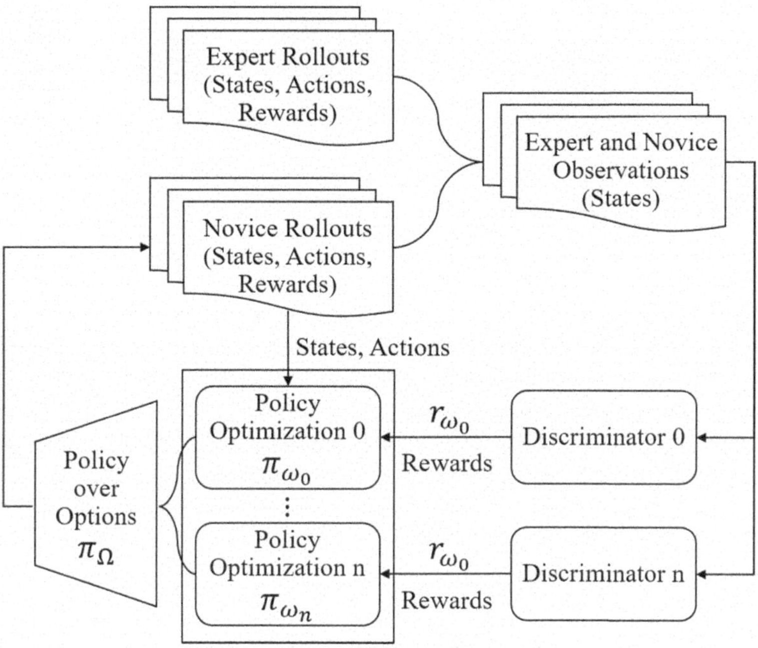

Fig. 8.7 The architecture of OptionGAN, adapted from Henderson et al. (2018)

The OptionGAN (Henderson et al. 2018) proposed by Henderson et. al. applies the option framework in hierarchical reinforcement learning (details in Chap. 10) to recover the joint reward-policy options with generative adversarial architecture using only observed states, as shown in Fig. 8.7. With the decomposition of policies, it is able to not only learn well on simple tasks, but also learn a general policy over options for complicated continuous control tasks.

One potential problem of IfO with above methods is that, even if the learned optimal policy is able to generate a state distribution that is very similar to the expert policy, it still does not mean that the actions are exactly the same for both imitation policy and the expert policy for all states. A simple example by Torabi et al. (2019d) would be, in a ring-like environment, two agents that move with the same speed but different directions (i.e., one clockwise and another one counter-clockwise) would result in each exhibiting the same state distribution even though their behaviors are opposite to one another (i.e., different action distributions given the states).

One way of solving above mismatch problem in action distributions is to feed a sequence of states instead of a single one to the discriminator, like proposed by Torabi et al. (2019b) and Torabi et al. (2018b), a similar algorithm but only with the difference that the discriminator considers state transitions, $\{(S_t, S_{t+1})\}$, as the input instead of single states. This changes the loss function of the discriminator to be

$$\mathbb{E}_{\mathcal{D}}[\nabla_\omega \log(D_\omega(S_t, S_{t+1}))] + \mathbb{E}_{\mathcal{D}^e}[\nabla_\omega \log(1 - D_\omega(S_t, S_{t+1}))] \tag{8.33}$$

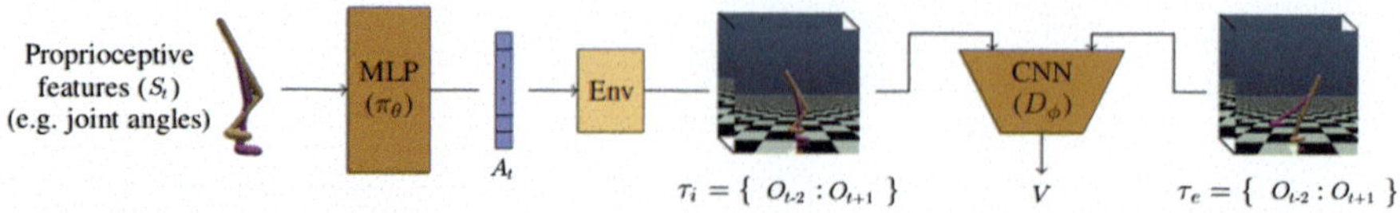

Fig. 8.8 Imitation learning from observations only, using the proprioceptive state. Figure is adapted from Torabi et al. (2019c)

where the state sequence can also be chosen to be longer than two in practice.

Another work by Torabi et al. (2019c) leverages the proprioceptive features as the state input for the policy instead of the observed images, to model the humans or animals proprioception-based control in reinforcement learning agents. Because of the low dimensions of the proprioceptive features, the policy can be represented by a simple multi-layer perceptron (MLP) instead of a convolutional neural network (CNN), while the discriminator still takes a sequence of observed images as inputs from both the explored samples and the expert demonstration, as shown in Fig. 8.8. The low-dimensional proprioceptive features make the learning process more efficient as well.

As mentioned in Chap. 7, the low sample efficiency is one of the key problems in present reinforcement learning algorithms, which also holds for the imitation learning and IfO areas. As the generative adversarial approaches are within the IRL domain, those methods introduced above can suffer from intensive computational cost as mentioned in Sect. 8.3. These adversarial imitation algorithms often require large numbers of demonstration examples and learning iterations to learn a policy imitating a demonstrator's behavior successfully. To further improve the sample efficiency of above methods, Torabi et al. (2019a) proposed to apply the linear quadratic regulators (LQR) (Tassa et al. 2012) as a trajectory-centric reinforcement learning method during the policy learning process, which has potential to make it possible to apply the algorithm for real-robot imitation learning.

The above works are mostly based on the basic assumptions that the demonstration data space and the imitators' learning space are consistent. However, when there is a mismatch between the two spaces, for example, the changes of viewpoint by placing the camera in different positions in the three-dimensional space for providing observation, the general imitation learning methods will have a degradation in performance. The difference of the spaces of demonstration and the imitator can be either in action space or the state space. For the difference in action spaces, Żołna et al. (2018) proposed to use pairs of states with random time gaps instead of consecutive states as the input of the discriminator, which can be regarded as dataset augmentation with noise for more robust and general performances. In their own experiments, it is indeed shown to improve the performances of imitator's policy with different action spaces from the demonstration. While for the difference of the state space, like the viewpoint changing mentioned above, Stadie et al. (2017) have proposed to apply a classifier to distinguish among samples of different viewpoints, with the output values of some initial layers in the discriminator as inputs. The proposed method leverages the idea of domain confusion to learn the domain agnostic features, where the domain indicates different viewpoints in this case. The confusion is maximized in the first layers of the discriminator (as a feature extractor) but minimized for the classifier, which also leverages the adversarial training framework. After training, the learned features of the extractor (first several layers of the discriminator) are invariant to the viewpoint.

There are also some other works in this field. Sun et al. (2019) proposed the first provably efficient algorithm in IfO, called Forward Adversarial Imitation Learning (FAIL), which can learn a near-optimal policy with the number of samples in a polynomial relationship with all relevant parameters but independent of the number of unique observations. The minimax game in FAIL learns a policy that matches the state distribution of the next state given the policies of the previous time steps. Recently, a method called Action-Guided Adversarial Imitation Learning (AGAIL) proposed by Sun and Ma (2019) tries to leverage the states and incomplete actions in demonstrations, which is a combination of IfO and traditional IL. The discriminator is used for discriminating single states, similar to the approach of Merel et al. (2017) described before. Additionally, a guided Q-network is employed to learn the true posterior of $p(a^e|a \sim \pi(s^e))$ in a supervised learning manner, where (s^e, a^e) denote samples of expert demonstration.

Reward-Engineering Methods The generative adversarial approach naturally provides the reward signals, from which the imitation policy can learn in a reinforcement learning manner. Apart from the generative adversarial approach, there are also methods with reward engineering for solving model-free IfO. Actually, the method RIDM in model-based IfO is a method with reward engineering mentioned in the previous section. The reward engineering indicates the need of manually designed reward functions for learning an imitation policy in a reinforcement learning manner with expert demonstrations. Reward engineering transfers the supervised learning approach of imitation learning into a reinforcement learning problem through formalizing the reward function for the reinforcement learning agent. What needs to be noticed is that the manually designed reward function does not have to be the true reward function leading to the expert policy, but more of an estimation from the demonstration dataset or prior knowledge about the tasks. For example, Kimura et al. (2018) proposed to use the Euclidean distance of the predicted next state by the predictor and the true next state by the demonstrator as the reward function. Then the reward function is used for learning an imitation policy in general reinforcement learning settings.

Another reward-engineering approach is called time-contrastive networks (TCNs) proposed by Sermanet et al. (2018) (Fig. 8.9). To handle the multi-viewpoint problem as mentioned before, which is important for learning from human behaviors, the TCN method learns a viewpoint invariant representation capturing the relationships among objects with the TCN network using several (two in the original paper) synchronous camera views from different perspectives. The adversarial training is therefore applied in the embedded representation space instead of the original state space in other IfO methods. The representation is

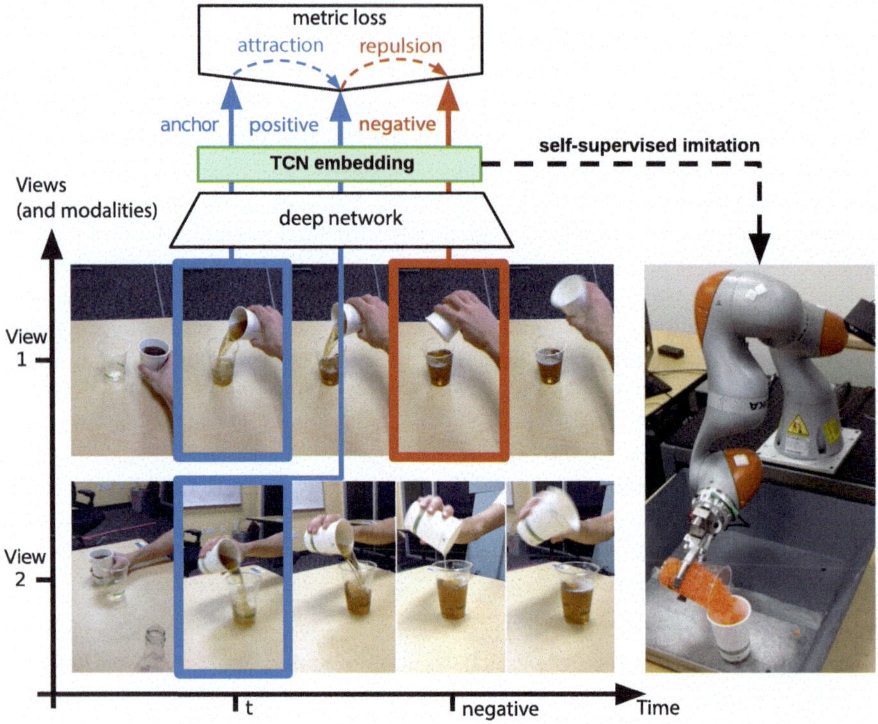

Fig. 8.9 The learning framework of time-contrastive network (TCN) with a triplet loss for observation embedding in self-supervised imitation learning from observations only. Figure is from Sermanet et al. (2018)

trained with a triplet loss with the TCN embedding network. The triplet loss is set to disperse the temporal neighbors of consecutive frames in the video demonstration with similar visual features but different actual dynamic states and also to attract those simultaneous frames from different viewpoints but with the same dynamic state in the embedding space. Therefore, the imitation policy can learn with unlabeled video of human demonstrations in a self-supervised learning manner. Similar as in Kimura's work, the reward function is defined to be the Euclidean distance between the state of demonstration and the state of the agent at each time step, but in the embedding space instead of the state space. The TCN is designed to work for single frame state embedding. Dwibedi et al. (2018) extended the work of TCN to multiple frames embedding for better representing the patterns in trajectory. Aytar et al. (2018) also took a similar approach, learning an embedding function for the YouTube video frames based on the demonstration, to solve the hard-exploration tasks like Montezuma's Revenge and Pitfall mentioned in the exploration challenge of Chap. 7. It can handle the small variance in domain like video artifacts and color changes. The measurement of closeness between the imitator's embedded states and some demonstrator's embedded states is also used as the reward function.

As mentioned in the previous section of the adversarial generative approach, a classifier can be employed to distinguish among observations from different viewpoints. A classifier can also be used to predict the order of frames in the demonstration as proposed by Goo and Niekum (2019), via a shuffle-and-learn (Misra et al. 2016) training manner. A reward function can be defined with respect to the learned classifier for training the imitator policy. Also in previous sections of the adversarial generative approach, the mismatch between the state spaces, like being caused by the viewpoints, can be handled with an invariant feature representation. However, instead of using the output values of the discriminator with demonstration states and the imitator's state as inputs, it can also train an imitation policy with a reward function defined to be the Euclidean distance between the two kinds of states in the representation space, as proposed by Gupta et al. (2017) and Liu et al. (2018).

8.4.3 Challenges of IfO

With the above mentioned methods developed in IfO, the agent can learn the policy from the observed states only, but still suffers from several problems as mentioned in the survey by Torabi et al. (2019d), which are listed as the challenges below:

- **Embodiment Mismatch**: The embodiment mismatch generally describes the differences of appearances (for visual-based control), dynamics, and other features between the imitator's domain and the demonstrator's domain. A typical example would be letting a robotic arm mimic the motion of a human's arm. Due to the significant differences in the controlling dynamics and perspectives of looking at the agents, the imitation learning process can be potentially very hard to conduct. Even determining if the robot is in the same state as the human's arm can be difficult. One way to solve this is to learn the hidden correspondences or latent representations that are invariant for the differences of the two domains, and conducting the imitation learning based on the correspondences or in that learned representation space. One IfO method developed to address this problem learns a correspondence between the embodiments using autoencoders in a supervised fashion (Gupta et al. 2017). The autoencoder is trained in such a way that the encoded representations are invariant with respect to the embodiment features. Another method learns the correspondence in an unsupervised fashion with a small amount of human supervision (Sermanet et al. 2018).
- **Viewpoint Difference**: As mentioned in several methods described above, like the TCN and some methods in model-based IfO, the difference of viewpoints can degrade the performance of the imitation policy significantly, for visual-based control with demonstration data provided by images or videos from a camera. Generally an encoding model for representing the states in a viewpoint invariant space is required as in Sieb et al. (2019), or a classifier for predicting the specific viewpoint for one frame as in Stadie et al. (2017). Another IfO approach that attempts to address this issue learns a context translation model to translate an

observation by predicting it in the target context (Liu et al. 2018). The translation is learned using data that consists of images of the target context and the source context, and the task is to translate the frame from the source context to that of the target. It would require the similar samples of the target context to be collected as in the source context.

8.5 Probabilistic Methods

Apart from the parameterized methods with deep neural networks (DNN), a variety of probabilistic inference methods are applied for imitation learning as well, especially in the robot motion domains, which include Gaussian mixture regression (GMR) (Calinon 2016), dynamical movement primitives (DMPs) (Pastor et al. 2009), probabilistic movement primitives (ProMPs) (Paraschos et al. 2013), kernelized movement primitives (KMP) (Huang et al. 2019), Gaussian process regression (GPR) (Schneider and Ertel 2010), and GMR-based GP process (Jaquier et al. 2019). As this book aims at introducing deep reinforcement learning with parameterization methods using DNN, we will only briefly discuss about probabilistic methods because the combination of probabilistic methods with DRL is non-trivial, unlike other previous approaches introduced in this chapter.

However, even if it is hard to apply the probabilistic methods on DRL tasks like using the supervised imitation learning as an initialization for reinforcement learning introduced in next section, probabilistic methods are still attractive to be investigated for imitation learning with several advantageous properties. Unlike the deterministic prediction results given by DNN, the covariance matrices of the prediction distributions computed by GMR, ProMPs, and KMP encode the variability of the predicted trajectory. This can be useful when applying the learned model on predicting or decision-making tasks where the belief of the prediction is also important, like ensuring the safety during robotic manipulation or vehicle driving cases. Apart from that, probabilistic methods usually have analytic solutions with the support of probabilistic theory, which is different from the "black-box" optimization process of DNN-based methods. This also makes probabilistic methods able to be solved within a short time when the amount of data is small. Probabilistic methods like GMR-based GP process have the quick adaptability for the unseen input datapoints, which will be discussed in the following sections. For probabilistic methods in IL, the dataset is considered to be provided in a labeled format with pairs of input and output, which is usually the set of state-action pairs $\{(s_i, a_i)|i = 0, \ldots, N\}$ for common reinforcement learning cases or time-state pairs $\{(t, S_t)|t = 0, \ldots, N\}$ (Jaquier et al. 2019) for the time-aligned demonstrations.

GMR-based GP regression is a combination of Gaussian mixture regression and Gaussian process regression. GMR exploits the Gaussian conditioning theorem to estimate the distribution of output data given input data. A Gaussian Mixture Model (GMM) is used to fit on the joint distribution of input and output datapoints with an Expectation Maximization (EM) algorithm. The conditional means and covariances

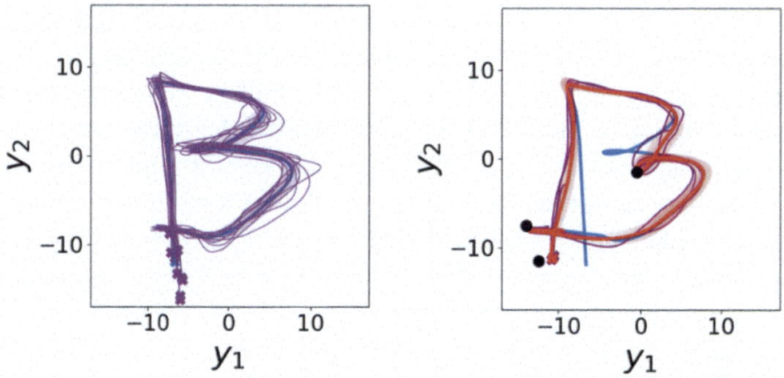

Fig. 8.10 GMR-based GP process for imitation learning. The left image shows the prior mean of the process and the sample trajectories in blue and purples lines, respectively. The right image shows the prior mean (same as the left one), sampled, and predicted trajectories in blue, pink, and red lines, respectively. The three black dots in the right image are observations. Figure is adapted from Jaquier et al. (2019)

given observed input can be solved in closed form, and the output can therefore be predicted via a linear combination of conditional expectations with the test input datapoints. GP aims at learning a deterministic input–output relationship, just like DNN's approach, based on a Gaussian prior over potential objective functions. The GMR-based GP is combined as a GP with its prior mean equal to the conditional mean of the GMR model, and with its kernel in the form of a sum of all separable kernels associated with the components of the corresponding GMM. This combination takes the advantage of the ability of GPs to encode various prior beliefs through the mean and kernel functions and allows the variability information retrieved by GMR to be encapsulated in the uncertainty estimated by the GP. When given new and unseen input observation points, the GMR-based GP method is able to quickly adapt to them and predict reasonable outputs as shown in Fig. 8.10. For a two-dimensional trajectories estimation process, the left image in Fig. 8.10 shows the given samples in purple lines, and the prior mean in blue line. The right image is the GMR-based GP process with 3 new observation points in black, and with the pink lines showing the sampled trajectories and red line as prediction. This method is testified to have a great performance on leveraging demonstrations but quickly adapting to new datapoints, which can be applied on manipulating the robot to avoid obstacles with demonstrations.

8.6 IL as Initialization for RL

The basic setup for applying imitation learning is to learn a policy without any reinforcement signals but only the demonstrations data, which means the learned policy through imitation learning is the final policy from the demonstrations. However, in practice, the policy learned from imitation learning is usually not general enough, especially for unseen cases. Therefore, we can leverage the imitation learning in the reinforcement learning process, which improves the learning efficiency of reinforcement learning. For example, a pre-trained policy using demonstration data can be used to initialize the policy in reinforcement learning. More about these approaches will be discussed later. Therefore, we do not require the policy from imitation learning to be optimal, but good enough with a relatively simple imitation learning process, like applying a supervised learning approach. So, we only choose some of the simple and straightforward methods described below as an initialization method for subsequent reinforcement learning processes. Those fancier techniques in imitation learning will provide a better initialization policy with no doubts, but may have drawbacks as longer pre-training time and so on.

Generally, the policy learned from imitating the demonstrations in a supervised manner, including the **BC, DAgger, Variational Dropout, and so on**, can be regarded as a good initialization for reinforcement learning policy, using methods like **policy replacement or residual policy learning** described in the following sections. We will experimentally show the improvement in reinforcement learning with the initialization policy trained using above mentioned supervised learning methods in the following sections.

In addition to the policy replacement approach for initialization of reinforcement learning, residual policy learning (Silver et al. 2018; Johannink et al. 2019) is another approach to realize initialization. It is based on good but imperfect controllers for robot manipulation tasks, and to learn a residual policy on top of that initial controller. For robot manipulation in real world, the initial controller could be a pre-trained policy in simulation; and for robot manipulation in simulation, the initial controller could be from the pre-trained supervised learning with expert trajectories as in Sect. 8.3.2.

The action in residual policy learning follows the combinatorial policy, which is the sum of the initial policy π_{ini} and the residual policy π_{res}:

$$a = \pi_{ini}(s) + \pi_{res}(s) \tag{8.34}$$

In this way, the residual policy learning is able to preserve the initialized policy performance to the best advantage.

Example: DDPG with Residual Policy Learning

We apply the DDPG algorithm for leveraging the demonstrations with residual policy learning. According to the residual policy learning, the actor's policy in DDPG consists of two parts: one is the pre-trained initialization policy, which will be fixed after initialization, and another one is the residual policy to be trained during the learning process. The initialization policy is pre-trained with the demonstration samples generated from inverse kinematics, which is the same as the policy replacement method. The pre-trained initialization policy only works for the actor part in DDPG. The process of applying residual policy learning in DDPG is as follows:

(1) Initialize all neural networks in DDPG with residual learning, including a general initialization of the critic, target critic, and an initialization with zero-valued final layers for the residual policy and the target residual policy, and an initialization with imitation learning for the policy and the corresponding target, totally six networks. Fix the initialized policy and its target, and start the training process.
(2) Let the agent interact with the environment, and the action value is the sum of the action values from the initialization policy and the residual policy: $a = a_{ini} + a_{res}$; store samples in the form of $(s, a_{res}, s', r, done)$.
(3) Draw samples $(s, a_{res}, s', r, \text{done})$ from the memory buffer, we have

$$Q_{target}(s, a_{res}) = r + \gamma Q^T\left(s, \pi^T_{res}(s)\right) \tag{8.35}$$

where Q^T, π^T_{res} denote the target critic and the target residual policy, respectively. The critic loss is $\text{MSE}(Q_{target}(s, a_{res}), Q(s, a_{res}))$. The objective for the actor is to maximize the action-value function of state s and action a_{res} as follows:

$$\max_\theta Q(s, a_{res}) = \max_\theta Q(s, \pi_{res}(s|\theta)) \tag{8.36}$$

which can be optimized via deterministic policy gradient.
(4) Repeat above steps (2) and (3) until the policy is converged or near optimal.

Compared with general DDPG algorithm, the difference of applying residual policy learning is just to learn action-value function and the policy with respect to the residual policy actions instead of the overall actions for the agent.

8.7 Other Approaches of Leveraging Demonstrations in RL

8.7.1 Feeding Demonstrations into Replay Buffer

Instead of pre-training a policy to initialize the reinforcement learning policy, deep Q-learning from demonstrations (DQfD) (Hester et al. 2018) leverages demonstrations through directly feeding those expert trajectories into memory buffer of off-policy reinforcement learning. It applies DQN for only discrete action space applications. DQfD uses a replay buffer initialized with all expert demonstrations and then keeps storing new samples in it. It applies the prioritized experience replay to sample the training batch from the replay buffer, and DQfD trains the policy using a combination of a supervised hinge loss for imitating the demonstrations and a general TD loss.

The approach of deep deterministic policy gradient from demonstrations (DDPGfD) (Večerík et al. 2017) is a method similar with the DQfD method as described above, but applies DDPG for continuous action space applications. DDPGfD leverages demonstrations through directly feeding those expert trajectories into memory of off-policy reinforcement learning (e.g., DDPG), to train the policy with both demonstrations and explorations. The prioritized experience replay (Schaul et al. 2015) is used as a natural balance of the two sources of training data. DDPGfD can work on solvable simple tasks for reinforcement learning, while learning from sparse rewards on harder task requires more active exploration during training.

Nair et al. (2018) proposed a method based on DQfD and DDPGfD to have better learning efficiency for hard tasks where further exploration based on demonstrations matters. The policy loss is a combination of policy gradient loss and the behavioral cloning loss, which gives the gradients as follows:

$$\lambda_1 \nabla_\theta J - \lambda_2 \nabla_\theta L_{BC} \tag{8.37}$$

where the J is the general reinforcement learning objective (maximized) and L_{BC} (minimized) is the behavior cloning loss as defined at the beginning of this chapter.

Moreover, the Q-filter technique is applied in this method, which requires the behavioral cloning loss to be only applied to states where the learned critic $Q(s, a)$ determines that the demonstrator action is better than the actor action:

$$L_{BC} = \sum_{i=1}^{N_D} ||\pi(s_i|\theta_\pi) - a_i||^2 \mathbb{1}_{Q(s_i,a_i)>Q(s_i,\pi(s_i))} \tag{8.38}$$

where the N_D is number of samples in demonstration dataset and (s_i, a_i) are sampled from the demonstration dataset. This ensures the policy to explore better actions other than being restricted by the demonstration data.

Using the same approach, QT-Opt (Kalashnikov et al. 2018) and Quantile QT-Opt (Bodnar et al. 2019) algorithms also apply a combination of on-policy buffer and an off-policy demonstration buffer to conduct off-line learning with actor-free CE method with DQN, which achieves the state-of-the-art performances in real-world robot learning tasks based on images.

8.7.2 *Normalized Actor-Critic*

Normalized actor-critic (NAC) (Gao et al. 2018) is another method for efficient reinforcement learning with demonstrations, and it pretrains a policy as initialization for a refinement reinforcement learning process. The key difference of NAC from other methods is that it uses exactly the same objective for the processes of pre-training an initialization policy with demonstrations and the refinement reinforcement learning process (not like a combination of supervised loss and reinforcement learning loss in DQfD, or two separate training processes with different loss in policy replacement and behavioral cloning methods), which makes NAC robust to suboptimal demonstrations data.

The NAC method is similar to methods of DDPGfD or DQfD, but trains the policy sequentially from demonstrations and samples from interactions instead of using samples from both sources at the same time.

8.7.3 Reward Shaping with Demonstrations

Reward shaping with demonstrations (Brys et al. 2015) is a method focusing on the initialization of value function instead of the action policy for reinforcement learning. It provides the agent an intermediate reward for enriching the sparse reward signals:

$$R_F(s, a, s') = R(s, a, s') + F^D(s, a, s') \tag{8.39}$$

where the shaping reward F^D from demonstrations D is defined with potential function ϕ in the following form to guarantee the convergence:

$$F^D(s, a, s', a') = \gamma \phi^D(s', a') - \phi^D(s, a) \tag{8.40}$$

and ϕ^D is defined as:

$$\phi^D(s, a) = \max_{(s^d, a)} e^{-\frac{1}{2}\left(s - s^d\right)^T \Sigma^{-1} \left(s - s^d\right)} \tag{8.41}$$

which is to maximize the value of the potential for the state s that is most similar as the demonstration state s^d. The optimized potential function is used to initialize the action-value function Q in reinforcement learning:

$$Q_0(s, a) = \phi^D(s, a) \tag{8.42}$$

The intuition of the reward shaping method is to bias the exploration in favor of those state-action pairs in demonstrations or close to those in demonstrations for accelerating the training process of reinforcement learning. Reward shaping provides a good approach of initialization for the value-evaluation function in reinforcement learning process.

Other methods like unsupervised perceptual rewards (Sermanet et al. 2016) also learn a dense and smooth reward functions with the demonstrations, using features in a pre-trained deep model.

8.8 Summary

Due to the low learning efficiency challenge of reinforcement learning as mentioned in Chap. 7, in this chapter, we introduce imitation learning (IL) as one potential solution leveraging the expert demonstration. The overall chapter is summarized into several main categories. The behavior cloning methods introduced in Sect. 8.2 are the most straightforward way of imitation learning in a supervised learning manner, which can be further combined with reinforcement learning like as an initialization introduced in Sect. 8.6. A more advanced way of combining the imitation learning with reinforcement learning is through IRL by recovering a reward function explicitly or implicitly from demonstration, as in Sect. 8.3. Methods like MaxEnt can explicitly learn the reward function but with heavy computation cost. Other methods in the generative adversarial approach like GAIL, GAN-GCL, AIRL learn in a more efficient way. Another problem is if the actions are missing in the demonstration dataset, like learning from the videos only, how to work properly with imitation learning? This falls into the category of IfO as in Sect. 8.4. Since the IfO problem is from another perspective, those methods mentioned before like BC, IRL can also be applied in IfO with proper modifications. The methods in IfO are generally summarized in model-based and model-free categories. The model-based method learns the dynamics model from samples, and it can actually recover the observation-only demonstration dataset with actions through leveraging the action-state relationship in the model, explicitly or implicitly. Then, the regular imitation learning methods can be applied if the actions are recovered explicitly. Methods like RIDM, BCO, ILPO, etc., fall into this model-based IfO category. For the model-free methods in IfO, either the reward engineering or generative adversarial approach can be applied for providing the reward function to enable reinforcement learning. Methods like OptionGAN, FAIL, AGAIL, and so on are in the category of generative adversarial IfO, while TCN and some other methods are in the category of reward engineering IfO. The two categories here in IfO actually also apply for general IL, like GAIL as a generative adversarial method and recently proposed contrastive forward dynamics (CFD) (Jeong et al. 2019) as a reward-engineering method for learning from demonstration with both observations and actions in IL. Then the probabilistic methods including GMR, GPR, and DMR-based GP are introduced as an alternative for general IL, with high-efficiency learning for relatively low-dimensional cases, as in Sect. 8.5. Finally some other approaches like DDPGfD and DQfD for feeding demonstration data into replay buffer in off-policy reinforcement learning and so on are introduced in Sect. 8.7. The research area of imitation learning is still very active as an efficient approach for solving learning problems, with an organic combination with reinforcement learning.

References

Abbeel P, Ng AY (2004) Apprenticeship learning via inverse reinforcement learning. In: Proceedings of the twenty-first international conference on machine learning. ACM, New York, p 1

Aytar Y, Pfaff T, Budden D, Paine T, Wang Z, de Freitas N (2018) Playing hard exploration games by watching YouTube. In: Advances in neural information processing systems, pp 2930–2941

Blau T, Ott L, Ramos F (2018) Improving reinforcement learning pre-training with variational dropout. In: 2018 IEEE/RSJ international conference on intelligent robots and systems (IROS). IEEE, Piscataway, pp 4115–4122

Bodnar C, Li A, Hausman K, Pastor P, Kalakrishnan M (2019) Quantile QT-Opt for risk-aware vision-based robotic grasping. Preprint. arXiv:191002787

Brys T, Harutyunyan A, Suay HB, Chernova S, Taylor ME, Nowé A (2015) Reinforcement learning from demonstration through shaping. In: Twenty-fourth international joint conference on artificial intelligence

Calinon S (2016) A tutorial on task-parameterized movement learning and retrieval. Intel Serv Robot 9(1):1–29

Duan Y, Andrychowicz M, Stadie B, Ho OJ, Schneider J, Sutskever I, Abbeel P, Zaremba W (2017) One-shot imitation learning. In: Advances in neural information processing systems, pp 1087–1098

Dwibedi D, Tompson J, Lynch C, Sermanet P (2018) Learning actionable representations from visual observations. In: 2018 IEEE/RSJ international conference on intelligent robots and systems (IROS). IEEE, Piscataway, pp 1577–1584

Edwards AD, Sahni H, Schroecker Y, Isbell CL (2018) Imitating latent policies from observation. Preprint. arXiv:180507914

Eysenbach B, Gupta A, Ibarz J, Levine S (2018) Diversity is all you need: learning skills without a reward function. Preprint. arXiv:180206070

Finn C, Christiano P, Abbeel P, Levine S (2016a) A connection between generative adversarial networks, inverse reinforcement learning, and energy-based models. Preprint. arXiv:161103852

Finn C, Levine S, Abbeel P (2016b) Guided cost learning: deep inverse optimal control via policy optimization. In: International conference on machine learning, pp 49–58

Fu J, Luo K, Levine S (2017) Learning robust rewards with adversarial inverse reinforcement learning. Preprint. arXiv:171011248

Gao Y, Lin J, Yu F, Levine S, Darrell T, et al (2018) Reinforcement learning from imperfect demonstrations. Preprint. arXiv:180205313

Goo W, Niekum S (2019) One-shot learning of multi-step tasks from observation via activity localization in auxiliary video. In: 2019 international conference on robotics and automation (ICRA). IEEE, Piscataway, pp 7755–7761

Goodfellow I, Pouget-Abadie J, Mirza M, Xu B, Warde-Farley D, Ozair S, Courville A, Bengio Y (2014) Generative adversarial nets. In: Proceedings of the neural information processing systems (Advances in neural information processing systems) conference

Guo X, Chang S, Yu M, Tesauro G, Campbell M (2019) Hybrid reinforcement learning with expert state sequences. Preprint. arXiv:190304110

Gupta A, Devin C, Liu Y, Abbeel P, Levine S (2017) Learning invariant feature spaces to transfer skills with reinforcement learning. Preprint. arXiv:170302949

Hanna JP, Stone P (2017) Grounded action transformation for robot learning in simulation. In: Thirty-first AAAI conference on artificial intelligence

Hausman K, Chebotar Y, Schaal S, Sukhatme G, Lim JJ (2017) Multi-modal imitation learning from unstructured demonstrations using generative adversarial nets. In: Advances in neural information processing systems, pp 1235–1245

Henderson P, Chang WD, Bacon PL, Meger D, Pineau J, Precup D (2018) OptionGAN: learning joint reward-policy options using generative adversarial inverse reinforcement learning. In: Thirty-second AAAI conference on artificial intelligence

Hester T, Vecerik M, Pietquin O, Lanctot M, Schaul T, Piot B, Horgan D, Quan J, Sendonaris A, Osband I, et al (2018) Deep Q-learning from demonstrations. In: Thirty-second AAAI conference on artificial intelligence

Ho J, Ermon S (2016) Generative adversarial imitation learning. In: Advances in neural information processing systems, pp 4565–4573

Huang Y, Rozo L, Silvério J, Caldwell DG (2019) Kernelized movement primitives. Inter J Robot Res 38(7):833–852

Jaquier N, Ginsbourger D, Calinon S (2019) Learning from demonstration with model-based Gaussian process. Preprint. arXiv:191005005

Jeong R, Aytar Y, Khosid D, Zhou Y, Kay J, Lampe T, Bousmalis K, Nori F (2019) Self-supervised sim-to-real adaptation for visual robotic manipulation. Preprint. arXiv:191009470

Johannink T, Bahl S, Nair A, Luo J, Kumar A, Loskyll M, Ojea JA, Solowjow E, Levine S (2019) Residual reinforcement learning for robot control. In: 2019 international conference on robotics and automation (ICRA). IEEE, Piscataway, pp 6023–6029

Kalashnikov D, Irpan A, Pastor P, Ibarz J, Herzog A, Jang E, Quillen D, Holly E, Kalakrishnan M, Vanhoucke V, et al (2018) QT-Opt: scalable deep reinforcement learning for vision-based robotic manipulation. Preprint. arXiv:180610293

Kimura D, Chaudhury S, Tachibana R, Dasgupta S (2018) Internal model from observations for reward shaping. Preprint. arXiv:180601267

Liu Y, Gupta A, Abbeel P, Levine S (2018) Imitation from observation: learning to imitate behaviors from raw video via context translation. In: 2018 IEEE international conference on robotics and automation (ICRA). IEEE, Piscataway, pp 1118–1125

Merel J, Tassa Y, Srinivasan S, Lemmon J, Wang Z, Wayne G, Heess N (2017) Learning human behaviors from motion capture by adversarial imitation. Preprint. arXiv:170702201

Misra I, Zitnick CL, Hebert M (2016) Shuffle and learn: unsupervised learning using temporal order verification. In: European conference on computer vision. Springer, Berlin, pp 527–544

Molchanov D, Ashukha A, Vetrov D (2017) Variational dropout sparsifies deep neural networks. In: Proceedings of the 34th international conference on machine learning, vol 70, JMLR.org, pp 2498–2507

Nair A, Chen D, Agrawal P, Isola P, Abbeel P, Malik J, Levine S (2017) Combining self-supervised learning and imitation for vision-based rope manipulation. In: 2017 IEEE international conference on robotics and automation (ICRA). IEEE, Piscataway, pp 2146–2153

Nair A, McGrew B, Andrychowicz M, Zaremba W, Abbeel P (2018) Overcoming exploration in reinforcement learning with demonstrations. In: 2018 IEEE international conference on robotics and automation (ICRA). IEEE, Piscataway, pp 6292–6299

Ng AY, Harada D, Russell S (1999) Policy invariance under reward transformations: theory and application to reward shaping. In: Proceedings of the international conference on machine learning (ICML), vol 99, pp 278–287

Ng AY, Russell SJ, et al (2000) Algorithms for inverse reinforcement learning. In: Proceedings of the international conference on machine learning (ICML), vol 1, p 2

Paraschos A, Daniel C, Peters JR, Neumann G (2013) Probabilistic movement primitives. In: Advances in neural information processing systems, pp 2616–2624

Pastor P, Hoffmann H, Asfour T, Schaal S (2009) Learning and generalization of motor skills by learning from demonstration. In: 2009 IEEE international conference on robotics and automation. IEEE, Piscataway, pp 763–768

Pathak D, Mahmoudieh P, Luo G, Agrawal P, Chen D, Shentu Y, Shelhamer E, Malik J, Efros AA, Darrell T (2018) Zero-shot visual imitation. In: Proceedings of the IEEE conference on computer vision and pattern recognition workshops, pp 2050–2053

Pavse BS, Torabi F, Hanna JP, Warnell G, Stone P (2019) RIDM: reinforced inverse dynamics modeling for learning from a single observed demonstration. Preprint. arXiv:190607372

Puterman ML (2014) Markov decision processes: discrete stochastic dynamic programming. Wiley, Hoboken

Ross S, Bagnell D (2010) Efficient reductions for imitation learning. In: Proceedings of the thirteenth international conference on artificial intelligence and statistics, pp 661–668

Ross S, Gordon G, Bagnell D (2011) A reduction of imitation learning and structured prediction to no-regret online learning. In: Proceedings of the fourteenth international conference on artificial intelligence and statistics, pp 627–635

Russell SJ (1998) Learning agents for uncertain environments. In: The 11th annual conference on computational learning theory, vol 98, pp 101–103

Schaul T, Quan J, Antonoglou I, Silver D (2015) Prioritized experience replay. In: International conference on learning representations

Schneider M, Ertel W (2010) Robot learning by demonstration with local Gaussian process regression. In: 2010 IEEE/RSJ international conference on intelligent robots and systems. IEEE, Piscataway, pp 255–260

Sermanet P, Xu K, Levine S (2016) Unsupervised perceptual rewards for imitation learning. Preprint. arXiv:161206699

Sermanet P, Lynch C, Chebotar Y, Hsu J, Jang E, Schaal S, Levine S, Brain G (2018) Time-contrastive networks: self-supervised learning from video. In: 2018 IEEE international conference on robotics and automation (ICRA). IEEE, Piscataway, pp 1134–1141

Sieb M, Xian Z, Huang A, Kroemer O, Fragkiadaki K (2019) Graph-structured visual imitation. Preprint. arXiv:190705518

Silver T, Allen K, Tenenbaum J, Kaelbling L (2018) Residual policy learning. Preprint. arXiv:181206298

Stadie BC, Abbeel P, Sutskever I (2017) Third-person imitation learning. Preprint. arXiv:170301703

Sun M, Ma X (2019) Adversarial imitation learning from incomplete demonstrations. Preprint. arXiv:190512310

Sun W, Vemula A, Boots B, Bagnell JA (2019) Provably efficient imitation learning from observation alone. Preprint. arXiv:190510948

Syed U, Bowling M, Schapire RE (2008) Apprenticeship learning using linear programming. In: Proceedings of the 25th international conference on machine learning. ACM, New York, pp 1032–1039

Tassa Y, Erez T, Todorov E (2012) Synthesis and stabilization of complex behaviors through online trajectory optimization. In: 2012 IEEE/RSJ international conference on intelligent robots and systems. IEEE, Piscataway, pp 4906–4913

Torabi F, Warnell G, Stone P (2018a) Behavioral cloning from observation. Preprint. arXiv:180501954

Torabi F, Warnell G, Stone P (2018b) Generative adversarial imitation from observation. Preprint. arXiv:180706158

Torabi F, Geiger S, Warnell G, Stone P (2019a) Sample-efficient adversarial imitation learning from observation. Preprint. arXiv:190607374

Torabi F, Warnell G, Stone P (2019b) Adversarial imitation learning from state-only demonstrations. In: Proceedings of the 18th international conference on autonomous agents and multiagent systems, international foundation for autonomous agents and multiagent systems, pp 2229–2231

Torabi F, Warnell G, Stone P (2019c) Imitation learning from video by leveraging proprioception. Preprint. arXiv:190509335

Torabi F, Warnell G, Stone P (2019d) Recent advances in imitation learning from observation. Preprint. arXiv:190513566

Večerík M, Hester T, Scholz J, Wang F, Pietquin O, Piot B, Heess N, Rothörl T, Lampe T, Riedmiller M (2017) Leveraging demonstrations for deep reinforcement learning on robotics problems with sparse rewards. Preprint. arXiv:170708817

Ziebart BD, Maas AL, Bagnell JA, Dey AK (2008) Maximum entropy inverse reinforcement learning. In: Proceedings of the AAAI conference on artificial intelligence, Chicago, vol 8, pp 1433–1438

Ziebart BD, Bagnell JA, Dey AK (2010) Modeling interaction via the principle of maximum causal entropy. In: Proceedings of the 27th international conference on international conference on machine learning

Żołna K, Rostamzadeh N, Bengio Y, Ahn S, Pinheiro PO (2018) Reinforced imitation learning from observations

Chapter 9
Integrating Learning and Planning

Huaqing Zhang, Ruitong Huang, and Shanghang Zhang

Abstract In this chapter, reinforcement learning is analyzed from the perspective of learning and planning. We initially introduce the concepts of model and model-based methods, with the highlight of advantages on model planning. In order to include the benefits of both model-based and model-free methods, we present the integration architecture combining learning and planning, with detailed illustration on Dyna-Q algorithm. Finally, for the integration of learning and planning, the simulation-based search applications are analyzed.

Keywords Model-based · Model-free · Dyna · Monte Carlo tree search · Temporal difference (TD) search

9.1 Introduction

In reinforcement learning, the agent is allowed to interact with the environment. Information collected during each round of interaction is regarded as the agent's experience and assists the agent to improve its policy. Generally, we refer learning as the policy improvement process based on actual experience during the interactions. As the simplest learning protocol, the direct policy learning is elaborated in Fig. 9.1. The agent initially takes actions in the environment following current policy. Based on the action and current state of the agent, the environment provides feedback of rewards, which let agent evaluate the performance of current policy and explore the policy improvement opportunities. However, direct policy learning is based on the experience of every single step of the action. Due to the uncertainty and randomness

H. Zhang
Google LLC, Mountain View, CA, USA

R. Huang
Borealis AI, Toronto, ON, Canada

S. Zhang (✉)
University of California, Berkeley, CA, USA

H. Dong et al. (eds.), *Deep Reinforcement Learning*,
https://doi.org/10.1007/978-981-15-4095-0_9

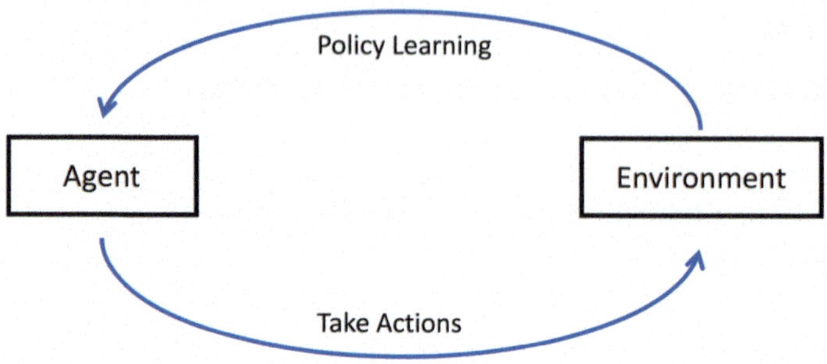

Fig. 9.1 Direct policy learning

of the environment, each experience may have large variance, which affects the learning speed and performance.

In order to improve the learning efficiency, for each episode of policy learning, it is beneficial to accumulate multiple rounds of interactions as the agent's experience. The interactions can be collected by performing roll-outs in the environment, which refer to forming a specific state-action-reward trajectory in the environment based on the current state and policy. In general model-free methods, the agent performs online roll-out in actual environment and combines the interactions for policy learning.

However, when the roll-out is applied online, generating experience with the environment is costly. In industrial scenarios, for example, some states may indicate system failure or engine explosion, which is dangerous to explore in actual situation for policy learning. Furthermore, the roll-out in the real environment has to be done sequentially. The failure on parallel computing results in low sample efficiency and slow learning speed. Therefore, for some scenarios, it is desirable to propose simulated environment and replace the real environment to generate experiences. We regard the roll-out from the simulated environment as planning, which can efficiently generate abundant simulated experiences with parallel computing for policy learning. In order to implement effective simulated environment for planning, the model-based methods are put forward and introduced in Sect. 9.2.

9.2 Model-Based Method

In order to apply planning, the concept of model is implemented between the agent and the environment (Kaiser et al. 2019). As shown in Fig. 9.2, when the agent stays in state S_t and takes actions A_t, the environment provides feedback rewards R_{t+1} and next state S_{t+1} information for the model. Based on the information collected from the experience between the agent and environment, the model representing the relations between S_{t+1} and (S_t, A_t) is called the transition model. The model representing the relations between R_{t+1} and (S_t, A_t) is called the reward model. When

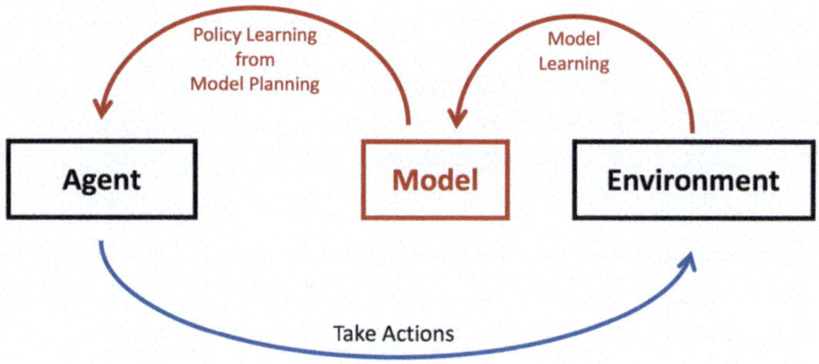

Fig. 9.2 Model-based reinforcement learning methods

the states are not fully represented by the observations, there is also the observation model $\mathcal{M}(O_t|S_t)$ and representation model $\mathcal{M}(S_{t+1}|S_t, A_t, O_{t+1})$ (Hafner et al. 2019), where the O_t is the corresponding observation of the state at time step t. For example, the images captured for representing the object motion are observations, as a function of the underlying dynamics state of the object. However, we will assume the state is fully observed and mainly focus on the transition model and the reward model in the following. The models can be formulated with two functions or distributions $\mathcal{F}_s$ and $\mathcal{F}_r$, to fit the relations, i.e.,

$$S_{t+1} \sim \mathcal{F}_s\left(S_t, A_t\right), \tag{9.1}$$

$$R_{t+1} = \mathcal{F}_r\left(S_t, A_t\right). \tag{9.2}$$

The model learning is a supervised regression learning process, which aims to build a visualized environment and function the same as the original one. Therefore, the model can be applied for planning by the agent with the insights on the true environment and help it perform policy learning.

For different application scenarios, the relations for model learning and policy learning may vary.

- **Direct Learning**: If the agent has already interacted with the environment multiple times following rule-based or expert knowledge, the collected experience can be directly applied for model learning first. When model is well learnt, the agent can regard the model as the environment and interact with it for policy learning.
- **Iterative Learning**: If the model is not well learnt, the model learning and policy learning can be performed iteratively. Based on limited experience from the interactions between the agent and the environment, some but limited insights about the environment can be learnt by the model. With the planning based on weakly learnt simulated model, the agent improves its policy a little and takes actions in real environment so as to provide further experience for model learning. With the number of iterations increasing, both the model learning

and policy learning gradually converge to the optimal results. Therefore, model learning and policy learning are able to assist each other and learn efficiently.

Accordingly, the model-based reinforcement learning establishes a model by learning from the real environment and performs planning with the agent for its policy learning. The advantages for the model learning can be depicted as follows.

- As the planning can be done with the interactions between the agent and the model, the agent does not need to take actions in the real environment for exploration and policy learning. Thus, for some environment that are costly on taking online actions, the model-based method can reduce the training time and guarantee the safety issues during the policy learning. For example, the real-world robot learning tasks require the robot manipulation in practice, like in Qt-opt (Kalashnikov et al. 2018) method, seven robots are collecting real-world samples days and nights for achieving the grasping task. A simulated environment (either learnt or manually engineered) can save large amounts of time and wear and tear of robots.
- When the policy learning is applied between the agent and the simulated model, the learning can be done in parallel. In the distributed system where there exist multiple learners contributing to the policy learning, each learner can interact with one model simulated from the environment. Therefore, the planning of multiple corresponding models can be independent with each other and does not affect the current state of the real environment. The parallelization on policy learning can improve the efficiency and scalability of the learning problem.

Nevertheless, considering the structure of model-based reinforcement learning, the weakness also exists.

- In model-based reinforcement learning, the performance of model learning will affect the policy learning results. For the complicated and dynamic environment scenarios, if the learnt model is unable to simulate the insights of the real environment, the agent actually interacts with the planning of a wrong or inaccurate model, which will further increase the error on the learnt policy.
- If there are some updates of adjustments on the environment, it takes several iterations for the model to learn the changes and further takes time for the agent to adjust the policy. Accordingly, the agent may have long delay to response and adapt to the changes on the online environment, which is not suitable for some applications with real-time requirements.

9.3 Integrated Architectures

Considering the pros and cons of model-free and model-based reinforcement learning methods, it is promising to combine the advantages of both model-based and model-free methods by integrating the learning and planning procedures. For different applications scenarios, the architectures integrating learning and planning are different.

Generally, in model-free method, the agent learns policies directly from real experience with the environment. No planning is adopted for policy improvement. In basic model-based method, model learning is firstly applied by the interactions between the agent and the environment. Based on the learnt model, the planning is applied and the policy can be iteratively learnt from the planning experience.

Since the model is implemented between the agent and the environment, for the policy learning of the agent, the experience can be classified into two categories.

- **Real experience**: In real experience, information is sampled directly from the interactions between the agent and the environment. Generally the real experience reflects the correct features of the environment, but it is costly and the states the hard to manually change or recover.
- **Simulated experience**: The simulated experience is generated by the planning of the model. The experience may not precisely provide the true features from the real environment, but the model is easy to manuscript and the model learning tries to minimizes the gap of mistakes.

For policy learning, if both real experience and simulated experience can be combined and considered simultaneously, we can take the advantages of both model-free and basic model-based methods, so as to improve the learning efficiency and accuracy. Therefore, the Dyna architecture is put forward (Sutton 1991). As shown in Fig. 9.3, based on the basic model-based method, during the policy learning, the agent updates the policy not only from the simulated experience with the learnt model, but considering the real experience with the real environment. Thus, in policy learning, the simulated experience can guarantee the quantity required from learning to reduce the variance of learning the features of the environment, while the real experience shows the correct features and dynamic changes from the environment, so as to reduce the bias through learning from the environment.

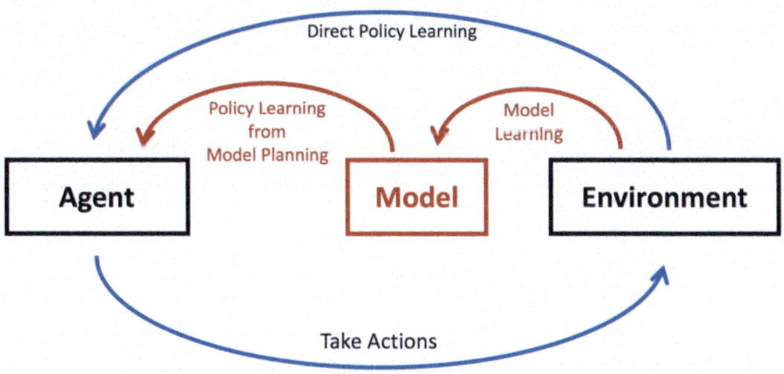

Fig. 9.3 Dyna architecture

Based on the architecture, the Dyna-Q algorithm is put forward and depicted in Algorithm 1. In the Dyna-Q learning, a Q table is established and maintained to instruct the actions of the agent. For each episode of learning, the Q table is learnt and updated from one-step action of the agent in the real environment. Moreover, the simulated model also learns from the real experience, and applies planning to generate n simulated experience for future learning. Accordingly, with the number of episode increasing, the Q table is learnt and converges to the correct results.

Algorithm 1 Dyna-Q

Initialize $Q(s, a)$ and $Model(s, a)$ for all $s \in \mathcal{S}$ and $a \in \mathcal{A}$
Do forever:
  (a) $s \leftarrow$ current (non-terminal) state
  (b) $a \leftarrow \epsilon$-greedy(s, Q)
  (c) Execute action a; Observe resultant reward r, Get next state s'
  (d) $Q(s, a) \leftarrow Q(s, a) + \alpha \left[r + \gamma \max_{a'} Q(s', a') - Q(s, a)\right]$
  (e) $Model(s, a) \leftarrow r, s'$
  (f) Repeat n times:
    $s \leftarrow$ random previously observed state
    $a \leftarrow$ random action previously taken in s
    $r, s' \leftarrow Model(s, a)$ random action previously taken in S
    $Q(s, a) \leftarrow Q(s, a) + \alpha \left[r + \gamma \max_{a'} Q(s', a') - Q(s, a)\right]$

9.4 Simulation-Based Search

In this section, we put more focus on the planning and introduce simulation-based search methods, which initialize at current state and take samples to roll out the trajectory. Accordingly, simulation-based search methods are generally the forward search paradigm using sample-based planning. The concept of forward search and sampling are further shown as follows.

- **Forward search**: Considering the planning of the problem, current state has more significance compared with all other states in Markov decision process (MDP). It is beneficial to view the MDP with finite choices in another perspective, which is a tree structure and the root is the current state. As shown in the Fig. 9.4, the forward search algorithm selects best action from the current state and look ahead for future planning in the branches of the tree structure.
- **Sampling**: When the planning is applied based on the MDP, from the current state, there may exist multiple options for the next states. Thus sampling is required in planning to randomly select the next state and continue to roll out the forward search. The randomness on the next state selection may follow some probabilities or distributions, which reflects the simulation policy adopted by the agent.

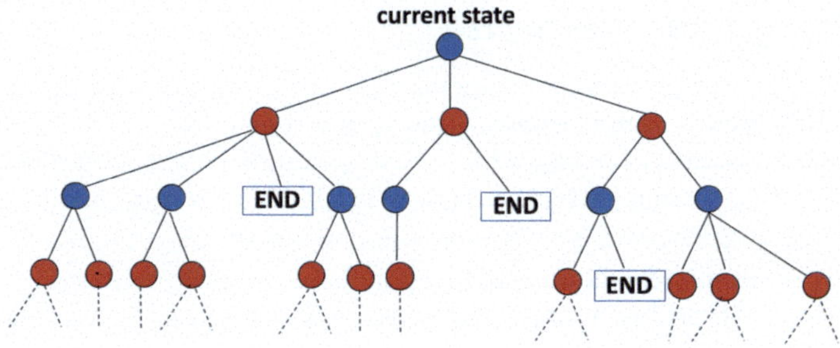

Fig. 9.4 Forward search

During the simulation-based search, the simulation policy is considered to instruct the planning direction. The simulation policy relates to the learning policy and assists on efficient planning to reflect the correct evaluations of the current policy of the agent.

For the rest of the section, we introduce different kinds of simulation-based search methods and combine with learning for problem-solving.

9.4.1 Simple Monte Carlo Search

When both the model $\mathcal{M}$ and policy π are fixed and initially provided, the simple Monte Carlo search can be applied to evaluate the performance of the action and update the learnt policy based on the experience. As shown in Algorithm 2, for each action a, $a \in \mathcal{A}$ applied on current state S_t, following the simulated policy π, K trajectories can be formulated and the total revenue from each trajectory is denoted as G_t^k. Based on the recorded trajectories, the performance by taking action A_t is evaluated as $Q(S_t, A_t)$. And based on the Q value of all actions, the optimal one is selected and the next state is determined.

Algorithm 2 Simple Monte Carlo search

Provided the model $\mathcal{M}$ and simulation policy π
for each action $a \in \mathcal{A}$ **do**
  for each episode $k \in \{1, 2, \ldots, K\}$ **do**
    Following the model $\mathcal{M}$ and simulation policy π, roll out in the environment started from current state S_t
    Record the trajectory as $\{S_t, a, R_{t+1}^k, S_{t+1}^k, A_{t+1}^k, R_{t+2}^k, \ldots S_T^k\}$
  end for
  Evaluate actions by mean return. $Q(S_t, a) = \frac{1}{K} \sum_{k=1}^{K} G_t^k$
end for
The learnt policy is to select current action with maximum Q value $A_t = \arg\max_{a \in \mathcal{A}} Q(S_t, a)$

9.4.2 Monte Carlo Tree Search

One clear drawback of simple Monte Carlo search is that the simulation policy π is prefixed, and, thus, never leverages the new information collected during planning. Monte Carlo Tree Search (MCTS) (Browne et al. 2012) is designed to overcome this drawback. In particular, MCTS proposes to maintain a search tree for keeping the collected information and improve the simulation policy gradually.

As shown in Algorithm 3, after sampling a trajectory from the current state S_t, MCTS updates the Q value for all the visited state and action pair (s, a) along the trajectory, similarly by the average reward of all the trajectories starting from (s, a). Then the simulation policy π on the nodes in the search tree is updated accordingly based on the new Q value in the tree. One example of updating π can be similar to the Q-learning, where at any node (state) s, π picks the optimal action based on the current Q with ϵ uniform exploration. When the simulation reaches a new state that is currently not in the tree, π sticks to the default policy like uniform exploration. The first new state in the trajectory will then be added into the search tree.[1] Such evaluation and policy improvement is repeated for every simulation until the simulation budget is reached. Lastly, the agent selects the action with maximum Q value at the current state S_t.

Algorithm 3 Monte Carlo tree search

Provided the model $\mathcal{M}$
Initialize simulation policy π
for each action $a \in \mathcal{A}$ **do**
  for each episode $k \in \{1, 2, \ldots, K\}$ **do**
    Following the model $\mathcal{M}$ and simulation policy π, roll out in the environment started from current state S_t
    Record the trajectory as $\{S_t, a, R_{t+1}, S_{t+1}, A_{t+1}, R_{t+2}, \ldots S_T\}$
    Update the Q value of every $(S_i, A_i), i = t, \ldots, T$ by mean return starting from (S_i, A_i) with $A_t = a$
    Update the simulation policy π according to the current Q values
  end for
end for
Output the action with maximum Q value at the current state, $A_t = \arg\max_{a \in \mathcal{A}} Q(S_t, a)$

9.4.3 TD Search

Apart from the Monte Carlo (MC) search methods, the temporal difference (TD) search can also be considered (Silver et al. 2012). Compared with the MC methods, the TD search does not require to roll out a trajectory to evaluate and update the

[1] Another option can be adding all the new nodes in the trajectory into the search tree.

current policy. Instead, in TD search, for each step of simulation, the policy is updated and instructs to select the action for the next state.

The TD search is applied in Dyna-2 algorithm (Silver et al. 2008), as depicted in Algorithm 4. In Dyna-2, the agent stores two sets of weights, denoted as long-term memory and short-term memory. Applying TD learning, the weights of short-term memory is updated with the simulated experience. The learnt $\overline{Q}$ with the weights of short-term memory further assists the agent to take actions in real environment. The higher-level TD learning is applied to update the weights of long-term memory. Finally, the learnt Q with weights θ is the learnt policy for the agent.

Algorithm 4 Dyna-2

function LEARNING
  Initialize $\mathcal{F}_s$ and $\mathcal{F}_r$
  $\theta \leftarrow 0$ # Initialize the weights of long-term memory
  loop
    $s \leftarrow S_0$
    $\overline{\theta} \leftarrow 0$ # Initialize the weights of short-term memory
    $z \leftarrow 0$ # Initialize eligibility trace
    SEARCH(s)
    $a \leftarrow \pi(s; \overline{Q})$ # Choose action based on polity related with $\overline{Q}$
    while s is not terminal **do**
      Execute a, observe reward r and next state s'
      $(\mathcal{F}_s, \mathcal{F}_r) \leftarrow UpdateModel(s, a, r, s')$
      SEARCH(s')
      $a' \leftarrow \pi(s'; \overline{Q})$ # Choose action applied in the next state s'
      $\delta \leftarrow r + Q(s', a') - Q(s, a)$ # Calculate TD-error
      $\theta \leftarrow \theta + \alpha(s, a)\delta z$ # Update weights of long-term memory
      $z \leftarrow \lambda z + \phi$ # Update eligibility trace
      $s \leftarrow s', a \leftarrow a'$
    end while
  end loop
end function

function SEARCH(s)
  while time available **do**
    $\overline{z} \leftarrow 0$ # Clear eligibility trace
    $a \leftarrow \overline{\pi}(s; \overline{Q})$ # Choose action based on polity related with $\overline{Q}$
    while s is not terminal **do**
      $s' \leftarrow \mathcal{F}_s(s, a)$ # Sample transition
      $r \leftarrow \mathcal{F}_r(s, a)$ # Sample reward
      $a' \leftarrow \overline{\pi}(s'; \overline{Q})$
      $\overline{\delta} \leftarrow R + \overline{Q}(s', a') - \overline{Q}(s, a)$ # Calculate TD-error
      $\overline{\theta} \leftarrow \overline{\theta} + \overline{\alpha}(s, a)\overline{\delta}\overline{z}$ # Update weights of short-term memory
      $\overline{z} \leftarrow \overline{\lambda}\overline{z} + \overline{\phi}$ # Update eligibility trance in short-term memory
      $s \leftarrow s', a \leftarrow a'$
    end while
  end while
end function

Compared with the MC methods, as the policy is updated for each step, TD search generally is more efficient. However, due to the frequent update, the search trends to reduce the variance but increase the bias for problem-solving.

References

Browne CB, Powley E, Whitehouse D, Lucas SM, Cowling PI, Rohlfshagen P, Tavener S, Perez D, Samothrakis S, Colton S (2012) A survey of Monte Carlo tree search methods. IEEE Trans Comput Intel AI Games 4(1):1–43

Hafner D, Lillicrap T, Ba J, Norouzi M (2019) Dream to control: learning behaviors by latent imagination. Preprint. arXiv:191201603

Kaiser L, Babaeizadeh M, Milos P, Osinski B, Campbell RH, Czechowski K, Erhan D, Finn C, Kozakowski P, Levine S, Mohiuddin A, Sepassi R, Tucker G, Michalewski H (2019) Model-based reinforcement learning for Atari. Preprint. arXiv:1903.00374

Kalashnikov D, Irpan A, Pastor P, Ibarz J, Herzog A, Jang E, Quillen D, Holly E, Kalakrishnan M, Vanhoucke V, et al (2018) Qt-opt: scalable deep reinforcement learning for vision-based robotic manipulation. Preprint. arXiv:180610293

Silver D, Sutton RS, Müller M (2008) Sample-based learning and search with permanent and transient memories. In: Proceedings of the 25th international conference on machine learning. ACM, New York, pp 968–975

Silver D, Sutton RS, Müller M (2012) Temporal-difference search in computer go. Mach Learn 87(2):183–219

Sutton RS (1991) Dyna, an integrated architecture for learning, planning, and reacting. ACM Sigart Bull 2(4):160–163

Chapter 10
Hierarchical Reinforcement Learning

Yanhua Huang

Abstract In this chapter, we introduce hierarchical reinforcement learning, which is a type of methods to improve the learning performance by constructing and leveraging the underlying structures of cognition and decision making process. Specifically, we first introduce the backgrounds and two primary categories of hierarchical reinforcement learning: options framework and feudal reinforcement learning. Then we have a detailed introduction of some typical algorithms in these categories, including strategic attentive writer, option-critic, and feudal networks, etc. Finally, we provide a summary of recent works on hierarchical reinforcement learning at the end of this chapter.

Keywords Hierarchical reinforcement learning · Options framework · Option-critic · Feudal reinforcement learning · Feudal networks

10.1 Introduction

Recently, deep reinforcement learning has achieved significant successes in many domains (Mnih et al. 2015; Schulman et al. 2015; Silver et al. 2016, 2017; Levine et al. 2018). Nevertheless, it is still a challenge for agents to learn long-term planning, especially in some environments with sparse rewards and long time horizon, such as Dota (OpenAI 2018) and StarCraft (Vinyals et al. 2019). Hierarchical reinforcement learning (HRL) provides a way for finding spatio-temporal abstractions and behavioral patterns of such complex control problems (Sutton et al. 1999; Dayan and Hinton 1993; Dietterich 2000; Dayan 1993; Kaelbling 1993; Parr and Russell 1998a; Vezhnevets et al. 2016; Barto and Mahadevan 2003; Bacon et al. 2017; Vezhnevets et al. 2017; Dietterich 1998; Nachum et al. 2018; Hausknecht 2000). Similar to the hierarchical structures of human cognition, HRL has the potential to abstract multi-level control where long-horizon planning and meta-

Y. Huang (✉)
Xiaohongshu Technology Co., Ltd., Shanghai, China

H. Dong et al. (eds.), *Deep Reinforcement Learning*,
https://doi.org/10.1007/978-981-15-4095-0_10

learning in higher-level guide the lower-level controllers. The modularization of hierarchical structures also allows transferability and interpretability, e.g., the skills about understanding the map and reaching beneficial states are commonly useful in games like grid-world (Tamar et al. 2016) or Doom (Kempka et al. 2016; Bhatti et al. 2016).

Most of the previous research on HRL began with four primary works: **options framework** (Sutton et al. 1999), **feudal reinforcement learning (FRL)** (Dayan and Hinton 1993), **MAXQ decomposition** (Dietterich 2000), and **hierarchical abstract machines (HAMs)** (Parr and Russell 1998b,a). The higher-level policy in the options framework switches lower-level policies at special time steps, to decompose the problems temporally. The senior controller in FRL agents proposes explicit goals, e.g., some particular states, for lower-level controllers to reach, to achieve a hierarchical decomposition of the state space. MAXQ also provides a state abstraction method by combining the solutions of decomposed sub-tasks with respect to the Q-value function. HAMs considers the learning process where actions that the agent can perform are constrained by hierarchies of finite state machines, to reduce the search space for large and complicated problems. In this chapter, we focus on recent works on applying deep learning to HRL. Specifically, we discuss two algorithms belonging to the options framework and FRL, respectively, and then provides a brief summary of deep HRL at the end of this chapter.

10.2 Options Framework

The options framework (Sutton et al. 1999; Hausknecht 2000) formalizes the idea of temporally extended actions. **Options**, also referred to as skills (Da Silva et al. 2012) or macro-actions (Hauskrecht et al. 1998; Vezhnevets et al. 2016), is a sub-policy with a termination condition, which observes the environment and outputs actions until the termination condition is satisfied. The termination condition is a kind of implicit term for splitting the trajectory into temporal parts, representing that the corresponding sub-policy has done its own work and the top-level policy-over-options needs to switch to another option. Given an MDP with states set $\mathcal{S}$ and actions set $\mathcal{A}$, an option $\omega \in \Omega$ is defined as a triple $(I_\omega, \pi_\omega, \beta_\omega)$ where $I_\omega \subseteq \mathcal{S}$ is an initiation set of states, $\pi_\omega : \mathcal{S} \times \mathcal{A} \to [0, 1]$ is an intra-option policy, and $\beta_\omega : \mathcal{S} \to [0, 1]$ is a termination function that provides stochastic terminal condition with Bernoulli distribution. An option ω is available for state s if and only if $s \in I_\omega$. An agent picks an option using its policy-over-options and subsequently follows it until termination, at which point the policy-over-options is queried again and the process continues. Note that if the option ω is taken, then actions are selected according to π_ω until the option terminates stochastically according to β_ω. For example, an option named open-the-door might consists of a policy for reaching, grasping, and turning the door knob, and a termination condition for determining the probability of the door being opened.

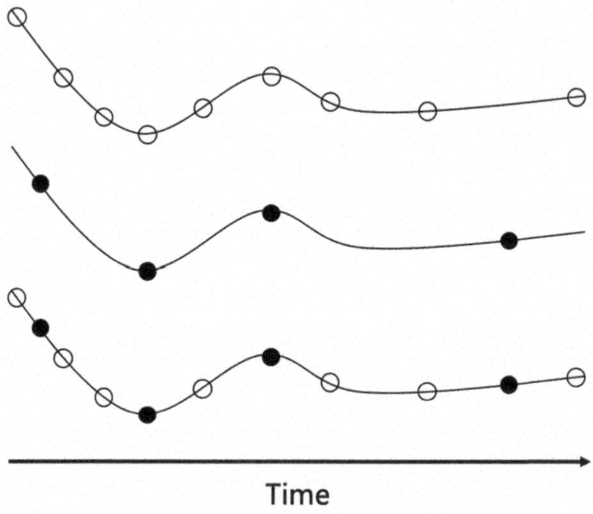

Fig. 10.1 Options under SMDP perspective, adapted from Sutton et al. (1999). Top: The state trajectory of a MDP. Middle: The state trajectory of a Semi-Markov Decision Process (SMDP). Bottom: The state trajectory of a MDP over two-level hierarchy. Filled circles represent SMDP decisions, while open circles are primitive steps within the corresponding option

In particular, an options framework consists of two-level hierarchy: each element in the bottom level is an option and the top level is a policy-over-options that picks an option at the beginning of episode or the termination of previous option. Policy-over-options learns from the reward signal given by the environment, while options can be learned with explicit sub-goals. For example, in the tabular case, each state can be viewed as a candidate of sub-goals (Wiering and Schmidhuber 1997; Schaul et al. 2015). Once the options are given, by treating them as actions, the top level can be learned using standard techniques. Recently, for complex environments like Minecraft and Atari, predefined sub-goals achieve promising performance with the combination of deep learning (Tessler et al. 2017; Kulkarni et al. 2016).

In options framework, top-level module learns a policy-over-options, and the bottom-level module learns policies to accomplish the objective of each option, which can be viewed as a decomposition of the Markov processes over temporal level, i.e., several time steps. The **semi-Markov decision process (SMDP)** provides a theoretical view of the options framework with the uncertainty in the time duration between actions (Sutton et al. 1999) as shown in Fig. 10.1. SMDP is a standard MDP with an additional element $\mathcal{F}$: $(\mathcal{S}, \mathcal{A}, \mathcal{P}, \mathcal{R}, \mathcal{F})$, where $\mathcal{F}(t|s, a)$ gives the probability that the transition time is t for state s in action a. Informally, the top-level control in options framework can be viewed as a policy over SMDP. For multi-level options, higher-level options represent a further temporally extended SMDP than the lower-level options below (Riemer et al. 2018).

It is demonstrated that handcrafted options can achieve significant performance with the combination of deep learning even in challenging environments like

Minecraft and Atari (Tessler et al. 2017; Kulkarni et al. 2016). However, the initiation set and termination condition is a restriction for the option framework. For example, a handcrafted policy π_ω for a mobile robot to dock with its battery charger might be defined only for states in which the battery charger is within sight. The termination condition indicates the probability of termination to be 1 when the robot is successfully docked or the state is outside of I_ω. As a consequence, discovering options autonomously has been the subject of HRL. We introduce two algorithms that propose to formulate option discovery as an optimization problem with solutions that are compatible with function approximation. The first one is a deep recurrent neural network named STRategic Attentive Writer (STRAW) that learns options with open-loop intra-option policies. The second one is the option-critic architecture that considers close-loop intra-option policies.

10.2.1 STRategic Attentive Writer (STRAW)

STRAW (Vezhnevets et al. 2016) is a novel deep recurrent neural network architecture that temporally abstracts commonly occurring sequences of actions, i.e., macro-actions, and learns a policy over them end-to-end. Note that macro-action is a particular option that is represented implicitly inside the neural network, where the action sequence (or distribution over them) is decided at the time that the macro-action is initiated. STRAW consists of two modules for short-term action distributions and long-term plan, respectively.

The first module translates environment observations into an **action-plan**—a state variable which represents an explicit stochastic plan of future actions. At time step t, the action-plan is represented by matrix $\boldsymbol{A} \in \mathbb{R}^{|\mathcal{A}|\times T}$ where T is the maximum time horizon of the plan. The τ-th column in $\boldsymbol{A}$ corresponds to the logits of actions at time step $t + \tau$.

The second module maintains **commitment-plan** by single row matrix $\boldsymbol{c}^t \in \mathbb{R}^{1\times T}$—a state variable that determines at which step the network terminates a macro-action and updates the action-plan. At time step t, the first element of $\boldsymbol{c}^{t-1}$ provides the parameter of the Bernoulli distribution for the terminate condition. During commitment, both the action-plan $\boldsymbol{A}^t$ and the commitment-plan $\boldsymbol{c}^t$ are rolled over to the next step by a time-shift operator ρ, where ρ shifts the given matrix by removing the first column and appending zero to the rear (Fig. 10.2).

Figure 10.2 shows an example of workflow with action-plans and commitment-plans in STRAW. To update these two types of plans, STRAW uses attentive writing technique (Gregor et al. 2015) over the temporal dimension, which allows the network focusing on the current part. This technique applies an array of Gaussian filters to plans along the temporal dimension. More precisely, for temporal size K, a grid of $|\mathcal{A}| \times K$ one-dimensional Gaussian filters is positioned on the plan by specifying the coordinates of the grid center and the stride between adjacent filters. Note that the stride is similar to the same term in CNN. Let ψ^A be the attention parameters for action-plan, i.e., grid position, stride, and standard deviation

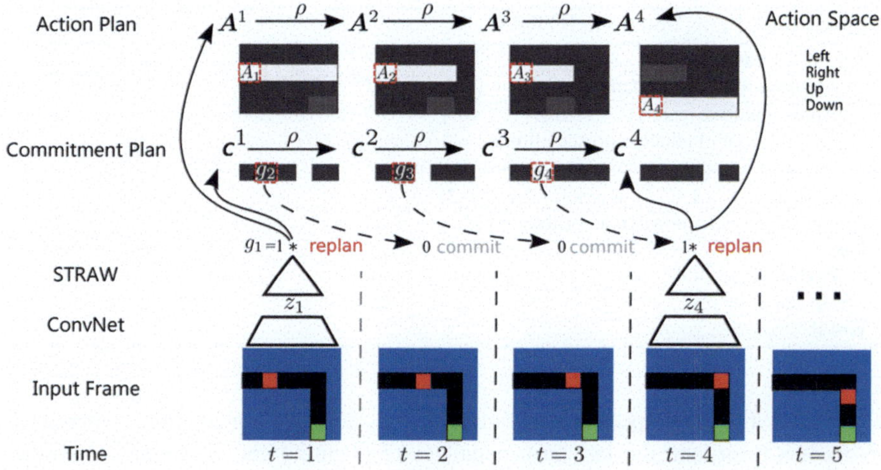

Fig. 10.2 The workflow of STRAW in a maze navigation game, adapted from Vezhnevets et al. (2016). The observations are raw pixels, where pixels with the colors blue, black, red, and green correspond to the wall, corridor, agent, and goal, respectively. The action space consists of moving to four positions. When $t = 1$, the frame's feature extracted by a conversational network is feed into STRAW. STRAW generates two plans immediately. During the successive 2 time steps, two plans are rolled over by ρ. And then the agent comes to the corner and the commitment-plan $\boldsymbol{c}^t$ gives a replan signal

of Gaussian filters. STRAW defines the attention operations as follows:

$$\boldsymbol{D} = \text{write}\big(p, \psi_t^A\big);\ \beta_t = \text{read}\big(\boldsymbol{A}^t, \psi_t^A\big), \tag{10.1}$$

where $p \in \mathbb{R}^{A\times K}$ is a patch of plan with temporal size K. The $write$ operation produces the smoothed plan $\boldsymbol{D}$ of the same size as $\boldsymbol{A}^t$, while the $read$ operation produces a read patch $\beta_t \in \mathbb{R}^{A\times K}$. Furthermore, given z_t as the feature representation of observation in time step t and applying similar attentive technique to commitment-plan, the algorithm for updating plans is shown in Algorithm 1, where f^{ψ}, f^A, and f^c are the linear functions, h is a multi-layer perceptron, $\boldsymbol{b} \in \mathbb{R}^{1\times T}$ is a bias filled with same scalar parameter b, and e is a scalar which is fixed to 40 in Vezhnevets et al. (2016) for frequently re-planning.

For further structured exploration, STRAW uses reparameterization trick on the diagonal Gaussian distribution: $Q(z_t|\zeta_t) = \mathcal{N}(\mu(\zeta_t), \sigma(\zeta_t))$, where ζ_t is the output of the feature extractor. The training loss of STRAW is defined as follows:

$$\mathcal{L} = \sum_{t=1}^{T}\Big(L\big(\boldsymbol{A}^t\big) + \alpha g_t \text{KL}(Q(z_t|\zeta_t)|P(z_t)) + \lambda \boldsymbol{c}_1^t\Big), \tag{10.2}$$

Algorithm 1 Plans update in STRAW

if $g_t = 1$ **then**
  Compute attention parameter of action-plan $\psi_t^A = f^{\psi}(z_t)$
  Apply attentive read: $\beta_t = read(\boldsymbol{A}^{t-1}, \psi_t^A)$
  Compute intermediate representation $\epsilon_t = h(\text{concat}(\beta_t, z_t))$
  Compute attention parameter of commitment-plan $\psi_t^c = f^c(\text{concat}(\psi_t^A, \epsilon_t))$
  Update $\boldsymbol{A}^t = \rho(\boldsymbol{A}^{t-1}) + \text{write}(f^A(\epsilon_t), \psi_t^A)$
  Update $\boldsymbol{c}_t = Sigmoid(\boldsymbol{b} + \text{write}(e, \psi_t^c))$
else
  Update $\boldsymbol{A}^t = \rho(\boldsymbol{A}^{t-1})$
  Update $\boldsymbol{c}_t = \rho(\boldsymbol{c}_{t-1})$
end if

where L is a domain specific loss function, e.g., negative log-likelihood of return, $P(z_t)$ is a prior, and the last term penalizes re-planning and encourages commitment.

Note that STRAW is a neural network architecture. For reinforcement learning tasks, a variety of learning algorithms are available. Vezhnevets et al. (2016) showed the performance on 2D mazes and Atari games with A3C (Mnih et al. 2016). 2D mazes are 2D grid-worlds with two types of cells—walls and corridors, where one of the corridor cells is chosen randomly as the goal. The agent fully observes the state of the maze and needs to reach the goal with structured exploration. In this task, Vezhnevets et al. (2016) showed that STRAW outperforms LSTM for the policy, and stays close to the optimal policy given by the Dijkstra algorithm. In the Atari domain, Vezhnevets et al. (2016) chose eight games that require some degree of planning and exploration, where STRAW and its variant reach higher scores on 6 out of 8 games than LSTM and simple feed-forward network.

10.2.2 The Option-Critic Architecture

The option-critic architecture (Bacon et al. 2017) extends policy gradient theorem to options, which provides a joint learning of options and policy-over-options in an end-to-end manner. It optimizes the discounted return directly. First consider the option-value function defined as follows:

$$Q_{\Omega}(s, \omega) = \sum_a \pi_{\omega}(a|s) Q_U(s, \omega, a), \tag{10.3}$$

where $Q_U : \mathcal{S} \times \Omega \times \mathcal{A} \rightarrow \mathbb{R}$ is the value of executing an action in the context of a state-option pair (s, ω):

$$Q_U(s, \omega, a) = R(s, a) + \gamma \sum_{s'} p(s'|s, a) U(\omega, s'), \tag{10.4}$$

where $U : \Omega \times \mathcal{S} \to \mathbb{R}$ is the value of executing ω upon entering a state s':

$$U(\omega, s') = (1 - \beta_\omega(s'))Q_\Omega(s', \omega) + \beta_\omega(s')V_\Omega(s'), \tag{10.5}$$

where $V_\Omega : \mathcal{S} \to \mathbb{R}$ is the optimal value function over options:

$$V_\Omega(s') = \max_{\omega \in \Omega} \mathbb{E}_\omega \left[\sum_{n=0}^{k-1} \gamma^n R_{t+n} + \gamma^k V_\Omega(S_{t+k}) | S_t = s' \right], \tag{10.6}$$

where k is the expected duration of ω when taken in state s'. Hence we can define $A_\Omega : \mathcal{S} \times \Omega \to \mathbb{R}$ as the advantage function over options by

$$A_\Omega(s, \omega) = Q_\Omega(s, \omega) - V_\Omega(s). \tag{10.7}$$

If option ω_t has been initiated or is executing at time step t in state S_t, by viewing state-option pairs as regular states in Markov chain, the probability of transitioning to (S_{t+1}, ω_{t+1}) in one step is:

$$\sum_a \pi_{\omega_t}(a|S_t)p(S_{t+1}|S_t, a)[(1 - \beta_{\omega_t}(S_{t+1}))\mathbf{1}_{\omega_t = \omega_{t+1}} + \beta_{\omega_t}(S_{t+1})\pi_\Omega(\omega_{t+1}|S_{t+1})]. \tag{10.8}$$

By assuming all options available everywhere, the transition above is a unique stationary distribution over state-option pairs.

The architecture of stochastic gradient descent algorithm for learning options is shown in Fig. 10.3, where the gradients are given by Theorems 10.1 and 10.2.

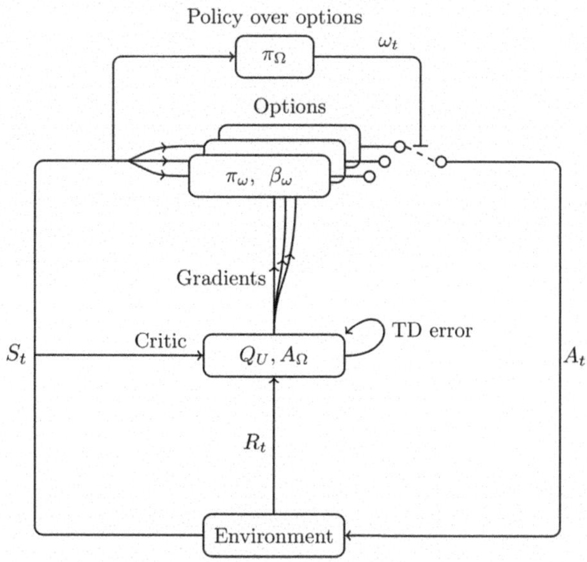

Fig. 10.3 The option-critic architecture, adapted from Bacon et al. (2017)

Moreover, Bacon et al. (2017) proposed to learn the values at a fast timescale while updating the intra-option policies and termination functions at a slower rate based on a two-timescale framework (Konda and Tsitsiklis 2000). We can see that in reference to the actor-critic architectures, intra-option policies, termination functions, and policy-over-options belong to the actor part while the critic consists of Q_U and A_Ω.

Theorem 10.1 (Intra-Option Policy Gradient Theorem (Bacon et al. 2017)) *Given a set of Markov options with stochastic intra-option policies differentiable in their parameters θ, the gradient of the expected discounted return with respect to θ and initial condition $(\hat{s}, \hat{\omega})$ is*

$$\sum_{s,\omega} \mu_\Omega(s, \omega | \hat{s}, \hat{\omega}) \sum_a \frac{\partial \pi_{\omega,\theta}(a|s)}{\partial \theta} Q_U(s, \omega, a), \tag{10.9}$$

where $\mu_\Omega(s, \omega|\hat{s}, \hat{\omega})$ is a discounted weighting of state-option pairs along trajectories starting from $(\hat{s}, \hat{\omega})$: $\sum_{t=0}^{\infty} \gamma^t p(S_t = s, \omega_t = \omega | S_0 = \hat{s}, \omega_0 = \hat{\omega})$.

Theorem 10.2 (Termination Gradient Theorem (Bacon et al. 2017)) *Given a set of Markov options with stochastic termination functions differentiable in their parameters φ, the gradient of the expected discounted return objective with respect to φ and the initial condition $(\hat{s}, \hat{\omega})$ is*

$$-\sum_{s',\omega} \mu_\Omega(s', \omega | \hat{s}, \hat{\omega}) \frac{\partial \beta_{\omega,\varphi}(s')}{\partial \varphi} A_\Omega(s', \omega), \tag{10.10}$$

where $\mu_\Omega(s', \omega|\hat{s}, \hat{\omega})$ is a discounted weighting of state-option pairs from $(\hat{s}, \hat{\omega})$, the same function as in the Intra-Option Policy Gradient Theorem.

Bacon et al. (2017) presented experiments in both discrete and continuous environments. In the discrete case, Bacon et al. (2017) trained on 4 Atari games in the Arcade Learning Environment (ALE) (Bellemare et al. 2013) with the same configuration as Mnih et al. (2015). The result shows that the option-critic was capable of learning structural options in all games. In the continuous domain, Bacon et al. (2017) selected the Pinball domain Konidaris and Barto (2009) where the agent controls a ball in a 2D maze with arbitrarily shaped polygons to a randomly generated target location. The trajectory learned by the option-critic shows the agent can achieve temporal abstraction.

10.3 Feudal Reinforcement Learning

Feudal reinforcement learning (FRL) (Dayan and Hinton 1993) considered a feudal control hierarchy in which managers have sub-managers who work for them and super-managers for whom they work. It mirrors the hierarchical aspects of a feudal fiefdom, where managers in each level can set tasks, reward, and punishment to their sub-managers. Notice that there are two key principles to guarantee the feudal rule: **Reward Hiding** and **Information Hiding**. Reward hiding means that sub-managers should obey their managers whether or not the command satisfies the super-managers. Information hiding means that sub-managers do not know their manager's task, while super-manager also do not know tasks the manager has set for the sub-managers. The top level of feudal agents, instead of learning a temporal decomposition of options like the options framework, decomposes the problem with respect to the state space by producing an explicit goal for the bottom-level policies. Such architecture allows reinforcement learning to scale to large domains with a clear division of labor among managerial levels.

Inspired by such decoupling learning, Vezhnevets et al. (2017) introduced a novel neural architecture called **FeUdal Networks** (**FuNs**) to discover sub-goals automatically, with soft condition about reward hiding and information hiding. It decouples end-to-end learning across multiple levels, which allows for utilizing different resolutions of time. Furthermore, hierarchical reinforcement learning with off-policy correction (**HIRO**) further improves sample efficiency with off-policy experience (Nachum et al. 2018). The experiments show that HIRO makes significant progresses and can solve exceedingly complex tasks that combine locomotion and rudimentary object interaction.

10.3.1 FeUdal Networks (FuNs)

A FeUdal Network (FuN) is a fully differentiable modular neural network for FRL with two modules: the manager and the worker. The manager sets goals at a lower temporal resolution in a latent state space, while the worker learns to achieve goals with intrinsic rewards. Figure 10.4 shows the architecture of FuN, where the forward process is described by the following equations:

$$z_t = f^{\text{Percept}}(S_t) \tag{10.11}$$

$$m_t = f^{\text{Mspace}}(z_t) \tag{10.12}$$

$$h_t^M, \hat{g}_t = f^{\text{Mrnn}}(m_t, h_{t-1}^M); \; g_t = \hat{g}_t / ||\hat{g}_t||; \tag{10.13}$$

$$w_r = \phi\left(\sum_{i=t-c}^{t} g_i\right) \tag{10.14}$$

$$h^W, U_t = f^{\text{Wrnn}}(z_t, h_{t-1}^W) \tag{10.15}$$

$$\pi_t = SoftMax(U_t w_t), \tag{10.16}$$

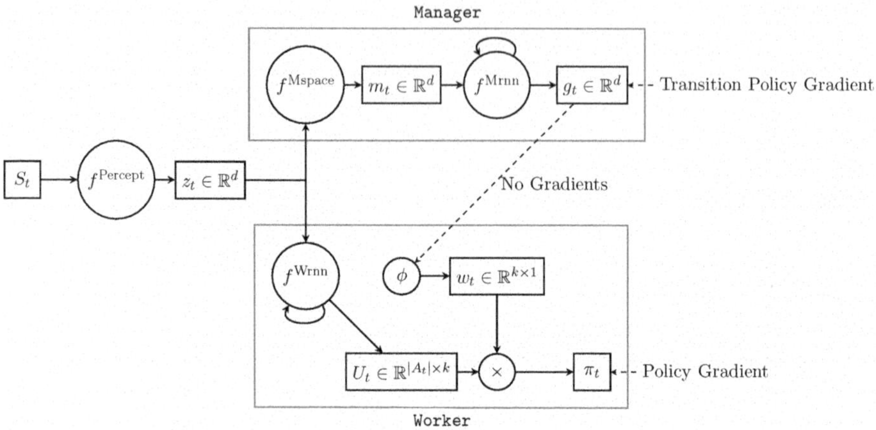

Fig. 10.4 The architecture of FuNs, adapted from Vezhnevets et al. (2017). Hyper-parameters k and d are specified to $k = 16 \ll d = 256$ in Vezhnevets et al. (2017)

where z_t is the representation of S_t, f^{Mspace} provides states m_t to the Manager, and g_t represents the goal outputted by the Manager. Note that following the two principles in FRL, there are no gradients propagated between Manager and Worker, but the perceptual module f^{Percept} that takes in observations is shared. Both f^{Mrnn} for Manager and f^{Wrnn} for Worker are recurrent modules, and f^{Mspace} is fully-connected. h^M and h^W correspond to the internal states of the manager and the worker, respectively. ϕ is a linear transform without biases that maps a goal g_t into an embedding vector w_t. U_t represents the embedding matrix of actions, outputting the logits of the worker's policy via a matrix product with w_t.

Consider the standard reinforcement learning setup that maximize the discounted return $G_t = \sum_{k=0}^{\infty} \gamma^k R_{t+k}$. A natural idea to learn the whole architecture is training end-to-end by a policy gradient algorithm because FuNs are fully differentiable. However, it will cause the gradients to propagate from the Worker to the Manager through the goal, which may make the goal just internal latent variables instead of hierarchical signals. As a consequence, FuNs train the Manager and the Worker independently. For the Manager, the update rule is followed by predicting the advantages directions:

$$\nabla g_t = \left(G_t - V_t^M(S_t, \theta)\right) \nabla_\theta d_{\cos}(m_{t+c} - m_t, g_t(\theta)), \tag{10.17}$$

where V_t^M is the Manager's value function and $d_{\cos}(\alpha, \beta) = \alpha^T \beta / (|\alpha||\beta|)$ is the cosine similarity. On the other hand, the Worker can be trained by any off-the-shelf deep reinforcement learning algorithm with the intrinsic reward defined as follows:

$$R_t^I = \frac{1}{c}\sum_{i=1}^{c} d_{\cos}(m_t - m_{t-i}, g_{t-i}), \tag{10.18}$$

where the directional shift in state space provides a structural invariance to the goal. In practice, FuNs soft the reward hiding condition in original FRL by training the Worker with $R_t + \alpha R_t^I$, where α is a hyper-parameter for regulating the influence of the intrinsic reward.

Vezhnevets et al. (2017) also provided a theoretical analysis about the training rule of the Manager. Consider a high-level policy-over-policies $o(S_t, \theta)$ that selects among several sub-policies with same fixed-duration c. For each sub-policy, the transition distribution $p(S_{t+c}|S_t, o)$ can be viewed as a transition policy $\pi^T(S_{t+c}|S_t, \theta)$. Similar to the SMDP perspective of the options framework, we can apply policy gradient theorem to $\pi^T(S_{t+c}|S_t, \theta)$ in the higher-level MDP

$$\nabla_\theta \pi^T(S_{t+c}|S_t, \theta) = \mathbb{E}\Big[\big(G_t - V_t^M(S_t, \theta)\big)\nabla_\theta \log p(S_{t+c}|S_t, o)\Big] \tag{10.19}$$

which is called **transition policy gradients**. With the assumption that direction $S_{t+c} - S_t$ follows the Mises-Fisher distribution, we would have $\log p(S_{t+c}|S_t, o) \propto d_{\cos}(S_{t+c} - S_t, g_t)$.

Besides, Vezhnevets et al. (2017) proposed the dilated LSTM for the Manager, which is analogous to dilated CNN for large receptive field without loss of resolution or coverage. Dilated LSTM maintains several internal LSTM cell states. At any time step, only one of the cell states is updated, and the output is the pooled result of the last c states that have been updated.

Note that similar to STRAW, FuN is also a neural network architecture for HRL. Vezhnevets et al. (2017) selected A3C as learning algorithms, and designed a series of experiments that showed the effectiveness of FuN over LSTM. First of all, Vezhnevets et al. (2017) presented the analysis of FuN on Montezuma's Revenge. Montezuma's Revenge is a difficult problem in Atari games for reinforcement learning agents, requiring multiple skills to avoid lethal traps and learn from sparse rewards. The experimental results showed that FuN achieves significant improvement in sample efficiency. Furthermore, Vezhnevets et al. (2017) also showed the improved performance on ten more Atari games, where FuN achieves significantly higher scores than Option-Critic. Likewise, Vezhnevets et al. (2017) used four different levels of DeepMind Lab 3D game platform (Beattie et al. 2016) to validate FuN. It demonstrates that FuN learns meaningful sub-policies, which are then efficiently integrated with memory to produce rewarding behavior.

10.3.2 Off-policy Correction

HRL methods propose to train multiple layers of policies to perform temporal and behavioral abstraction. In previous sections, we discussed STRAW and FuNs for learning a hierarchy of policies within neural architectures, and Option-Critic for learning both the internal policies and the termination conditions of options end-to-end. There are number of issues remained in HRL, e.g., generality, transferability, and sample efficiency. In this section, we discuss HIerarchical Reinforcement learning with Off-policy correction (HIRO) (Nachum et al. 2018) which gives a generally applicable and data-efficient method for training HRL agents.

For generality, HIRO considers the scheme that higher-level controllers supervise lower-level controllers by proposing some goals automatically. More precisely, at each time step t, HIRO drives the agent with a goal g_t. Given a user-specific parameter c, if t is a multiple of c, the goal g_t is produced by the higher-level policy μ^h, otherwise g_t is provided by the goal transition function h: $g_t = h(S_{t-1}, g_{t-1}, S_t)$ with previous goal g_{t-1}. Similar to FuNs, the goal refers to the higher-level decision that contains information about desired positions and orientations. The experiments found that, instead of representing the goal within an embedding space, using the raw observation directly is more effective by HIRO. Note that we can design the intrinsic reward and goal transition function by domain knowledge in specific tasks. Specifically, in the simplest case, the intrinsic reward is defined by

$$R_t^I = -||S_t + g_t - S_{t+1}||_2, \quad (10.20)$$

and the goal transition function is defined by

$$h(S_{t-1}, g_{t-1}, S_t) = S_{t-1} + g_{t-1} - S_t \quad (10.21)$$

to maintain the goal directions.

For data efficiency, HIRO extends the off-policy technique to both higher and lower-level training. HIRO lets the lower-level policy μ^l store the experience $(S_t, g_t, A_t, R_t^I, S_{t+1}, h(S_t, g_t, S_{t+1}))$ and train these policies with arbitrary off-policy algorithm by viewing g_t as an additional input into the models. For the higher-level policy, the transition tuples $(S_{t:t+c}, g_{t:t+c}, A_{t:t+c}, R_{t:t+c}, S_{t+c})$ (':' means slice in Python, do not contain the last element) can also be trained with an arbitrary off-policy algorithm by viewing g_t as an action and accumulating $R_{t:t+c}$ as a reward. However, the transitions obtained from past lower-level controllers do not accurately reflect the actions. To tackle this issue, HIRO proposed using the re-label technique to correcting high-level transitions. The old transition $(S_t, g_t, \sum R_{t:t+c}, S_{t+c})$ will be re-labeled with a different goal $\hat{g}_t$ such that $\hat{g}_t$ maximizes the probability $\mu^l(A_{t:t+c}|S_{t:t+c}, \hat{g}_{t:t+c})$, where $\hat{g}_{t+1:t+c}$ is computed by the goal transition function h. With stochastic behavior policies, the log probability

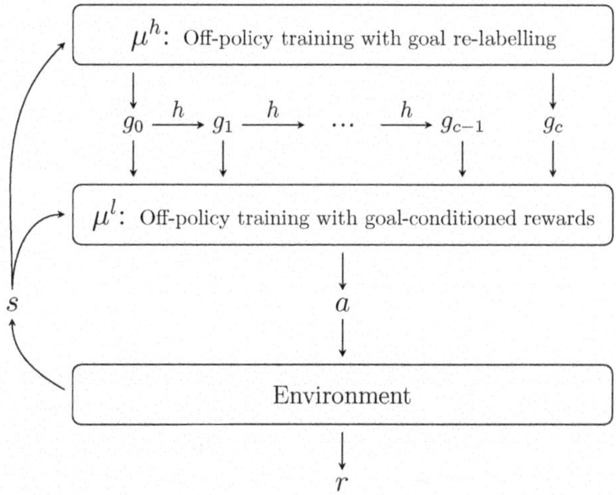

Fig. 10.5 The architecture of HIRO, adapted from Nachum et al. (2018). The lower-policy takes in higher-level goals and interacts with the environment directly, where goals are generated by higher-level policy or goal transition function

$\log \mu^l(A_{t:t+c}|S_{t:t+c}, \hat{g}_{t:t+c})$ can be computed by

$$\log \mu^l(A_{t:t+c}|S_{t:t+c}, \hat{g}_{t:t+c}) \propto -\frac{1}{2} \sum_{i=t}^{t+c-1} ||A_t - \mu^l(S_i, \hat{g}_i)||_2^2 + \text{const.} \qquad (10.22)$$

In practice, HIRO chooses the goal that maximizes the log probability above from a candidate goal set that includes the original goal, a goal corresponding to the difference $S_{t+c} - S_t$ and goals sampled from a diagonal Gaussian distribution with each mean item equivalent to the element in the vector of $S_{t+c} - S_t$ randomly, where the minus is an element-wise operator.

The full architecture of HIRO is shown in Fig. 10.5. Nachum et al. (2018) validated HIRO on four challenging tasks in Duan et al. (2016). The experiments show that off-policy correction provides a significant benefit and relabelling in lower-level controllers provides an initial speed-up in training.

10.4 Other Works

In this section, we provide a brief summary of recent work in HRL. Figure 10.6 shows two kinds of taxonomy. According to the reward signal of low-level policies, there are typically two categories of these works. The first proposes to learn low-level policies end-to-end directly from environment, such as STRAW (Vezhnevets et al. 2016) and the option-critic architecture (Bacon et al. 2017) introduced before. The second argues that learning from auxiliary rewards can achieve better hierarchy, such as the FuNs (Vezhnevets et al. 2017) and HIRO (Nachum et al. 2018) introduced before.

- **High-level abstracted objectives**
  - **State abstraction**
  - **Temporal abstraction**
  - **State space decomposition**
- **Low-level reward signal**
  - **Learning from environment**
  - **Using auxiluary reward**

Fig. 10.6 Two perspectives of HRL algorithms

Generally, the first approach can gain more effectiveness from end-to-end learning. Most of the work in this branch focus on options. For option discovery, both the STRAW (Vezhnevets et al. 2016) and the option-critic architecture (Bacon et al. 2017) can be viewed as top-down approaches, where the observed rewards are used to climb gradients. Machado et al. (2017) instead introduced a bottom-up method that utilized representations of the environment captured by Proto-value functions (PVFs) under a graph Laplacian framework, which provided task-independent options with theoretical foundations. Riemer et al. (2018) extended option-critic architecture and derived policy gradient theorems for a deep hierarchy of options. The resultant hierarchical option-critic obtains efficient empirical performances on both discrete and continuous environments. Harutyunyan et al. (2018) proposed to improve the termination condition by decoupling it into the behavior and target terminations as off-policy learning, which showed faster convergence by their experiments. Motivated by the SMDP view of options, Sharma et al. (2017) proposed Fine Grained Action Repetition (FiGAR) that learns to acquire the ability to predict the number of time steps for which an action chosen for execution is to be repeated. In addition, it is straightforward for combining meta-learning with this end-to-end approach to form a hierarchical architecture. Frans et al. (2017) developed a meta-learning algorithm for improving sample efficiency on unseen tasks by sharing the hierarchical structure of primitives-policies and achieve significant results on 3D humanoid robots. However, it is still an issue to scale this approach to complex domains because of the sole dependency on final tasks (Nachum et al. 2018; Bacon et al. 2017; Frans et al. 2017).

The second approach instead uses auxiliary rewards. Both FuNs (Vezhnevets et al. 2017) and HIRO (Nachum et al. 2018) construct the intrinsic reward with goal-oriented directions for lower-level policies. There are many other works focusing on goal-oriented rewards that make performance progress on a range of domains. Universal value function approximators (UVFAs) (Schaul et al. 2015) generalize value functions over goals. Levy et al. (2018) further introduced hindsight goal transitions that extends the idea of hindsight experience replay (Andrychowicz et al. 2017), which achieves significant stability. Kulkarni et al. (2016) introduced h-DQN that learns hierarchical action-value functions at different time scales, where the action-value function in the top level learns policy-over-options and the action-value function in the lower level learns to satisfy given sub-goals. Another method to construct handcrafted auxiliary rewards by taking the advantages of domain knowledge. Heess et al. (2016) introduced an architecture for locomotion tasks by first pre-training on several related simple tasks. Tessler et al. (2017) proposed a lifelong learning system for the Minecraft domain that selectively transfers learned skills to new tasks. Florensa et al. (2017) introduced stochastic neural network architecture that learns higher-level policies with pre-trained skills, which requires minimal domain knowledge about the downstream tasks and utilizes the transferability of learned skills. However, both goal-oriented rewards and handcrafted rewards are difficult to scale to tasks in other domain naively, such as pixel-wise observations.

We can also understand HRL algorithms through the perspective of abstracted objectives. Options framework typically learns temporal abstraction while FuNs consider state abstraction. HIRO can be viewed as both considering state abstraction and temporal abstraction, where goals provide state direction and goal transition function models temporal information. For temporal abstraction, contrast to options framework, Haarnoja et al. (2018) used graphical models to achieve another hierarchical idea that in hierarchy, each layer attempts to solve the current task directly if it is not fully successful. This makes the job easier for the layer above it. Besides state abstraction and temporal abstraction, Mnih et al. (2014) provided a state space decomposition method by taking the advantage of an attention mechanism. More precisely, this work adds a visual attention mechanism to the state space before selecting actions, where the attention achieves a high-level planning in state space (Sahni et al. 2017; Schulman 2016). For selecting abstraction objects, the main objective is to answer: how does a higher policy guide the lower-policies? For a domain with adequate prior knowledge, a skills combination learning through meta-learning may achieve better performance. For long-term planning, temporal abstraction is necessary in the highest level.

As we saw, HRL is still an advanced topic of reinforcement learning with many issues to be addressed. Recall that the motivation of HRL is to achieve hierarchical abstraction to improve sample efficiency and reuse learned skills for long time horizon problems. The empirical results have shown hierarchical architecture brings advantages but there is not enough evidence that it does achieve hierarchical abstraction or just more efficient exploration (Nachum et al. 2018). Additional future works on probabilistic planning, hierarchies over other reinforcement learning domain, and theoretical guarantees may provide breakthroughs.

References

Andrychowicz M, Wolski F, Ray A, Schneider J, Fong R, Welinder P, McGrew B, Tobin J, Abbeel OP, Zaremba W (2017) Hindsight experience replay. In: Advances in neural information processing systems, pp 5048–5058

Bacon PL, Harb J, Precup D (2017) The option-critic architecture. In: Thirty-first AAAI conference on artificial intelligence

Barto AG, Mahadevan S (2003) Recent advances in hierarchical reinforcement learning. Discrete Event Dyn Syst 13(1–2):41–77

Beattie C, Leibo JZ, Teplyashin D, Ward T, Wainwright M, Küttler H, Lefrancq A, Green S, Valdés V, Sadik A, et al (2016) DeepMind lab. Preprint. arXiv:161203801
Bellemare MG, Naddaf Y, Veness J, Bowling M (2013) The arcade learning environment: an evaluation platform for general agents. J Artif Intell Res 47:253–279
Bhatti S, Desmaison A, Miksik O, Nardelli N, Siddharth N, Torr PH (2016) Playing doom with slam-augmented deep reinforcement learning. Preprint. arXiv:161200380
Da Silva B, Konidaris G, Barto A (2012) Learning parameterized skills. Preprint. arXiv:12066398
Dayan P (1993) Improving generalization for temporal difference learning: the successor representation. Neural Comput 5(4):613–624
Dayan P, Hinton GE (1993) Feudal reinforcement learning. In: Advances in neural information processing systems, pp 271–278
Dietterich TG (1998) The MAXQ method for hierarchical reinforcement learning. In: Proceedings of the international conference on machine learning (ICML), vol 98, Citeseer, pp 118–126
Dietterich TG (2000) Hierarchical reinforcement learning with the MAXQ value function decomposition. J Artif Intell Res 13:227–303
Duan Y, Chen X, Houthooft R, Schulman J, Abbeel P (2016) Benchmarking deep reinforcement learning for continuous control. In: International conference on machine learning, pp 1329–1338
Florensa C, Duan Y, Abbeel P (2017) Stochastic neural networks for hierarchical reinforcement learning. Preprint. arXiv:170403012
Frans K, Ho J, Chen X, Abbeel P, Schulman J (2017) Meta learning shared hierarchies. Preprint. arXiv:171009767
Gregor K, Danihelka I, Graves A, Rezende DJ, Wierstra D (2015) Stochastic backpropagation and approximate inference in deep generative models. In: Proceedings of the international conference on machine learning (ICML)
Haarnoja T, Hartikainen K, Abbeel P, Levine S (2018) Latent space policies for hierarchical reinforcement learning. Preprint. arXiv:180402808
Harutyunyan A, Vrancx P, Bacon PL, Precup D, Nowe A (2018) Learning with options that terminate off-policy. In: Thirty-second AAAI conference on artificial intelligence
Hausknecht MJ (2000) Temporal abstraction in reinforcement learning. PhD thesis
Hauskrecht M, Meuleau N, Kaelbling LP, Dean T, Boutilier C (1998) Hierarchical solution of Markov decision processes using macro-actions. In: Proceedings of the fourteenth conference on Uncertainty in artificial intelligence. Morgan Kaufmann Publishers, Burlington, pp 220–229
Heess N, Wayne G, Tassa Y, Lillicrap T, Riedmiller M, Silver D (2016) Learning and transfer of modulated locomotor controllers. Preprint. arXiv:161005182
Kaelbling LP (1993) Hierarchical learning in stochastic domains: preliminary results. In: Proceedings of the tenth international conference on machine learning (ICML), vol 951, pp 167–173
Kempka M, Wydmuch M, Runc G, Toczek J, Jaśkowski W (2016) ViZDoom: A Doom-based AI research platform for visual reinforcement learning. In: 2016 IEEE conference on computational intelligence and games (CIG). IEEE, Piscataway, pp 1–8
Konda VR, Tsitsiklis JN (2000) Actor-critic algorithms. In: Advances in neural information processing systems, pp 1008–1014
Konidaris G, Barto AG (2009) Skill discovery in continuous reinforcement learning domains using skill chaining. In: Advances in neural information processing systems, pp 1015–1023
Kulkarni TD, Narasimhan K, Saeedi A, Tenenbaum J (2016) Hierarchical deep reinforcement learning: integrating temporal abstraction and intrinsic motivation. In: Advances in neural information processing systems, pp 3675–3683
Levine S, Pastor P, Krizhevsky A, Ibarz J, Quillen D (2018) Learning hand-eye coordination for robotic grasping with deep learning and large-scale data collection. Int J Robot Res 37(4–5):421–436
Levy A, Platt R, Saenko K (2018) Hierarchical reinforcement learning with hindsight. Preprint. arXiv:180508180

Machado MC, Bellemare MG, Bowling M (2017) A Laplacian framework for option discovery in reinforcement learning. In: Proceedings of the 34th international conference on machine learning, vol 70, JMLR.org, pp 2295–2304
Mnih V, Heess N, Graves A, et al (2014) Recurrent models of visual attention. In: Advances in neural information processing systems, pp 2204–2212
Mnih V, Kavukcuoglu K, Silver D, Rusu AA, Veness J, Bellemare MG, Graves A, Riedmiller M, Fidjeland AK, Ostrovski G, Petersen S, Beattie C, Sadik A, Antonoglou I, King H, Kumaran D, Wierstra D, Legg S, Hassabis D (2015) Human-level control through deep reinforcement learning. Nature 518:529–533
Mnih V, Badia AP, Mirza M, Graves A, Lillicrap T, Harley T, Silver D, Kavukcuoglu K (2016) Asynchronous methods for deep reinforcement learning. In: International conference on machine learning (ICML), pp 1928–1937
Nachum O, Gu SS, Lee H, Levine S (2018) Data-efficient hierarchical reinforcement learning. In: Advances in neural information processing systems, pp 3303–3313
OpenAI (2018) Openai five. https://blog.openai.com/openai-five/
Parr R, Russell SJ (1998a) Reinforcement learning with hierarchies of machines. In: Advances in neural information processing systems, pp 1043–1049
Parr RE, Russell S (1998b) Hierarchical control and learning for Markov decision processes. University of California, Berkeley
Riemer M, Liu M, Tesauro G (2018) Learning abstract options. In: Advances in neural information processing systems, pp 10424–10434
Sahni H, Kumar S, Tejani F, Schroecker Y, Isbell C (2017) State space decomposition and subgoal creation for transfer in deep reinforcement learning. Preprint. arXiv:170508997
Schaul T, Horgan D, Gregor K, Silver D (2015) Universal value function approximators. In: International conference on machine learning, pp 1312–1320
Schulman J (2016) Optimizing expectations: from deep reinforcement learning to stochastic computation graphs. PhD thesis, UC Berkeley
Schulman J, Levine S, Abbeel P, Jordan M, Moritz P (2015) Trust region policy optimization. In: International conference on machine learning (ICML), pp 1889–1897
Sharma S, Lakshminarayanan AS, Ravindran B (2017) Learning to repeat: fine grained action repetition for deep reinforcement learning. Preprint. arXiv:170206054
Silver D, Huang A, Maddison CJ, Guez A, Sifre L, Van Den Driessche G, Schrittwieser J, Antonoglou I, Panneershelvam V, Lanctot M, et al (2016) Mastering the game of go with deep neural networks and tree search. Nature 529:484–489
Silver D, Hubert T, Schrittwieser J, Antonoglou I, Lai M, Guez A, Lanctot M, Sifre L, Kumaran D, Graepel T, et al (2017) Mastering chess and shogi by self-play with a general reinforcement learning algorithm. Preprint. arXiv:171201815
Sutton RS, Precup D, Singh S (1999) Between MDPs and semi-MDPs: a framework for temporal abstraction in reinforcement learning. Artif Intell 112(1–2):181–211
Tamar A, Wu Y, Thomas G, Levine S, Abbeel P (2016) Value iteration networks. In: Advances in neural information processing systems, pp 2154–2162
Tessler C, Givony S, Zahavy T, Mankowitz DJ, Mannor S (2017) A deep hierarchical approach to lifelong learning in minecraft. In: Thirty-first AAAI conference on artificial intelligence
Vezhnevets A, Mnih V, Osindero S, Graves A, Vinyals O, Agapiou J, et al (2016) Strategic attentive writer for learning macro-actions. In: Advances in neural information processing systems, pp 3486–3494
Vezhnevets AS, Osindero S, Schaul T, Heess N, Jaderberg M, Silver D, Kavukcuoglu K (2017) Feudal networks for hierarchical reinforcement learning. In: Proceedings of the 34th international conference on machine learning, vol 70, JMLR.org, pp 3540–3549
Vinyals O, Babuschkin I, Czarnecki WM, Mathieu M, Dudzik A, Chung J, Choi DH, Powell R, Ewalds T, Georgiev P, et al (2019) Grandmaster level in StarCraft II using multi-agent reinforcement learning. Nature 575(7782):350–354
Wiering M, Schmidhuber J (1997) HQ-learning. Adapt Behav 6(2):219–246

Chapter 11
Multi-Agent Reinforcement Learning

Huaqing Zhang and Shanghang Zhang

Abstract In reinforcement learning, complicated applications require involving multiple agents to handle different kinds of tasks simultaneously. However, increasing the number of agents brings in the challenges on managing the interactions among them. In this chapter, according to the optimization problem for each agent, equilibrium concepts are put forward to regulate the distributive behaviors of multiple agents. We further analyze the cooperative and competitive relations among the agents in various scenarios, combining with typical multi-agent reinforcement learning algorithms. Based on all kinds of interactions, a game theoretical framework is finalized for general modeling in multi-agent scenarios. Analyzing the optimization and equilibrium situation for each component of the framework, the optimal multi-agent reinforcement learning policy for each agent can be guided and explored.

Keywords Multi-agent reinforcement learning · Equilibrium · Game theory · Zero-sum game · Chicken dare game · Stackelberg game

11.1 Introduction

Basic reinforcement learning is powerful on reinforcing one agent to behave outstandingly based on the rules and rewards reflected by the environment. However, for various applications in artificial intelligence, when the environment is in large-scale and the tasks are complicated, not only do we expect one single agent to make smart actions, but also we hope there exist a group of agents who can communicate and make decisions with each other. Accordingly, we need to propose and apply learning strategies for each agent. Considering the interactions among

H. Zhang
Google LLC, Mountain View, CA, USA

S. Zhang (✉)
University of California, Berkeley, CA, USA

H. Dong et al. (eds.), *Deep Reinforcement Learning*,
https://doi.org/10.1007/978-981-15-4095-0_11

multiple agents, multi-agent reinforcement learning is put forwards to achieve the expectation.

For clear analysis and better understanding, we set the basic component of multi-agent learning as agent, policy, and utility, which are further elaborated as follows.

- **Agent**: We define the agent as the autonomous individual, which can independently interact with the environment and take its own strategy based on the observation of others' behaviors, aiming to achieve maximum revenue or minimum loss for itself. In the considered scenarios, there exists multiple agents. When the number of agent equals to one. The multi-agent reinforcement learning is identical to regular reinforcement learning scenarios.
- **Policy**: Each agent follows its own policy in the multi-agent reinforcement learning. The policy is normally designed to maximize the revenue and minimize the cost of the agent while it is affected by the environment and the policies of other agents.
- **Utility**: Each agent has unique utility, considering its requirements and dependencies with the environment and other agents. The utility is defined as the revenue minus the cost of the agent based on different objectives. In multi-agent scenarios, each agent aims to maximize its own utility through learning from the environment and other agents.

Accordingly, in the multi-agent reinforcement learning, agents are assigned with their own utility functions. Based on the observation and experience through interactions, each agent performs policy learning autonomously, aiming to optimize its own utility value, without considering the utilities of other agents'. Therefore, there may exist competitions or co-operations through the interactions with all other agents. Considering different kinds of interactions among multiple agents, game theoretical analysis is commonly applied as a powerful tool for decision making (Fudenberg and Tirole 1991). Based on different scenarios, the games can be fitted into different categories, listed as follows.

- **Static Game**: The static game is the simplest form to model the interactions of agents. In the static game, single decision is required by each agent. As each agent only acts once, unexpected cheating and betraying in the static game can be profitable. Thus, each agent is required to carefully predict the strategies of the other agents so as to act smartly to gain high utility.
- **Repeated Game**: The repeated game refers to the situation where all agents can take actions repeatedly based on the same state for multiple iterations. The overall utility of each agent is the summation of the discounted utility for each iteration of the game. Due to the repeated actions of all agent, the cheating and betraying during the interactions can cause penalty or revenge from other agents in future iterations. Thus, the repeated game avoids the malicious behaviors of the agents and generally improve the total utilities for all agents.

- **Stochastic Game**: The stochastic game (or Markov game) can be regarded as the MDP with multiple agents or the repeated game with multiple states. The game models the iterated interactions of multiple agents in general scenarios, where for each iteration, each agent is at different states and tries to gain high utility based on the observation and predication of other agents.

In the chapter, based on the fundamental reinforcement learning for single agent, we focus more on the relations between agents, seeking equilibrium scenarios where each agent is able to achieve high and stable utilities.

11.2 Optimization and Equilibrium

As each agent aims to maximize its own utility, multi-agent reinforcement learning can be considered as solving optimization problem for each agent. Suppose there are m agents, $\mathcal{X} = \mathcal{X}_1 \times \mathcal{X}_2 \times, \ldots, \times \mathcal{X}_m$ refers to the policy space of all agents and $\mathbf{u} = (u_1(\mathbf{x}), \ldots, u_m(\mathbf{x}))$ represents the lists of utility profile of all agents with policy profile $\mathbf{x}$, where $\mathbf{x} \in \mathcal{X}$. Accordingly, each agent i, $\forall i \in \{1, 2, \ldots, m\}$, requires to maximize its own utility considering on the behaviors of others. For multi-agent reinforcement learning, the task is generally solving multiple optimization problems simultaneously or sequentially so as to make sure each agent is able to get high utility.

As utility of each agent can be affected by the policies of all other agents, it is required to seek a stable policy profile for all agents so that no agents are willing to deviate from the current policy for higher utilities. Therefore, the equilibrium concept in multi-agent reinforcement learning is put forward. For better understanding and analysis, without loss of generality, we introduce different equilibrium concepts based on an intuitive chicken dare game. The chicken dare game is a static game scenario related with the interactions between two agents. Both agents are able to choose "chicken" (short as "C") or "dare" (short as "D") as its action independently with each other. Based on different actions of both agents, the utilities are shown in Fig. 11.1. When both agents choose "D," both dare and receive lowest utility 0. When one agent chooses "D" and the other chooses "C," the one who dares receives largest utility 7 while the one who chickens still gets relatively low utility 3. When both agents choose "C," both agents chicken and can get relatively high rewards 5.

Fig. 11.1 Chicken dare game

	C	D
C	5, 5	3, 6
D	6, 3	0, 0

11.2.1 Nash Equilibrium

Following the chicken dare game (Rapoport and Chammah 1966) in Fig. 11.1, we set the rule that both agents are required to make actions simultaneously. When both agents choose "C," each of them would like to switch to "D" to gain higher utility based on the assumption that its opponent will not change its action. When both agents switch their actions to "D," both of them will receive 0 utility and would definitely switch the actions back to "C" for higher utility. Nevertheless, when one agent chooses the action "C" while the other "D," assuming the component would not switch its action, each agent cannot switch actions anymore to gain higher utility. Therefore, we call the scenarios when one agent chooses "C" and the other agent chooses "D" as Nash equilibrium (Nash et al. 1950), which are formally defined as follows.

Definition 11.1 Let $(\mathcal{X}, \mathbf{u})$ denotes the static scenario with m agents. $\mathcal{X} = \mathcal{X}_1 \times \mathcal{X}_2 \times, \ldots, \times \mathcal{X}_m$ refers to the policy space of all agents and $\mathbf{u} = (u_1(\mathbf{x}), \ldots, u_m(\mathbf{x}))$ is the utility profile of all agents with policy profile $\mathbf{x}$, where $\mathbf{x} \in \mathcal{X}$. Let x_i be a policy of agent i, $\mathbf{x}_{-i}$ be the policies of other agents except for agent i. The policy $x^* \in \mathcal{X}$ is able to achieve the Nash equilibrium if $\forall i, x_i \in \mathcal{X}_i$,

$$u_i(x_i^*, \mathbf{x}_{-i}^*) \geq u_i(x_i, \mathbf{x}_{-i}^*). \tag{11.1}$$

Pure Strategy Nash Equilibrium

As shown in the definition, in the static scenario of multi-agent reinforcement learning, each agent is required to determine one action at each timestamp. When the actions of the other agents are fixed and each agent cannot deviate its current action for higher utility, all agents achieve the pure strategy Nash equilibrium. In the chicken dare game, there exist two pure strategy Nash equilibrium solutions, where one agent behaves chicken and the other behaves dare. The pure strategy Nash equilibrium may not always exists, as the pure action of each agent may let the other agents deviate from its current behavior.

Mixed Strategy Nash Equilibrium

Moreover, each agent can set up a policy, where each action is chosen with probabilities at each timestamp. The policies of the agents bring in randomness and uncertainty for the interactions. Thus, the agents can adjust their policies considering its effect for other agents and the mixed strategy Nash equilibrium always exists. Take the chicken dare game as an example, we suppose the probability for agent 1 to behave chicken is p, then the probability to behave dare is $1 - p$. Therefore, in order to guarantee the policy of agent 1 will not cause bias on the

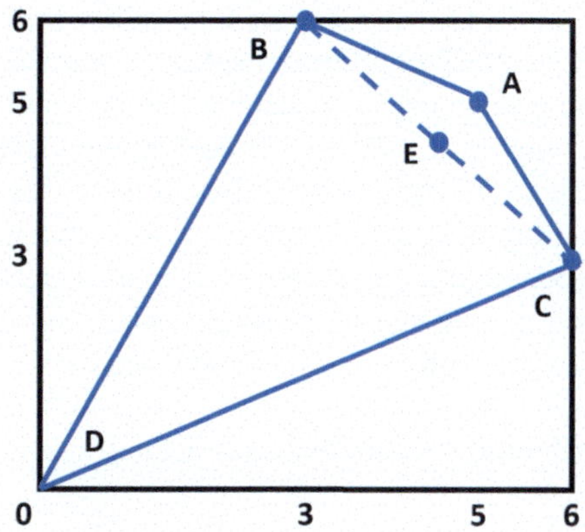

Fig. 11.2 Nash equilibrium in chicken dare game

decision making for agent 2. The following equation exists:

$$5p + 3(1 - p) = 6p + 0(1 - p). \quad (11.2)$$

Thus, we get $p = 0.75$, and the policy for both agents are the same, namely to choose action "C" with probability 0.75, and to choose action "D" with probability 0.25. With the policy, both agents achieve the Nash Equilibrium too, and the expected utility for each agent is 4.5.

Based on the above, we further show the results in Fig. 11.2, where the X axis is the utility of agent 1, and the Y axis is the utility of agent 2. Based on the relations of both agents' utilities in Fig. 11.1, the point A refers to the result when both agents act as "C." The point B denotes the result when agent 1 acts as "C" and agent 2 acts as "D." The point C represents the result when agent 1 acts as "D" and agent 2 acts as "C." The point D indicates the result when both agents act as "D." Therefore, whatever the policies of each agent, the result falls in the region $ABDC$. And point B and point C are the pure strategy Nash Equilibrium with determined actions. The middle point E of the line BC is the mixed strategy Nash Equilibrium. The total utilities for both agents equals 9 for all Nash equilibrium solutions.

11.2.2 Correlated Equilibrium

In Nash equilibrium solution, the total utility of both agents is 9, which is less than the maximum value 10. However, both agents are required to choose "C" at the same time to achieve the maximum value 10, which is unstable in distributed fashion. Therefore, correlation concept is put forwards among agents to further improve total utility and guarantee the stability of the solution at the same time.

In the chicken dare game, we set the probability distribution for both agents to choose "CC" (the first action is for agent 1 and the second action is for agent 2), "CD," "DC," and "DD" as $\mathbf{v}$. When both agents correlate with each other and set

$\mathbf{v} = [1/3, 1/3, 1/3, 0]$, the total utility for both agents are 9.3333, which is larger than the Nash equilibrium solution. Moreover, when one agent choose "C," as the agent knows its component will follow the probability distribution in correlation, its component will takes mixed strategy and choose "C" with probability 0.5 and "D" with probability 0.5. Thus, if the agent continues to choose "C," it can receive the utility of $0.5 * 5 + 0.5 * 3 = 4$. If the agent switches its action and assumes its component fixes its current policy, it can receive the utility of $0.5 * 6 + 0.5 * 0 = 3$, which is less than 4. Similarly when the agent chooses "D," its component will follow the correlation and choose action "C" with probability 1. Therefore, the agent cannot switch its action to "C" to gain higher utility. Accordingly, the probability distribution $\mathbf{v}$ lets both agents achieve correlated equilibrium, with the definition as follows.

Definition 11.2 Correlated equilibrium (Aumann 1987) can be achieved by any probability distribution $\mathbf{v}$ satisfying,

$$\sum_{\mathbf{x}_{-i} \in \mathcal{X}_{-i}} v(x_i^*, \mathbf{x}_{-i})[u_i(x_i^*, \mathbf{x}_{-i}) - u_i(x_i, \mathbf{x}_{-i})] \geqslant 0, \forall x_i \in \mathcal{X}_i, \tag{11.3}$$

where $\mathcal{X}_i$ is the policy space of the agent i and $\mathcal{X}_{-i}$ denotes the policy space of all the agents except agent i.

Therefore, as long as both agents follow the correlated probability distribution, each agent cannot deviate with current policy for higher utility. We further depict correlated equilibrium as point F in Fig. 11.3. Moreover, for all points in the region ABC, as long as it satisfies relations in (11.3), it can be called correlated equilibrium solutions.

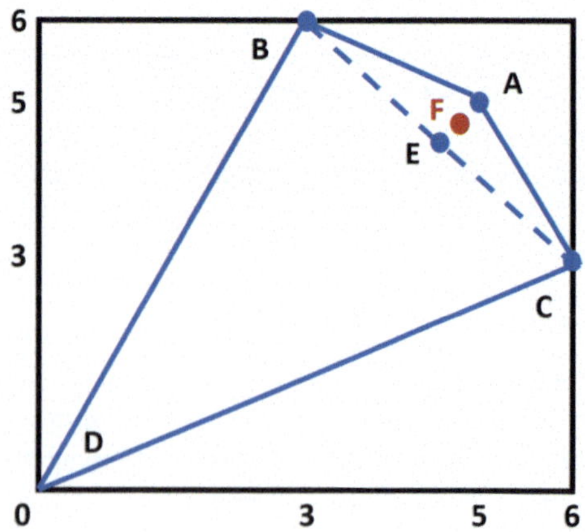

Fig. 11.3 Correlated equilibrium in chicken dare game

11.2.3 Stackelberg Equilibrium

Apart from the simultaneous scenarios, both agents may also take actions sequentially. In sequential scenarios, the agents are divided into leaders and followers, where the leaders act first and the followers act correspondingly (Bjorn and Vuong 1985). Accordingly, the first-mover advantage exists, where the leaders are able to predict the corresponding reactions of followers and take actions for high utilities. In the chicken dare game, if we assume the agent 1 as leader and agent 2 as follower, the agent 1 can choose action "D," since when agent 1 choose "D," the agent 2 will definitely choose "C" for higher utility. Thus, the utility for the agent 1 can achieve the maximum value 6 for itself and both agents achieve the stackelberg equilibrium (Zhang et al. 2018), which can be defined as follows.

Definition 11.3 Let $((\mathcal{X}, \boldsymbol{\Pi}), (g, f))$ be the general sequential scenario with m leaders and n followers. $\mathcal{X} = \mathcal{X}_1 \times \mathcal{X}_2 \times, \ldots, \times \mathcal{X}_m$ and $\boldsymbol{\Pi} = \boldsymbol{\Pi}_1 \times \boldsymbol{\Pi}_2 \times, \ldots, \times \boldsymbol{\Pi}_n$ are the policy space of all leaders and all followers, respectively. $g = (g_1(\mathbf{x}), \ldots, g_m(\mathbf{x}))$ is the utility function of leaders for $\mathbf{x} \in \mathcal{X}$, and $f = (f_1(\boldsymbol{\pi}), \ldots, f_n(\boldsymbol{\pi}))$ is the utility function of followers for $\boldsymbol{\pi} \in \boldsymbol{\Pi}$. Let x_i be the policy of leader i, $\mathbf{x}_{-i}$ be policies of all leaders except for leader i, π_j be the policy of follower j, and $\boldsymbol{\pi}_{-j}$ be policies of all other followers except for leader j. The policies of $x^* \in \mathcal{X}$ and $\boldsymbol{\pi}^* \in \boldsymbol{\Pi}$ can achieve the stackelberg equilibrium of the multi-leader multi-follower scenario if $\forall i, \forall j$ $x_i \in \mathcal{X}_i, \pi_j \in \boldsymbol{\Pi}_j$,

$$g_i\left(x_i^*, \mathbf{x}_{-i}^*, \boldsymbol{\pi}^*\right) \geq g_i\left(x_i, \mathbf{x}_{-i}^*, \boldsymbol{\pi}^*\right) \geq g_i\left(x_i, \mathbf{x}_{-i}, \boldsymbol{\pi}^*\right), \tag{11.4}$$

$$f_j\left(\mathbf{x}, \boldsymbol{\pi}_j^*, \boldsymbol{\pi}_{-j}^*\right) \geq f_j\left(\mathbf{x}, \boldsymbol{\pi}_j, \boldsymbol{\pi}_{-j}^*\right). \tag{11.5}$$

11.3 Competition and Cooperation

In the last section, we take an example of chicken dare static game to introduce the optimization and equilibrium concepts. Moreover, the relation among multiple agents varies for different applications. In this section, we will further analyze the competitive and cooperative relations among multiple agents in distributed fashion. Without specific explanation, we consider the scenario where there are m agents, $\mathcal{X} = \mathcal{X}_1 \times \mathcal{X}_2 \times, \ldots, \times \mathcal{X}_m$ refers to the policy space of all agents and $\mathbf{u} = (u_1(\mathbf{x}), \ldots, u_m(\mathbf{x}))$ is the utility profile of all agents with policy profile $\mathbf{x}$, where $\mathbf{x} \in \mathcal{X}$.

11.3.1 Cooperation

When multiple agents cooperate with each other, in most times, the total utilities will be higher than the utilities of all agents without cooperation. Moreover, in distributed network, each agent only considers its own utility. Accordingly, in order to include the agent in the cooperated coalition, the agent is required to receive higher utility compared with its non-cooperated behaviors. The optimization problem for agent i, $\forall i \in \{1, 2, \dots, m\}$ can be formulated as following:

$$\begin{aligned} &\max_{x_i} \ \textstyle\sum_{k=1}^{k=m} u_k(x_k|\mathbf{x}_{-k}), \\ &s.t. \ \ u_i(x_i^*|\mathbf{x}_{-i}^*) \geq u_i(x_i|\mathbf{x}_{-i}^*). \end{aligned} \tag{11.6}$$

11.3.2 Zero-Sum Game

Zero-sum game (VINCENT 1974) is frequently adopted in multiple applications. For simplicity, we suppose there are two agents, and each agent can choose to take action "A" or "B." The utility function is shown in Fig. 11.4, where we can observe the total utility for each situation equal to zero. Accordingly, for general zero-sum problem, each agent is required to maximize its own utility based on the prediction that its utility is minimized by its opponents at the same time. The optimization problem for agent i, $\forall i \in \{1, 2, \dots, m\}$ can be summarized as follows:

$$\max_{x_i} \ \min_{\mathbf{x}_{-i}} \ u_i. \tag{11.7}$$

In Littman (1994), the authors analyze a simplified football competition and model it as a zero-sum game. Generally, in the game, there are two agents, each agent tries to maximize its own utility while minimize the utility of its component. Thus, for the agent i, its optimization problem can be represented as

$$\max_{\pi_i} \ \min_{\mathbf{a}_{-i}} \ \textstyle\sum_{a_i} Q(s, a_i, \mathbf{a}_{-i})\pi_i, \tag{11.8}$$

where π_i is the strategy for the agent i and a_i is the actual action of the agent i based on the strategy π_i. In the game, the agent i tries to maximize its value function, while its component tries to minimize the value function by taking the action $\mathbf{a}_{-i}$.

Fig. 11.4 Zero-sum game

	A	B
A	1, -1	-1, 1
B	-1, 1	1, -1

11.3.3 Simultaneous Competition

Apart from the zero-sum game, there are many applications requiring general simultaneous competition for multiple agents. In simultaneous competition, all agents are required to take actions at the same time. The optimization problem for agent i, $\forall i \in \{1, 2, \ldots, m\}$ can be summarized as follows:

$$\max_{x_i} \quad u_i(\mathbf{x_i}|\mathbf{x_{-i}}). \tag{11.9}$$

In Hu and Wellman (1998), the general Q-learning is put forward to solve the competitions among multiple agents. The algorithm is illustrated in 1. Based on the experience of interactions, each agent i maintains a Q table to instruct its policies π_i and update with the following function:

$$Q_i(s, a_i, \mathbf{a}_{-i}) = (1 - \alpha_i)Q_i(s, a_i, \mathbf{a}_{-i}) + \alpha_i\big[r_i + \gamma\pi_i(s')Q_i\big(s', a'_i, \mathbf{a}'_{-i}\big)\boldsymbol{\pi}_{-i}(s')\big]. \tag{11.10}$$

In the multi-agent scenarios, as the update of Q table requires the policies of the other agents $\boldsymbol{\pi}_{-i}$, the agent i is also required to maintain an estimated Q table for all other agents. According to the prediction of other agents' policies $\boldsymbol{\pi}_{-i}$, the agent i aims to set up the policy π_i so that $(\pi_i, \boldsymbol{\pi}_{-i})$ achieves the mixed strategy Nash equilibrium.

Algorithm 1 Multi-agent general Q learning

```
Set initial values for Q table Q_i(s, a_i, a_{-i}) = 1, ∀i ∈ {1, 2, ..., m}.
for episode = 1 to M do
    Set initial state s = S_0.
    for step = 1 to T do
        Each agent i chooses action a_i based on π_i(s), which is a mixed Nash equilibrium strategy
        based on Q values of all agents.
        Observe experience (s, a_i, a_{-i}, r_i, s') and apply it to update Q_i value
        Update the state s = s'.
    end for
end for
```

Except for the basic Q learning, other deep reinforcement learning approaches can also be explored considering the interactions among agents. The multi-agent deep deterministic policy gradient (MADDPG) (Lowe et al. 2017), developed from the single-agent deep deterministic policy gradient (DDPG) algorithm, provides strategies for each agent in the simultaneous competition scenario. In MADDPG, as shown in Algorithm 2, each agent is allocated with a den-centralized actor, which suggests the agent to take actions. On the other side, the critic is centralized and maintains the Q value related with action profile of all agents.

Algorithm 2 Multi-agent deep deterministic policy gradient (MADDPG)

for episode = 1 to M **do**
  Set initial state $s = S_0$.
  for step = 1 to T **do**
    Each agent i chooses action a_i based on current policy π_{θ_i}.
    The actions of all agents $\mathbf{a} = (a_1, a_2, \ldots, a_m)$ are executed simultaneously.
    Store $(s, \mathbf{a}, r, s')$ in replay buffer $\mathcal{M}$
    Update the state $s = s'$.
    for agent i = 1 to m **do**
      Sample a batch of previous experience from replay buffer $\mathcal{M}$.
      Calculate the gradients and update the weights for both actor and critic network.
    end for
  end for
end for

Specifically, the gradient of the expected return for each actor i can be denoted as

$$\nabla_{\theta_i^\pi} J(\pi_i) = \mathbb{E}\left[\nabla_{\theta_i^\pi} \boldsymbol{\pi}_i\left(o_i|\theta_i^\pi\right)\nabla_{a_i} Q_i^\pi\left(o_1, \ldots, o_m, a_1, \ldots, a_m|\theta_i^Q\right)\right], \tag{11.11}$$

where $o_1, \ldots, o_m$ is the observations of m agents, respectively. Parameterized by θ_i^π, $\boldsymbol{\pi}_i$ is the deterministic policy for agent i satisfying $a_i = \boldsymbol{\pi}_i(o_i)$.

Correspondingly, the loss function of the critic for the agent i is the TD-error of the Q value, such as

$$\mathcal{L}_i = \mathbb{E}\left[\left(Q_i^\pi\left(o_1, \ldots, o_m, a_1, \ldots, a_m|\theta_i^Q\right) - r_i - \gamma Q_i^{\pi'}\left(o_1', \ldots, o_m', a_1', \ldots, a_m'|\theta_i^{Q'}\right)\right)^2\right], \tag{11.12}$$

where $\theta_i^{Q'}$ is the delayed parameter for Q prediction. $\boldsymbol{\pi}'$ refers to the target policies with delayed parameters $\theta_i^{\pi'}$.

11.3.4 Sequential Competition

For some applications, different types of agents may have different priorities when take actions. Thus, the agents in competition take actions sequentially and the agents act first will have first-mover advantage. Generally we suppose $((\mathcal{X}, \boldsymbol{\Pi}), (g, f))$ as the general sequential scenario with m leaders and n followers. $\mathcal{X} = \mathcal{X}_1 \times \mathcal{X}_2 \times, \ldots, \times \mathcal{X}_m$ and $\boldsymbol{\Pi} = \boldsymbol{\Pi}_1 \times \boldsymbol{\Pi}_2 \times, \ldots, \times \boldsymbol{\Pi}_n$ are the policy space of all leaders and all followers, respectively. $g = (g_1(\mathbf{x}), \ldots, g_m(\mathbf{x}))$ is the utility function of leaders for $\mathbf{x} \in \mathcal{X}$, and $f = (f_1(\boldsymbol{\pi}), \ldots, f_n(\boldsymbol{\pi}))$ is the utility function of followers for $\boldsymbol{\pi} \in \boldsymbol{\Pi}$. Accordingly, the optimization problem for follower j, $\forall j \in \{1, 2, \ldots, n\}$ is

$$\max \quad f_j(\pi_j | \boldsymbol{\pi}_{-j}, \boldsymbol{x}). \tag{11.13}$$

The optimization problem for leader i, $\forall i \in \{1, 2, \ldots, m\}$ can be depicted as

$$\begin{aligned} &\max \quad g_i(x_i | \mathbf{x}_{-i}, \boldsymbol{\pi}), \\ &s.t. \quad \pi_j = \arg\max \quad f_j(\pi_j | \boldsymbol{\pi}_{-j}, \boldsymbol{x}), \quad \forall j \in \{1, 2, \ldots, n\}. \end{aligned} \tag{11.14}$$

11.4 Game Theoretical Framework

Based on the analysis on the relationships of multiple agents, we summarize a general game theoretical framework in Fig. 11.5. In the framework, we suppose it is an iterative scenario where all the agents are able to take actions during each time interval. Within the same time interval, we further classify the agents into multiple levels. The agents at top level act first. Based on the observation of the actions in top levels, the agents in lower levels behave correspondingly. Moreover, within each level, there are multiple agents take actions simultaneously. Accordingly, the stackelberg equilibrium is expected between each two levels and the Nash equilibrium or correlated equilibrium is expected for all agents within one level.

The game theoretical framework can be regarded as a general structure to deal with all kinds of multi-agent reinforcement learning problems. For further tests and evaluations, various multi-agent platforms have been put forward. For example, AlphaStar is the platform for simulating the behaviors of multiple agents in StarCraft video game. Multi-agent connected autonomous driving (MACAD) platform (Palanisamy 2019) is provided for learning and adapting the driving environment. Google research football (Kurach et al. 2019) is the platform to

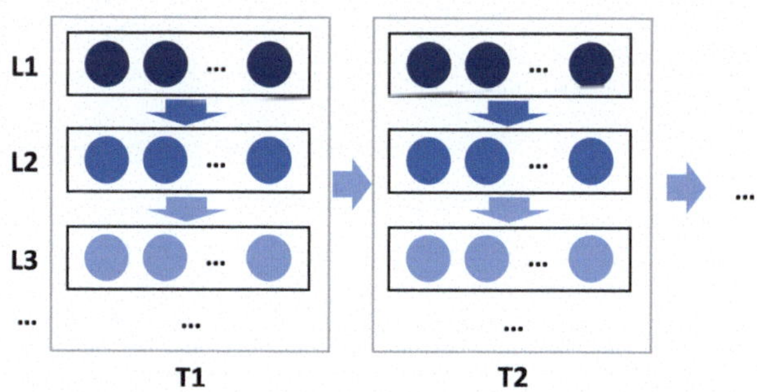

Fig. 11.5 Game theoretical framework

simulate football games for multiple autonomous agents. Based on the multi-agent platforms in different kinds of scenarios, detailed game theoretical framework for multi-agent reinforcement learning can be proposed and analyzed for optimal strategies.

References

Aumann RJ (1987) Correlated equilibrium as an expression of Bayesian rationality. Econometrica J Econ Soc 55:1–18

Bjorn PA, Vuong QH (1985) Econometric modeling of a stackelberg game with an application to labor force participation. Working Papers no. 577, California Institute of Technology, Division of the Humanities and Social Sciences. https://ideas.repec.org/p/clt/sswopa/577.html

Fudenberg D, Tirole J (1991) Game theory, 1991. Cambridge, Mass. 393(12):80

Hu J, Wellman MP (1998) Multiagent reinforcement learning: theoretical framework and an algorithm. In: International conference on robotics and automation (ICRA)

Kurach K, Raichuk A, Stańczyk P, Zając M, Bachem O, Espeholt L, Riquelme C, Vincent D, Michalski M, Bousquet O, Gelly S (2019) Google research football: a novel reinforcement learning environment. Preprint. arXiv:1907.11180

Littman ML (1994) Markov games as a framework for multi-agent reinforcement learning. In: Proceedings of the international conference on machine learning (ICML), pp 157–163. https://doi.org/10.1016/b978-1-55860-335-6.50027-1

Lowe R, Wu Y, Tamar A, Harb J, Abbeel OP, Mordatch I (2017) Multi-agent actor-critic for mixed cooperative-competitive environments. In: Advances in neural information processing systems

Nash JF, et al (1950) Equilibrium points in n-person games. Proc Natl Acad Sci 36(1):48–49

Palanisamy P (2019) Multi-agent connected autonomous driving using deep reinforcement learning. Preprint. arXiv:1911.04175

Rapoport A, Chammah AM (1966) The game of chicken. Am Behav Sci 10(3):10–28

Vincent P (1974) Learning the optimal strategy in a zero-sum game. Econometrica 42(5):885–891

Zhang H, Khairy S, Cai LX, Han Z (2018) Resource allocation in unlicensed long term evolution HetNets. Springer, Berlin

Chapter 12
Parallel Computing

Huaqing Zhang and Tianyang Yu

Abstract Due to the low sample efficiency of reinforcement learning, parallel computing is an efficient solution to speed up the training process and improve the performance. In this chapter, we introduce the framework applying parallel computation in reinforcement learning. Based on different scenarios, we firstly analyze the synchronous and asynchronous communication and elaborate parallel communication in different network typologies. Taking the advantage of parallel computing, classic distributed reinforcement learning algorithms are depicted and compared, followed by summaries of fundamental components in the distributed computing architecture.

Keywords Distributed computing · Asynchronous advantage actor-critic · Hybrid GPU/CPU A3C · Importance weighted actor-learner architecture (IMPALA) · Scalable efficient deep-RL (SEED) · Distributed proximal policy optimization (DPPO) · Ape-X · Retrace-actor (Reactor) · Recurrent replay distributed DQN (R2D2)

12.1 Introduction

In deep reinforcement learning, large amounts of data is required for model training. Take OpenAI Five (OpenAI et al. 2019) as an example, batches of around two million frames are applied for training every 2 s, so as to let the agents learn and behave smartly in the Dota game. Moreover, from the optimization perspective, large batch size can reduce variance especially for policy gradient methods. However, due to the sequential interactions between the agent and the environment, the reinforcement learning algorithm suffers from the low sample efficiency, result in the unsatisfied

H. Zhang (✉)
Google LLC, Mountain View, CA, USA

T. Yu
Nanchang University, Nanchang, China

H. Dong et al. (eds.), *Deep Reinforcement Learning*,
https://doi.org/10.1007/978-981-15-4095-0_12

training performance and slow convergence speed. Parallel computing, referring to the simultaneous computation on separated but independent tasks, is implemented as an efficient solution. Generally, the parallelization can be considered from the following two perspectives:

- **Parallel Computation**: Data computation is the core procedure to perform feature engineering, modeling learning, and performance evaluations. The computation is taken over by computing unit, which can be combined and extended to different scales. Within each level, the performance of computation can be regarded into two perspectives. One is focusing multiple computing units on one task. The other is to map multiple computing units to multiple tasks and apply computation in parallel fashion. Compared with the above computation strategies, with increasing number of computing units applied on one task, the efficiency to finish the task increases but soon converges due to some bottleneck processes. In deep reinforcement learning, when the computing resources are sufficiently provided, in order to further improve the efficiency, it is beneficial to separate the task into multiple independent sub-tasks, each allocated to efficient amounts of computing resources.
- **Parallel Transmission**: When sufficient computing resources are provided, how to manage the data transmission between the computing resources may become the bottleneck to solve the problem. Generally different data transmission network typologies are put forward for different applications to avoid the transmission redundancy, to balance the transmission loads and to reduce the transmission delay. In parallel computing, as there are multiple processes or threads finishing different tasks at the same time, it is challenging to manage the data traffic and guarantee the transmission efficiency in the network with limited communication bandwidth.

In a supervised setting, one simple way to speed up the learning is to process many different input samples as once. However, in deep reinforcement learning, this is not possible because we have to let the agent and environment interact with each other sequentially to obtain all required information. What we should follow in deep reinforcement learning instead is to apply parallelization on different trajectories or batches when updating weights in deep policy and value network. In this chapter, we analyze the parallelization of deep reinforcement learning in the perspective of data computation and data transmission. We further enumerate significant distributed computing algorithms and show the general distributed computing architecture to be applied for large-scale deep reinforcement learning problems.

12.2 Synchronization and Asynchronization

In parallel computing, most of the common data transmission methods apply star topology, which is composed of one master node and multiple slave nodes. The master node generally manages the data information for the problem. It applies data distribution and collection with each slave node. Based on the accumulated data, the general network parameters are learnt and updated. Each slave node, on the other hand, receives the allocated data from the master node, performs data computation and submits its computing results back to the master node. As there are multiple slave nodes working at the same time, under the management of the master node, the data computation can be done in parallel to finish a large-scale problem in cooperation.

The star topology is widely considered in solving deep reinforcement learning applications. The parallel version of the actor-critic method, for example, usually adopts one master node and multiple slave nodes. Each slave node maintains a deep policy network, which has the same structure as all other slave nodes and the master node. Therefore, the slave nodes can be initialized by copying the weights of the policy network from their master node. Then it can independently interact with the environment for exploration. After several rounds of interactions, the slave node communicates with the master node and sends the information related with the weights of networks. The information can be single-step exploration experience, trajectory exploration experience, buffered exploration experiences with priority information, computed gradients of the network parameters, etc., based on different detailed architecture. Accumulating the feedback and experience from each slave node, the master node can update network parameters and further announce its updated weights to slave nodes for their next round of explorations.

The star topology clearly separates the tasks and accelerates the policy learning with the parallel computing among slave nodes. However, with different computation power, each slave node may explore and collect experience with different time schedules. Then how to determine the pattern for data communication varies for different problems and system architectures, which generally is classified into synchronous communication and asynchronous communication.

The synchronous communication pattern is shown in Fig. 12.1, where the red bar is the time applied for data communication among the nodes and the blue bar is the time for computation within the node. It is noticed that the time for communication falls into the same time intervals for all slave nodes. For the master node, it has same time intervals to communicate with all slave nodes. However, for the slave nodes, the ones computing faster have to wait until all other slower ones finish the

Fig. 12.1 Synchronization

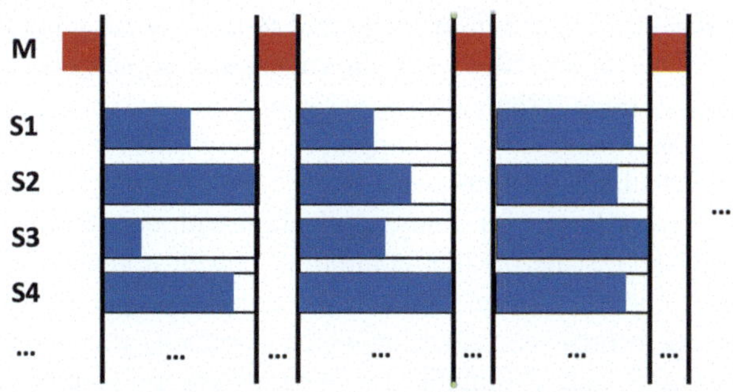

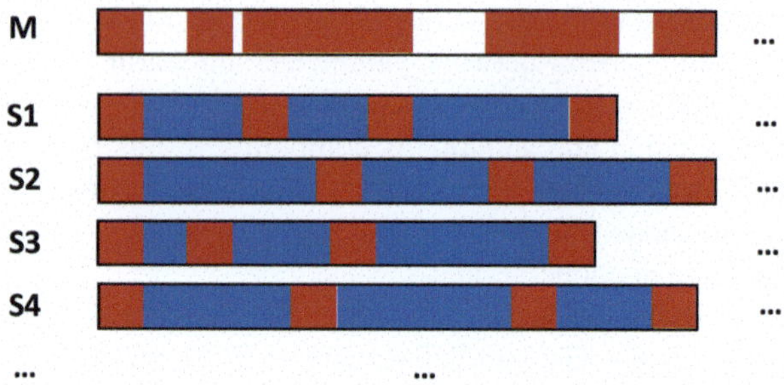

Fig. 12.2 Asychronization

computation within the round. Thus, synchronous communication is more organized for the master node to collect and analyze the computation results from the slave nodes, but there are lots of computing resources wasted on slave nodes due to the waiting for synchronization.

In order to avoid the waiting time for slave node and improve the efficiency to apply computing resources, asynchronous communication is put forward correspondingly. As depicted in Fig. 12.2, each slave node is able to submit the information to the master node as long as it finishes the training task within one round, and the master node collects the information and synchronizes with the slave node whenever the slave node finishes. Accordingly, the data communication to different slave nodes is performed within different time intervals. The master node may require communication with different slave nodes from time to time, but the computing resource are fully adopted for model training for each slave node.

12.3 Parallel Communication and Networking

The star topology is a centralized way to apply parallel computation, where the master node is able to manage and maintain the system to guarantee that all distributed tasks are well-organized. On the other side, the master node is also the weakness part of the system. In order to guarantee high performance, the master node is required to be much more efficient on information processing compared with the slave nodes. Secondly, the data transmission bandwidth towards the master node also requires to be sufficient so as to avoid delay for the data computation for all slave nodes. Moreover, the robustness of the system is highly dependent on the master node. Whatever issues causing the breakdown event on the master node, the whole system stops working even though the computing resources on slave nodes are available and sufficient.

Accordingly, for many application with demanding requirements on robustness and large-scale parallel computing power, a general distributed data computation and communication structure is necessary. We assume there are multiple independent processes, each of which maintains its own deep reinforcement learning network and communicates with others frequently from data synchronization.

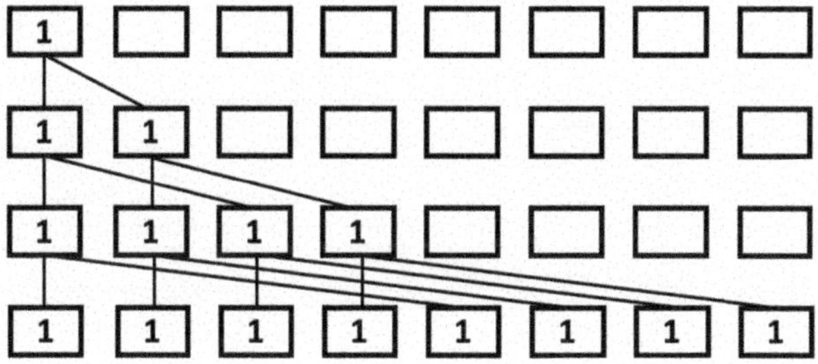

Fig. 12.3 Tree-structured communication

As each process requires to exchange the information with all other processes, when the number of processes increasing, the communication cost exponentially increases. In order to reduce the redundant communication and achieve efficiency on information synchronization, inter-process communication (IPC) are referred with message passing interfaces (MPI). Generally MPI provides basic interfaces for message sending, broadcasting, and receiving for each process. Based on the standard, different communication structures are further provided to improve the communication efficiency. The following takes some communication structures as examples for insights.

- **Tree-Structured Communication**: Assume there are N processors in the system. When a process would like to broadcast its information to all other $N - 1$ processors, it can follow a tree-structure as shown in Fig. 12.3. In the tree-structured communication, the processor first communicates with its $m - 1$ neighboring processes. Then in the next iterations, all neighbors receiving the information will further communicate to $m - 1$ new different processors in parallel to expand to all other processors. Accordingly, with increasing parallel communication, it takes $\lceil \log_m N \rceil$ iterations to broadcast the information to all processors and the information sender processor only requires to send $(m - 1)\lceil \log_m N \rceil$ times. Compared with the method to send its information to all other processors, the tree-structured communication reduces the sending times for each processor but increases the iterations to apply parallel communication.
- **Butterfly Communication**: When all N processors need to broadcast their information to all other processors simultaneously, each processor can follow the tree-structured communication and formulate the butterfly communication structure. In butterfly communication, as shown in Fig. 12.4, each processor first sends its information to its neighbors, who will further accumulate and forward the information to all other processors. As each node can collect and accumulate the information before transmitting to all other processors, the efficiency is further improved in distributed fashion. In general, it takes $\lceil \log_m N \rceil$ iterations to broadcast the information to all processors and each processor only requires to send $(m - 1)\lceil \log_m N \rceil$ times. Moreover, whenever there is a node breakdown in the middle of communication, all other nodes are still able to continue and let the information synchronized on all other processors.

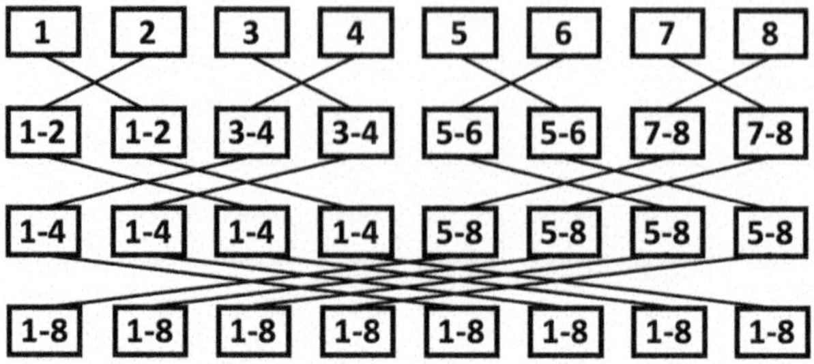

Fig. 12.4 Butterfly communication

Based on different communication structures, the parallel computation and transmission for reinforcement learning algorithm can be widely diverse and flexible. For different applications, the system architecture can be different to improve the parallelization and efficiency. In the next section, we will further summarize the general distributed computing architecture in deep reinforcement learning.

12.4 Distributed Reinforcement Learning Algorithms

12.4.1 Asynchronous Advantage Actor-Critic

Asynchronous advantage actor-critic (A3C) (Mnih et al. 2016) is the distributed algorithm derived from the advantage actor-critic (A2C) method. As shown in Fig. 12.5, there are multiple actor-learners interacting with separated but identical environments by applying A2C algorithm. Each actor-learner maintains a policy network and a value network to make smart actions. For the initialization and synchronization of the network parameters for all actor-learners, a parameter server is established, supporting asynchronous communications with all actor-learners.

From the perspective of each actor-learner, we elaborate the learning algorithm in Algorithm 1. For each learning episode, each actor-learner initially obtains the network parameters from the parameter server asynchronously. Based on the synchronized policy network, the actor-learner chooses actions and interacts with the environment for at most $t_{\max}$ steps. The explored experience is collected to train the policy and value network, generating the accumulated gradients θ and $d\theta_v$, respectively. After $T_{\max}$ steps of exploration, the actor-learner reports the accumulated gradients to the parameter server and updates the general network parameters θ and θ_v asynchronously.

Algorithm 1 Asynchronous advantage actor-critic (Actor-Learner)

Hyperparameters: Total number of steps T_{max}. Maximum steps for each episode t_{max}.
Initialize step counter $t = 1$.
while $T \leq T_{max}$ **do**
  Reset gradients: $d\theta = 0$ and $d\theta_v = 0$.
  Sync with parameter server to obtain network parameters $\theta' = \theta$ and $\theta'_v = \theta_v$.
  $t_{start} = t$
  Set starting state S_t for the episode
  while Reach terminal state **or** $t - t_{start} == t_{max}$ **do**
    Choose action a_t based on policy $\pi(S_t|\theta')$
    Act in the environment and receive rewards R_t and next state S_{t+1}
    $t = t + 1, T = T + 1$
  end while
  if Reach terminal state **then**
    $R = 0$
  else
    $R = V(S_t|\theta'_v)$
  end if
  for $i = t - 1, t - 2, \ldots, t_{start}$ **do**
    Update discounted rewards $R = R_i + \gamma R$
    Accumulate gradients wrt θ', $d\theta = d\theta + \nabla_{\theta'} \log \pi(S_i|\theta')(R - V(S_i|\theta'_v))$
    Accumulate gradients wrt θ'_v, $d\theta_v = d\theta_v + \partial(R - V(S_i|\theta'_v))^2/\partial\theta'_v$
  end for
  Asynchronously update θ with $d\theta$ and θ_v with $d\theta_v$.
end while

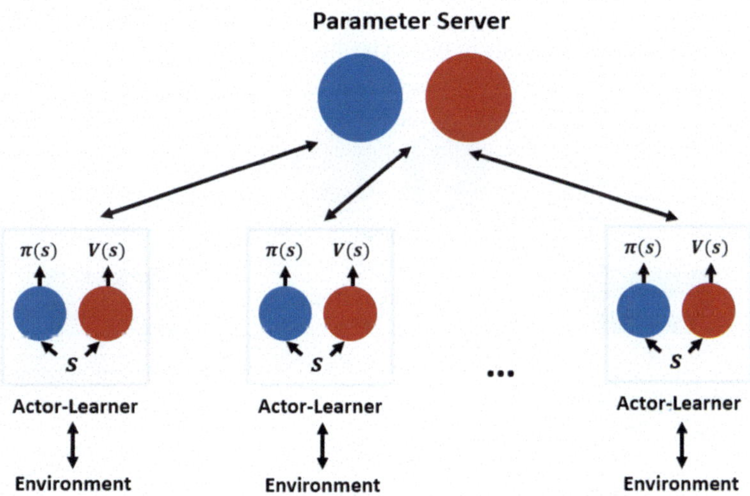

Fig. 12.5 A3C architecture

12.4.2 Hybrid GPU/CPU A3C

In order to better leverage the GPU's computational power and improve the computation efficiency, A3C architecture is further optimized to the hybrid GPU/CPU A3C (GA3C) (Babaeizadeh et al. 2017). As depicted in Fig. 12.6, from the environments or simulators to the learning model, there exist components of agent, predictor, and trainer. The functionality for each component is shown as follows.

- **Agent**: There are multiple agents interacting with their simulated environments, respectively. Each agent does not need to maintain a policy network for decision making. Instead, based on the current state S_t, the agent pushes one request to the prediction queue and lets the predictor assist to choose actions from the general policy network. After the action A_t is taken and the reward R_t and next state S_{t+1} are provided from the environment, the agent submits the experience (S_t, A_t, R_t, S_{t+1}) to the training queue for the model training.
- **Predictor**: The predictor collects the requests from the agent in the prediction queue, batches the requests, and sends to the general policy network for model inference. The batched input data for model inference takes the advantage of the parallel computation in GPU, improving the computation efficiency of the learning model. Based on the number of requests, multiple predictors with multiple prediction queues are supported to balance the trade-off of computation latency and computation efficiency.
- **Trainer**: Receiving the experiences from multiple agents, the trainer collects the data from the training queue, batches the training data, and sends to the general policy and value network for model training. The batched model training improves the computation efficiency with GPU and moreover reduces the variance and fluctuations in model training.

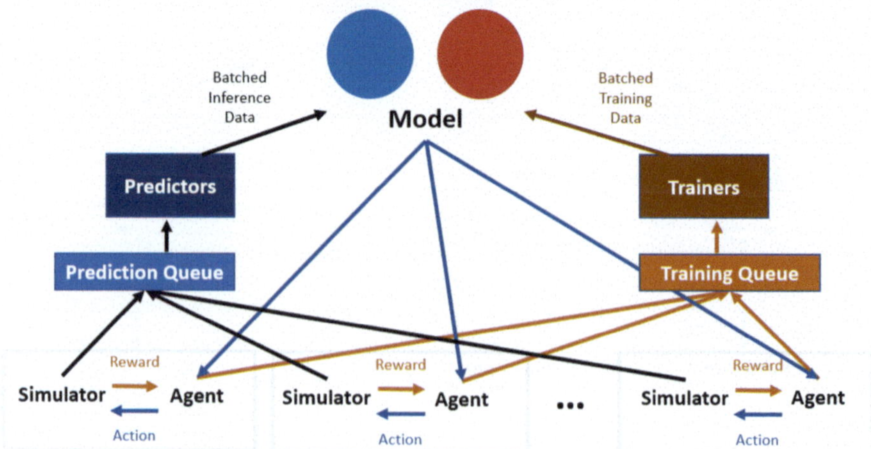

Fig. 12.6 GA3C architecture

12.4.3 Distributed Proximal Policy Optimization

Distributed Proximal Policy Optimization (DPPO) is a distributed version of the PPO algorithm. As depicted in Fig. 12.7, the algorithm includes the chief as the parameter server and workers the same as actor-learners in A3C. It distributes data collection and gradient calculation over multiple workers, which greatly reduces the learning time. Periodically, the chief updates parameters after averaging gradients passed by workers, and then passes the latest parameters to workers synchronously.

The pseudocode of the DPPO algorithm is provided in Algorithms 2, 3, and 4, corresponding to one chief and two different workers. Workers can be one of the two versions of PPO algorithm: PPO-Penalty and PPO-Clip. This section provides the corresponding two DPPO algorithms: DPPO-Penalty and DPPO-Clip. The only difference is the way in which the workers calculate the gradients, while the chief part is the same, as shown in the pseudocode.

The chief collects gradients from workers and update parameters. As shown in Algorithm 2, during each iteration, the chief waits for at least $(W - D)$ gradients from workers and updates with the averaged gradients. The latest parameters are returned to workers to continue the sampling and gradients-calculation process. At each iteration, M and B sub-iterations are performed on actor and critic, respectively.

Workers collect data samples and calculate gradients, then pass the gradients to the chief. Algorithms 3 and 4 have a similar process, except for the methods of calculating the policy gradient. At each iteration, the worker first collects a bunch of data $\mathcal{D}_k$, calculates $\hat{G}_t$ and $\hat{A}_t$, stores π_θ as π_{old}, and then performs M and B sub-iterations on actor and critic, respectively.

In DPPO-Clip, the parameter λ is also shared across workers, but its updates are determined based on local average KL divergence. Other statistical values for

Fig. 12.7 DPPO architecture

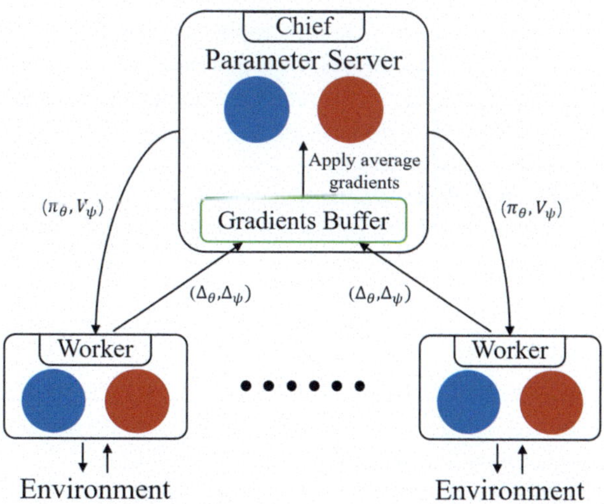

Algorithm 2 DPPO (chief)

Hyperparameters: the number of workers W, threshold for numbers of gradients available workers D, the number of sub-iterations M, B
Input: initial global policy parameters θ, initial global value function parameters ϕ
for k = 0, 1, 2, ... **do**
  for $m \in \{1, \ldots, M\}$ **do**
    Wait until at least $W - D$ gradients wrt. θ are available average gradients and update global θ
  end for
  for $b \in \{1, \ldots, B\}$ **do**
    Wait until at least $W - D$ gradients wrt. ϕ are available average gradients and update global ϕ
  end for
end for

normalization in data collection are also recommended to be shared among workers, like means and standard deviations of observations, rewards, and advantages. An additional penalty term is also adopted in DPPO-Clip when the KL divergence exceeds the valid change. Early stopping is also used during each sub-iteration on the actor to improve stability.

12.4.4 IMPALA and SEED

Based on the advantage actor-critic (A2C) learning algorithm, the Importance Weighted Actor-Learner Architecture (IMPALA) (Espeholt et al. 2018) applies the trajectory experiences of the agents as the communication information for distributed computation. As shown in Fig. 12.8, the IMPALA architecture is composed of actors and learners, with the detailed introductions as follows.

- **Actor**: Within each actor, a replicated policy network interacts with a simulated environment and stores the experience into the buffer. After certain number of interactions, each actor sends the trajectory of stored experiences to the learners and receive the updates of policy network parameters from the learners in a synchronized fashion.
- **Learner**: When interacting with the actor, the learner receives the trajectory experiences of the actors and applies it for model training. The value approximation at state S_T is defined as the n-step V-trace target, as follows:

$$\text{Target} = V(S_T) + \textstyle\sum_{t=T}^{T+n-1} \gamma^{t-T} \left(\Pi_{i=T}^{t-1} c_i\right) \delta_t V, \tag{12.1}$$

  where $\delta_t V = \rho_t (R_t + \gamma V(S_{t+1}) - V(S_t))$ is the temporal difference. $\rho_t = \min(\bar{\rho}, \frac{\pi(S_t)}{\mu(S_t)})$. $c_i = \min(\bar{c}, \frac{\pi(S_i)}{\mu(S_i)})$. π is the learner policy, which is averagely several updates ahead of the actor's policy μ.

Algorithm 3 DPPO (PPO-Penalty worker)

Hyperparameters: KL penalty coefficient λ, adaptive parameters $a = 1.5, b = 2$, the number of sub-iterations M, B
Input: initial local policy parameters θ, initial local value function parameters ϕ
for k = 0, 1, 2, ... **do**
  Collect set of trajectories $\mathcal{D}_k = \{\tau_i\}$ by running policy π_θ in the environment
  Compute rewards-to-go $\hat{G}_t$
  Compute advantage estimates, $\hat{A}_t$(using any method of advantage estimation) based on the current value function V_{ϕ_k}
  Store partial trajectory information
  $\pi_{\text{old}} \leftarrow \pi_\theta$
  for $m \in \{1, \ldots, M\}$ **do**

$$J_{PPO}(\theta) = \sum_{t=1}^{T} \frac{\pi_\theta(A_t|S_t)}{\pi_{\text{old}}(A_t|S_t)} \hat{A}_t - \lambda \text{KL}[\pi_{\text{old}}|\pi_\theta] - \xi \max(0, \text{KL}[\pi_{\text{old}}|\pi_\theta] - 2\text{KL}_{\text{target}})^2$$

    if $\text{KL}[\pi_{\text{old}}|\pi_\theta] > 4\text{KL}_{\text{target}}$ **then**
      break and continue with next outer iteration $k + 1$
    end if
    Compute $\nabla_\theta J_{PPO}$
    send gradient wrt. θ to chief
    wait until gradient accepted or dropped; update parameters
  end for
  for $b \in \{1, \ldots, B\}$ **do**
    $L_{BL}(\phi) = -\sum_{t=1}^{T}(\hat{G}_t - V_\phi(S_t))^2$
    Compute $\nabla_\phi L_{BL}$
    send gradient wrt. ϕ to chief
    wait until gradient accepted or dropped; update parameters
  end for
  Compute $d = \hat{\mathbb{E}}_t\,[\text{KL}[\pi_{\text{old}}(\cdot|S_t), \pi_\theta(\cdot|S_t)]]$
  if $d < d_{\text{target}}/a$ **then**
    $\lambda \leftarrow \lambda/b$
  else if $d > d_{\text{target}} \times a$ **then**
    $\lambda \leftarrow \lambda \times b$
  end if
end for

Moreover, there can be multiple learners, separated as worker learners and master learner. Each learner interacts with different actors and finishes model training independently. Periodically, all worker learners communicate with the master learner with learning gradients and the master announces the update of the network parameters synchronously.

The Scalable, Efficient, Deep-RL (SEED) architecture (Espeholt et al. 2019) is closely related with the IMPALA. The key difference is that the inference policy network is moved from the actor to the learner, which reduces the computation requirement for the actor and decreases the communication latency. The detailed SEED architecture is shown in Fig. 12.9. As each actor implements one or multiple

Algorithm 4 DPPO (PPO-Clip worker)

Hyperparameters: clip factor ϵ, the number of sub-iterations M, B
Input: initial local policy parameters θ, initial local value function parameters ϕ
for k = 0, 1, 2, ... **do**
  Collect set of trajectories $\mathcal{D}_k = \{\tau_i\}$ by running policy π_θ in the environment
  Compute rewards-to-go $\hat{G}_t$
  Compute advantage estimates, $\hat{A}_t$(using any method of advantage estimation) based on the current value function V_{ϕ_k}
  Store partial trajectory information
  $\pi_{\text{old}} \leftarrow \pi_\theta$
  for $m \in \{1, \ldots, M\}$ **do**
    Update the policy by maximizing the PPO-Clip objective:

$$J_{PPO}(\theta) = \frac{1}{|\mathcal{D}_k|T} \sum_{\tau \in D_k} \sum_{t=0}^{T} \min\left(\frac{\pi_\theta(A_t|S_t)}{\pi_{\text{old}}(A_t|S_t)}\hat{A}_t, \text{clip}(\frac{\pi(A_t|S_t)}{\pi_{\text{old}}(A_t|S_t)}, 1-\epsilon, 1+\epsilon)\hat{A}_t\right)$$

    Compute $\nabla_\theta J_{PPO}$
    send gradient wrt. θ to chief
    wait until gradient accepted or dropped; update parameters
  end for
  for $b \in \{1, \ldots, B\}$ **do**
    Fit value function by regression on mean-squared error:

$$L_{BL}(\phi) = -\frac{1}{|\mathcal{D}_k|T} \sum_{\tau \in \mathcal{D}_k} \sum_{t=0}^{T} \left(V_\phi(S_t) - \hat{G}_t\right)^2$$

    typically via some gradient descent algorithm
    send gradient wrt. ϕ to chief
    wait until gradient accepted or dropped; update parameters
  end for
end for

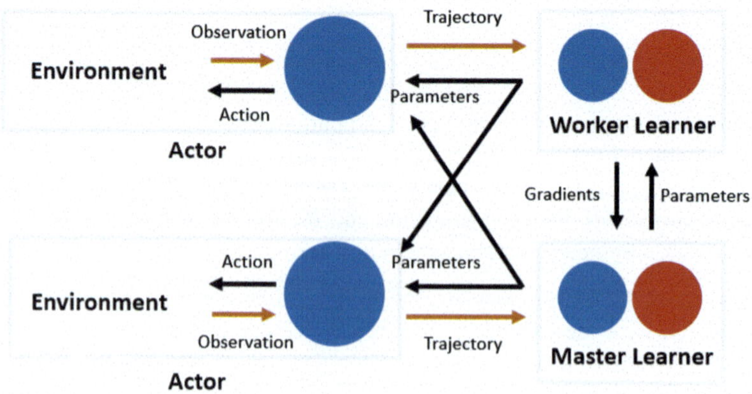

Fig. 12.8 IMPALA architecture

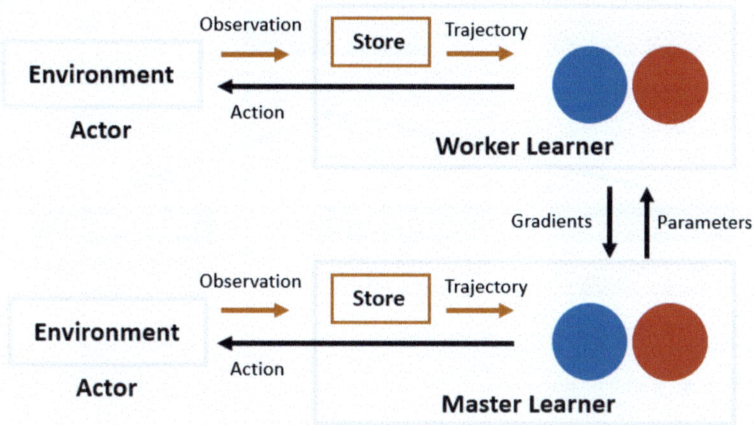

Fig. 12.9 SEED architecture

environments only, all kinds of machines with weak computation power can be regarded as the actor in the architecture. Based on the instructed actions from the learner, the actor provides one-step feedback experience to the learner and the experience is stored in the experience buffer within the learner. After several iterations, the trajectory data is applied to model training, where V-trace target in Eq. (12.1) is also applied as the value approximation.

12.4.5 Ape-X, Reactor, and R2D2

In distributed network, considering the multiple interactions between agents and environments, it is scalable and beneficial to extend prioritized experience replay to the architecture. Ape-X (Horgan et al. 2018) is the typical distributed architecture including prioritized experience replay. As shown in Fig. 12.10, there exist multiple independent actors. Within each actor, an agent interacting with an environment with the guidance from the policy network. Based on the experience collected from multiple actors, the learner train the network parameters and learn the optimal policy. Most importantly, apart from the actor and learner, there exists a replay buffer collecting the experience from actors, updating the priorities of each experience entry and batching the prioritized ones to the learner for model training. The batched prioritized experiences improve the computation efficiency and model learning performance.

The algorithms from the perspective of each actor are elaborated in Algorithm 5. Each actor initially synchronizes with the learner on network parameters. The updated parameters then instruct the agent to interact with the environment. Receiving the feedback from the environment, the actor calculates the priorities of the explored experience and sends both the data and priority information to the replay buffer.

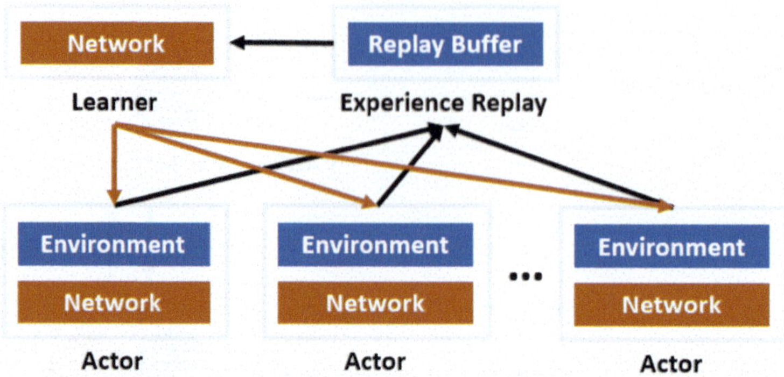

Fig. 12.10 Ape-X architecture

Algorithm 5 Ape-X (Actor)

Hyperparameters: Send to replay with batch size B in local buffer. Number of iterations T
Sync with learner to obtain latest network parameters θ_0.
Get initial state S_0 from environment.
for t $= 0, 1, 2, \ldots, T-1$ **do**
  Choose action A_t based on policy $\pi(S_t|\theta_t)$
  Add experience (S_t, A_t, R_t, S_{t+1}) to the local buffer
  if The local buffer reaches its size requirements B **then**
    Get buffered data with batch size B
    Calculate the priorities p of the buffered data.
    Send the batched buffered data and its priorities to the replay
  end if
  Periodically sync and update the latest network parameters θ_t
end for

When the replay buffer collects certain amount of experiences from the actors, the learner interacts with the replay buffer for learning. The algorithm from the perspective of the learner is shown in Algorithm 6. For each episode of model learning, the learner firstly samples prioritized batch of experience data from the replay buffer. Each data entry is represented as (i, d), where i denotes the index of the data and d is the detailed information of the experience including state, action, reward, and next state. The batched data is employed to train network parameters of the learner, which will periodically synchronize with network parameters of all actors. After model training, the priorities of the sampled data are adjusted and updated in replay buffer. Due to the limits of the replay buffer size, periodically, the data with low priorities will be removed in replay buffer.

Following the general architecture, Ape-X DQN and Ape-X DPG are proposed when the learning model follows the DQN and DPG algorithm, respectively. In Ape-X DQN, the Q-network exists in the learner and all actors. The actions for the actor are guided with the Q values from the network. In Ape-X DPG, both policy

Algorithm 6 Ape-X (Learner)

Hyperparameters: Number of learning episodes T.
Initialize the network parameters θ_0.
for $t = 1, 2, 3, \dots, T$ **do**
  Sample a prioritized batch of data (i, d) from replay
  Applying training with the batched data
  Update network parameters to θ_t
  Calculate the priorities p for batched data d
  Update the priorities p for data with index i on replay
  Periodically remove data with low priorities in replay
end for

network and value network exist in learner, while each actor only replicates the policy network to instruct the actions.

Built upon the prioritized distributed replay, Retrace-Actor (Reactor) (Gruslys et al. 2017) is further put forward based on the actor-critic architecture. Instead of the single experience, the sequence of experiences are pushed into the buffer and distributional Retrace(λ) algorithm is implemented to update the estimation of Q values. In the perspective of the neural network, the LSTM network is added in both policy and value network for better model learning.

Similarly, Recurrent Replay Distributed DQN (R2D2) (Kapturowski et al. 2018) applies fixed-length sequence of experiences in prioritized distributed replay. Developed from the DQN, R2D2 implements LSTM layer in the network and train the LSTM from replay with stored states.

12.4.6 Gorila

Implemented from the Deep Q-Network algorithm, general reinforcement learning architecture (Gorila) (Nair et al. 2015) is depicted in Fig. 12.11. Synchronizing the parameters of deep Q-network from the parameter server, the actors interact with the environment based on the policy instructed by the deep Q-network. The experiences received from the interactions are instantly forward to the replay buffer. The replay buffer stores and maintains the collected experience from all actors. Fetching the batched experience data from the replay buffer, the learner applies model learning and calculates the gradients of the Q-network. Within the learner, there are one learning Q-network and one target Q-network to calculate the TD-error. The learning Q-network sync with the parameter sever for each step of learning, while the target Q-network sync with the parameter server every N steps. Periodically, the parameter server receives the gradients from the learner and updates the network parameters for future explorations.

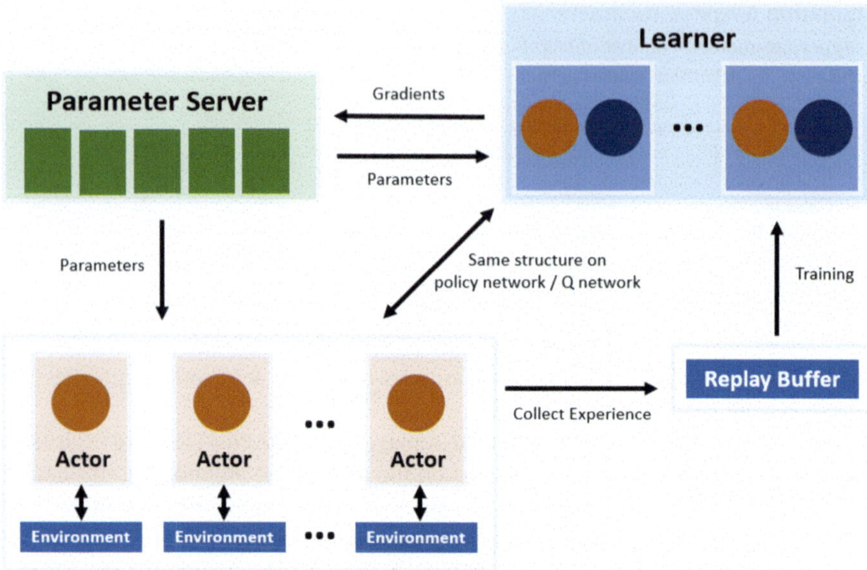

Fig. 12.11 Gorila architecture

12.5 Distributed Computing Architecture

Based on the basic patterns and structures in parallel computing, in distributed reinforcement learning, the large-scale parallel computing architecture can be further explored and investigated. Generally, the system can be composed of the following basic components:

- **Environments**: The environment is the component the agent interact with. In large-scale parallel computing of deep reinforcement learning, the environment can have multiple replicas which are mapped to different replicas of actors to gain experience in parallel fashion. Moreover, in model-based reinforcement learning, the model can also be regarded as simulated environment in the system to help learning in parallelization.
- **Actors**: The actor concept in the system refers to the component directly interacting with the environment. There can be multiple actors mapping to single or multiple real or simulated environments, and each actor is able to make actions independently with each other in the assigned environment. The action is determined by its own or shared policy network or Q-network from the parameter servers or its corresponding learners. With multiple actions applied sequentially in the environment, trajectories are formed, which will be pushed into the replay memory buffers or directly fed into the learners. As interactions between actor and environment can be costly in time, the parallelization on actors can improve

the speed to generate experiences and contributes to the training performance for learners.

- **Replay Memory Buffers**: The replay memory buffer is the component to collect the planning trajectories from actors and provide to learners for policy learning or Q learning. As the memory buffer requires to perform fast data writing, shuffling, and data reading, the storage structure should also support in dynamic and parallel fashion. Moreover, as most learners reply on the data in replay memory buffer for training, it is recommended to allocate reply memory buffers closely connected with learners, so as to guarantee the learning efficiency.
- **Learners**: The learners are the key component for deep reinforcement learning. Based on different deep reinforcement learning algorithms, the structure for each learner will be different. Normally, each learner maintains a policy network or Q-network and trains the deep network weights based on the actors' experience in replay memory buffers. Before and after training, the learner will communicate with parameter servers for synchronization on deep network weights or its learning gradients. Either synchronous or asynchronous communication approach can be applied between multiple learners and parameter servers.
- **Parameter Servers**: The parameter server is the component to collect all information from the learners and maintain the weights of policy network or Q-network in general. The parameter server will periodically synchronize with all learners for weight updates and assist all learners to start learning based on the training results from other learners. Moreover, the parameter server can instruct the actors for its interaction with the environments. In large-scale deep reinforcement learning system, in order to guarantee robustness and efficiency for data interactions from parameter servers to learners and actors, the parameter severs can also be in different kinds of structures and the communication among parameter servers can be centralized or distributed.

The general computing architecture is combined based on the above components. As parallel computing is applicable within each component, the structure can be flexible and easily adapted for the requirements from problems.

References

Babaeizadeh M, Frosio I, Tyree S, Clemons J, Kautz J (2017) Reinforcement learning through asynchronous advantage actor-critic on a GPU. In: International conference on learning representations

Espeholt L, Soyer H, Munos R, Simonyan K, Mnih V, Ward T, Doron Y, Firoiu V, Harley T, Dunning I, et al (2018) IMPALA: scalable distributed deep-RL with importance weighted actor-learner architectures. Preprint. arXiv:180201561

Espeholt L, Marinier R, Stanczyk P, Wang K, Michalski M (2019) SEED RL: scalable and efficient Deep-RL with accelerated central inference. Preprint. arXiv:191006591

Gruslys A, Dabney W, Azar MG, Piot B, Bellemare M, Munos R (2017) The reactor: a fast and sample-efficient actor-critic agent for reinforcement learning. Preprint. arXiv:1704.04651

Horgan D, Quan J, Budden D, Barth-Maron G, Hessel M, van Hasselt H, Silver D (2018) Distributed prioritized experience replay. Preprint. arXiv:1803.00933

Kapturowski S, Ostrovski G, Quan J, Munos R, Dabney W (2019) Recurrent experience replay in distributed reinforcement learning. In: International Conference on Learning Representations (ICLR).

Mnih V, Badia AP, Mirza M, Graves A, Lillicrap T, Harley T, Silver D, Kavukcuoglu K (2016) Asynchronous methods for deep reinforcement learning. In: International conference on machine learning (ICML), pp 1928–1937

Nair A, Srinivasan P, Blackwell S, Alcicek C, Fearon R, Maria AD, Panneershelvam V, Suleyman M, Beattie C, Petersen S, Legg S, Mnih V, Kavukcuoglu K, Silver D (2015) Massively parallel methods for deep reinforcement learning. Preprint. arXiv:1507.04296

OpenAI: Berner C, Brockman G, Chan B, Cheung V, Dębiak P, Dennison C, Farhi D, Fischer Q, Hashme S, Hesse C, Józefowicz R, Gray S, Olsson C, Pachocki J, Petrov M, de Oliveira Pinto HP, Raiman J, Salimans T, Schlatter J, Schneider J, Sidor S, Sutskever I, Tang J, Wolski F, Zhang S (2019) Dota 2 with large scale deep reinforcement learning. Preprint. arXiv:1912.06680

Part III
Applications

Zihan Ding
e-mail: zhding@mail.ustc.edu.cn

To help the readers deeply understand DRL and quickly apply those widely used techniques in practice, the following chapters introduce five selected applications or large-scale projects, including the learning to run challenge, image enhancement, AlphaZero, robot learning in simulation, and multi-agent reinforcement learning with Arena platform. The applications are selected to cover as much usage as possible, for helping the readers to understand the details and tricks of implementation in different scenarios. Table 1 lists the applications along with the algorithms, type of policy, action space, and observation, which we believe would be useful for you to find the application that can be reused or extended to other applications.

The related codes with environment supports and algorithm implementations are released in the following link: https://github.com/deep-reinforcement-learning-book.

Table 1 Summary of application projects in the book

Applications	Algorithms	Action space	Observation
Learning to run	SAC	Continuous	Continuous
Image enhancement	PPO	Discrete	Image features
AlphaZero	Monte Carlo tree search	Discrete	Binary Chessboard matrix
Robot learning	SAC	Continuous	Continuous
Multi-agent RL	MADDPG, etc.	Any	Any

Chapter 13
Learning to Run

Zihan Ding and Hao Dong

Abstract In this chapter, we provide a practical project for readers to have some hands-on experiences of deep reinforcement learning applications, in which we adopt one challenge hosted by CrowdAI and NIPS (now NeurIPS) 2017: `Learning to Run`. The environment has a 41-dimension state space and 18-dimension action space, both continuous, which is a moderately large-scale environment for novices to gain some experiences. We provide a soft actor-critic solution for the task, as well as some tricks applied for boosting performances. The environment and code are available at https://github.com/deep-reinforcement-learning-book/Chapter13-Learning-to-Run.

Keywords Learning to run · Deep reinforcement learning · Soft actor-critic · Parallel training

13.1 NIPS 2017 Challenge: Learning to Run

13.1.1 Introduction of the Environment

`Learning to Run` is a competition hosted by CrowdAI and NIPS 2017, which attracts a lot of reinforcement learning researchers to participate in. In this task, the participants are required to develop a controller to enable a physiological human model to navigate a complex obstacle course as quickly as possible. It provides a human musculoskeletal model and a physics-based simulation environment. To model physics and biomechanics and navigate it with reinforcement learning agent,

Z. Ding (✉)
Imperial College London, London, UK
e-mail: zhding@mail.ustc.edu.cn

H. Dong
Peking University, Beijing, China
e-mail: hao.dong@pku.edu.cn

H. Dong et al. (eds.), *Deep Reinforcement Learning*,
https://doi.org/10.1007/978-981-15-4095-0_13

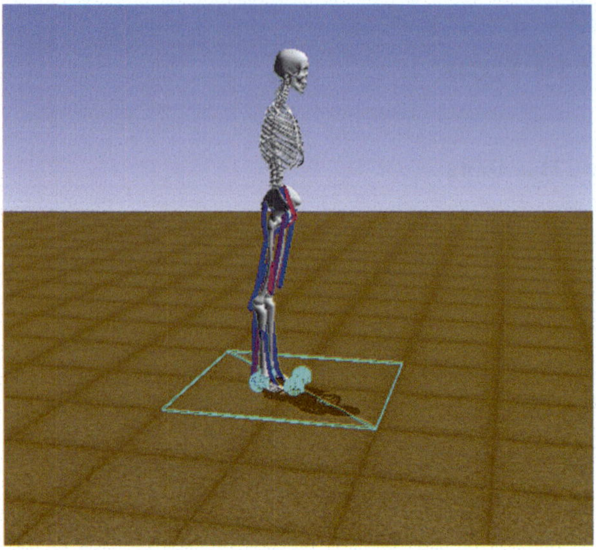

Fig. 13.1 Scene of NIPS 2017 Challenge: Learning to Run environment

the environment *osim-rl* based on *OpenSim* library is provided, where *OpenSim* is a standard biomechanical physics environment for musculoskeletal simulations. The environment scene containing the agent is shown in Fig. 13.1.

The environment incorporates a musculoskeletal model that includes body segments for each leg, a pelvis segment, and a single segment to represent the upper half of the body (trunk, head, arms). The segments are connected with joints (e.g., knee and hip) and the motion of these joints is controlled by the excitation of muscles. The muscles in the model have complex paths (e.g., muscles can cross more than one joint and there are redundant muscles). The muscle actuators themselves are also highly nonlinear. The agent operates in a 2D world. To make it easier to understand and operate, the environment used in our project is simplified compared with the original one used in the challenge, so there may be some slight differences from the official documents of the competition. All components of the skeleton model are shown as follows.

- The observation contains 41 values:
  - position of the pelvis (rotation, x, y)
  - velocity of the pelvis (rotation, x, y)
  - rotation of each ankle, knee, and hip (6 values)
  - angular velocity of each ankle, knee, and hip (6 values)
  - position of the center of mass (2 values)
  - velocity of the center of mass (2 values)
  - positions (x, y) of head, pelvis, torso, left and right toes, left and right talus (14 values)

  - strength of left and right psoas: value is 1.0 for difficulty level lower than 2 (a parameter of environment, default), otherwise a random normal variable with mean 1.0 and standard deviation 0.1 fixed for the entire simulation. (Note: in our simplified environment, these strength values are all set to be 0.0)
  - next obstacle: x-distance from the pelvis, y-position of the center relative to the ground, radius. (Note: in our simplified environment there are not obstacles, so these values are all set to be 0.0)

- The action contains 18 scalar values for representing the actuation of 18 muscles (9 per leg):
  - hamstrings,
  - biceps femoris,
  - gluteus maximus,
  - iliopsoas,
  - rectus femoris,
  - vastus,
  - gastrocnemius,
  - soleus,
  - tibialis anterior.

- The reward function:
  - The reward is computed as a change in position of the pelvis along the x-axis minus the penalty for the use of ligaments.

- Other details:
  - The "done" signal indicates if the move was the last step of the environment. This happens if either 1000 iterations were reached or the pelvis height is below 0.65 m.

As we can see from the above environment descriptions, the environment in the competition is a quite complicated task with both high-dimensional observation space and action space, compared with other games in OpenAI Gym[1] or DeepMind Control Suite.[2] Therefore, some specific techniques are required for solving the task with a good performance, as well as a relatively short period of training. We will introduce those specific methods and parallel training framework applied for solving the task. Since we provide a duplicate of the game environment and the solutions in the repository[3] together with the book, we also recommend the readers to take a hands-on practice with the project.

---

[1]OpenAI Gym website: https://gym.openai.com/.

[2]DeepMind Control Suite environments: https://github.com/deepmind/dm_control.

[3]https://github.com/deep-reinforcement-learning-book.

13.1.2 Installation

The environment can be installed with following commands according to the official repository[4]:

1. Create a conda environment (named opensim-rl) with the OpenSim package.

```
conda create -n opensim-rl -c kidzik opensim python=3.6.1
```

2. Activate the conda environment we just created.
   On Windows, run:

```
activate opensim-rl
```

   On Linux/OSX, run:

```
source activate opensim-rl
```

   You need to type in the above command every time you open a new terminal.
3. Install our Python reinforcement learning environment.

```
conda install -c conda-forge lapack git
pip install osim-rl
```

The challenges are hosted every year after 2017, though, the original `Learning to Run` environment has been deprecated as the update of the environment package. However, we still choose to employ the original environment of 2017 version for its simplicity here. We provide a repository for hosting the environment of 2017 version in our project:

```
git clone
    https://github.com/deep-reinforcement-learning-book/Chapter13
    -Learning-to-Run.git
```

Our codes of the reinforcement learning algorithms and environment wrappers are also provided in the above repository.

Through the above steps, we have completed the installation of the environment. You can verify the success of the installation with the following comments.

```
python -c "import opensim"
```

If it runs smoothly, the installation is successful. Otherwise, solutions can be found at: http://osim-rl.stanford.edu/docs/faq/

[4]https://github.com/stanfordnmbl/osim-rl.

To execute 200 iterations of the simulation with random rollout, we can enter the Python interpreter and run the following (on Linux):

```
from osim.env import RunEnv # load package
env = RunEnv(visualize=True) # initialize environment
observation = env.reset(difficulty = 0) # reset the environment
for i in range(200): # rollout
    observation, reward, done, info =
        env.step(env.action_space.sample())
```

This environment is user-friendly as it has already been formalized as an OpenAI Gym game, with a pre-defined reward function. Our task is to derive a function which takes the current state observation (a 41 dimensional vector) and returns the muscle excitation action (18 dimensional vector) in a way that maximizes the reward. As mentioned previously, the reward function is defined as the change of the x-axis of pelvis during each iteration minus the magnitude of the ligament forces used in that iteration, which tries to encourage the agent to move forward with minimal loss in its body.

13.2 Training an Agent to Run

In order to solve this task and obtain a great performance, a bunch of tricks are needed in the implementation of the training framework, including:

- a parallel training framework for balancing CPU and GPU resources;
- reward scaling;
- exponential linear unit (ELU) activation function;
- layer normalization;
- action repetition;
- update repetition;
- observation normalization and action discretization can be useful but we do not use it in our solution;
- data augmentation w.r.t. the symmetry of the agent's legs is also helpful but we do not use it in our solution.

We will introduce those methods one by one in the following sections. Note that the last two tricks are also potentially useful according to the experiments and reports of teams participated in the competition, but we do not apply them in our solution as they are more of task-specific methods and less general to be applied in other tasks. However, it is worth knowing that the observation normalization, action discretization, and data augmentation can accelerate the learning process in general cases.

A typical drawback of this environment is its slow simulation speed. It takes at least dozens of seconds to finish the simulation of a single episode on a general

CPU. In order to learn a policy efficiently, we need to parallelize the sampling and training process.

13.2.1 Parallel Training

There are at least two reasons to conduct parallel training for this task. The first one is the low simulation speed of the `Learning to Run` environment as described above. It takes at least dozens of seconds to finish the simulation of a single episode. The second one would be its high intrinsic complexity. Based on the experience of the authors, the environment needs at least hundreds of CPU/GPU hours with general model-free reinforcement learning algorithms like deep deterministic policy gradient (DDPG) or soft actor-critic (SAC) to obtain a relatively good policy. Therefore, a training framework with multiprocessing across multiple GPUs is needed.

Due to the high complexity of the `Learning to Run` environment, the training process needs to be implemented in a parallel and distributed manner with multiple CPUs and GPUs. Moreover, the balance between CPUs and GPUs is also critical for this task, as the sampling process through interaction with environments is usually on CPUs and the back-propagation training process is usually on GPUs. The overall training efficiency satisfies the buckets effect in practice. The parallel training for balancing the CPUs and GPUs is also discussed in Chaps. 12 and 18. Here we apply one of the solutions for this task.

As shown in Fig. 13.2, the training function in normal single-process deep reinforcement learning is handled with a single process, which is usually not capable of using the computational resources to the best, especially when there are multiple CPUs and multiple GPUs.

Figure 13.3 shows a parallel framework for deploying off-policy deep reinforcement learning algorithms on multiple CPUs and multiple GPUs, where a single agent and single environment are wrapped into a "worker" to run with a single process. Multiple workers may share the same GPU, because a single worker may not be able to fully occupy the GPU memory. In this setting, the number of processes and number of workers sharing the same GPU can be manually configured to ensure a maximum usage of all computational resources in learning.

Our project provides a highly parallelized SAC algorithm following the above framework for handling this task using multiprocessing and multiple-GPU computation. Because the memory for multiple processes are not shared across each other,

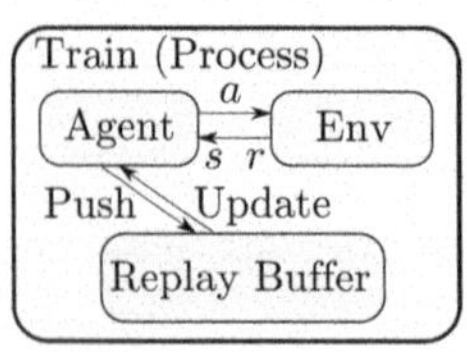

Fig. 13.2 Single-process training in off-policy deep reinforcement learning: only one process for sampling and policy training

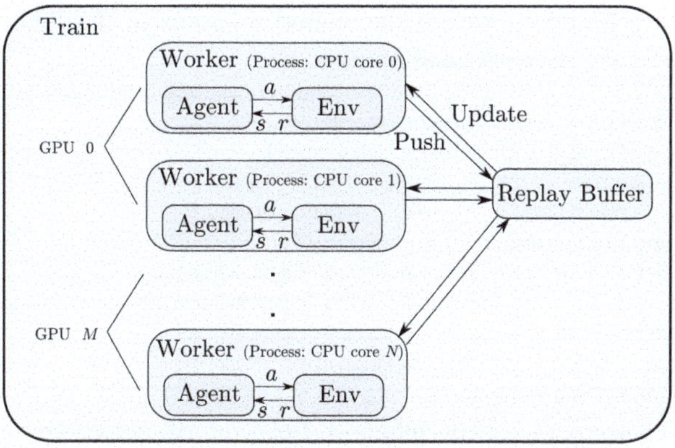

Fig. 13.3 A parallel training framework for off-policy deep reinforcement learning. Each worker contains an agent for interaction with the environment, and the policy is trained on distributed GPUs

specific modules are needed for handling the communication and parameter sharing functions. In the code, the replay buffer is shared with `multiprocessing` module within Python, and the networks and parameters updated during the training process are shared with the PyTorch `multiprocessing` module on Linux machine.

In practice, although we apply several workers with each containing an agent, the networks within the agents are actually shared across workers, therefore only one set of networks (for a single agent) are maintained actually. PyTorch `nn.Module` can handle the case when multiple processes try to update the network parameters with the shared memory. As Adam optimizer also has some statistical values maintained during training, we also share those values across processes with the following `ShareParameters` function:

```
def ShareParameters(adamoptim):
  ''' share parameters of Adam optimizers for multiprocessing '''
   for group in adamoptim.param_groups:
       for p in group['params']:
           state = adamoptim.state[p]
           # initialize: have to initialize here, or else cannot
                find
           state['step'] = 0
           state['exp_avg'] = torch.zeros_like(p.data)
           state['exp_avg_sq'] = torch.zeros_like(p.data)

           # share in memory
           state['exp_avg'].share_memory_()
           state['exp_avg_sq'].share_memory_()
```

In the training function, we set the shared modules in the SAC algorithm as follows, including the networks and optimizers:

```
sac_trainer.soft_q_net1.share_memory()
sac_trainer.soft_q_net2.share_memory()
sac_trainer.target_soft_q_net1.share_memory()
sac_trainer.target_soft_q_net2.share_memory()
sac_trainer.policy_net.share_memory()
ShareParameters(sac_trainer.soft_q_optimizer1)
ShareParameters(sac_trainer.soft_q_optimizer2)
ShareParameters(sac_trainer.policy_optimizer)
ShareParameters(sac_trainer.alpha_optimizer)
```

The `share_memory()` is an inherent function of networks inherited from `nn.Module` in PyTorch. We can share the entropy factor as well, but not do it here. The "forkserver" start method is used with Python 3 for CUDA subprocesses as follows in our code:

```
torch.multiprocessing.set_start_method('forkserver', force=True)
```

The replay buffer can be shared with `multiprocessing` of Python:

```
from multiprocessing.managers import BaseManager

replay_buffer_size = 1e6
BaseManager.register('ReplayBuffer', ReplayBuffer)
manager = BaseManager()
manager.start()
replay_buffer = manager.ReplayBuffer(replay_buffer_size) # share
    the replay buffer through manager
```

Run the following commands in the cloned file to start training (cannot work on Windows 10 for parallel training due to the "forkserver" start method):

```
python sac_learn.py --train
```

We can also test the trained model with:

```
python sac_learn.py --test
```

13.2.2 Tricks

However, even with above well paralleled framework we still cannot achieve a great performance in this task. Due to the complexity of the task and the non-linearity of the deep learning models, local optimums and non-smooth or even non-differentiable curvatures on the loss surface can trap the optimization process (for policy or value functions) easily. Fine-tuning strategy is usually needed in deep reinforcement learning methods, especially for complicated tasks like `Learning to Run`. So below we introduce some tricks we applied in our method for solving this task in a more efficient and stable way:

- Reward scaling: reward scaling is just following the normal scaling routine, which is dividing the reward values by the standard deviation of sampled batch during training. Reward scaling is a common technique in reinforcement learning to make the training process stable and therefore accelerate the convergence. As reported in the follow-up paper of SAC algorithm (Haarnoja et al. 2018), the maximum entropy reinforcement learning algorithms can be sensitive on the scaling of reward function, which is different from other conventional reinforcement learning algorithms. Therefore, the authors of SAC add a gradient-based temperature tuning module for the entropy regularization term, which significantly alleviates the difficulty of hyperparameter tuning in practice.
- The exponential linear unit (ELU) (Clevert et al. 2015) activation function is used instead of rectified linear unit (ReLU) (Agarap 2018): to leverage the faster learning and more generalized learning performance, we use ELU as activation functions for the policy networks. The ELU function is defined as:

$$f(x) = \begin{cases} x, & \text{if } x > 0 \\ \alpha \exp(x - 1), & \text{if } x \leq 0 \end{cases} \tag{13.1}$$

  Comparison of ELU and ReLU is shown in Fig. 13.4. Compared with ReLU, ELUs have negative values which allows them to push mean unit activations closer to zero like batch normalization but with lower computational complexity. The mean shifts toward zero speed up learning by bringing the normal gradient closer to the unit natural gradient because of a reduced bias shift effect.
- Layer normalization: we also use layer normalization (Ba et al. 2016) for each hidden layer in value networks and policy networks. Compared with batch normalization, layer normalization computes the mean and variance used for normalization from all of the summed inputs to the neurons in a layer on a single training case. The individual adaptive bias and gain are given to each neuron, and they are applied after the normalization but before the non-linearity. This method will help with accelerating the training speed in practice.
- Action repetition: we use a common trick called action repetition (or called frame-skipping) in our training process, to speed up the wall-clock training time. Original paper of DQN applies both frame-skipping and pixel-wise max operator for image-based learning on Atari 2600 Games. If we define the original observation for a single frame is o_i with i indicating its frame index, the inputs in original DQN paper are stacked 4 frames after maximization over two consecutive non-skipped frames $[\max(o_{i-1}, o_i), \max(o_{i+3}, o_{i+4}), \max(o_{i+7}, o_{i+8}), \max(o_{i+11}, o_{i+12})]$ with frame-skipping rate of 4 (actually 2, 3, or 4 for different games). Actions are repeated for those skipped frames. The max is applied pixel-wise on image observations and rewards are summed over all skipped

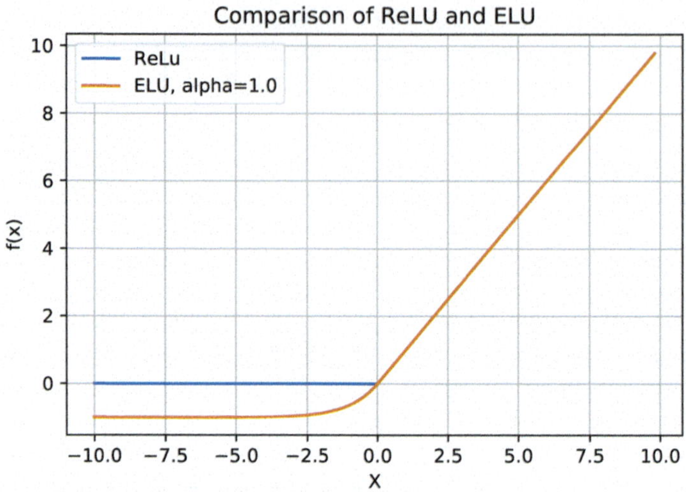

Fig. 13.4 Comparison of ReLU and ELU activation functions. ELU is differentiable at zero

and non-skipped frames. The original frame-skipping scheme in DQN increases the stochasticity as well as accelerating sampling. However, in our task, we use a different setting without max operator and stacked frames: the action is simply repeated over skipped frames, and all samples for each skipped and non-skipped frame are stored in the replay buffer. We use an action repetition rate of 3 in practice. This reduces the inference time of forwarding the policy during interaction with the environment.

- Update repetition: we also use the trick of repeatedly updating the policy during training with a smaller learning rate. So the policy is updated with a repetition rate of 3 on the same batch of samples.

13.2.3 Learning Results

With above settings and tricks applied in SAC algorithm, the agent is able to learn the skill of running in a human way for quite a long distance with 3 days of training time on 4-GPUs and 56-CPUs machine, as shown in Fig. 13.5. The learning curve is shown in Fig. 13.6, with both the original reward values and a moving-average smoothed curve in an increasing manner. The vertical axis is the cumulative reward in an episode, indicating the distance and gesture of running for the agent.

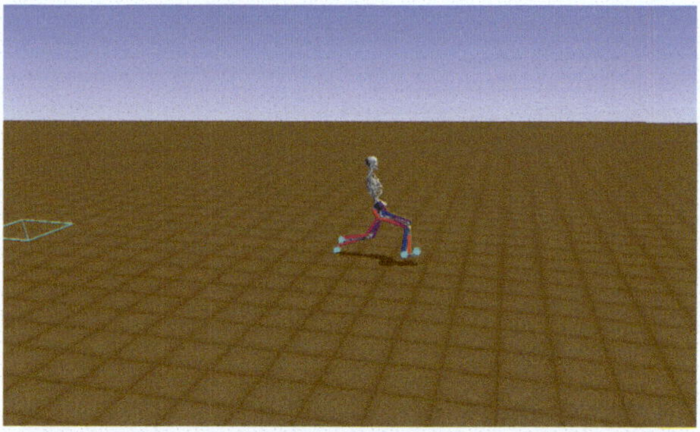

Fig. 13.5 The final performance of the running agent in task `Learning to Run`

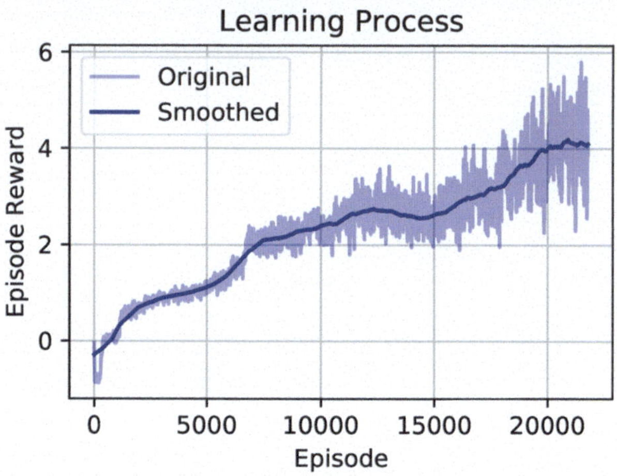

Fig. 13.6 The learning process of task `Learning to Run`

References

Agarap AF (2018) Deep learning using rectified linear units (ReLU). Preprint. arXiv:1803.08375

Ba JL, Kiros JR, Hinton GE (2016) Layer normalization. Preprint. arXiv:1607.06450

Clevert DA, Unterthiner T, Hochreiter S (2015) Fast and accurate deep network learning by exponential linear units (ELUs). Preprint. arXiv:1511.07289

Haarnoja T, Zhou A, Hartikainen K, Tucker G, Ha S, Tan J, Kumar V, Zhu H, Gupta A, Abbeel P, et al (2018) Soft actor-critic algorithms and applications. Preprint. arXiv:181205905

Chapter 14
Robust Image Enhancement

Yanhua Huang

Abstract Deep generative models such as GAN and Unet have achieved significant progress over classic methods in several computer vision tasks like super-resolution and segmentation. However, such learning-based methods lack robustness and interpretability, which limits their applications in real-world situations. In this chapter, we discuss a robust way for image enhancement that can combine a number of interpretable techniques through deep reinforcement learning. We first present some background about image enhancement. Then we formulate the image enhancement as a pipeline modeled by MDP. Finally, we show how to implement an agent on this MDP with PPO algorithm. The experimental environment is constructed by a real-world dataset that contains 5000 photographs with both the raw images and adjusted versions by experts. Codes are available at: https://github.com/deep-reinforcement-learning-book/Chapter14-Robust-Image-Enhancement.

Keywords Image processing · Image enhancement · Robust learning

14.1 Image Enhancement

Image enhancement belongs to image processing techniques. Its principal objective is to make the processed images more suitable for the needs of various applications. Typical image enhancement techniques contain denoising, deblurring, and brightness improvement. Real-world images always need multiple image enhancement techniques. Figure 14.1 shows an enhancement pipeline that consists of brightness improvements and denoising. Professional photo editing software, such as Adobe Photoshop, allows powerful image retouching but is not efficient and requires expertise in photo editing for users. In large-scale situations like recommendation systems, the subjective quality of images is vital for user experience, where an automatic image enhancement method that satisfies effectiveness, robustness, and

Y. Huang (✉)
Xiaohongshu Technology Co., Ltd., Shanghai, China

H. Dong et al. (eds.), *Deep Reinforcement Learning*,
https://doi.org/10.1007/978-981-15-4095-0_14

Fig. 14.1 An example of image enhancement pipeline. The raw image in the left is underexposed with JPEG compression noise

efficiency is needed. In particular, robustness is the most important condition, especially in user-generated content platforms, e.g., Facebook and Twitter, even if 1% of enhancement results are bad it will hurt millions of users.

Unlike image classification or segmentation that has a unique ground truth, the training data of image enhancement relies on human experts. As a result, no large-scale public dataset for image enhancement is available. Classical methods are mainly based on gamma correction and histogram equalization that enhance the image with the help of prior expert knowledge. These methods do not require a large amount of data either. Gamma correction takes advantage of nonlinearity in human perception such as our capacity to perceive light and color (Poynton 2012). Histogram equalization achieves the idea that allows areas of lower local contrast to gain a higher contrast for better distribution on the pixel histogram, which is useful when backgrounds and foregrounds are both bright or both dark such as X-ray images. Although these methods are fast and simple, the lack of consideration of contextual information limits their performance.

Recently, learning-based methods, which try to approximate the mapping from the input image to the desired pixel values with CNN, have achieved great success (Bychkovsky et al. 2011; Ulyanov et al. 2018; Kupyn et al. 2018; Wang et al. 2019). However, such methods are not without issues. First of all, it is hard to train a comprehensive neural network that can handle multiple enhancement situations. Besides, pixel-to-pixel mapping lacks robustness, e.g., it does not perform very well when dealing with some detailed information such as hair and characters (Zhang et al. 2019; Nataraj et al. 2019). Some researchers have proposed to apply deep reinforcement learning to image enhancement by formulating the enhancement procedure as a sequence of iterative decision-making problems to address the challenges above (Yu et al. 2018; Park et al. 2018; Furuta et al. 2019). In this chapter, we follow these methods and propose a new MDP formulation for image enhancement. We demonstrate our approach on a dataset containing 5000 pairs of images with code examples, for providing a quick hands-on learning process.

Before discussing the algorithm, we introduce two Python libraries *Pillow* (Clark 2015) and *scikit-image* (Van der Walt et al. 2014) that provide a number of friendly interfaces to implement image enhancement. One can install them directly from PyPI as follows:

```
pip install Pillow
pip install scikit-image
```

Here is an example code for contrast adjustment by Pillow's sub-module ImageEnhance.

```
from PIL import ImageEnhance

def adjust_contrast(image_rgb, contrast_factor):
    """Adjust contrast
    Args:
        image_rgb (PIL.Image): RGB image
        contrast_factor (float): color balance factor range from 0
            to 1.
    Return:
        PIL.Image object
    """
    enhancer = ImageEnhance.Contrast(image_rgb)
    return enhancer.enhance(contrast_factor)
```

14.2 Reinforcement Learning for Robust Processing

When applying reinforcement learning to image enhancement, one needs to first consider how to construct an MDP in this domain. An idea that naturally emerges is to consider processing pixels to be states and different image enhancement technologies to be actions in the context of reinforcement learning. This formulation provides a combination method of several controllable primary enhancers to achieve robust and effective results. In this section, we discuss such a reinforcement learning-based color enhancement method. For simplicity, we only take global enhancement actions. Note that it is natural to adapt to general enhancement algorithms by adding region proposal modules (Ren et al. 2015).

Suppose that the training dataset contains N pairs of RGB images $\{(l_i, h_i)\}_{i=1}^N$ where l_i is the low-quality raw image and h_i is the high-quality retouched image. In order to maintain the data distribution, the initial state S_0 should be sampled from $\{l_i\}_{i=1}^N$ uniformly. In each step, the agent takes a predefined action such as contrast adjustment with a certain factor and then applies it to the current state. Note that the current state and selected action fully determine the transition, i.e., no environment uncertainty exists. Following previous works (Park et al. 2018; Furuta et al. 2019), we use the improvement on CIELAB color space as the transition reward function:

$$||L(h) - L(S_t)||_2^2 - ||L(h) - L(S_{t+1})||_2^2 \tag{14.1}$$

where h is the corresponding high-quality image of S_0 and L maps images from RGB color space to CIELAB color space.

Another important thing is the terminal condition during learning and evaluation. Unlike reinforcement learning applications on games where the terminal state can be determined by the environment, agents in image enhancement need to decide an exit time by themselves. Park et al. (2018) proposed a DQN-based agent that exits when all predicted Q-values are negative. However, the overestimation problem of function approximation in Q-learning might lead to less robust results during inference. We address this issue by training an explicit policy and adding a "NO-OP" action to represent the exit choice. Table 14.1 lists all predefined actions, where the action with index 0 represents "NO-OP."

Training a convolutional neural network from scratch needs a large amount of retouched image pairs. Instead of using raw image states as observations, we consider the activation of the last convolutional layer in ResNet50 pre-trained on the ILSVRC classification dataset (Russakovsky et al. 2015), which is a significant deep feature that improves many other visual recognition tasks (Ren et al. 2016; Redmon et al. 2016). Inspired by previous work (Park et al. 2018; Lee et al. 2005), we further consider the histogram information when constructing observations. Specifically, we calculate the histogram statistics of the state in RGB color space over ranges $(0, 255)$, $(0, 255)$, $(0, 255)$, and CIELAB color space over ranges $(0, 100)$, $(-60, 60)$, $(-60, 60)$. These three features are concatenated as $2048 + 2000$ dimensional observations. We select PPO (Schulman et al. 2017) as the policy optimization algorithm. PPO is an actor-critic method that achieves significant results on a number of tasks. The network consists of three parts: three-layers feature extractor serving as a backbone, one-layer actor, and one-layer critic. All layers are fully connected, where the outputs of the layers in feature extractor are 2048, 512, and 128 units with ReLU activation, respectively.

Table 14.1 The action set for global color enhancement

Index	Description
0	No operation
1	Contrast $\times 0.95$
2	Contrast $\times 1.05$
3	Saturation $\times 0.95$
4	Saturation $\times 1.05$
5	Brightness $\times 0.95$
6	Brightness $\times 1.05$
7	Red and green $\times 0.95$
8	Red and green $\times 1.05$
9	Green and blue $\times 0.95$
10	Green and blue $\times 1.05$
11	Red and blue $\times 0.95$
12	Red and blue $\times 1.05$

Table 14.2 Hyper-parameters of PPO for image enhancement

Hyper-parameter	Value
Optimizer	Adam
Learning rate	1e−5
Clip norm	1.0
GAE λ	0.95
Episodes per iter	4
Optimization per iter	2
Max iter	10,000
Entropy factor	1e−2
Reward scale	0.1
Reward clip	[−1, 1]
γ	0.95

We evaluated our method on the MIT-Adobe FiveK (Bychkovsky et al. 2011) dataset including 5000 raw images, each with five retouched images produced by different experts (A/B/C/D/E). Following previous work (Park et al. 2018; Wang et al. 2019), we only use the retouched images by Expert C, which randomly selected 4500 images for training and the rest 500 images for testing. The raw images are DNG format while the retouched images are TIFF format. We convert all of them to JPEG format with quality 100 and color space sRGB by Adobe Lightroom. For efficient training, we resized images such that the maximal side consists of 512 pixels for each image. Hyper-parameters are provided in Table 14.2.

From now on, we demonstrate how to implement the algorithm above. First of all, we need to construct an environment object.

```
class Env(object):
   """Training env wrapper of image processing RL problem"""
   def __init__(self, src, max_episode_length=20,
       reward_scale=0.1):
      """
      Args:
         src (list[str, str]): list of raw and retouched path,
             initial
                             state will sample from it uniformly
         max_episode_length (int): max number of actions can be
             taken
      """
      self._src = src
      self._backbone = backbone
      self._preprocess = preprocess
      self._rgb_state = None
      self._lab_state = None
      self._target_lab = None
      self._current_diff = None
      self._count = 0
      self._max_episode_length = max_episode_length
      self._reward_scale = reward_scale
      self._info = dict()
```

With the ResNet API from TensorFlow, we build the observation by function `_state_feature` as follows:

```
backbone = tf.keras.applications.ResNet50(include_top=False,
    pooling='avg')
preprocess = tf.keras.applications.resnet50.preprocess_input

def get_lab_hist(lab):
   """Get hist of lab image"""
   lab = lab.reshape(-1, 3)
   hist, _ = np.histogramdd(lab, bins=(10, 10, 10),
                        range=((0, 100), (-60, 60), (-60, 60)))
   return hist.reshape(1, 1000) / 1000.0

def get_rgb_hist(lab):
   """Get hist of lab image"""
   lab = lab.reshape(-1, 3)
   hist, _ = np.histogramdd(lab, bins=(10, 10, 10),
                        range=((0, 255), (0, 255), (0, 255)))
   return hist.reshape(1, 1000) / 1000.0

def _state_feature(self):
   s = self._preprocess(self._rgb_state)
   s = tf.expand_dims(s, axis=0)
   context = self._backbone(s).numpy().astype('float32')
   hist_rgb = get_rgb_hist(self._rgb_state).astype('float32')
   hist_lab = get_lab_hist(self._lab_state).astype('float32')
   return np.concatenate([context, hist_rgb, hist_lab], 1)
```

Then we define the transition function `_transit` following Table 14.2, and implement reward function `_reward` with Eq. (14.1), to construct same interfaces as OpenAI Gym (Brockman et al. 2016):

```
def step(self, action):
   """One step"""
   self._count += 1
   self._rgb_state = self._transit(action)
   self._lab_state = rgb2lab(self._rgb_state)
   reward = self._reward()
   done = self._count >= self._max_episode_length or action == 0
   return self._state_feature(), reward, done, self._info

def reset(self):
   """Reset"""
   self._count = 0
   raw, retouched = map(Image.open, random.choice(self._src))
   self._rgb_state = np.asarray(raw)
   self._lab_state = rgb2lab(self._rgb_state)
   self._target_lab = rgb2lab(np.asarray(retouched))
   self._current_diff = self._diff(self._lab_state)
   self._info['max_reward'] = self._current_diff
   return self._state_feature()
```

In contrast to the implementation in Sect. 5.10.6, we apply the PPO (Schulman et al. 2017) algorithm in the discrete case. Note that we use LogSoftmax as the activation function in the actor network, which provides better numerical stability when calculating the surrogate objective. For the PPO agent, we first define its initialization and act function:

```
class Agent(object):
    """PPO Agent"""
    def __init__(self, feature, actor, critic, optimizer,
                 epsilon=0.1, gamma=0.95, c1=1.0, c2=1e-4,
                     gae_lambda=0.95):
        """
        Args:
            feature (tf.keras.Model): backbone of actor and critic
            actor (tf.keras.Model): actor network
            critic (tf.keras.Model): critic network
            optimizer (tf.keras.optimizers.Optimizer): optimizer
                for NNs
            epsilon (float): epsilon in clip
            gamma (float): reward discount
            c1 (float): factor of value loss
            c2 (float): factor of entropy
        """
        self.feature, self.actor, self.critic = feature, actor,
            critic
        self.optimizer = optimizer

        self._epsilon = epsilon
        self.gamma = gamma
        self._c1 = c1
        self._c2 = c2
        self.gae_lambda = gae_lambda

    def act(self, state, greedy=False):
        """
        Args:
            state (numpy.array): 1 * 4048
            greedy (bool): whether select action greedily

        Returns:
            action (int): selected action
            logprob (float): log prob of the selected action
            value (float): value of the current state
        """
        feature = self.feature(state)
        logprob = self.actor(feature)
        if greedy:
            action = tf.argmax(logprob[0]).numpy()
            return action, 0, 0
        else:
            value = self.critic(feature)
            logprob = logprob[0].numpy()
```

```
        action = np.random.choice(range(len(logprob)),
            p=np.exp(logprob))
        return action, logprob[action], value.numpy()[0, 0]
```

During sampling, we record the trajectories with the GAE (Schulman et al. 2015) algorithm

```
def sample(self, env, sample_episodes, greedy=False):
    """ Sample trajectories from given env
    Args:
        env: environment
        sample_episodes (int): how many episodes will be sampled
        greedy (bool): whether select action greedily
    """
    trajectories = [] # s, a, r, logp
    e_reward = 0
    e_reward_max = 0
    for _ in range(sample_episodes):
        s = env.reset()
        values = []
        while True:
            a, logp, v = self.act(s, greedy)
            s_, r, done, info = env.step(a)
            e_reward += r
            values.append(v)
            trajectories.append([s, a, r, logp, v])
            s = s_
            if done:
                e_reward_max += info['max_reward']
                break
        episode_len = len(values)
        gae = np.empty(episode_len)
        reward = trajectories[-1][2]
        gae[-1] = last_gae = reward - values[-1]
        for i in range(1, episode_len):
            reward = trajectories[-i - 1][2]
            delta = reward + self.gamma * values[-i] - values[-i -
                1]
            gae[-i - 1] = last_gae = \
                delta + self.gamma * self.gae_lambda * last_gae
        for i in range(episode_len):
            trajectories[-(episode_len - i)][2] = gae[i] + values[i]
    e_reward /= sample_episodes
    e_reward_max /= sample_episodes
    return trajectories, e_reward, e_reward_max
```

Finally, the optimization part is provided as follows:

```
def _train_func(self, b_s, b_a, b_r, b_logp_old, b_v_old):
    all_params = self.feature.trainable_weights + \
                self.actor.trainable_weights + \
                self.critic.trainable_weights
    with tf.GradientTape() as tape:
```

```
        b_feature = self.feature(b_s)
        b_logp, b_v = self.actor(b_feature), self.critic(b_feature)

        entropy = -tf.reduce_mean(
            tf.reduce_sum(b_logp * tf.exp(b_logp), axis=-1))
        b_logp = tf.gather(b_logp, b_a, axis=-1, batch_dims=1)
        adv = b_r - b_v_old
        adv = (adv - tf.reduce_mean(adv)) /
              (tf.math.reduce_std(adv) + 1e-8)

        c_b_v = b_v_old + tf.clip_by_value(b_v - b_v_old,
                                           -self._epsilon, self._epsilon)
        vloss = 0.5 * tf.reduce_max(tf.stack(
            [tf.pow(b_v - b_r, 2), tf.pow(c_b_v - b_r, 2)],
                axis=1), axis=1)
        vloss = tf.reduce_mean(vloss)

        ratio = tf.exp(b_logp - b_logp_old)
        clipped_ratio = tf.clip_by_value(
            ratio, 1 - self._epsilon, 1 + self._epsilon)
        pgloss = -tf.reduce_mean(tf.reduce_min(tf.stack(
            [clipped_ratio * adv, ratio * adv], axis=1), axis=1))

        total_loss = pgloss + self._c1 * vloss - self._c2 * entropy
    grad = tape.gradient(total_loss, all_params)
    self.optimizer.apply_gradients(zip(grad, all_params))
    return entropy

def optimize(self, trajectories, opt_iter):
    """ Optimize based on given trajectories """
    b_s, b_a, b_r, b_logp_old, b_v_old = zip(*trajectories)
    b_s = np.concatenate(b_s, 0)
    b_a = np.expand_dims(np.array(b_a, np.int64), 1)
    b_r = np.expand_dims(np.array(b_r, np.float32), 1)
    b_logp_old = np.expand_dims(np.array(b_logp_old, np.float32),
        1)
    b_v_old = np.expand_dims(np.array(b_v_old, np.float32), 1)
    b_s, b_a, b_r, b_logp_old, b_v_old = map(
        tf.convert_to_tensor, [b_s, b_a, b_r, b_logp_old, b_v_old])
    for _ in range(opt_iter):
        entropy = self._train_func(b_s, b_a, b_r, b_logp_old,
            b_v_old)

    return entropy.numpy()
```

where the value loss clipping and advantage normalization are followed by Dhariwal et al. (2017). Figure 14.2 shows an example result.

Fig. 14.2 An example result of global enhancement on the MIT-Adobe FiveK dataset. The global brightness is increased while some areas like sky in the upper right corner need local enhancement

References

Brockman G, Cheung V, Pettersson L, Schneider J, Schulman J, Tang J, Zaremba W (2016) OpenAI gym. Preprint. arXiv:160601540

Bychkovsky V, Paris S, Chan E, Durand F (2011) Learning photographic global tonal adjustment with a database of input/output image pairs. In: Conference on computer vision and pattern recognition 2011. IEEE, Piscataway, pp 97–104

Clark A (2015) Pillow (PIL fork) documentation. https://github.com/python-pillow/Pillow

Dhariwal P, Hesse C, Klimov O, Nichol A, Plappert M, Radford A, Schulman J, Sidor S, Wu Y, Zhokhov P (2017) OpenAI baselines. GitHub, GitHub repository

Furuta R, Inoue N, Yamasaki T (2019) Fully convolutional network with multi-step reinforcement learning for image processing. In: Proceedings of the AAAI conference on artificial intelligence, vol 33, pp 3598–3605

Kupyn O, Budzan V, Mykhailych M, Mishkin D, Matas J (2018) DeblurGAN: Blind motion deblurring using conditional adversarial networks. In: Proceedings of the IEEE conference on computer vision and pattern recognition, pp 8183–8192

Lee S, Xin J, Westland S (2005) Evaluation of image similarity by histogram intersection. Color research & application: endorsed by inter-society color council, the colour group (Great Britain), Canadian society for color, color science association of Japan, Dutch society for the study of color, the Swedish colour centre foundation, colour society of Australia, centre. Français de la Couleur 30(4):265–274

Nataraj L, Mohammed TM, Manjunath B, Chandrasekaran S, Flenner A, Bappy JH, Roy-Chowdhury AK (2019) Detecting GAN generated fake images using co-occurrence matrices. J Electron Imaging 2019:532-1

Park J, Lee JY, Yoo D, So Kweon I (2018) Distort-and-recover: color enhancement using deep reinforcement learning. In: Proceedings of the IEEE conference on computer vision and pattern recognition, pp 5928–5936

Poynton C (2012) Digital video and HD: algorithms and interfaces. Elsevier, Amsterdam

Redmon J, Divvala S, Girshick R, Farhadi A (2016) You only look once: unified, real-time object detection. In: Proceedings of the IEEE conference on computer vision and pattern recognition, pp 779–788

Ren S, He K, Girshick R, Sun J (2015) Faster R-CNN: towards real-time object detection with region proposal networks. In: Advances in neural information processing systems, pp 91–99

Ren S, He K, Girshick R, Zhang X, Sun J (2016) Object detection networks on convolutional feature maps. IEEE Trans Pattern Anal Mach Intell 39(7):1476–1481

Russakovsky O, Deng J, Su H, Krause J, Satheesh S, Ma S, Huang Z, Karpathy A, Khosla A, Bernstein M (2015) ImageNet large scale visual recognition challenge. Int J Comput Vision 115(3):211–252

Schulman J, Moritz P, Levine S, Jordan M, Abbeel P (2015) High-dimensional continuous control using generalized advantage estimation. Preprint. arXiv:150602438

Schulman J, Wolski F, Dhariwal P, Radford A, Klimov O (2017) Proximal policy optimization algorithms. Preprint. arXiv:170706347

Ulyanov D, Vedaldi A, Lempitsky V (2018) Deep image prior. In: Proceedings of the IEEE conference on computer vision and pattern recognition, pp 9446–9454

Van der Walt S, Schönberger JL, Nunez-Iglesias J, Boulogne F, Warner JD, Yager N, Gouillart E, Yu T (2014) Scikit-image: image processing in python. PeerJ 2:e453

Wang R, Zhang Q, Fu CW, Shen X, Zheng WS, Jia J (2019) Underexposed photo enhancement using deep illumination estimation. In: Proceedings of the IEEE conference on computer vision and pattern recognition, pp 6849–6857

Yu K, Dong C, Lin L, Change Loy C (2018) Crafting a toolchain for image restoration by deep reinforcement learning. In: Proceedings of the IEEE conference on computer vision and pattern recognition, pp 2443–2452

Zhang S, Zhen A, Stevenson RL (2019) GAN based image deblurring using dark channel prior. Preprint. arXiv:190300107

Chapter 15
AlphaZero

Hongming Zhang and Tianyang Yu

Abstract In this chapter, we introduce combinatorial games such as chess and Go and take Gomoku as an example to introduce the AlphaZero algorithm, a general algorithm that has achieved superhuman performance in many challenging games. This chapter is divided into three parts: the first part introduces the concept of combinatorial games, the second part introduces the family of algorithms known as Monte Carlo Tree Search, and the third part takes Gomoku as the game environment to demonstrate the details of the AlphaZero algorithm, which combines Monte Carlo Tree Search and deep reinforcement learning from self-play.

Keywords AlphaZero · Monte Carlo Tree Search · Upper confidence bounds for trees · Self-play · Deep reinforcement learning · Deep neural network

15.1 Introduction

The AlphaZero (Silver et al. 2018, 2017a) algorithm is a generalized version of the AlphaGo Zero (Silver et al. 2017b) algorithm, which has achieved superhuman performance in the game of Go. Different from the initial AlphaGo (Silver et al. 2016) series such as AlphaGo Fan (defeated Fan Hui), AlphaGo Lee (defeated Lee Sedol), and AlphaGo Master (defeated Jie Ke), the AlphaZero algorithm is based solely on reinforcement learning from games via self-play (play against itself). It starts with random plays and without supervised learning from human expert games. There are two key parts in AlphaZero: (1) Monte Carlo Tree Search is used in self-play to collect data; (2) a deep neural network is used to learn from the data and predict the move selections and state values in Monte Carlo

H. Zhang (✉)
Peking University, Beijing, China
e-mail: zhanghongming@pku.edu.cn

T. Yu
Nanchang University, Nanchang, China

H. Dong et al. (eds.), *Deep Reinforcement Learning*,
https://doi.org/10.1007/978-981-15-4095-0_15

Tree Search. This general algorithm is suitable not only for the game of Go, but it also defeated world champion programs in the games of chess and shogi, which indicates its generality. In this chapter, we first introduce the concept of combinatorial games (including Go, chess, Gomoku, etc.) and implement the free-style Gomoku as an example, then explain the concrete steps in Monte Carlo Tree Search, and finally take Gomoku as the game environment to demonstrate the details of the AlphaZero algorithm. To help the readers' understanding, we provide the implementation of Gomoku game and the AlphaZero algorithm in the link: https://github.com/deep-reinforcement-learning-book/Chapter15-AlphaZero.

15.2 Combinatorial Games

Combinatorial game theory (CGT) (Albert et al. 2007) is a branch of research in mathematics and theoretical computer science that typically studies sequential games with perfect information. These types of games usually have the following characteristics:

- The game contains two players (Go, chess). One player's games (Sudoku, Solitaire) can also be regarded as combinatorial games that played between the game designer and the player. Games with more players are not considered combinatorial due to the social aspect of coalitions that may arise during play (Browne et al. 2012).
- The game does not contain any chance factors that will influence the outcome of the game such as a dice, etc.
- The game offers perfect information (Muthoo 1996), which means each player is perfectly informed of all the events that have previously occurred.
- The players perform actions in a turn-based manner, and the action space and state space are both finite.
- The game ends with finite time steps, results are divided into win and loss, and some games have a draw.

Many of the combinatorial games (Albert et al. 2007) including single-player games like Sudoku and Solitaire, two-player games like Hex, Go, and chess are classical problems for computer scientists to solve. Since IBM's DEEP BLUE (Campbell et al. 2002; Hsu 1999) beat grandmaster Gary Kasparov on chess, Go has become the standard yardstick for AI. Besides, there are also many other games like Othello, Amazons, Khet, Shogi, Chinese Checkers, Connect Four, Gomoku, etc., attracting people to find solutions with computing machines.

We take Gomoku as an example of combinatorial games and show some details about the code of Gomoku. Gomoku is also called "Gobang" or "five-in-a-row." It begins with an empty board and ends with a row of five stones (in a horizontal, vertical, or diagonal line) that indicates one player wins, or otherwise a draw. There

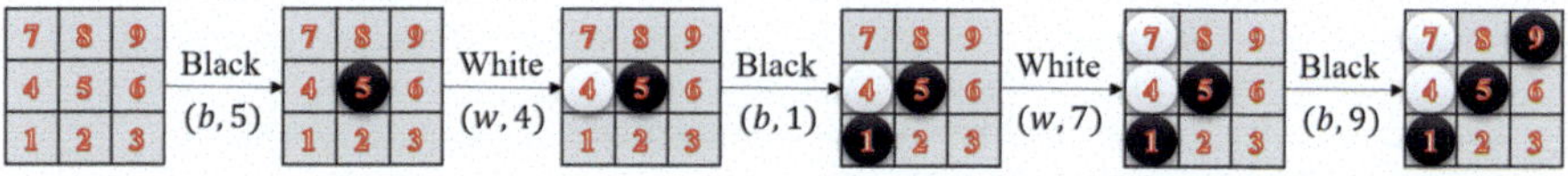

Fig. 15.1 A sequence of Gomoku gaming process on a 3 × 3 board. The "b" is shorten for "the black player" and the "w" is shorten for "the white player." $(b, 5)$ means the black player puts the stone at the position 5. The black player wins the game in the end

are various sets of rules and most variations are based on either free-style Gomoku or standard Gomoku. Free-style Gomoku simply requires a row of five or more stones for a win. Standard Gomoku requires a row of exactly five stones for a win: rows of six or more, called overlines, do not count. In this game, we choose the free-style Gomoku as a demonstration.

We further simplify the board to have a size of 3 × 3 as an example for demonstration here. A row of three stones indicates winning (we may call it "three-in-a-row"). A sequence of game steps on this board are shown in Fig. 15.1.

The red numbers on the board are used for indexing different possible positions on the board, which can be used to represent the choice for each move. White and black circles are the game stones of two players. The gaming process can be represented as a sequence: $((b, 5), (w, 4), (b, 1), (w, 7), (b, 9))$, where the "$b$" stands for "the black player" and the "w" stands for "the white player". The black player has a row of three stones in the end which means he wins the game, as shown in the last board state in Fig. 15.1. Recall the definition we mentioned before, this simplified Gomoku (or "three-in-a-row") satisfies all characteristics of combinatorial games: the game contains two players; the game does not contain any chance factors; the game offers perfect information; the players perform actions in a turn-based manner; the game ends with finite time steps.

Here, we implement the free-style Gomoku as an example:

- Define the game as `Board` class with the rule of the game implemented as some functions. Though we illustrated the rule of Gomoku with a simplified version before, we can define a standard five-in-a-row by passing the `n_in_row` variable with value 5.

```
class Board(object):
    '''
    board for the game
    '''
    def __init__(self, width, height, n_in_row): ... #
        Initialization function
    def move_to_location(self, move): ... # Transfer move
        denotation
    def location_to_move(self, location): ... # Transfer move
        denotation
    def do_move(self, move): ... # Update each move and exchange
        the current player
```

```
def has_a_winner(self): ... # The rule of Gomoku, to judge if
    someone wins
def current_state(self): ... # Generate board states as
    inputs of the network
...
```

- Each action on the board is denoted by a number like pictured in Fig. 15.1. This way makes it more convenient to build tree nodes in MCTS. But it is inconvenient to identify whether there is a row of five stones. So, we define the transition functions between coordinates and numbers. Coordinates are used to judge whether a player has a row of five stones, and numbers are used to build tree nodes in MCTS.

```
def move_to_location(self, move):
    '''
    Transfer number to coordinate
    3*3 board's moves like:
    6 7 8
    3 4 5
    0 1 2
    and move 5's location is (1,2)
    '''
    h = move // self.width
    w = move % self.width
    return [h, w]

def location_to_move(self, location):
    '''
    Transfer coordinate to number
    '''
    if len(location) != 2:
        return -1
    h = location[0]
    w = location[1]
    move = h * self.width + w
    if move not in range(self.width * self.height):
        return -1
    return move
```

- To find out if a player wins, a function is needed to judge if there is a line of five stones in a row or in a column or in a diagonal. The function `has_a_winner()` is shown:

```
def has_a_winner(self):
    '''
    Judge if there's a 5-in-a-row, and which player if so
    '''
    width = self.width
    height = self.height
```

```
        states = self.states
        n = self.n_in_row

        moved = list(set(range(width * height)) -
            set(self.availables))
        # Moves have been played
        if len(moved) < self.n_in_row + 2:
            # Too few moves to get 5-in-a-row
            return False, -1

        for m in moved:
            h, w = self.move_to_location(m)
            player = states[m]

            if (w in range(width - n + 1) and
                    len(set(states.get(i, -1) for i in range(m, m +
                        n))) == 1):
                # Judge if there's a 5-in-a-row in a line
                return True, player

            if (h in range(height - n + 1) and
                    len(set(states.get(i, -1) for i in range(m, m + n
                        * width, width))) == 1):
                # Judge if there's a 5-in-a-row in a column
                return True, player

            if (w in range(width - n + 1) and h in range(height - n
                + 1) and
                    len(set(states.get(i, -1) for i in range(m, m + n
                        * (width + 1), width + 1))) == 1):
                # Judge if there's a 5-in-a-row in a top right
                    diagonal
                return True, player

            if (w in range(n - 1, width) and h in range(height - n
                + 1) and
                    len(set(states.get(i, -1) for i in range(m, m + n
                        * (width - 1), width - 1))) == 1):
                # Judge if there's a 5-in-a-row in a top left
                    diagonal
                return True, player

        return False, -1
```

15.3 Monte Carlo Tree Search

Monte Carlo Tree Search (MCTS) (Browne et al. 2012) is a method for finding optimal decisions in a given domain by taking random samples in the decision space and building a search tree according to the results. This method started a revolution

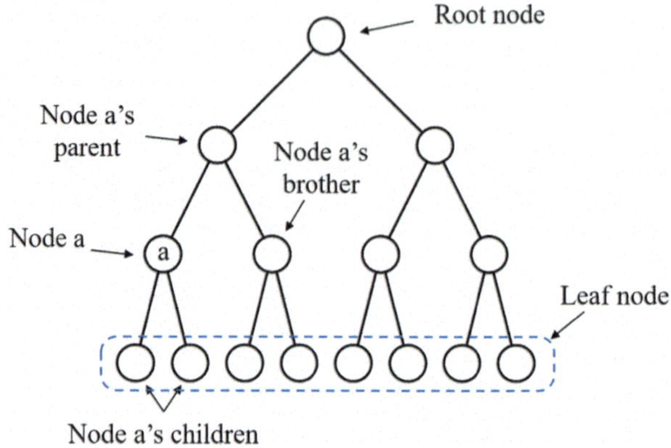

Fig. 15.2 A tree structure

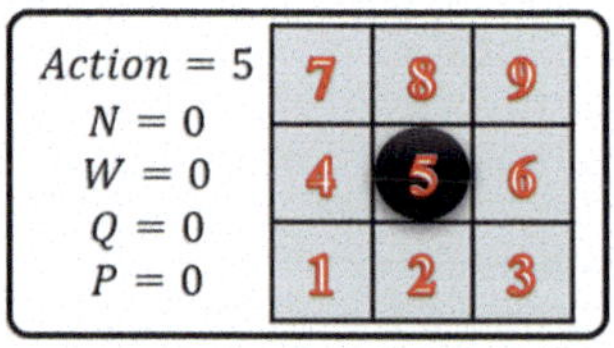

Fig. 15.3 A node with the state that a black stone on the board at the position 5. The action can be represented as $(b, 5)$ with $Action = 5$ by the black player. $N = 0$ means the visit count of this node is 0. W means the total rewards, Q means the average reward, P means the probability of taking the $Action = 5$. As the node has not been visited, these values are initialized with 0

in the fields of games and planning problems and pushed the performance such as computer Go to ever high levels.

Monte Carlo Tree Search includes two parts, the tree and the search methods. The tree is a kind of data structure (Fig. 15.2), it contains nodes connected by edges. Some important concepts include the root node, the parent and the child, the leaf node, etc. The node at the top of the tree is called the root node; the node above the current node is its parent, and the node below the current node is its child; a node that does not have child node is called a leaf node. Usually, besides the state and actions, the node in MCTS contains information about the visit count and average reward. In the AlphaZero algorithm, the tree also contains the probability distribution of the action for each state.

In the AlphaZero algorithm, each node contains the following information that is used to design search methods. (The node is illustrated in Fig. 15.3.)

- $Action$: action that is taken and reaches this node.
- N: the node's visit count. Initial value is zero, means the node has not been visited.
- W: the node's total reward which is used to calculate the average reward. Its initial value is zero.

- Q: the node's average reward which indicates the state value of the node. Its initial value is zero.
- P: the probability of taking the $Action$. It is the output by the policy network with the input of its parent node's state.

Before we go ahead, we have to mention an important aspect. Because the game is played by two players, there are two perspectives in a tree and the information at a node is either from the perspective of the black player or the perspective of the white player. For example, in Fig. 15.3, this node has a black stone on the board, so it should be the white player's turn to play the next move at this node. However, the information at this node is from the perspective of its parent node, which is the black player. It is the parent node that expands its child nodes and chooses an action to reach one of these child nodes, so the $Action$, N, W, Q, P at this node are all used by the black player. The black player takes $Action = 5$ to reach this node, and the information at the present time is $N = 0$, $W = 0$, $Q = 0$, $P = 0$. It is very important to have a clear understanding of the perspective of each node; otherwise, it will be confusing when we do the *backup* step in MCTS.

With the tree, the search methods aim to explore the decision space and estimate the state-action value function $Q^{\pi}(s, a)$ of the root node by some heuristic methods. Exploring from the root node to leaf nodes, obtaining the average reward of each action over multiple explorations, and finding the optimal trajectory in the tree. The action-value function without a discount factor can be expressed as follows (Couetoux et al. 2011):

$$Q^{\pi}(s, a) = \mathbb{E}_{\pi}\left[\sum_{h=0}^{T-1} P(S_{h+1}|S_h, A_h)R(S_{h+1}|S_h, A_h)|S_0 = s, A_0 = a, A_h = \pi(S_h)\right]. \tag{15.1}$$

$Q^{\pi}(s, a)$ denotes the value function, i.e., the expected reward gathered when executing action a in state s and following policy π for all subsequent actions until arriving at a terminal state.

Usually, there are four steps in each tree search iteration: *select, expand, simulate, backup* (all these steps are conducted in the tree search, which indicates there is not a real move on the board).

- *Select*: actions are selected from the root node by a particular policy until reaching a leaf node.
- *Expand*: child nodes are added to the current leaf node.
- *Simulate*: a policy such as a random policy is used to simulate the process of playing stones until the end of the game to get a result: win, loss, or draw. Reward is given from the game and usually $+1$ for a win, -1 for a loss, 0 for a draw.
- *Backup*: the simulating result is backpropagated to update the information at each node that is selected in this iteration.

The most commonly used tree search algorithm is the Upper Confidence Bound in Tree (UCT) (Kocsis and Szepesvári 2006), which handles the exploration versus exploitation trade-off well. The Upper Confidence Bound (UCB) (Auer et al. 2002) (discussed in Chap. 2 Sect. 2.2.2) algorithm is a classical strategy to solve the multi-armed bandit problem, in which the agent has to choose among various gambling machines at each time step to maximize its payoffs. UCB chooses the next action at time t by the following policy:

$$A_t = \arg\max_a \left[Q_t(a) + c\sqrt{\frac{\ln t}{N_t(a)}} \right]. \tag{15.2}$$

$Q_t(a)$ is the estimated action value which determines the exploitation. The square-root term determines the exploration. $N_t(a)$ is the number of times action a has been selected till time t and c is a positive real number that determines how much exploration needs to be done. There are some variations in the family of UCB algorithms, like UCB1, UCB1-NORMAL, UCB1-TUNED, and UCB2 (Auer et al. 2002).

UCT is the version that implements UCB1 in the Tree. Each time, a child node with the largest UCT value is selected which is defined as follows:

$$\text{UCT} = \overline{X}_j + C_p\sqrt{\frac{2\ln n}{n_j}}. \tag{15.3}$$

Here, n is the number of visits to the current node, n_j is the number of visits to its child node j, and $C_p > 0$ is a constant that controls the exploration based on specific problems. The average reward term $\overline{X}_j$ encourages the exploitation of higher-reward actions, while the square-root term $\sqrt{\frac{2\ln n}{n_j}}$ encourages the exploration of less-visited actions.

UCT algorithm addresses the exploration versus exploitation trade-off in each state of the tree search space and has revolutionized quite a few large-scale RL problems such as board game Hex, Go, and real-time games like Atari. Levente Kocsis and Csaba Szepesv́ari (Kocsis and Szepesvári 2006) proved that: Consider a finite-horizon MDP with rewards scaled to lie in the interval [0, 1]. Let the horizon of the MDP be D, and the number of actions per state be K. Consider algorithm UCT such that the bias terms of UCB1 are multiplied by D. Then the bias of the estimated expected payoff, $\overline{X}_n$, is $O(\frac{\log n}{n})$. Further, the failure probability at the root converges to zero at a polynomial rate as the number of episodes grows to infinity. It indicates that, as the number of explored samples increases, the UCT algorithm can guarantee the tree search converges to the optimal solution.

In the AlphaZero algorithm, the step of *simulate* is discarded and the deep neural network is used to output the result directly. So, there are three key steps as follows, and the process is pictured in Fig. 15.4.

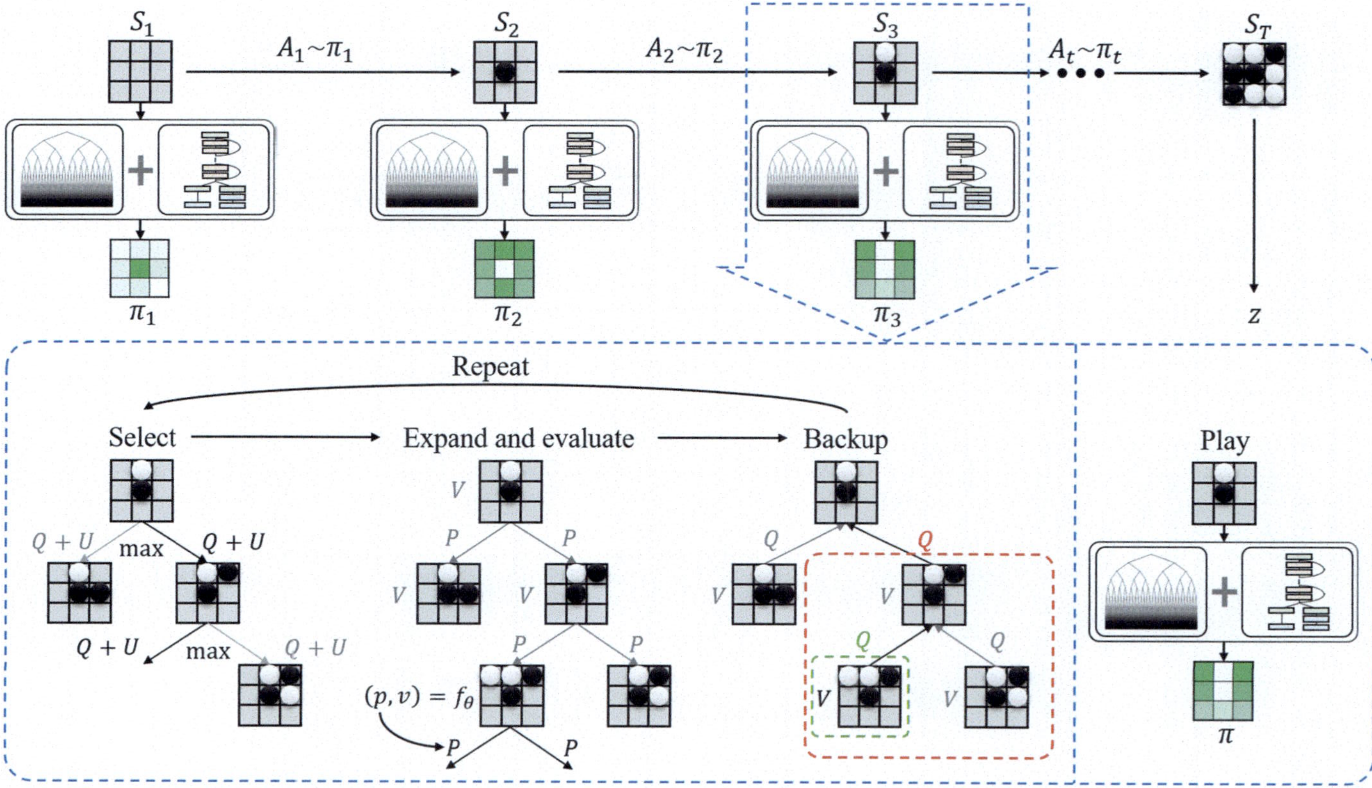

Fig. 15.4 MCTS in the AlphaZerc algorithm. Each time before a stone is placed on the real board, MCTS will be repeated for many times. It first selects actions from the root node and reaches a leaf node, then the leaf node is expanded and evaluated, finally the *backup* step is taken to update nodes' information

- *Select*: actions are selected from the root node by a particular policy until reaching a leaf node.
- *Expand and evaluate*: child nodes are added to the current leaf node. The probability of each action and the value of the state are evaluated directly by the policy network and the value network. A threshold value is usually applied to judge if the node is supposed to be expanded, for saving the computing resources without losing the effectiveness of the algorithm. In our implementation, we ignore this threshold and expand the node as long as it reaches a leaf node.
- *Backup*: each time after expanding and evaluating, the result is backpropagated to update the information at the nodes that are selected in the current iteration. If the leaf node is not the terminal of the game, the result cannot be given by the game and it is output by the deep neural network; otherwise, it is given by the game directly.

In the step of *select*, the action is selected by the formula $a = \arg\max_a(Q(s,a) + U(s,a))$. $Q(s,a) = \frac{W}{N}$ encourages the exploitation of higher-reward actions. $U(s,a) = c_{puct}P(s,a)\frac{\sqrt{\sum_b N(s,b)}}{1+N(s,a)}$ encourages the exploration of less-visited actions, c_{puct} is a parameter determining the exploration scale, which is 5 in the AlphaZero algorithm.

In the step of *expand and evaluate*, the policy network outputs a probability $p(s,a)$ for each action, the value network outputs a value v indicating the state value for the current state s. The $p(s,a)$ is used to calculate $U(s,a)$ in the step of *select*, $U(s,a) = c_{puct}P(s,a)\frac{\sqrt{\sum_b N(s,b)}}{1+N(s,a)}$. The v is used to calculate W in the step of *backup*, $W(s,a) = W(s,a) + v$. The probabilities and values output by the neural networks may not be accurate at the beginning, but they will be more accurate gradually during the training process.

In the step of *backup*, information at each selected node is updated, $N(s,a) = N(s,a) + 1$, $W(s,a) = W(s,a) + v$, $Q(s,a) = \frac{W(s,a)}{N(s,a)}$.

Some core code is shown as follows:

- The process of Monte Carlo Tree Search is defined as a `MCTS` class, it contains the whole tree and the tree search function `_playout()`:

```
class MCTS(object):
    '''
    An implementation of Monte Carlo Tree Search.
    '''
    def __init__(self, policy_value_fn,action_fc,evaluation_fc,
        is_selfplay,c_puct, n_playout): ... # Init function
    def _playout(self, state): ... # The process of tree search
```

- The node in the tree is defined as a `TreeNode` class with above steps: *select, expand and evaluate, backup*.

```
class TreeNode(object):
    '''
    A node in the tree.
    Each node keeps track of its own value Q, prior
        probability P, and
    its visit-count-adjusted prior score u.
    '''
    def __init__(self, parent, prior_p): ... # Init function
    def select(self, c_puct): ... # Choose action
    def expand(self, action_priors, add_noise): ...# Expand
        the node, evaluate each action and state
    def update(self, move): ... # Update node after expanding
        and evaluating
    ...
```

- The function `select()` corresponds to the step of *select*:

```
def select(self, c_puct):
    '''
    Select an action among children that gives maximum action
        value Q plus bonus u(P).
    Return: A tuple of (action, next_node)
    '''
    return max(self._children.items(),
               key=lambda act_node: act_node[1].get_value(c_puct))
```

- The function `expand()` here corresponds to the step of *expand and evaluate*. Besides, we add some Dirichlet noises at the nodes for random exploration:

```
def expand(self, action_priors, add_noise):
    '''
    Expand tree by creating new children.
    action_priors: a list of tuples of actions and their prior
        probability
       according to the policy function.
    '''
    if add_noise:
       action_priors = list(action_priors)
       length = len(action_priors)
       dirichlet_noise = np.random.dirichlet(0.3 *
           np.ones(length))
       for i in range(length):
          if action_priors[i][0] not in self._children:
             self._children[action_priors[i][0]] =
                 TreeNode(self,
```

```
                0.75 * action_priors[i][1] + 0.25 *
                    dirichlet_noise[i])
    else:
        for action, prob in action_priors:
            if action not in self._children:
                self._children[action] = TreeNode(self, prob)
```

- The function `update_recursive()` corresponds to the step of *backup*:

```
def update_recursive(self, leaf_value):
    '''
    Like a call to update(), but applied recursively for all
        ancestors.
    '''
    # If it is not root, this node's parent should be updated
        first.
    if self._parent:
        self._parent.update_recursive(-leaf_value)
        # Every step for recursive update.
        # We should change the perspective by the way of taking
            the opposite value
    self.update(leaf_value)

def update(self, leaf_value):
    '''
    Update node values from leaf evaluation.
    leaf_value: the value of subtree evaluation from the
        current player's
      perspective.
    '''
    self._n_visits += 1
    # update visit count
    self._Q += 1.0 * (leaf_value - self._Q) / self._n_visits
    # Update Q, a running average of values for all visits.
    # there is just: (v-Q)/(n+1)+Q =
        (v-Q+(n+1)*Q)/(n+1)=(v+n*Q)/(n+1)
```

- The tree search function `_playout()` in the class `MCTS` executes the three steps iteratively: *select, expand and evaluate, backup*.

```
def _playout(self, state):
    '''
    Run a single MCTS from the root to the leaf, get a value at
    the leaf and propagate it back through its parents.
    '''
    node = self._root
    # Select
    while(1):
```

```
            if node.is_leaf():
                break
            action, node = node.select(self._c_puct)
            state.do_move(action)
        # Evaluate and expand
        action_probs, leaf_value =
            self._policy_value_fn(state,self._action_fc,self._evaluation_fc)
        end, winner = state.game_end()
        if not end:
            node.expand(action_probs,add_noise=self._is_selfplay)
        else:
            if winner == -1: # draw
                leaf_value = 0.0
            else:
                leaf_value = (
                    1.0 if winner == state.get_current_player() else -1.0
                )
        # Backup
        node.update_recursive(-leaf_value)
```

15.4 AlphaZero: A General Algorithm for Board Games

Generally, the AlphaZero algorithm is suitable for all kinds of combinatorial games, such as Go, chess, Shogi, and so on. Here, we take Gomoku with the free-style rule described in Sect. 15.2 as an example, to introduce the details of the AlphaZero algorithm. Gomoku is a turn-based game with simple rules, which is suitable for demonstration as the game itself is not the focus. We further simplify the game board to have size 3×3 as an example, and a row of three stones indicates winning as we mentioned before. In addition, it is necessary to mention that the AlphaZero algorithm generalizes the AlphaGo Zero algorithm, so the two algorithms are very similar. In our implementation, we compare both methods.

We will demonstrate the details of the AlphaZero algorithm in this section for better understanding. The whole algorithm can be split into two parts: (1) self-play process using Monte Carlo Tree Search for data collection; (2) the deep neural network for training. The schematic is shown in Fig. 15.5.

First, we execute the self-play process with MCTS for collecting data. In order to demonstrate the tree search to the end of the game with relatively short paragraphs, we assume the game begins from the state as shown in Fig. 15.6 (normally, the game is started from an empty board). It is the white player's turn to play a stone now.

We build a tree from this node and start the MCTS process following the three steps: *select, expand and evaluate, backup*. Now, there is only one node in the tree. This node is a root node since it is at the top of the tree, and it is also a leaf node as it does not have child node. It means we have reached a leaf node and the step of *select* is done. So, the node needs to be expanded, it turns to the second step: *expand and evaluate*. In Fig. 15.7, the node is expanded, and the probability of each action is given by the policy network, with the input of the current board state.

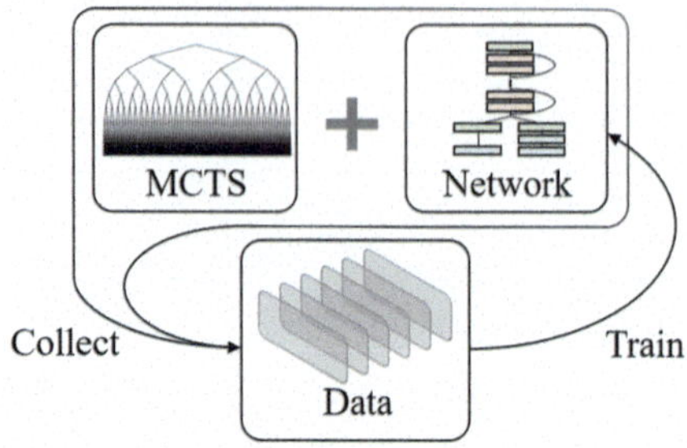

Fig. 15.5 Algorithm schematic. In the AlphaZero algorithm, a cycle is formed among MCTS, the data, and the network. MCTS is used to generate data, the data is used to improve the precision of the network, the more accurate network is used in MCTS to generate higher quality data

Fig. 15.6 Board state. The size of the board is 3 × 3. In this state, it is the white player's turn to play

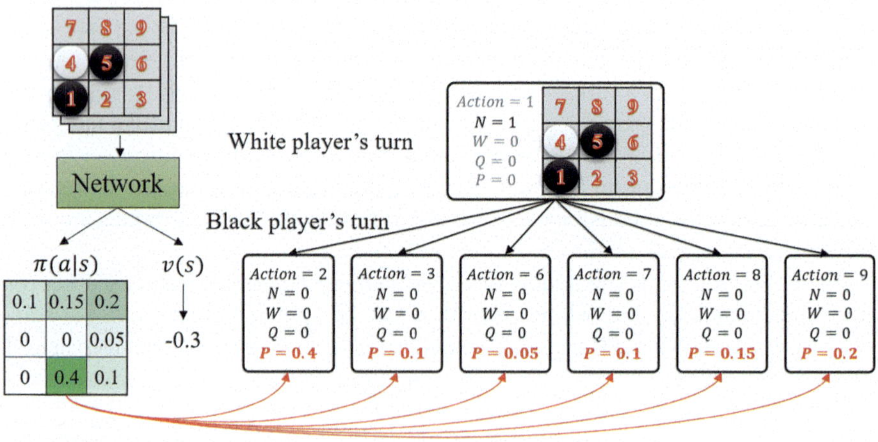

Fig. 15.7 *Expand and evaluate* at the root node. All the available actions are expanded, and the probabilities $\pi(a|s)$ are given by the network

The last step is *backup*. As it is a root node now, we do not need to backup W and Q (used to judge if we should go to this node) and only need to update the visit count N. $N = 0$ is changed to $N = 1$, and the tree search iteration has been finished once.

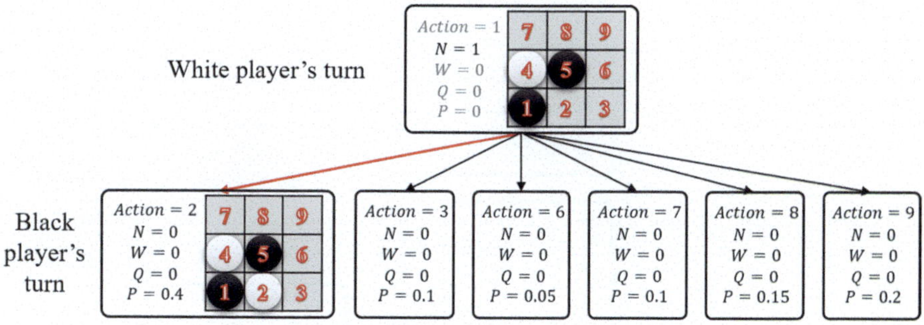

$$a = argmax_a(Q(s,a) + U(s,a))$$

$$\text{Exploration}: U(s,a) = c_{puct}P(s,a)\frac{\sqrt{\sum_b N(s,b)}}{1+N(s,a)}$$

$$\text{Exploitation}: Q(s,a) = \frac{W}{N}$$

Fig. 15.8 *Select* at the root node. The white player selects $Action = 2$ $(w, 2)$ and reaches a leaf node. In this node, it is the black player's turn to play

At each time we execute a tree search iteration, we will begin from the root node. So, the second tree search process also starts from the root node. This time, the root node has child nodes, which means it is not a leaf node. The action is selected by the formula $a = \arg\max_a(Q(s,a)+U(s,a))$, $Q(s,a) = \frac{W}{N}$, $U(s,a) = c_{puct}P(s,a)\frac{\sqrt{\sum_b N(s,b)}}{1+N(s,a)}$. Here, the action chosen by the white player is 2 $(w, 2)$, and it reaches a new node. This new node is a leaf node and it is the black player's turn at this node (Fig. 15.8).

We expand and evaluate this leaf node. It is the same as the first time: the node is expanded, and the probability of each action is given by the policy network (Fig. 15.9).

Then, it is time to backup. Now there are two nodes; we first update the current node and then the preceding node. Both of them follow the same rules for updating in the *backup* step: $N(s,a) = N(s,a) + 1$, $W(s,a) = W(s,a) + v(s)$, $Q(s,a) = \frac{W(s,a)}{N(s,a)}$. It should be noted that there are two perspectives in the tree: black and white. We should be careful about the perspectives and always update the value in the current player's perspective. For example, in Fig. 15.10, the value from the value network is $v(s) = -0.1$, it is from the perspective of the black player. When updating the information that belongs to the white player, it needs to be reversed, i.e., $v(s) = 0.1$. So, we get $N = 1$, $W = 0.1$, $Q = 0.1$.

Then we return to its parent node. Like in the first tree search process, as the current state is a root node, we do not need to backup W and Q here and only need to update the visit time N. So, let $N = 2$ and we have done another tree search as shown in Fig. 15.11.

The third tree search process also starts from the root node. With the formula $a = \arg\max_a(Q(s,a) + U(s,a))$ and the current information in the tree, the action chosen by the white player is 2 $(w, 2)$, and the black player chooses the action

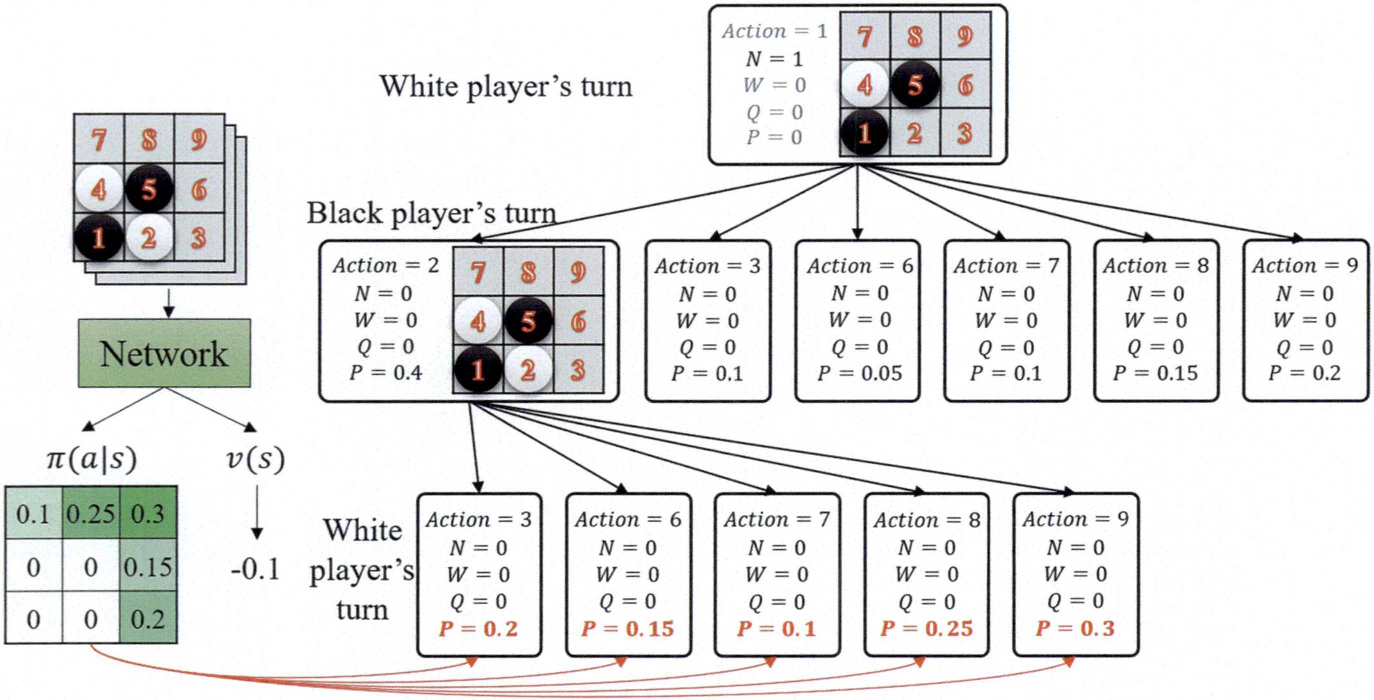

Fig. 15.9 *Expand and evaluate* at the new node. All the available actions are expanded, and the probabilities $\pi(a|s)$ are given by the network

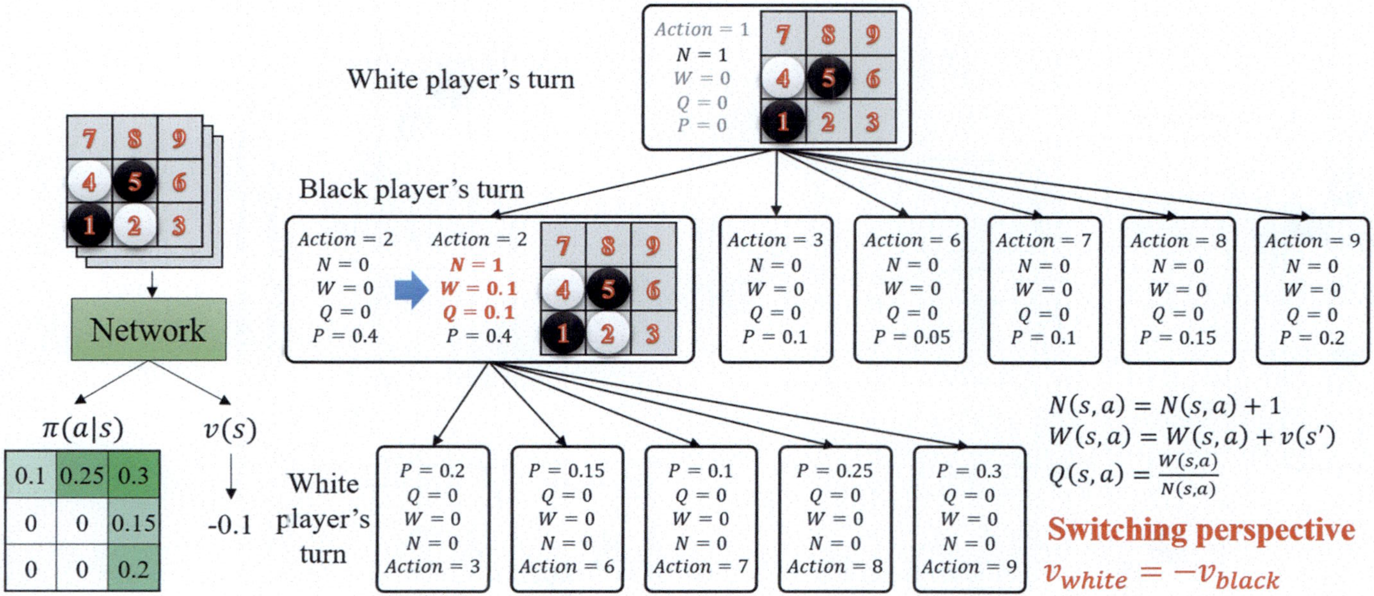

Fig. 15.10 *Backup* at the new node. The information at the current node is updated and Q should be updated in the white player's perspective: $N = 1$, $W = 0.1$, $Q = 0.1$

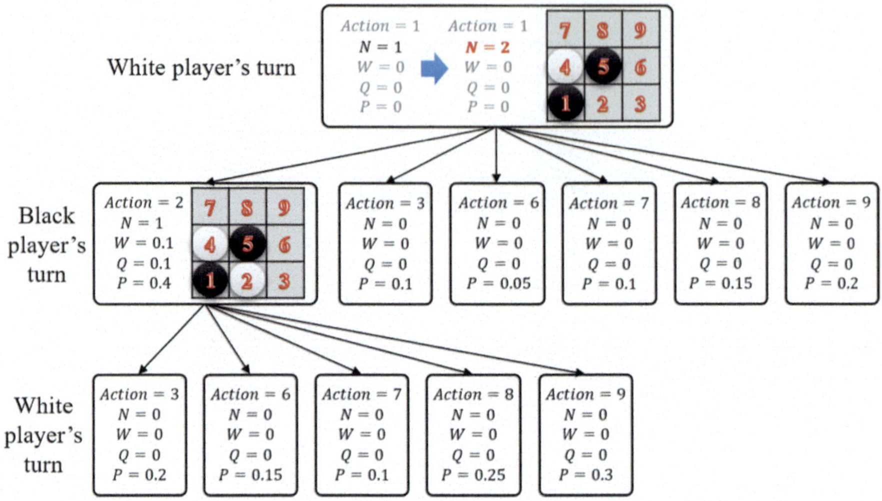

Fig. 15.11 *Backup* at the root node. N is updated to 2, W and Q are not needed to update

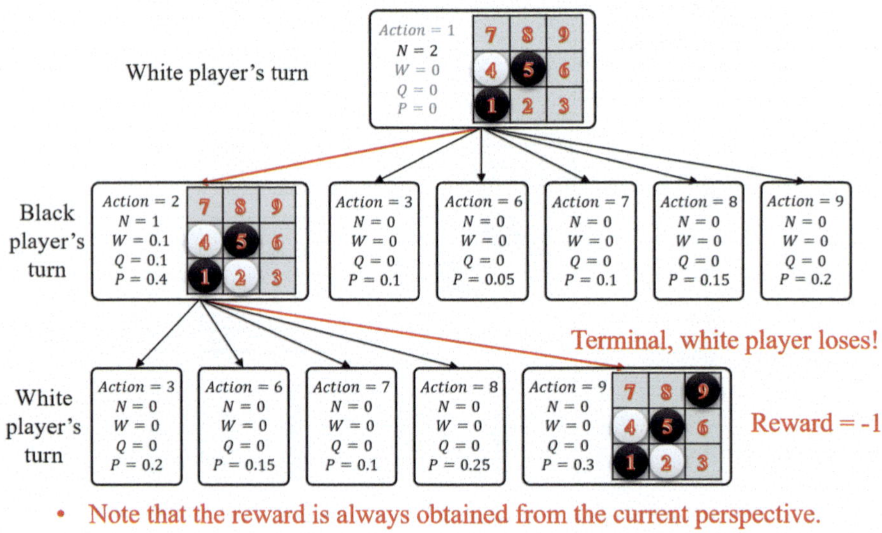

Fig. 15.12 *Expand and evaluate* at the terminal node. Since the game is end at this node, no node will be expanded and the reward can be obtained from the game directly. So, the policy network and the value network are not used here

9 $(b, 9)$. As shown in Fig. 15.12, the *select* step leads to a new node that is the end of the game. This time, for the step of *expand and evaluate*, the node will not be expanded and the value v can be obtained from the game directly. So, the value network is not used to evaluate the state and the policy network is not used to output the probability of each action.

Then, it is time to backup and there are three nodes in the trajectory. As we mentioned before, the nodes are updated recursively from the leaf node to the root node, $N(s,a) = N(s,a) + 1$, $W(s,a) = W(s,a) + v(s)$, $Q(s,a) = \frac{W(s,a)}{N(s,a)}$. Moreover, the perspective of each node should also be switched, which means $v_{white} = -v_{black}$. In this game, the black player chooses 9 $(b, 9)$ and reaches a new node. Now it is the white player's turn to choose an action, but unluckily the game is over and the white player loses the game. So the $reward = -1$ is from the perspective of the white player, that is to say $v_{white} = -1$. When we update the information at this node, as we mentioned before, these information is used by the black player to choose $Action = 9$ to reach this node, so the value for this node should be $v_{black} = -v_{white} = 1$, and other information is based on it, $N = 1$, $W = 1$, $Q = 1$. The information at other nodes are updated in the same way.

After the *backup* step, the renewed tree is shown in Fig. 15.13. The root node has been visited three times and the information at each visited node is updated.

We have demonstrated three times MCTS above. After the tree search is conducted 400 times as shown in Fig. 15.14 (in the AlphaGo Zero algorithm, the number is 1600; in the AlphaZero algorithm, the number is 800), the tree has grown much larger and the estimated values are more accurate.

After the MCTS, a stone can be placed on the real board now. The action is chosen by calculating the probability related to the visit count rather than the outputs of the policy network: $\pi(a|s) = \frac{N(s,a)^{1/\tau}}{N(s)^{1/\tau - 1}} = \frac{N(s,a)^{1/\tau}}{\sum_b N(s,b)^{1/\tau}}$, where $\tau \to 0$ is a

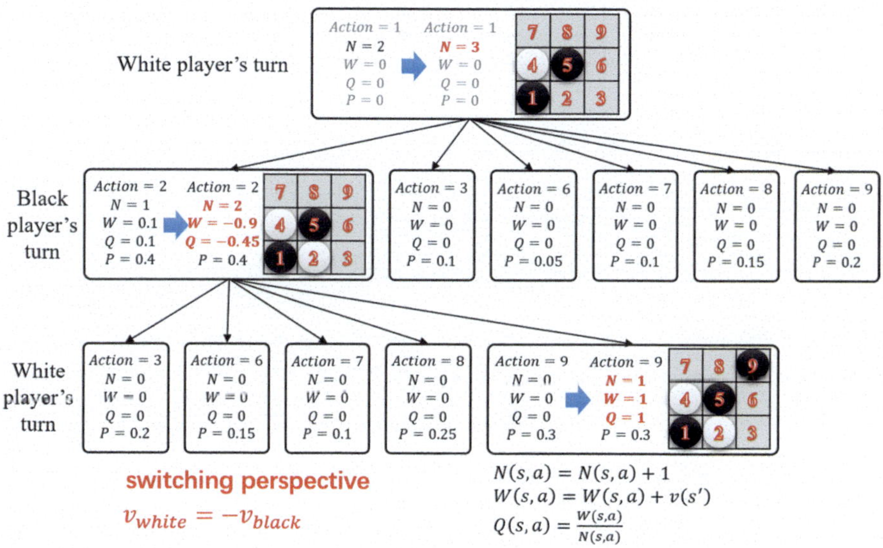

Fig. 15.13 The tree after the *backup* step. In the process of the third MCTS, the *backup* step updates the information of the three visited nodes recursively. Be careful, because two players are in a single tree and $v_{white} = -v_{black}$, the information should be updated from the right perspective

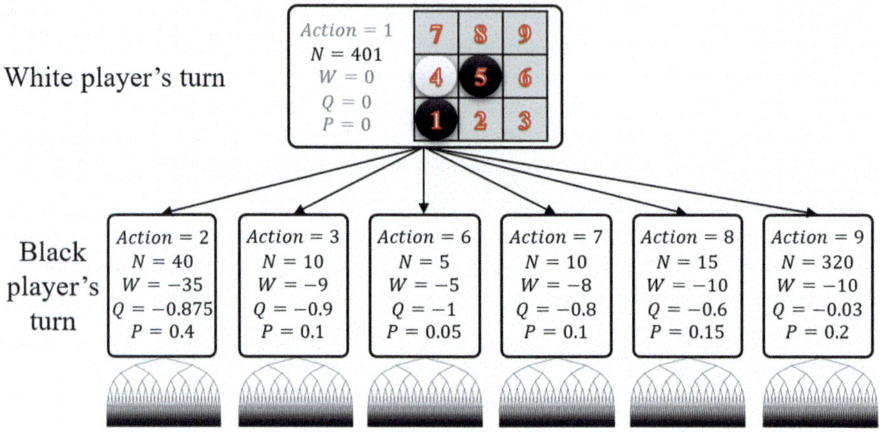

Fig. 15.14 The whole tree after searching 400 times. As the first MCTS process is started from the step of *expand and evaluate* and it does not select a child node, the sum of the visit counts of its child nodes is 400 and the root node's visit count is 401

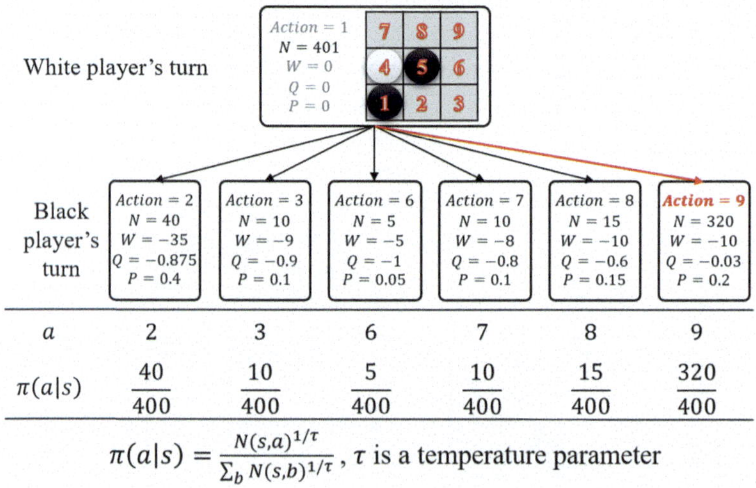

Fig. 15.15 Play a move on the board. After searching 400 times, an action is selected according to $\pi(a|s) = \frac{N(s,a)^{1/\tau}}{\sum_b N(s,b)^{1/\tau}}$. Here, the white player selects 9 $(w, 9)$

temperature parameter, $b \in A$ denotes the available action at state s. Here, the selected action is 9 $(w, 9)$ as shown in Fig. 15.15.

The temperature parameter is used to control the exploration. If $\tau = 1$, it selects moves proportionally to their visit counts in MCTS, this means a high-level exploration and ensures a diverse set of positions are encountered. If $\tau \to 0$, it means a low exploration and selects the move with maximum visit count. In the AlphaZero and AlphaGo Zero algorithm, when they perform the self-play process to collect data, the temperature is set to $\tau = 1$ for the first 30 moves (12 moves in our implementation), and $\tau \to 0$ for the remainder of the game. When playing a real game with an opponent, the temperature is set to $\tau \to 0$ all the time.

The white stone has been put at the position 9 $(w, 9)$ on the board, so the root node in the tree will be changed to the child node and MCTS will go on from this new root node. Other sibling nodes and its parent node will be discarded to prune the tree and save memory (Fig. 15.16).

When the game is over on the game board, we get the data and the result (Fig. 15.17).

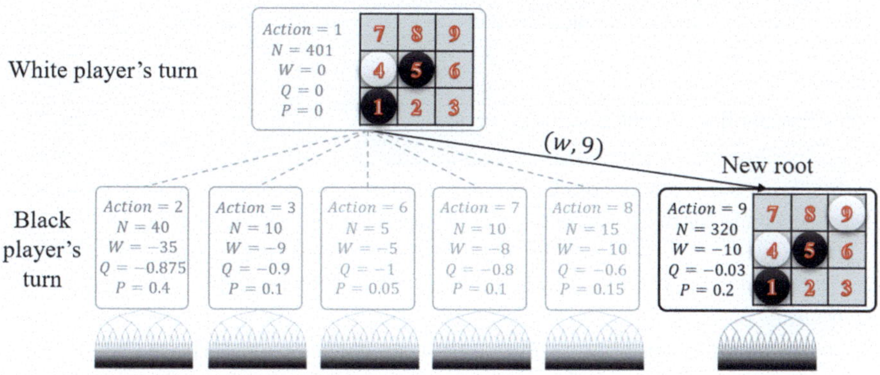

Fig. 15.16 The new root node. The nodes that under the new root node will be maintained, and other nodes will be discarded

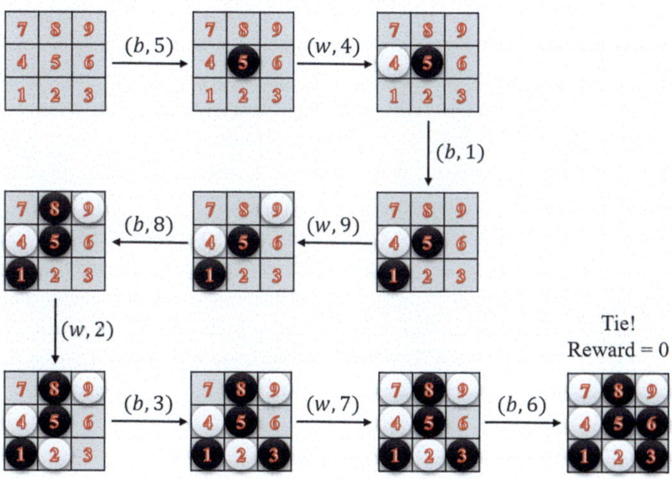

Fig. 15.17 The game data. All states in the game will be saved and labeled with probabilities $\pi(a|s)$ and value $v(s)$

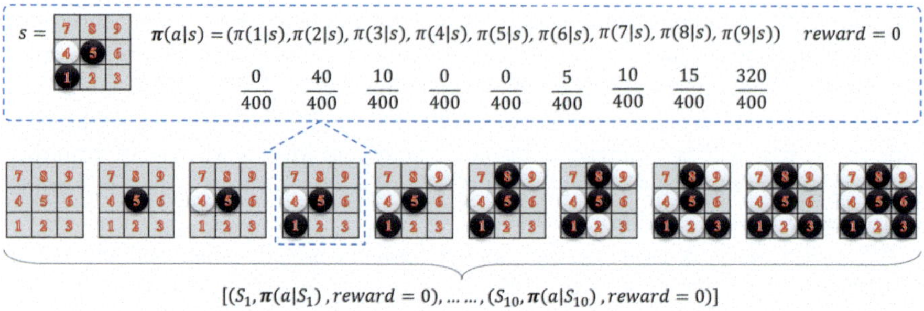

Fig. 15.18 Data with labels. The probability of the action is according to $\pi(a|s) = \frac{N(s,a)^{1/\tau}}{\sum_b N(s,b)^{1/\tau}}$, and the value $v(s)$ is from the result of the game: +1 for a win, −1 for a loss, 0 for a draw

The probability for each action is: $\pi(a|s) = \frac{N(s,a)^{1/\tau}}{\sum_b N(s,b)^{1/\tau}}, \tau = 1$. Be careful that the probabilities here are related to visit counts; this is the key point that combines the MCTS self-play and policy network training. As the result of this game is a draw, the labels for the value network here are 0 for all these data (Fig. 15.18).

Now, the data is available by the self-play process using Monte Carlo Tree Search, the next part is to apply the deep neural network for data training. In the training process, the data is first transformed into some feature planes. Each feature plane composes binary values indicating the presence of the player's stones, with one set of planes for the current player and another set of planes for the opponent's stones. These planes are concatenated together in order with history features included. Then, we augment the data in the same way as in the AlphaGo Zero algorithm: Since the rules of Go and Gomoku are both invariant to rotation and reflection, the data are augmented by rotation and reflection before training. And during the MCTS, board positions are randomly rotated or reflected before being evaluated by the neural network, which can average different biases. However, in the AlphaZero algorithm, the trick is not used since some games' rules are not invariant to rotation and reflection. With more and more sampled data collected, the network is trained to be more accurate for estimation.

We use `ResNet` (He et al. 2016) as the network structure (Fig. 15.19) which is the same as the AlphaGo Zero algorithm. The network's inputs are the game states, the outputs are the probabilities of actions and the values of the states. The network can be denoted as $(\boldsymbol{p}, v) = f_\theta(s)$ and the data is $(s, \boldsymbol{\pi}, r)$. The loss function l combines the cross-entropy losses over the actions' probability distribution, mean-squared error over the state value, and the L2 weight regularization over the parameters. The concrete formula is $l = (r - v)^2 - \boldsymbol{\pi}^T \log \boldsymbol{p} + c||\theta||^2$, where c is a parameter controlling the level of regularization.

There are some details about the timing of updating the model. In the AlphaGo Zero algorithm, the new model will play with the current best model for 400 games; if the new one wins by a margin of 55%, it becomes a better model. By contrast, in the version of the AlphaZero algorithm, it does not play against previous models and the parameters are updated continuously. These are alternative approaches with

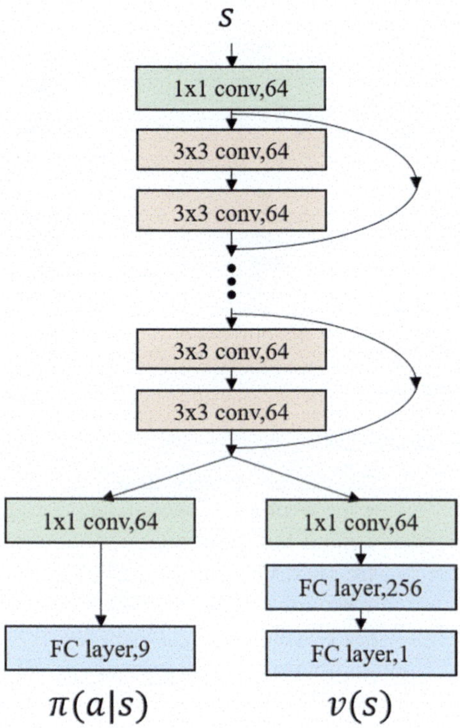

Fig. 15.19 Network structure. The structure is the same with the AlphaGo Zero algorithm. `ResNet` is used as the backbone and two heads output the probability distribution and state value separately

different performances, and we implement our code in the AlphaGo Zero's manner in this project to make the training process more stable. Besides, if you want to train it faster, you can collect data in a parallel way using multi-process and do the tree search process asynchronously as the original paper does. In the parallel way (Fig. 15.20), there are many processes executing at the same time: the parameters of the neural network are trained with the latest self-play data, the new self-play data are generated from the best model, the latest model is continually evaluated (in AlphaGo Zero), all these components are executed simultaneously.

Then, with a stream of new data saved in a buffer and with this neural network to learn a more accurate policy head and value head, we will get a powerful Gomoku AI after iterating the MCTS and network training for sufficient times.

We finally trained a Gomoku model in a parallel way with the free-style rule (a row of five or more stones for a win) on a 11×11 board. Some specific parameters are presented in Table 15.1. A model on a 15×15 board is also trained successfully with the same parameters, which indicates the generality and stability of the AlphaZero algorithm.

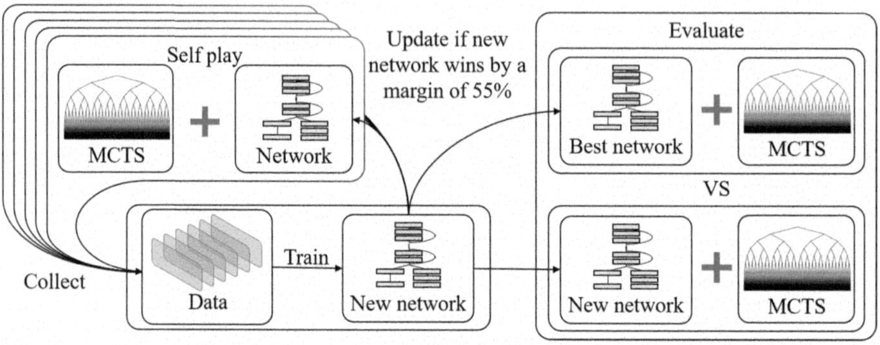

Fig. 15.20 Parallel training structure

Table 15.1 Comparison of parameters

Parameters setting	Gomoku	AlphaGo Zero	AlphaZero
c_{puct}	5	5	5
MCTS times	400	1600	800
Residual blocks	19	19/39	19/39
Batch size	512	2048	4096
Learning rate	0.001	Annealed	Annealed
Optimizer	Adam	SGD with momentum	SGD with momentum
Dirichlet noise	0.3	0.03	0.03
Weight of noise	0.25	0.25	0.25
$\tau = 1$ for the first n moves	12	30	30

References

Albert M, Nowakowski R, Wolfe D (2007) Lessons in play: an introduction to combinatorial game theory. CRC Press, Boca Raton

Auer P, Cesa-Bianchi N, Fischer P (2002) Finite-time analysis of the multiarmed bandit problem. Mach Learn 47(2–3):235–256

Browne CB, Powley E, Whitehouse D, Lucas SM, Colton S (2012) A survey of Monte Carlo tree search methods. IEEE Trans Comput Intell Ai Games 4(1):1–43

Campbell M, Hoane Jr AJ, Hsu FH (2002) Deep blue. Artif. Intell. 134(1–2):57–83

Couetoux A, Milone M, Brendel M, Doghmen H, Sebag M, Teytaud O (2011) Continuous rapid action value estimates. In: Asian conference on machine learning, pp 19–31

He K, Zhang X, Ren S, Sun J (2016) Deep residual learning for image recognition. In: Proceedings of the IEEE conference on computer vision and pattern recognition, pp 770–778

Hsu Fh (1999) IBM's deep blue chess grandmaster chips. IEEE Micro 19(2):70–81

Kocsis L, Szepesvári C (2006) Bandit based Monte-Carlo planning. In: European conference on machine learning. Springer, Berlin, pp 282–293

Muthoo RBA (1996) A course in game theory by Martin J. Osborne; Ariel Rubinstein. Economica 63(249):164–165

Osborne MJ, Rubinstein A (1994) A course in game theory. MIT press

Silver D, Huang A, Maddison CJ, Guez A, Sifre L, Van Den Driessche G, Schrittwieser J, Antonoglou I, Panneershelvam V, Lanctot M, et al (2016) Mastering the game of go with deep neural networks and tree search. Nature 529:484

Silver D, Hubert T, Schrittwieser J, Antonoglou I, Lai M, Guez A, Lanctot M, Sifre L, Kumaran D, Graepel T, et al (2017a) Mastering chess and shogi by self-play with a general reinforcement learning algorithm. Preprint. arXiv:171201815

Silver D, Schrittwieser J, Simonyan K, Antonoglou I, Huang A, Guez A, Hubert T, Baker L, Lai M, Bolton A, Chen Y, Lillicrap T, Hui F, Sifre L, van den Driessche G, Graepel T, Hassabis D (2017b) Mastering the game of go without human knowledge. Nature 550(7676):354

Silver D, Hubert T, Schrittwieser J, Antonoglou I, Lai M, Guez A, Lanctot M, Sifre L, Kumaran D, Graepel T, et al (2018) A general reinforcement learning algorithm that masters chess, shogi, and Go through self-play. Science 362(6419):1140–1144

Chapter 16
Robot Learning in Simulation

Zihan Ding and Hao Dong

Abstract This chapter introduces a hands-on project for robot learning in simulation, including the process of setting up a task with a robot arm for objects grasping in CoppeliaSim and the deep reinforcement learning solution with soft actor-critic algorithm. The effects of different reward functions are also shown in the experimental sections, which testifies the importance of auxiliary dense rewards for solving a hard-to-explore task like the robot grasping ones. Brief discussions on robot learning applications, sim-to-real transfer, other robot learning projects and simulators are also provided at the end of this chapter.

Keywords Robot learning · Deep reinforcement learning · Simulation · Dense reward · Parallel training · Soft actor-critic · Domain randomization

Deep reinforcement learning methods have many potential real-world application scenarios, and robotic control is one of the most exciting areas. Although deep reinforcement learning methods have already been able to solve most of the simple games like in OpenAI Gym as described in previous Chaps. 4, 5, and 6, at present, we may not expect that the deep reinforcement learning methods can be applied in robotics control as a completely alternative approach of traditional control methods with inverse kinematics or proportional-integral-derivative (PID) controller. However, deep reinforcement learning algorithms can be applied in some specific situations or as a combinatorial approach with traditional control methods, especially for highly complicated systems or dexterous manipulations (Akkaya et al. 2019; Andrychowicz et al. 2018).

Z. Ding (✉)
Imperial College London, London, UK
e-mail: zhding@mail.ustc.edu.cn

H. Dong
Peking University, Beijing, China
e-mail: hao.dong@pku.edu.cn

H. Dong et al. (eds.), *Deep Reinforcement Learning*,
https://doi.org/10.1007/978-981-15-4095-0_16

Fig. 16.1 The scene of a robotic hand for solving the Rubik's cube. Figure is adapted from Akkaya et al. (2019)

In most cases, the dynamic process of robotic control can be approximated with Markov process well, which makes it an ideal experimental site for deep reinforcement learning, in both simulation and real world. Moreover, the great potential of deep reinforcement learning for robotic control in the real world drives high-tech companies like DeepMind and OpenAI to dedicate in this research area. Recently, OpenAI even solved the Rubik's cube with a single five-fingered humanoid robotic hand (Akkaya et al. 2019), using automatic domain randomization technique for sim-to-real transfer, as shown in Fig. 16.1. Other companies also start to investigate the usage of robotic arms in the distribution process of warehouse logistics, even through directly training the robot in real world (Korenkevych et al. 2019).

However, due to the sample inefficiency and the safety issues when applying deep reinforcement learning algorithms in the real world, one can find it hard to train the reinforcement learning policy directly in the real world for complicated robotic systems or for dexterous manipulations (Akkaya et al. 2019). Training in simulation and transferring the policy into real-world applications later on or leveraging human expert demonstrations are potential approaches for satisfying the computational and safety requirements in robot learning tasks. The simulators for robots have been widely developed for decades, including, DART, CoppeliaSim (previously called V-REP for version 3.6.2 and before) (Rohmer et al. 2013), MuJoCo, Gazebo, etc., which will be discussed in the last section of this chapter. Most of these simulators have Python counterparts for the convenience of applying reinforcement learning control policies and other numerical operations.

Learning in simulation is meaningful for at least two aspects. First, the simulated environments can be used as the testbed for proposed learning methods or frameworks (including but not limited to reinforcement learning area), especially for large-scale real-world applications, e.g. robotic learning tasks. Learning in simulation serves as a verification process for new methods before being applied in real-world scenarios. Second, learning in simulation is an indispensable step for achieving the real-world tasks via the sim-to-real transfer approach, to reduce the time consumption and mechanical loss.

In this chapter, we will introduce the process of applying deep reinforcement learning algorithms on a simple robotic object-grasping task in simulation using CoppeliaSim (V-REP) simulator and its Python wrapper: PyRep (James et al. 2019a). We also release the code of task descriptions and deep reinforcement learning algorithms for the project described in this chapter,[1] for the convenience of learning and understanding.

Since there is already another application working on large-scale high-dimensional continuous control as in Chap. 13, the robot learning task in this chapter will focus more on different aspects of applying deep reinforcement learning methods in practice, including how to build up an environment for achieving a certain task with reinforcement learning, how to design the reward function for assisting reinforcement learning and achieving the final goal of the task in the end, etc, to provide the readers a better understanding of applying reinforcement learning not only in training but also in the designing of learning environments.

16.1 Robotics Simulation

The first step we need to do is to set up an environment containing: a robotic arm and the objects it interacts with in simulation, which is supposed to follow realistic physical dynamics. However, here we need to emphasize that, a realistic simulation does not mean the policy learned in simulation can be directly applied in real world with good performance. The "realistic" evaluation can be achieved in a variety of specific forms, but only one of it could match the exact real-world setting. For example, the lighting conditions can give different kinds of shadow effects on the objects, all looking realistic, but one of them matches the real case and slight differences in the appearance may lead to significantly different actions in reality, due to the sensitivity of deep neural networks. To solve this kind of problem in the simulation-to-reality transfer process, a variety of methods including domain randomization (Andrychowicz et al. 2018) or dynamics randomization (Peng et al. 2018) are applied, which will be discussed later in this chapter.

There are several simulators for robotics, including CoppeliaSim (V-REP), MuJoCo, Unity, and so on. Original CoppeliaSim (V-REP) software using common interfaces of C++ and Lua languages, only partial functions are implemented with Python. However, it is better to use Python interfaces for the convenience of applying reinforcement learning algorithms. Luckily, we have PyRep package developed for bringing CoppeliaSim (V-REP) to deep robot learning. In this project, we choose to use CoppeliaSim (V-REP), with its package PyRep using the Python interfaces.

[1]Code link:https://github.com/deep-reinforcement-learning-book/Chapter16-Robot-Learning-in-Simulation.

We will demonstrate a basic process of setting up a robot learning task in this section.

16.1.1 Install CoppeliaSim and PyRep

The CoppeliaSim (V-REP) software can be downloaded from its official website,[2] and we need the version of CoppeliaSim (V-REP) to be 3.6.2 (which can be found at the website[3]) to be compatible with the PyRep during the period of writing this book. It can be installed directly through unzipping the downloaded file. Note that a version of CoppeliaSim (V-REP) higher than 3.6.2 may not be compatible with the other modules in this project.

After installing CoppeliaSim (V-REP), we can install a forked stable version of PyRep with the following steps on the repository website (`https://github.com/deep-reinforcement-learning-book/PyRep`):

```
git clone
    https://github.com/deep-reinforcement-learning-book/PyRep.git
pip3 install -r requirements.txt
python3 setup.py install --user
# change the path to the installation path of VREP on your local
    machine
export VREP_ROOT=EDIT/ME/PATH/TO/V-REP/INSTALL/DIR
export LD_LIBRARY_PATH=$LD_LIBRARY_PATH:$VREP_ROOT
export QT_QPA_PLATFORM_PLUGIN_PATH=$VREP_ROOT
source ~/.bashrc
```

Remember to change the path of V-REP in the above script of `VREP_ROOT`.

16.1.2 Git Clone Our Project

Our project for deep reinforcement learning on robotics learning tasks can be downloaded here:

```
git clone https://github.com/deep-reinforcement-learning-book/
    Chapter16-Robot-Learning-in-Simulation.git
```

It contains the robotic parts (arms, grippers) and other objects we needed, built-up scenes for robot grasping tasks, deep reinforcement learning algorithms for training an agent-control policy, etc. The robot grasping scene built in this project is

[2]http://www.coppeliarobotics.com.

[3]http://www.coppeliarobotics.com/previousVersions.

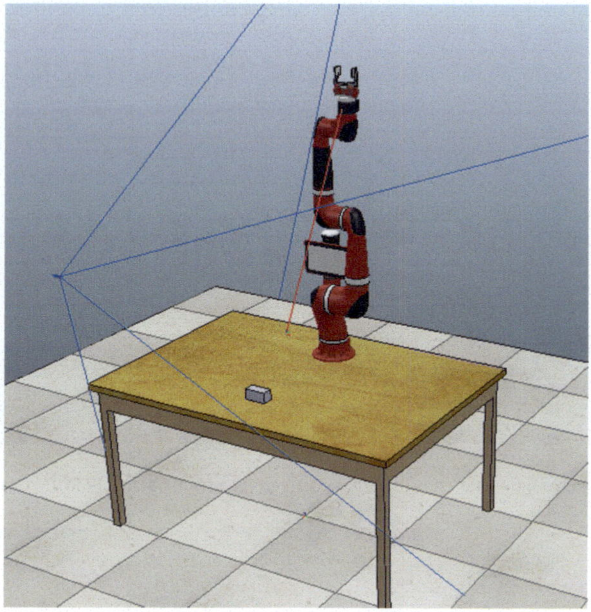

Fig. 16.2 The scene of *Grasping* task in CoppeliaSim (V-REP)

visualized in Fig. 16.2. We will demonstrate how to construct such a scene with the basic components in the following several sections.

16.1.3 Assemble the Robot

We use the robotic arm named *Rethink Sawyer* with a gripper at its end, and the gripper we use in this example is the *BraxterGripper*. Official PyRep package provides a variety of robot arms and grippers, which can be used to assemble and build the task scenes you want. Here we assemble the gripper onto the robot arm as an example, as shown in Fig. 16.3.

In our git folder, we drag the `./hands/BaxterGripper.ttm` and `./arms/Sawyer.ttm` into a new scene opened in CoppeliaSim (V-REP). We choose the gripper and Ctrl+Left Click (on Mouse) the end joint of *Sawyer* (i.e., `Sawyer_wrist_connector`, which is a force sensor in CoppeliaSim (V-REP) and could be used for connecting different objects), then click assemble button as in Fig. 16.4. There are different types of connector provided in CoppeliaSim (V-REP), here the force sensor is only one of them with broken potentials when the true forces applied on the joints are larger than the threshold. Another thing is, we should not use "group/merge" here, in order to control the gripper separately from the robot arm. More details about how to connect and combine multiple bodies can be referred to the website of CoppeliaSim (V-REP). After we finish the above process, the scene hierarchy will be the same as displayed in Fig. 16.5.

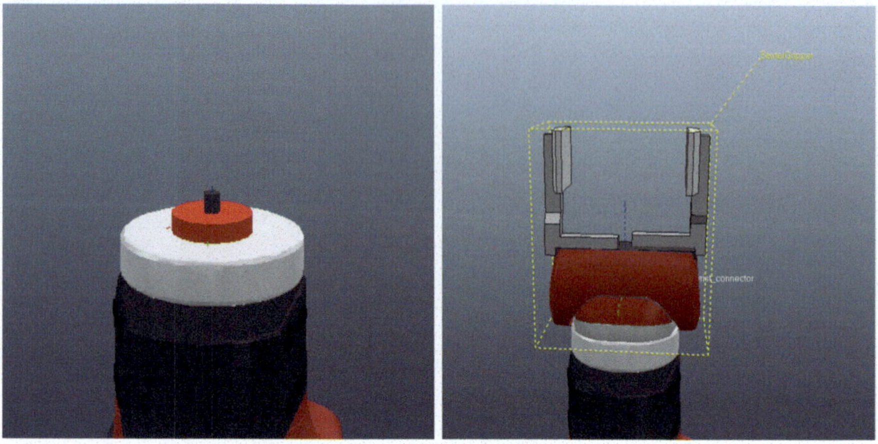

Fig. 16.3 The end of *Sawyer* arm (left) and the assembled *BaxterGripper* (right)

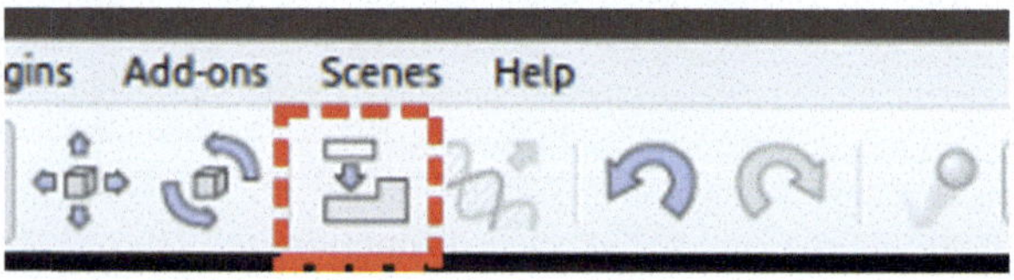

Fig. 16.4 The "assemble" button in CoppeliaSim (V-REP)

Set Up the Learning Environment

A pre-built environment is provided in `./scenes/sawyer_reacher_rl.ttt` with CoppeliaSim (V-REP), as shown in Fig. 16.2. In order to build such a scene, we need to add some other objects in the current scene, which only contains the robot arm and a gripper at present.

First, we add a target object with `Add -> Primitive shape -> Cuboid`, resize its shape as we want and rename it as "target." We need to double click the icon prior to the "target," and choose `Common -> Renderable` to make the object visible for the vision sensor.

After those steps, we need to add a vision sensor for providing us a customized view of the scene. This vision sensor can keep capturing images of the view during the simulation process, which is necessary if we use image-based control (otherwise not necessary). We can get images as the return of a simulation step only if we enable such a vision sensor in the scene. In order to set such a scene, click `Add -> Vision sensor -> perspective type`, and then right click the mouse on the scene, choose `Add -> Floating view`. Then first click the `Vision_sensor` we just create, and right click the mouse on the opened floating view, choose `View -> associate view with selected vision sensor`. We manually set the position and rotation of the added vision sensor, and get the scene after setting up as in Fig. 16.6.

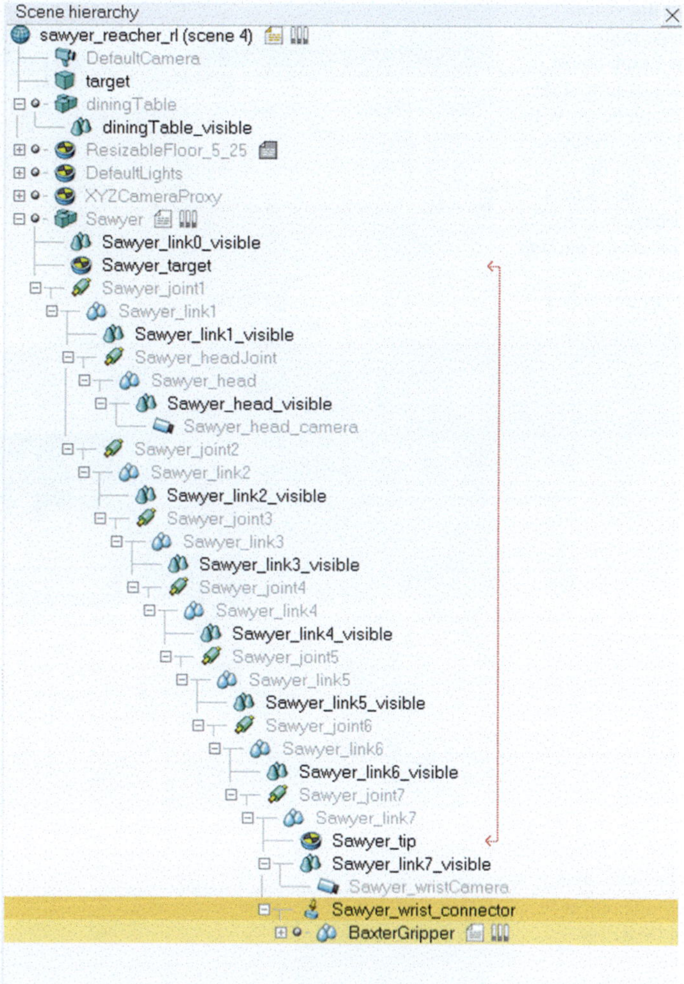

Fig. 16.5 The hierarchy of the task scene in CoppeliaSim (V-REP), with all physical models including the *Sawyer* robot arm. The red arrow indicates the inverse kinematics chain for end-position control mode. The black fonts represent the objects visible in the scene, while the gray fonts indicate the virtual objects that are not visible

Then, we drag in an object `table.ttt` from `./objects` in the project folder. We manually set the position of the robot arm with gripper and the target cuboid on the surface of the table as in Fig. 16.7 by clicking the button of `Object/item shift`.

Above is the setup process of the environment scene, which provides us the visible entities in the task. The dynamics of those entities will follow the simulated rules of the physical simulator. Apart from that, we also need to define

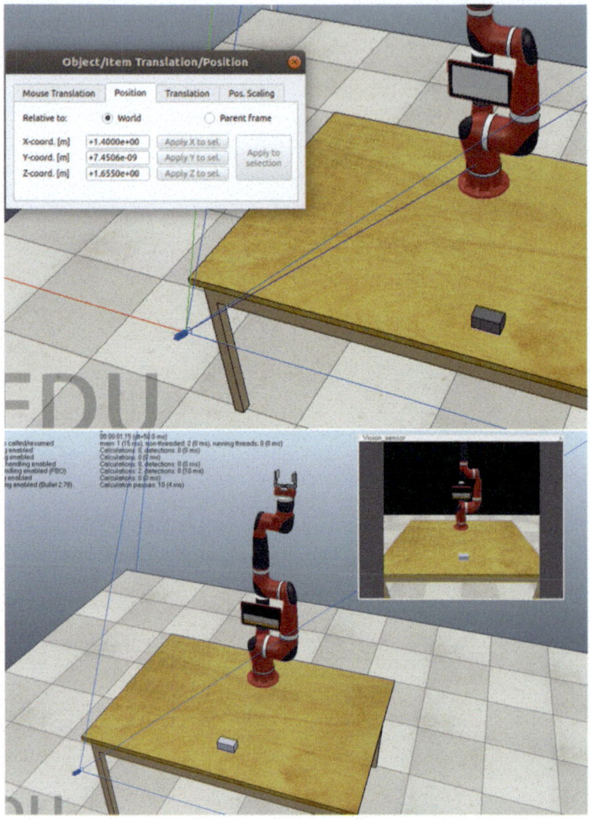

Fig. 16.6 Set the vision sensor in CoppeliaSim (V-REP). The graph above is for setting the position of the camera; the graph below with a small window on the upper right corner is from the camera placed. The camera can provide image observation for each time step if being called by the image-based control policy

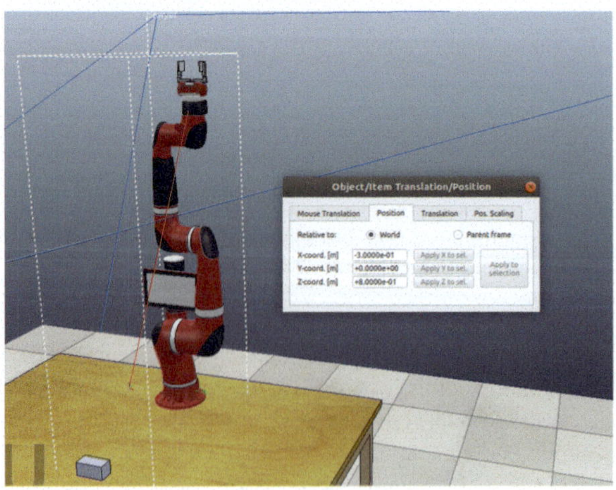

Fig. 16.7 Manually change the position of the object in CoppeliaSim (V-REP)

the control process and reward functions of the environment, which generally includes the constraints of object movement (for locomotion tasks mainly), the starting and ending procedures for an episode during training, initialization conditions, the observation formats, and so on. In our git folder, we provide the file `sawyer_grasp_env_boundingbox.py` for achieving these functions in the built scene, and it is written with APIs close to the OpenAI Gym environments, for the convenience of applying reinforcement learning algorithms to control it later. The scene we built above is static itself, and the control script provides the functions to define the dynamics (apart from those defined in the physical simulator) for it. For the robot grasping task here, we use the control mechanism of forward kinematics (control by joints velocity directly) for the robotic arm, and we also have another scene with inverse kinematics control (control by the end position of robotic arm) with different configurations. Inverse kinematics usually requires to solve the inverse Jacobian matrix of the function describing the relationship of joint angles and end position of the robot arm, which is supported by PyRep as well. More details about inverse kinematics settings are beyond the scope of this book. The scripts in our provided example code for defining the dynamics and controlling the robot support both of the controlling mechanisms.

Note In practice, when you are trying to build your own robotic models or to assemble different parts for other customized robotic arms, you need to be careful of the order and dependency relationships of different modules on the robotic arm. This is concerned with some requirements on the dynamic and static parts when building a model in CoppeliaSim (V-REP) (e.g., the `Sawyer_tip` for inverse kinematics is a static component). More details are provided at website.[4]

After setting up the environment scene in CoppeliaSim (V-REP), we need to write a control script with PyRep package for defining the dynamics and reward function of the environment. The code for defining the environment is provided in our repository. We will introduce the functions and modules in the project with the following sections.

Modules in Environment Script

Import the necessary packages and set global variables needed afterward:

```
from os.path import dirname, join, abspath
from pyrep import PyRep
from pyrep.robots.arms.sawyer import Sawyer
from pyrep.robots.end_effectors.baxter_gripper import
    BaxterGripper
from pyrep.objects.proximity_sensor import ProximitySensor
from pyrep.objects.vision_sensor import VisionSensor
from pyrep.objects.shape import Shape
from pyrep.objects.dummy import Dummy
from pyrep.const import JointType, JointMode
```

[4]http://www.coppeliarobotics.com/helpFiles/en/designingDynamicSimulations.htm.

```
import numpy as np
import matplotlib.pyplot as plt
import math

POS_MIN, POS_MAX = [0.1, -0.3, 1.], [0.45, 0.3, 1.] # valid
    position range of target object
```

The overall structure of the class defining the environment of robot grasping task is shown below, and all functions defined in the class are shortened here, which will be expanded later.

```
class GraspEnv(object):
    ''' Sawyer robot grasping a cuboid '''
    def __init__(self, headless, control_mode='joint_velocity'):
        '''
        parameters:
        :headless: bool, if True, no visualization, else with
            visualization.
        :control mode: str, 'end_position' or 'joint_velocity'.
        '''
        ...

    def _get_state(self):
        '''
        Return state containing arm joint angles/velocities &
            target position.
        '''
        ...

    def _is_holding(self):
        '''
         Return the state of holding the target or not, return
             bool.
        '''
        ...

    def _move(self, action, bounding_offset=0.15,
         step_factor=0.2, max_itr=20, max_error=0.05,
         rotation_norm =5.):
        '''
        Move the tip according to the action with inverse
            kinematics for 'end_position' control mode. Inverse
            kinematics mode control is achieved through setting
            the tip target instead of using .solve_ik(), because
            sometimes the .solve_ik() does not function correctly.
        Mode: close-loop proportional control, using inverse
            kinematics.

        parameters:
        :bounding_offset: offset of bounding box outside the valid
            target position range, as valid and safe range of
            action
```

```
        :step_factor: small step factor multiplied on the
            difference of current and desired positions, i.e.
            proportional factor
        :max_itr: maximum moving iterations
        :max_error: upper bound of distance error for movement at
            each call
        :rotation_norm: factor for normalization of rotation
            values, since the actions are of the same scale for
            each dimension
        '''
        ...

    def reinit(self):
        '''
        Reinitialize the environment, e.g. when the gripper is
            broken during exploration.
        '''
        ...

    def reset(self, random_target=False):
        '''
        Reset the gripper position and the target position.
        '''
        ...

    def step(self, action):
        '''
        Move the robot arm according to the action.
        If control_mode=='joint_velocity', action is 7 dim of
            joint velocity values + 1 dim rotation of gripper;
        if control_mode=='end_position', action is 3 dim of tip
            (end of robot arm) position values + 1 dim rotation of
            gripper;
        '''
        ...

    def shutdown(self):
        ''' Close the simulator '''
        ...
```

The first step is to initialize the environment, including setting the public variables, launching and setting up the scene, setting the proxy variables for the counterparts in the scene, which is defined as the `__init__()` function as follows.

```
    def __init__(self, headless, control_mode='joint_velocity'):
        '''
        parameters:
        :headless: bool, if True, no visualization, else with
            visualization.
        :control mode: str, 'end_position' or 'joint_velocity'.
        '''
        # set public variables
```

```
self.headless = headless # if headless is True, no
    visualization
self.reward_offset = 10.0 # reward value of grasping the
    object
self.reward_range = self.reward_offset # reward range
self.penalty_offset = 1. # penalty value for undesired
    cases
self.fall_down_offset = 0.1 # distance for judging the
    target object fall off the table
self.metadata=[] # gym env argument
self.control_mode = control_mode # the control mode of
    robotic arm: 'end_position' or 'joint_velocity'
```

The second part of the `__init__()` function is to launch and set up the scene, and set the proxy variables for the counterparts in the scene:

```
self.pr = PyRep() # call the PyRep
if control_mode == 'end_position': # the control mode with
    all joints in inverse kinematics mode
   SCENE_FILE = join(dirname(abspath(__file__)),
       './scenes/sawyer_reacher_rl_new_ik.ttt') # scene
       with joints controlled by ik (inverse kinematics)
elif control_mode == 'joint_velocity': # the control mode
    with all joints in force/torch mode for forward
    kinematics
   SCENE_FILE = join(dirname(abspath(__file__)),
       './scenes/sawyer_reacher_rl_new.ttt') # scene with
       joints controlled by forward kinematics
self.pr.launch(SCENE_FILE, headless=headless) # launch the
    scene, headless means no visualization
self.pr.start()  # start the scene
self.agent = Sawyer() # get the robot arm in the scene
self.gripper = BaxterGripper() # get the gripper in the
    scene
self.gripper_left_pad = Shape('BaxterGripper_leftPad') #
    the left pad on the gripper finger
self.proximity_sensor =
    ProximitySensor('BaxterGripper_attachProxSensor') #
    need the name of the sensor here
self.vision_sensor = VisionSensor('Vision_sensor') # need
    the name of the sensor here
self.table = Shape('diningTable') # the table in the scene
    for checking collision
if control_mode == 'end_position': # control the robot arm
    by the position of its end using inverse kinematics
   self.agent.set_control_loop_enabled(True) # if false,
       inverse kinematics won't work
   self.action_space = np.zeros(4) # 3 DOF end position
       control + 1 DOF rotation of gripper
elif control_mode == 'joint_velocity': # control the robot
    arm by directly setting velocity values on each joint,
    using forward kinematics
   self.agent.set_control_loop_enabled(False)
```

```
            self.action_space = np.zeros(7) # 7 DOF velocity
                control, no need for extra control of end rotation,
                the 7th joint controls it.
        else:
            raise NotImplementedError
        self.observation_space = np.zeros(17) # scalar positions
            and scalar velocities of 7 joints + 3-dimensional
            position of the target
        self.agent.set_motor_locked_at_zero_velocity(True)
        self.target = Shape('target') # get the target object
        self.agent_ee_tip = self.agent.get_tip() # a part of robot
            as the end of inverse kinematics chain for controlling
        self.tip_target = Dummy('Sawyer_target') # the target
            point of the tip (end of the robot arm) to move towards
        self.tip_pos = self.agent_ee_tip.get_position() # tip
            x,y,z position
```

The third part of the `__init__()` function is to set a proper initial robot gesture or tip position:

```
        if control_mode == 'end_position':
            initial_pos = [0.3, 0.1, 0.9]
            self.tip_target.set_position(initial_pos) # set target position
            # one big step for rotation setting is enough, with
                reset_dynamics=True, set the rotation instantaneously
            self.tip_target.set_orientation([0,np.pi,np.pi/2],
                reset_dynamics=True) # first two dimensions along x and y
                axis make gripper face downwards
            self.initial_tip_positions = self.initial_target_positions =
                initial_pos
        elif control_mode == 'joint_velocity':
            self.initial_joint_positions = [0.0, -1.4, 0.7, 2.5, 3.0, -0.5,
                4.1] # a proper initial gesture
            self.agent.set_joint_positions(self.initial_joint_positions)
        self.pr.step()
```

A function to get the observed state is as follows, including the joint positions and velocities, and the 3-dimensional position of the target object, totally 17 dimensions.

```
    def _get_state(self):
        '''
        Return state containing arm joint positions/velocities &
            target position.
        '''
        return np.array(self.agent.get_joint_positions() + # list,
            dim=7
                self.agent.get_joint_velocities() + # list, dim=7
                self.target.get_position()) # list, dim=3
```

A function to determine whether the object is grasped by the gripper is defined as `_is_holding()`, which applies the collision detection on the pad of the gripper and the proximity sensor for determining if the object is within the gripper.

```
def _is_holding(self):
    '''
     Return the state of holding the target or not, return
         bool.
    '''
    # Note that the collision check is not always accurate,
    # for continuous collision frames, maybe only the first
        4-5 frames of collision can be detected.
    pad_collide_object =
        self.gripper_left_pad.check_collision(self.target)
    if pad_collide_object and
        self.proximity_sensor.is_detected(self.target)==True:
       return True
    else:
       return False
```

A function for moving the end effector of the robot arm in inverse kinematics mode operation within a valid range ("bounding box") is provided as `_move()`. A tip is applied in PyRep at the end of the robot arm for achieving the end effector control with inverse kinematics through setting up the tip position and orientation. By calling the `pr.step()` function, the inverse kinematics for controlling robot joint movements will be solved automatically within PyRep. Due to the inaccuracy of the single big-step control, here we decompose the transition movement of the action into small steps and take a feedback control loop with maximum iteration and maximum tolerated error to conduct the small-step actions.

```
def _move(self, action, bounding_offset=0.15, step_factor=0.2,
     max_itr=20, max_error=0.05, rotation_norm =5.):
    '''
    Move the end effector on robot arm according to the action with
         inverse kinematics for 'end_position' control mode.
    Inverse kinematics mode control is achieved through setting the tip
         target instead of using .solve_ik(), because sometimes the
         .solve_ik() does not function correctly.
    Mode: a close-loop proportional control, using inverse kinematics.
    -------------------------------------------------------------------
    parameters:
    :bounding_offset: offset of a bounding box outside the valid target
         position range, as valid and safe range for restricting the
         potential action
    :step_factor: small step factor mulitplied on the difference of
         current and desired position, i.e. proportional factor
    :max_itr: maximum moving iterations
    :max_error: upper bound of distance error for movement at each call
    :rotation_norm: factor for normalization of rotation values, since the
         actions are of the same scale for each dimension
    '''
    pos=self.gripper.get_position()

    # check whether state+action will be within of the bounding box, if
         so, move normally; otherwise the action is not conducted.
    # i.e. x_min < x < x_max and y_min < y < y_max and z > z_min
```

```
if pos[0]+action[0]>POS_MIN[0]-bounding_offset and
    pos[0]+action[0]<POS_MAX[0]+bounding_offset \
  and pos[1]+action[1] > POS_MIN[1]-bounding_offset and
      pos[1]+action[1] < POS_MAX[1]+2*bounding_offset \
  and pos[2]+action[2] > POS_MIN[2]-2*bounding_offset: # larger
      offset in z axis

  # there is a mismatch between the object set_orientation() and
      get_orientation():
  # the (x,y,z) in set_orientation() will be (y,x,-z) in
      get_orientation().
  ori_z=-self.agent_ee_tip.get_orientation()[2] # the minus is
      because the mismatch between the set_orientation() and
      get_orientation()
  target_pos =
      np.array(self.agent_ee_tip.get_position())+np.array(action[:3])
  diff=1 # intialization
  itr=0
  while np.sum(np.abs(diff))>max_error and itr<max_itr:
    itr+=1
    # set pos in small step
    cur_pos = self.agent_ee_tip.get_position()
    diff=target_pos-cur_pos # difference of current and target
        position, close-loop control
    pos = cur_pos+step_factor*diff # step small step according to
        current difference, to prevent that ik cannot be solved
    self.tip_target.set_position(pos.tolist())
    self.pr.step() # every time when setting target tip, need to
        call simulation step to achieve it

  # one big step for z-rotation is enough, but small error still
      exists due to the ik solver
  ori_z+=rotation_norm*action[3] # normalize the rotation values,
      because usually the same action range is used in policy for
      both rotation and position
  self.tip_target.set_orientation([0, np.pi, ori_z]) # make gripper
      face downwards and rotate ori_z along z axis
  self.pr.step() # simulation step

else:
  print("Potential Movement Out of the Bounding Box!")
  pass # no action if potentially moving out of the bounding box
```

A function for re-initializing the scene is provided.

```
def reinit(self):
    '''
    Reinitialize the environment, e.g. when the gripper is
        broken during exploration.
    '''
    self.shutdown() # shutdown the original env first
    self.__init__(self.headless) # initialize with the same
        headless mode
```

A function to reset the target and the robot arm in the scene is as following.

```
def reset(self, random_target=False):
    '''
    Reset the gripper position and the target position.
    '''
    # set target object
    if random_target: # randomize
        pos = list(np.random.uniform(POS_MIN, POS_MAX)) # sample from
             uniform dist. in valid range
        self.target.set_position(pos) # random position
    else: # non-randomize
        self.target.set_position(self.initial_target_positions) # fixed
             position
    self.target.set_orientation([0,0,0])
    self.pr.step()

    # set end position to be initialized
    if self.control_mode == 'end_position': # JointMode.IK
        self.agent.set_control_loop_enabled(True) # ik mode
        self.tip_target.set_position(self.initial_tip_positions) #
             cannot set joint positions directly due to ik mode or
             force/torch mode is on
        self.pr.step()
        # prevent the stuck cases. as using ik for moving, stucking can
             make ik unsolvable therefore not reset correctly, so
        # some random actions are taken when the desired position is
             not reached.
        itr=0
        max_itr=10
        while np.sum(np.abs(np.array(self.agent_ee_tip.get_position()-
             np.array(self.initial_tip_positions))))>0.1 and itr<max_itr:
            itr+=1
            self.step(np.random.uniform(-0.2,0.2,4)) # take random
                 actions for preventing the stuck cases
            self.pr.step()

    elif self.control_mode == 'joint_velocity': # JointMode.FORCE
        self.agent.set_joint_positions(self.initial_joint_positions)
        self.pr.step()

    # set collidable, for collision detection
    self.gripper_left_pad.set_collidable(True) # set the pad on the
         gripper to be collidable, so as to check collision
    self.target.set_collidable(True)
    # open the gripper if it's not fully open
    if np.sum(self.gripper.get_open_amount())<1.5:
        self.gripper.actuate(1, velocity=0.5)
        self.pr.step()

    return self._get_state() # return current state of the environment
```

The `step()` function of the environment usually applied in other environments (OpenAI Gym, etc) with an action as input is as follows. If the robot is controlled with `end_position` mode using inverse kinematics, it needs to call the `_move()` function defined previously to conduct the action; if the robot is controlled with `joint_velocity` mode using forward kinematics, the joint positions can be set directly.

```
def step(self, action):
    '''
    Move the robot arm according to the action.
    If control_mode=='joint_velocity', action is 7 dim of
        joint velocity values + 1 dim rotation of gripper;
    if control_mode=='end_position', action is 3 dim of tip
        (end of robot arm) position values + 1 dim rotation of
        gripper;
    '''
    # initialization
    done=False # episode finishes
    reward=0
    hold_flag=False # holding the object or not
    if self.control_mode == 'end_position':
       if action is None or action.shape[0]!=4: # check if
           action is valid
          print('No actions or wrong action dimensions!')
          action = list(np.random.uniform(-0.1, 0.1, 4)) #
              random
       self._move(action)

    elif self.control_mode == 'joint_velocity':
       if action is None or action.shape[0]!=7: # check if
           action is valid
          print('No actions or wrong action dimensions!')
          action = list(np.random.uniform(-0.1, 0.1, 7)) #
              random
       self.agent.set_joint_target_velocities(action) #
           Execute action on arm
       self.pr.step()

    else:
       raise NotImplementedError
```

Apart from moving the robot arm, the reward function, absorbing state, `done`, and other information like the flag for object holding state are also needed to be handled in the `step()` function, which is provided as follows. The reward for successfully grasping the object is given by a positive offset value, while the penalty for the object to be pushed off the table is a negative offset of the same value, which forms a sparse reward and is potentially very hard to learn for the agent. So we add distance penalty from the end effector to the target object for assisting learning, as well as the penalty for the collision between the gripper and the table to avoid gripper damage, which forms a dense reward. However, we need to know that the dense reward can potentially have divergence with the final goal of the task, which is to let the robot grasp the target object. Since the distance penalty is proportional to the distance between the gripper and the object center, it will promote the gripper to be as close to the object center as possible, which may not even be a proper gesture to grasp the object. More discussions about the divergence between the reward function and the goal of the reinforcement learning task are provided in Chap. 18. Because of that, we also augment the reward function with an offset displacement of the target position above the target object rather than at the center of the object, as discussed in later sections.

```
ax, ay, az = self.gripper.get_position()
if math.isnan(ax): # capture the broken gripper cases
    during exploration
   print('Gripper position is nan.')
   self.reinit()
   done=True
tx, ty, tz = self.target.get_position()
sqr_distance = (ax - tx) ** 2 + (ay - ty) ** 2 + (az - tz)
    ** 2 # squared distance between the gripper and the
    target object

# close the gripper if it's close enough to the object and
    the object is detected with the proximity sensor
if sqr_distance<0.1 and
    self.proximity_sensor.is_detected(self.target)== True:
   # make sure the gripper is open before grasping
   self.gripper.actuate(1, velocity=0.5)
   self.pr.step()
   self.gripper.actuate(0, velocity=0.5) # if done, close
       the hand, 0 for close and 1 for open; velocity 0.5
       ensures the gripper to close with in one frame
   self.pr.step() # Step the physics simulation

   if self._is_holding():
      reward += self.reward_offset # extra reward for
          grasping the object
      done=True
      hold_flag = True
   else:
      self.gripper.actuate(1, velocity=0.5)
      self.pr.step()
elif np.sum(self.gripper.get_open_amount())<1.5: # if
    gripper is closed (not fully open) due to collision or
    others, open it; get_open_amount() return list of
    gripper joint values
   self.gripper.actuate(1, velocity=0.5)
   self.pr.step()
else:
   pass

# the base reward is negative distance to target
reward -= np.sqrt(sqr_distance)

# case when the object fall off the table
```

```
        if tz <
            self.initial_target_positions[2]-self.fall_down_offset:
            done = True
            reward = -self.reward_offset

        # Penalty for collision with the table
        if self.gripper_left_pad.check_collision(self.table):
            reward -= self.penalty_offset
            #print('Penalize collision with table.')

        if math.isnan(reward): # capture the cases of numerical
            problem
            reward = 0.

        return self._get_state(), reward, done, {'finished':
            hold_flag}
```

The function for closing the environment is relatively simple.

```
    def shutdown(self):
        ''' Close the simulator '''
        self.pr.stop()
        self.pr.shutdown()
```

In the following experiments, we only use the very primary settings of the above grasping task: the initial position of the target object is fixed; the initialization of the robot joint positions is properly chosen to prevent more complicated gestures of robot; the robot is controlled with forward kinematics mode with joint velocities; and the robot is controlled with numerical states including joint positions, joint velocities, and target positions as observations. But the readers are free to play around with more complicated settings like using inverse kinematics mode for controlling the end position of the robot, using visual-based control from raw images or combining it with partial numerical states, using less information as observations, or setting the tasks to be harder and more complicated, etc.

In the project file, the environment of *Sawyer* grasping task can be tested with:

```
python sawyer_grasp_env_boundingbox.py
```

16.2 Reinforcement Learning for Robotics Tasks

The above robot learning environment based on forward kinematics control has a 7-dimension continuous action space for joint velocities, and a 17-dimension continuous state space, which is a relatively complicated environment compared with examples in previous chapters like Chaps. 5 and 6. Moreover, the robot simulation system is complicated as well, which makes the sampling process require considerable amounts of time. It is hard to train a relatively good policy using a single-thread/process training framework within a short time. In practice, the bottleneck for the speed of policy learning for the robot learning task mainly lies in the simulation process in CoppeliaSim (V-REP), which makes the whole learning process very inefficient if there is only one process for sampling. The paralleled off-line training framework can improve the sampling speed for the task.

In this project, we use the paralleled soft actor-critic (SAC) algorithm, which follows the same parallel framework as in the previous project of Chap. 13. The detailed introduction of SAC algorithm is provided in Chap. 6 with both theories and implementations, so here we only briefly describes the advantages of SAC algorithm as a reason for our choice. As an off-policy learning algorithm, SAC with diagonal Gaussian policies can handle high-dimensional continuous action spaces, and it is more stable and less sensitive to hyperparameters than other algorithms like deep deterministic policy gradient in training, especially with the adaptive learning of the entropy factor (Haarnoja et al. 2018). It also leverages the soft Q-learning with entropy regularization, which helps to boost explorations during training for hard tasks like the robot grasping task here. Moreover, due to the off-policy learning manner of SAC, it can be easily modified into a paralleled version in practice.

Even with the paralleled sampling processes, it can still be hard for the robot to explore a good gesture for grasping the object with dense reward above, and even harder for the sparse reward only. To further accelerating the learning process, we augment the reward function with heuristics. First, as the target object is a cuboid with its length longer than the gripper opening length, so the gripper can only grasp the object with its direction vertical to the length direction, as well as facing downwards. Therefore, we add an extra reward penalty as follows:

```
# Augmented reward for orientation: better grasping gesture if the
    gripper has vertical orientation to the target object.
# Note: the frame of gripper has a difference of pi/2 in z
    orientation as the frame of target.
desired_orientation = np.concatenate(([np.pi, 0],
    [self.target.get_orientation()[2]])) # gripper vertical to
    target in z and facing downwards,
rotation_penalty =
    -np.sum(np.abs(np.array(self.agent_ee_tip.get_orientation())
    -desired_orientation))
rotation_norm = 0.02
reward += rotation_norm*rotation_penalty
```

Secondly, as mentioned above, the negative distance between the gripper and the object as part of the reward function may not lead to an optimal grasping gesture. So the second augmentation on the reward function is an offset distance above the center of the target object as the zero-penalty point, which modifies the distance term to be:

```
    offset=0.08 # augmented reward: offset of target position
        above the target object
    sqr_distance = (ax - tx) ** 2 + (ay - ty) ** 2 + (az -
        (tz+offset)) ** 2 # squared distance between the
        gripper and the target object
```

With above two augmentations in the reward function, the learning performance is further improved compared with the original dense or sparse reward cases, as shown in Fig. 16.8.

The reward engineering is one way of effectively combining prior knowledge from humans for assisting learning in practice, which may be opposite to the research pursuits. From the perspective of research, people may focus more on reducing the amount of reward engineering and other human efforts in assisting intelligent agent's learning, and improve methods for achieving more intelligent and automatic learning process. On the contrary, if the target is to achieve the task in practice, some aids can be helpful. Apart from reward engineering, learning from expert demonstrations is another way of effectively improve the learning performances in practice, as detailed in Chap. 8.

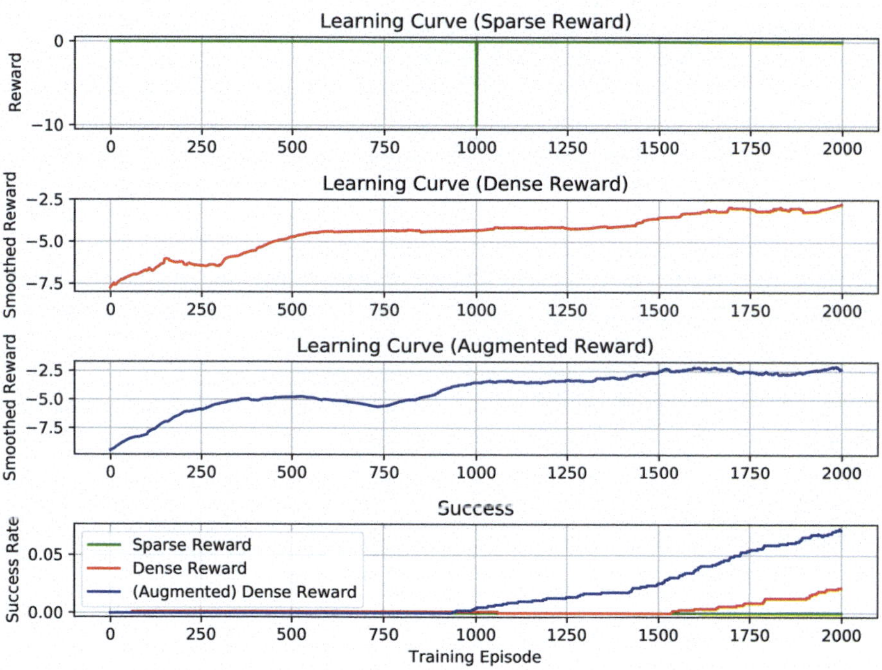

Fig. 16.8 The learning performance of *Sawyer* robot grasping task with parallel training of SAC algorithm, with different reward functions

16.2.1 Parallel Training

The CoppeliaSim (V-REP) requires an individual process for each simulation environment. Therefore, in order to speed up the sampling process, we have to set up multi-processing instead of multi-threading for collecting samples in parallel. A multi-processing version of SAC algorithm implemented with PyTorch is provided in our code repository. The training and testing process can be started by simply run:

```
# training
python sac_learn.py --train
# testing
python sac_learn.py --test
```

In this code, the interaction process with the environment is achieved with multiple processes, and each process contains one environment.

16.2.2 Learning Performance

We experimentally test the algorithm on the *Sawyer* grasping task, and a list of hyperparameters during training is shown in Table 16.1. The learning performances are shown in Fig. 16.8, with three different types of reward functions. The minus ten reward in the sparse reward case in Fig. 16.8 is caused by the penalty of the object falling off the table. Different reward functions will give different scales of reward, which may not be a fair comparison purely from the learning curves by reward. Therefore, in addition to the (smoothed) episode reward, we also display the success rate of grasping during the learning process. As the training proceeds, we can clearly see the success event is happening more and more frequently, which indicates an improvement of grasping skills for the robot. The augmented reward

Table 16.1 Hyperparameters of SAC

Parameter	Value
Optimizer	Adam (Kingma and Ba 2014)
Learning rate	$3 \cdot 10^{-4}$
Reward discount (γ)	0.99
Number of workers	6
Hidden layers (policy)	4
Number of hidden units (policy)	512
Hidden layers (Q-network)	3
Number of hidden units (Q-network)	512
Batch size	128
Target entropy	− Action dimension
Buffer size	$1 \cdot 10^6$

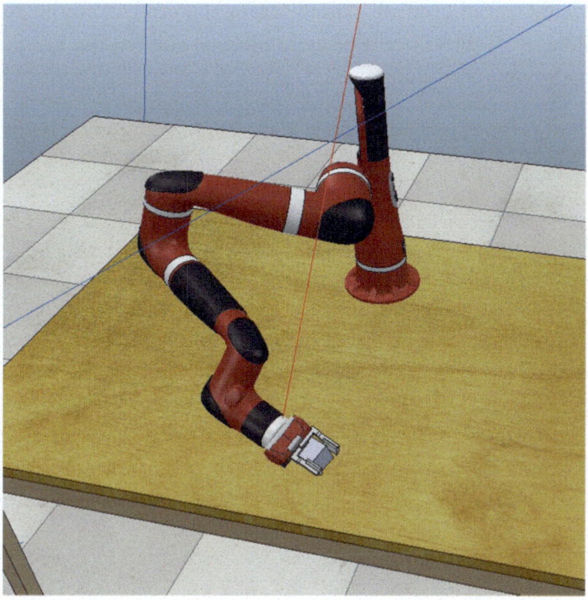

Fig. 16.9 The *Sawyer* in simulation is grasping the object with deep reinforcement learning policy after training

function accelerates the learning process significantly compared with the original dense reward, and the sparse reward makes it almost impossible to explore and learn to grasp the object.

After thousands of episodes of training, the robot is already able to grasp the target object with a fixed position, although in an inelegant gesture with not so high success rate, as shown in Fig. 16.9. The overall learning process is conducted from scratch without any demonstration or pre-training for this example.

16.2.3 Domain Randomization

When we apply the policy trained in simulation in reality, it is common to find that the policy does not work in practice due to the differences between real-world dynamics and the dynamic process we set in simulation. Domain randomization is a method for improving the generalization ability of the policy when we try to transfer the policy learned in simulation into real-world scenarios.

Domain randomization can be achieved through randomizing the physical parameters in the environment, including the parameters for determining the physical dynamics of the robotic arm, objects, and their interactions in the scene. More specifically, randomizing the dynamic parameters is called dynamics randomization (Peng et al. 2018), for example, the mass of the object, the friction of the joints on robot arms, the friction between the object and the table, and so on. Moreover, the color, light conditions, and textures can be randomized if using visual-based control, which learns the controlling agent according to the observed images of the robot. For example, we can set the color of the object in PyRep with following commands:

```
self.target.set_color(np.random.uniform(low=0, high=1,
    size=3).tolist()) # set [r,g,b] 3 channel values of
    the target object color
```

Other physical parameters in the simulation can be changed accordingly, which is beyond the scope of this chapter. When training the agent, we can set these parameters to be random for each episode or dozens of episodes during the whole training process. Moreover, it is important to make sure that the randomization ranges of dynamics parameters and other characteristics in simulation can cover the real dynamic process in reality, so as to mitigate the reality gaps.

Domain randomization is only one potential approach for mitigating the reality gap for sim-to-real transfer, and above visual feature randomization with PyRep in CoppeliaSim is just a very simple example. A detailed description of other sim-to-real methods is provided in Chap. 7.

16.2.4 Robot Learning Benchmark

In the above sections, we show how to build a simple scene for robot grasping task and apply a reinforcement learning algorithm to solve it. Recently, James et al. (2019b) proposes *RLBench* package (https://github.com/stepjam/RLBench) as a large-scale benchmark and learning environment featuring 100 unique, hand-design tasks, tailored to facilitate research in a number of vision-guided manipulation research areas, not only in reinforcement learning, but also in imitation learning, multi-task learning, geometric computer vision, and few-shot learning. As shown in Fig. 16.10,[5] *RLBench* is built on PyRep used in previous sections, and it contains 100 basic tasks for robotic manipulation including grasping, moving, stacking, and other various manipulations, which are common to see in daily life. It also supports customized task settings with a simple configuration process. Different learning methods including reinforcement learning can be employed for solving these tasks.

The robotic grasping task in previous sections provides a typical framework for robotic learning in simulation with CoppeliaSim (V-REP), which is also applicable to *RLBench* package. They both contains at least three key components: (1) the task scene built in CoppeliaSim (V-REP), (2) the scripts for defining the dynamics of the environment, including `reset()` and `step()` functions, and (3) the scripts providing a learning agent with algorithms like reinforcement learning. *RLBench* follows this building framework but with a hierarchical structure for constructing general tasks.

[5]Figure source: https://github.com/stepjam/RLBench.

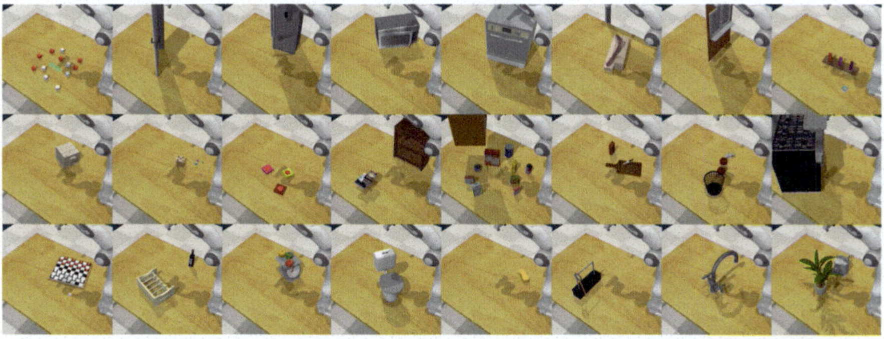

Fig. 16.10 Tasks defined in *RLBench* for robot learning

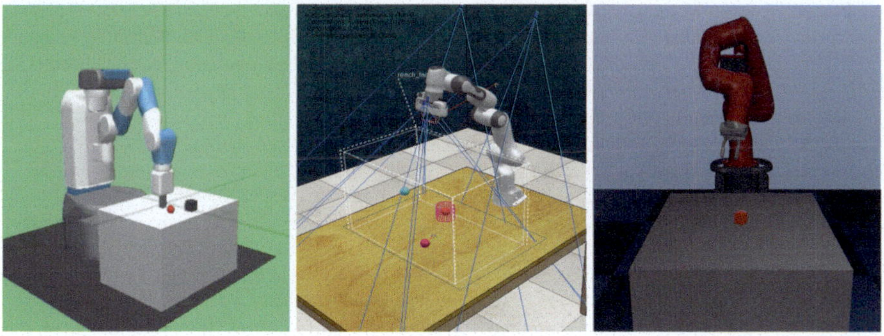

Fig. 16.11 Robot learning tasks: (1) *FetchPush* environment in OpenAI Gym (left); (2) Goal-reaching task with PyRep (middle); (3) *SawyerLift* task in RoboSuite (right)

The *RLBench* package can be installed with the following commands (if you have already installed PyRep):

```
git clone https://github.com/stepjam/RLBench.git
pip3 install -r requirements.txt
python3 setup.py install --user
```

16.2.5 Other Simulators

Notice that there are a bunch of different simulation software for robot learning, as shown in Fig. 16.11, including OpenAI Gym, CoppeliaSim (V-REP/PyRep) (Rohmer et al. 2013; James et al. 2019a), MuJoCo (Todorov et al. 2012), Gazebo, Bullet/PyBullet (Coumans et al. 2013; Coumans and Bai 2016), Webots (Michel 2004), Unity 3D, NVIDIA Isaac SDK, etc. These toolkits or platforms have different characteristics for different applications in practice. For example, the OpenAI

Gym robotics environment is a relatively simple environment for fast verification of proposed methods, the CoppeliaSim and Unity 3D are both built on physics simulators with relatively good rendering effects, and the MuJoCo has more realistic and accurate physics engine, which can benefit the sim-to-real transfer, Isaac SDK is a relatively new software (released in 2019) with strong supports in deep learning algorithms and applications, as well as photo-realistic rendering based on Unity 3D, etc.

References

Akkaya I, Andrychowicz M, Chociej M, Litwin M, McGrew B, Petron A, Paino A, Plappert M, Powell G, Ribas R, et al. (2019) Solving Rubik's cube with a robot hand. arXiv:191007113

Andrychowicz M, Baker B, Chociej M, Jozefowicz R, McGrew B, Pachocki J, Petron A, Plappert M, Powell G, Ray A, et al. (2018) Learning dexterous in-hand manipulation. arXiv:180800177

Coumans E, Bai Y (2016) Pybullet, a python module for physics simulation for games, robotics and machine learning. http://pybullet.org

Coumans E, et al. (2013) Bullet physics library. Open source 15(49):5. bulletphysics.org

Haarnoja T, Zhou A, Hartikainen K, Tucker G, Ha S, Tan J, Kumar V, Zhu H, Gupta A, Abbeel P, et al. (2018) Soft actor-critic algorithms and applications. arXiv:181205905

James S, Freese M, Davison AJ (2019a) PyRep: bringing V-REP to deep robot learning. arXiv:190611176

James S, Ma Z, Arrojo DR, Davison AJ (2019b) RLBench: the robot learning benchmark and learning environment. arXiv:190912271

Kingma D, Ba J (2014) Adam: a method for stochastic optimization. In: Proceedings of the international conference on learning representations (ICLR)

Korenkevych D, Mahmood AR, Vasan G, Bergstra J (2019) Autoregressive policies for continuous control deep reinforcement learning. arXiv:190311524

Michel O (2004) Cyberbotics Ltd. Webots™: professional mobile robot simulation. Int J Adv Robot Syst 1(1):5

Peng XB, Andrychowicz M, Zaremba W, Abbeel P (2018) Sim-to-real transfer of robotic control with dynamics randomization. In: 2018 IEEE international conference on robotics and automation (ICRA). IEEE, Piscataway, pp 1–8

Rohmer E, Singh SP, Freese M (2013) V-rep: a versatile and scalable robot simulation framework. In: 2013 IEEE/RSJ international conference on intelligent robots and systems. IEEE, Piscataway, pp 1321–1326

Todorov E, Erez T, Tassa Y (2012) MuJoCo: a physics engine for model-based control. In: Proceedings of the International Conference on Intelligent Robots and Systems (IROS). IEEE, Piscataway

Chapter 17
Arena Platform for Multi-Agent Reinforcement Learning

Zihan Ding

Abstract In this chapter, we introduce a project named *Arena* for multi-agent reinforcement learning research. The hands-on instructions are provided in this chapter for building games with *Arena* toolkit, including a single agent game and a simple two-agent game with different reward schemes. The reward scheme in *Arena* is a way to specify the social structure among multiple agents, which contains social relationships of non-learnable, isolated, competitive, collaborative, and mixed types. Different reward schemes can be applied at the same time in a hierarchical structure in one game scene, together with the individual-to-group structure for physical units, to describe the complex relationships in multi-agent systems comprehensively. Moreover, we also show the process of applying the baseline in *Arena*, which provides several implemented multi-agent reinforcement learning algorithms as a benchmark. Through this project, we want to provide the readers with a useful tool for investigating multi-agent intelligence with customized game environments and multi-agent reinforcement learning algorithms.

Keywords Multi-agent reinforcement learning · Learning environment · Toolkit · Competitive · Collaborative · Social relationship

In this chapter, we will introduce a powerful toolkit for multi-agent reinforcement learning (MARL): *Arena* (Song et al. 2019). *Arena* is a general evaluation platform for multi-agent intelligence based on Unity, with learning environments of diverse logic and representations, as well as easy configurations on complex social tree relationships between multiple agents. *Arena* also contains the implementation of state-of-the-art deep multi-agent reinforcement learning algorithm baselines, which can help the users to quickly testify the built-up environments. Generally, *Arena* is a building toolkit for researchers to easily invent and build unexplored multi-agent problems with customized game environments. The official website of *Arena*

Z. Ding (✉)
Imperial College London, London, UK
e-mail: zhding@mail.ustc.edu.cn

H. Dong et al. (eds.), *Deep Reinforcement Learning*,
https://doi.org/10.1007/978-981-15-4095-0_17

is: https://sites.google.com/view/arena-unity/home. *Arena* focuses on first-person or third-person action games, leveraging the fantastic rendering effects of Unity. Recently published open-sourced project OpenSpiel (https://github.com/deepmind/open_spiel) by DeepMind, focuses on multi-agent board or card games.

There are two major modules in *Arena*: (1). the Building Toolkit, which is used to quickly build multi-agent environments with customized characteristics; (2). the Baselines, for testing the built-up environments with MARL algorithms. We will start by building the environments in *Arena*.

17.1 Intallation

Unity ML-agents toolkit is a prerequisite for *Arena*, which needs to be installed before applying *Arena*. The complete process of installing *Arena* follows the official website of Building Toolkit[1] and Baseline.[2]

Note that if you are running on a remote server without a graphical user interface (e.g., X-Server) or you cannot get access to the X-Server, you will need to set up the virtual display following instructions in Sect. 17.3.1 or guidelines on the official website of *Arena*.

After installation, we can find that in the `Arena-BuildingToolkit/Assets/ArenaSDK/GameSet/` file of *Arena* folder, there are dozens of built-in games with both continuous and discrete action spaces. They are pre-designed as examples for using *Arena*, you can read the scripts of all those games for better understanding of how *Arena* works. All games and abstraction layers share one Unity project. Each game is held in an independent folder, with the game's name as the folder name. The folder `ArenaSDK` holds all the abstraction layers and shared code, assets, and utilities. Code style is kept as consistent as possible to the Unity ML-agents toolkit.

17.2 Build Game with *Arena*

We will go through making a multi-agent game with the *Arena* Building Toolkit. It will not require much coding work with many off-the-shelf assets and multi-agent features managed by *Arena*. Before you start, we are expecting you to have some basic knowledge about Unity. Therefore, you are recommended to finish the roll-a-ball tutorial (https://learn.unity.com/project/roll-a-ball-tutorial) to learn all the basic concepts of Unity.

[1]Building Toolkit: https://github.com/YuhangSong/Arena-BuildingToolkit.

[2]Baseline: https://github.com/YuhangSong/Arena-Baselines.

Fig. 17.1 Built-in games in *Arena* file

In order to use *Arena*, run Unity, choose open project, and select the cloned or downloaded "Arena-BuildingToolkit" file. The opening process may take some time for the first time.

We can see dozens of built-in games in the *Arena* folder, as shown in Fig. 17.1. They are pre-designed as examples for using *Arena*, you can read the scripts of all those games for better understanding of how *Arena* works. We will provide basic instructions on building these games in the following sections.

17.2.1 Simple One-Player Game

We start by building a basic game environment with only one player in it:

- Create a folder to host your game. In this part, we create a folder named "1P" for only one-player game.
- On the left-side "Hierarchy" window, we delete the original **Main Camera** and **Directional Light** in it. Drag the prefab **GlobalManager** built in *Arena* folder `Assets/ArenaSDK/SharedPrefabs` as shown in Fig. 17.2, to the left-side "Hierarchy" window, as shown in Fig. 17.3. Note that the prefabs are useful and shared components in Unity for using any built-in objects with a simple dragging operation. **GlobalManager** in *Arena* manages the whole game, so all the other components need to be attached under it.
- Next we need to place a playground for the agent to play on, we find the prefabs in *Arena* called **PlayGroundWithDeadWalls**, and attach it to the child called **World** of **GlobalManager**. The **GlobalManager** also has another child **TopDownCamera** for providing an overall view of the game. This step is shown in Fig. 17.4.
- Similar as above, we need to attach a **BasicAgent** from *Arena* prefabs and attach it to the **GlobalManager** as shown in Fig. 17.5. So now we have one agent on the playground in the scene, and we can manually drag the agent to a proper position,

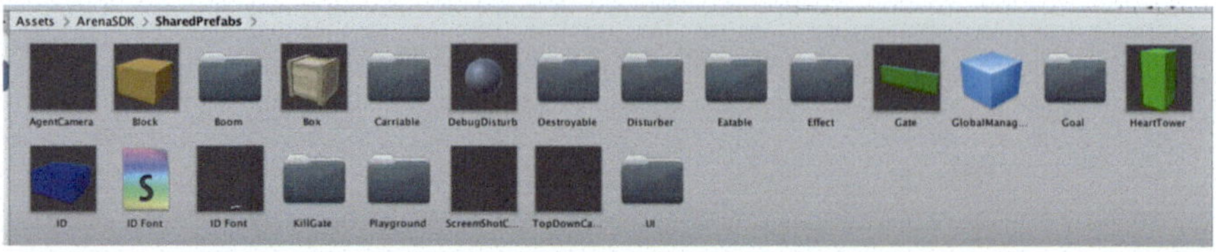

Fig. 17.2 The *Arena* built-in prefabs

Fig. 17.3 Drag the **GlobalManager** in *Arena* prefabs to the "Hierarchy" window of current game

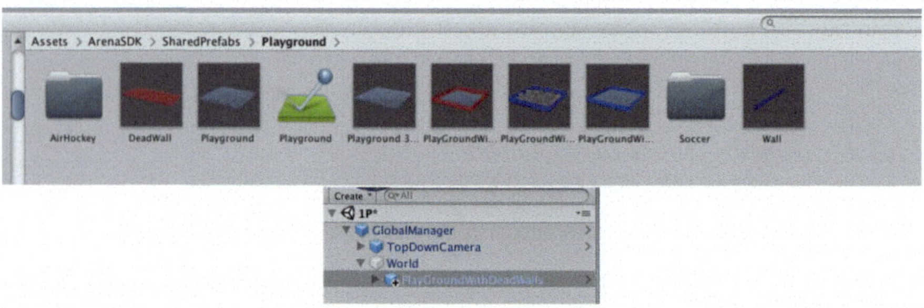

Fig. 17.4 Choose a playground in *Arena* prefabs and attach it to the child of the **GlobalManager**

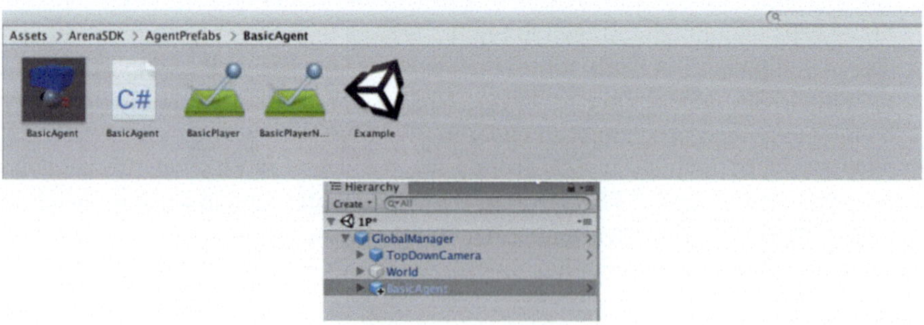

Fig. 17.5 Choose and attach a **BasicAgent** in *Arena* prefabs and attach it to the child of the **GlobalManager**

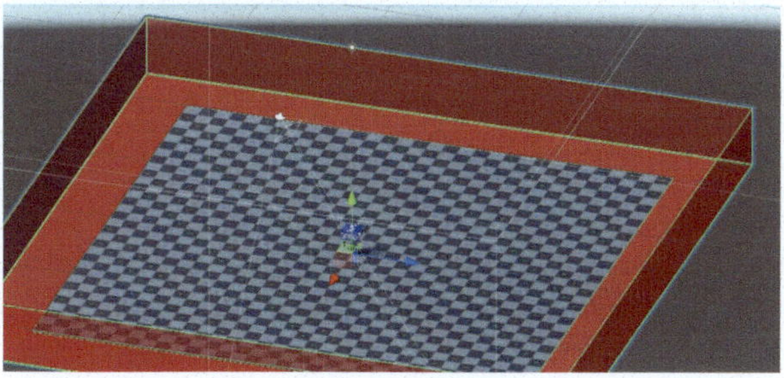

Fig. 17.6 The scene with a single agent on the playground

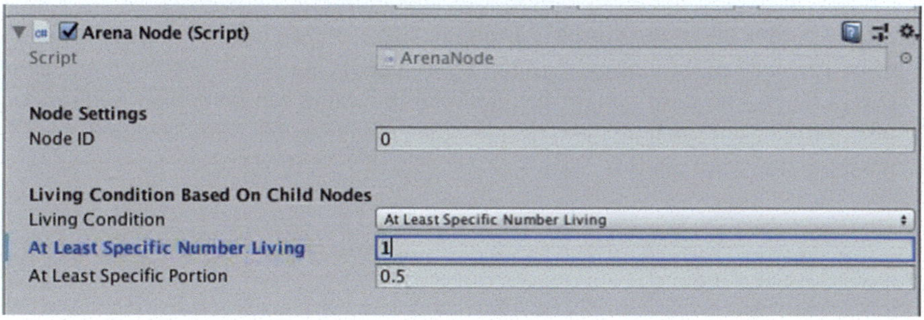

Fig. 17.7 Configure the game settings of a single-player game

as in Fig. 17.6. The values of the x, y, and z coordinates of position and rotation will be displayed on the **Transform** property of the agent.

- In order to make the game work normally, we also need to configure the game parameters as shown in Fig. 17.7. Here we only need to change the **Living Condition Based On Child Nodes** of the **GlobalManager**. The **Living Condition** is chosen to be **At Least Specific Number Living** and the **At Least Specific Number Living** value is set to be 1. As we only have one agent in this game, the above settings ensure that whenever the number of agents under **GlobalManager** is smaller than one, the game episode will end and restart. Now we can press the **Play** button to play the game and operate the agent with keys "W, A, S, D." As on the edges of the playground are the "dead walls," whenever the agent touches it, it will die and the game will restart. There are lots of other properties if you apply the **BasicAgent**, including different **Actions Settings**, **Reward Functions** (for reinforcement learning), etc. You can play around with them (only the **Actions Settings** are valid in this simple game) to get familiar with *Arena* Building Toolkit.

17.2.2 *Simple Two-Player Game with Reward Scheme*

In this section, we will introduce how to deploy more than one agent in the game environment with a social tree.

- First, let us start from the above single-player game. If we choose the **Global-Manager** or the **BasicAgent**, we will find that for both objects there is a script called **Arena Node (Script)** as shown in Figs. 17.8 and 17.9, which is a basic concept used to define the social relationships in *Arena* games. Descriptions about **Arena Node** will be provided in this section.
- We choose **BasicAgent** built before and duplicate it by pressing Ctrl+C and Ctrl+V in the left hierarchy window, as shown in Fig. 17.10. Now we have two **Arena Node**s under the **Global Manager**, therefore we need to set the **Node ID** to be 1 instead of 0 for any one of the **BasicAgent**s in order to discriminate them (Fig. 17.11). The positions of the agents in the scene can be moved to a proper position, so as to separate them because the two agents are initialized at the same position after duplication.

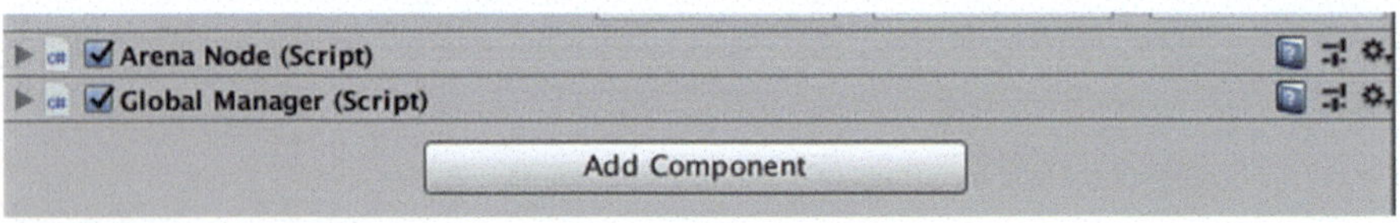

Fig. 17.8 The **Arena Node (Script)** exists in **GlobalManager**

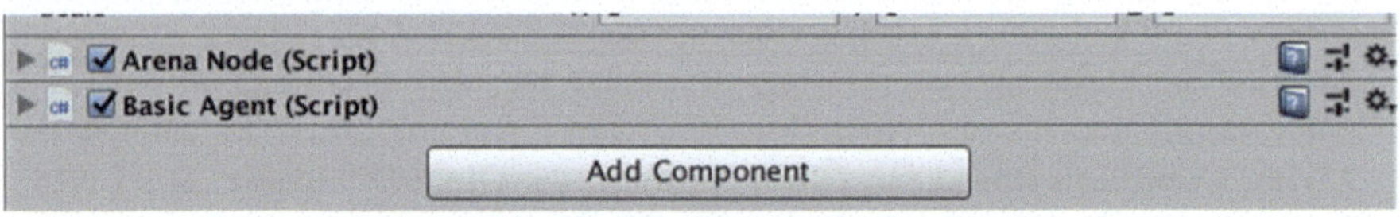

Fig. 17.9 The **Arena Node (Script)** exists in **BasicAgent**

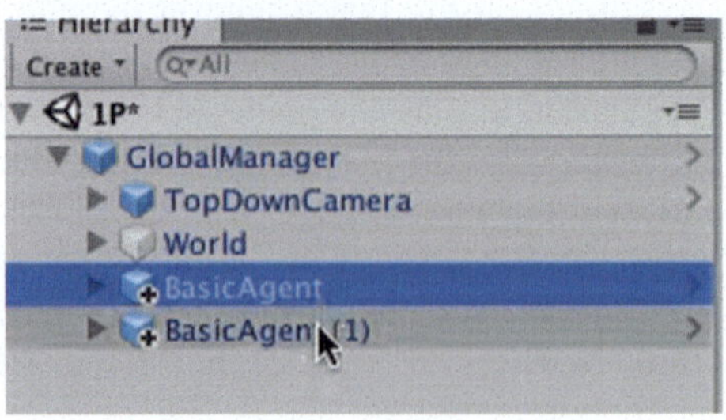

Fig. 17.10 Duplicate the **BasicAgent** under the **GlobalManager**

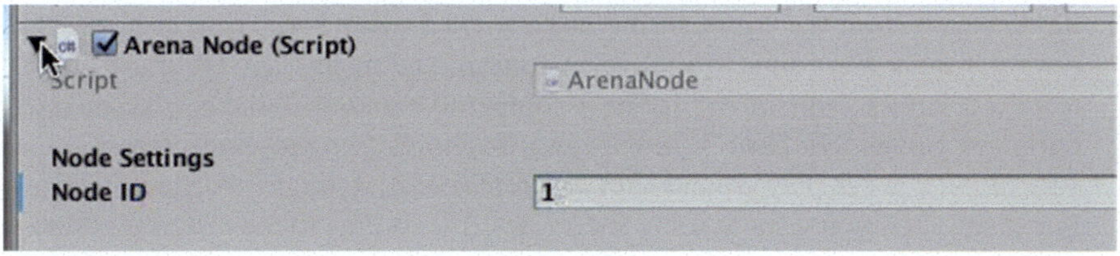

Fig. 17.11 Change node ID to be different from each other when there are multiple nodes under **GlobalManager**

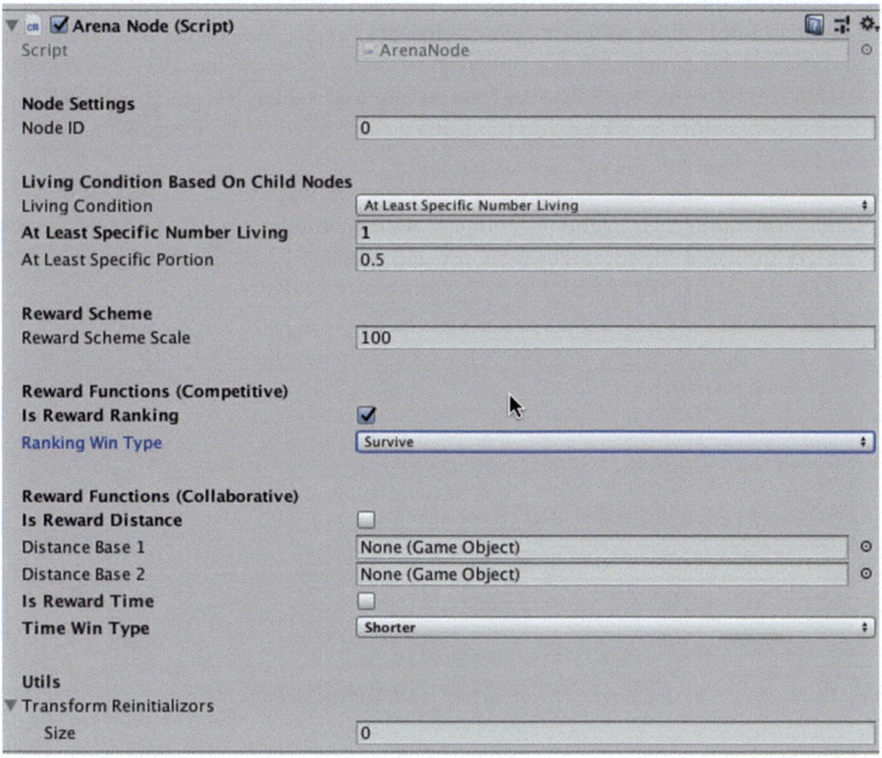

Fig. 17.12 Set the reward scheme under **GlobalManager**

- Next we choose **GlobalManager, Arena Node (Script)**, and we can set the reward functions for the game, as shown in Fig. 17.12. We click the **Is Reward Ranking**, which is a competitive reward function for agents under **GlobalManager**. We also choose **Ranking Win Type** to be **Survive**, which means the agent died at the last place (surviving in the end) will get a positive reward. If you select **Depart**, the reward will be given to the agent died in the first place. We also need to unclick **Is Reward Distance** (which gives dense reward values according to the distance from the agent to the target) and **Is Reward Time** (which gives reward values according to the living time of the agent). Above are different reward schemes built in *Arena*, in a competitive and/or collaborative manner. Different games will have different reward settings to represent different social structures. You can play around with different reward settings for different games. For example, if you want to set a dense reward according to the distance between the agent and the target for a task like reaching the target, you need to click **Is Reward Distance**, as well as dragging a target object to the **Target** blank to make it work.

- We need to set the **At Least Specific Number Living** to be 2 under the **Living Condition Based On Child Nodes** of **GlobalManager**, as shown in Fig. 17.13. So that only when at least two agents are living, the game will continue; otherwise the game will end and restart. Now we click the **Play** button, the game should work normally. As long as one agent died, the game will end and rewards/penalties will be given to agents as displayed in the **Console**, as shown in Fig. 17.14.
- Next we will make the game more complex, we want to have two teams with each containing two agents to compete with each other. So, first, we create an empty object at the hierarchy window and name it "2 Player Team". Then we attach the **Arena Node** script to it, as shown in Fig. 17.15.

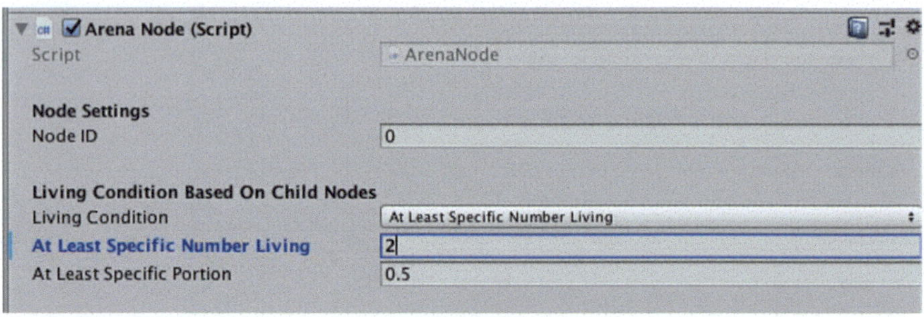

Fig. 17.13 Set the least number of living agents in **GlobalManager**

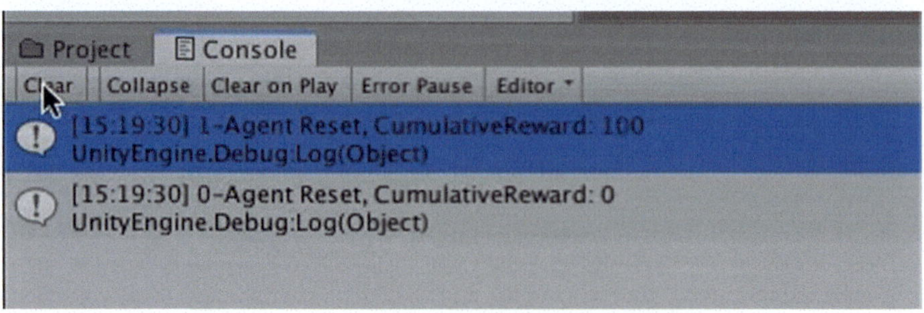

Fig. 17.14 The rewards given to each agent are displayed at the **Console**

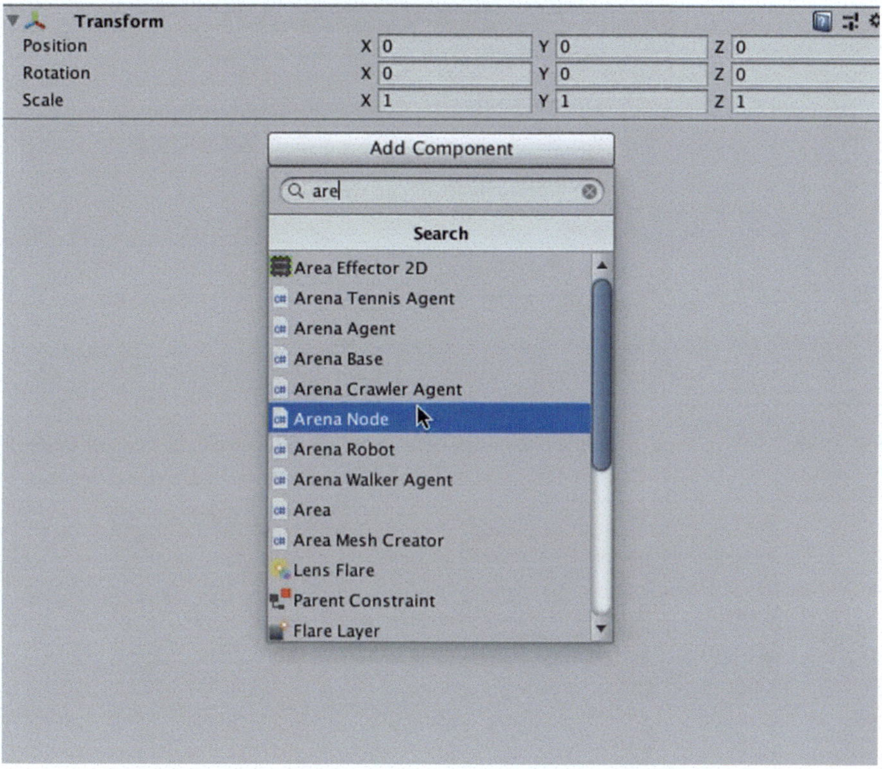

Fig. 17.15 Attach the **Arena Node** script to the team object

- Now we drag the two previous **BasicAgent** to the new-built team object **2 Player Team**. Then we duplicate the **2 Player Team**, change the **Node ID** of the second team object to be 1 instead of 0. Now we have a structure of teams and agents as shown in Fig. 17.16. If we click the **Play** button now, we shall see two teams with two agents each in the scene as Fig. 17.17.
- As the **At Least Specific Number Living** of **GlobalManager** is set to be 2, any team's death will cause the game to end. Also as the **At Least Specific Number Living** of **2 Player Team** is 1 as default, only when both two agents died at the same team will cause the team to die. We can also set the game logic to be different, if we set the **Living Condition** of the **2 Player Team** to be **All Living**, then any agent died in a team will cause the team to die, and therefore end the game. From above, you can see that with the social tree structure like **GlobalManager->Team->Agent**, *Arena* can basically support any kinds of social relationships with the **Arena Node** through defining the living and reward schemes.

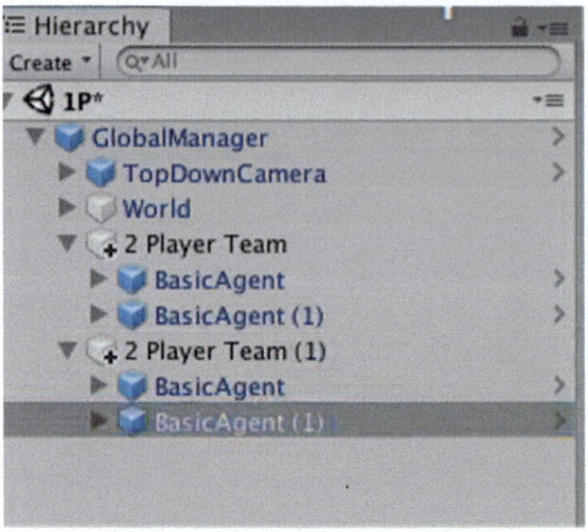

Fig. 17.16 The hierarchy structure of two teams with two agents each under the **GlobalManager** in *Arena*

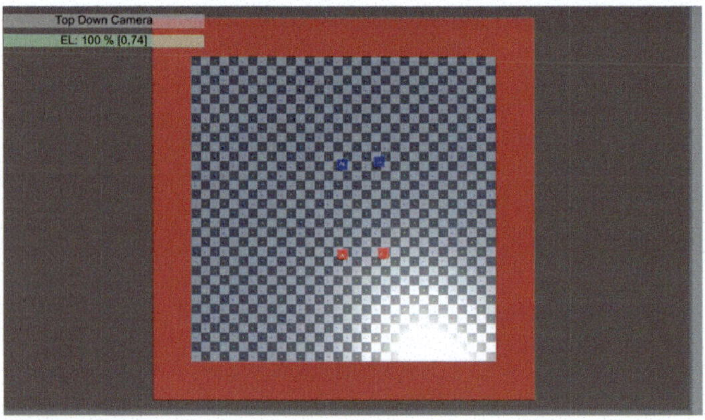

Fig. 17.17 The game scene of two teams with two agents each during the play

17.2.3 Advanced Settings

Reward Scheme

To construct complex social tree relationship, there are five basic multi-agent reward schemes (BMaRSs) in *Arena* that define different social paradigms at each node in the social tree, including: **non-learnable (NL), isolated (IS), competitive (CP), collaborative (CL), competitive and collaborative mixed (CC)**. Specifically, each BMaRS is a restriction applied to the reward function, so it corresponds to a batch of reward functions that can lead to a specific social paradigm. For each BMaRS, *Arena* provides multiple ready-to-use reward functions (sparse and dense), simplifying the construction of games with complex social relationships. Apart from providing reward functions, *Arena* also offers a verification option for customized reward functions, so that one can make sure that the programmed reward functions lie in one of the BMaRSs and produce a specific social paradigm. We will introduce these five different reward schemes in detail.

First we need to give some preliminaries. We consider a Markov game as defined in the basics of RL, consisting of multiple agents $x \in \mathcal{X}$, a finite global state space $s_t \in \mathcal{S}$, a finite action space $a_{x,t} \in \mathcal{A}_x$ for each agent x, and a bounded-step reward space $r_{x,t} \in \mathbb{R}$ for each agent x. For the environment, it consists of a transition function $g : \mathcal{S} \times \left\{\mathcal{A}_x : x \in \mathcal{X}\right\} \rightarrow \mathcal{S}$, which is a stochastic (due to the stochasticity of Unity simulator) function $s_{t+1} \sim g\big(s_{t+1}|(s_t, \{a_{x,t} : x \in \mathcal{X})\}\big)$, a reward function for each agent $f_x : \mathcal{S} \times \left\{\mathcal{A}_x : x \in \mathcal{X}\right\} \rightarrow \mathbb{R}$, which is a deterministic function $r_{x,t+1} = f_x\big(s_t, \{a_{x,t} : x \in \mathcal{X}\}\big)$, a joint reward function $f = \{f_x : x \in \mathcal{X}\}$, and episode reward $R_x^f = \sum_{t=1}^{T} r_{x,t}$ for each agent x under the joint reward function f. For the agent, *Arena* considers that it observes $s_{x,t} \in \mathcal{S}_x$, where $\mathcal{S}_x$ consists of a part of the information from the global state space $\mathcal{S}$. Therefore, there is a policy $\pi_x : \mathcal{S}_x \rightarrow \mathcal{A}_x$, which is a stochastic function $a_{x,t} \sim \pi_x(s_{x,t})$. Besides, *Arena* considers that agent x can take a policy π_x from a set of policies Π_x. *Arena* assumes that the random seed of all sampling operations is k, which is sampled from the whole seed space $\mathcal{K}$.

The definitions of different BMaRSs employ the basic concepts including agents $\{x : x \in \mathcal{X}\}$, policies $\{\pi_x : \Pi_x\}$, agent rewards $\{R_x^f : x \in \mathcal{X}\}$, and joint reward functions $\mathcal{F} = \{f : \cdot\}$ on population $\mathcal{X}$. The five different BMaRSs in *Arena* are defined in the following way:

1. **Non-learnable (NL)** BMaRSs ($\mathcal{F}^{NL}$) are a set of joint reward functions f as follows:

$$\mathcal{F}^{NL} = \left\{f : \forall k \in \mathcal{K}, \forall x \in \mathcal{X}, \forall \pi_x \in \Pi_x, \partial R_x^f / \partial \pi_x = \mathbf{0}\right\}, \tag{17.1}$$

   where $\mathbf{0}$ is a zero matrix of the same size and shape as the parameter space that defines π_x. Intuitively, $\mathcal{F}^{NL}$ means R_x^f for any agent $x \in \mathcal{X}$ is not optimizable by improving its policy π_x.
2. **Isolated (IS)** BMaRSs ($\mathcal{F}^{IS}$) are a set of joint reward functions f as follows:

$$\mathcal{F}^{IS} = \left\{ f : f \notin \mathcal{F}^{NL} \text{ and } \forall k \in \mathcal{K}, \forall x \in \mathcal{X}, \forall x' \in \mathcal{X} \setminus \{x\}, \right. \\ \left. \forall \pi_x \in \Pi_x, \forall \pi_{x'} \in \Pi_{x'}, \frac{\partial R_x^f}{\partial \pi_{x'}} = \mathbf{0} \right\}, \tag{17.2}$$

   where "\" is the set difference. Intuitively, $\mathcal{F}^{IS}$ means that the episode reward R_x^f received by any agent $x \in \mathcal{X}$ is not related to any policy $\pi_{x'}$ taken by any other agent $x' \in \mathcal{X} \setminus \{x\}$. Reward functions f_x in f of $\mathcal{F}^{IS}$ are often called *internal reward functions* in other multi-agent approaches (Hendtlass 2004; Jaderberg et al. 2018; Bansal et al. 2018), meaning that apart from the reward functions applied at a population level (such as win/lost), which are too sparse to learn, there are also reward functions directing the learning process towards receiving the population-level rewards, yet are more frequently available, i.e., more dense (Singh et al. 2009, 2010; Heess et al. 2017). $\mathcal{F}^{IS}$ is especially practical when the agent is a robot requiring continuous control of applying force on each of its joints, which means basic motor skills (such as moving) need to be learned before generating population-level intelligence. Therefore, *Arena* provides f in $\mathcal{F}^{IS}$ of: energy cost, punishment of applying a big force, encouragement of keeping a steady velocity, and moving distance towards the target.

3. **Competitive (CP)** BMaRSs ($\mathcal{F}^{CP}$) are inspired by Cai and Daskalakis (2011) and defined as

$$\mathcal{F}^{CP} = \Bigg\{ f : f \notin \mathcal{F}^{NL} \cup \mathcal{F}^{IS} \text{ and } \forall k \in \mathcal{K}, \forall x \in \mathcal{X},$$

$$\forall \pi_x \in \Pi_x, \forall \pi_{x'} \in \Pi_{x'}, \frac{\partial \int_{x' \in \mathcal{X}} R^f_{x'} dx'}{\partial \pi_x} = \mathbf{0} \Bigg\}, \tag{17.3}$$

which intuitively means that for any agent $x \in \mathcal{X}$, taking any possible policy $\pi_x \in \Pi_x$, the sum of the episode reward of all agents will not change. If the episode length is 1, it expresses a classic multi-player zero-sum game (Cai and Daskalakis 2011).

Useful examples of f within $\mathcal{F}^{CP}$ are: (1) agents fight for a limited amount of resources that are always exhausted at the end of the episode, and the agent is rewarded for the amount of resources that it gained; and (2) fight till death, and the reward is given based on the order of death (the reward can also be based on the reversed order, so that the one departing the game first receives the highest reward, such as in some poker games, the one who first discards all cards wins). *Rock, Paper, and Scissors* in normal-form game (Myerson 2013) and *Cyclic Game* in Balduzzi et al. (2019) are both special cases of $\mathcal{F}^{CP}$;

4. **Collaborative (CL)** BMaRSs ($\mathcal{F}^{CL}$) are inspired by Cai and Daskalakis (2011) and defined as

$$\mathcal{F}^{CL} = \Bigg\{ f : f \notin \mathcal{F}^{NL} \cup \mathcal{F}^{IS} \text{ and } \forall k \in \mathcal{K}, \forall x \in \mathcal{X},$$

$$\forall x' \in \mathcal{X} \setminus \{x\}, \forall \pi_x \in \Pi_x, \forall \pi_{x'} \in \Pi_{x'}, \frac{\partial R^f_{x'}}{\partial R^f_x} \geq 0 \Bigg\}, \tag{17.4}$$

which intuitively means that there is no conflict of interest ($\partial R^f_{x'} / \partial R^f_x < 0$) for any pair of agents (x', x). Besides, since $f \notin \mathcal{F}^{NL} \cup \mathcal{F}^{IS}$, there is at least one pair of agents (x, x') that makes $\partial R^f_{x'} / \partial R^f_x > 0$. This indicates that this pair of agents shares a common interest, so that improving R^f_x for agent x means improving $R^f_{x'}$ for agent x'. The most common example of f within $\mathcal{F}^{CL}$ is that f_x for all $x \in \mathcal{X}$ is identical, such as the moving distance of an object that can be pushed forward by the joint effort of multiple agents, or the alive duration of the population (as long as there is at least one agent alive in the population, the population is alive). Therefore, *Arena* provides f in $\mathcal{F}^{CL}$: living time of the team (both positive and negative, since some games require the team to survive as long as possible, while other games require the team to depart as early as possible, such as poker).

5. **Competitive and Collaborative Mixed (CC)** BMaRSs ($\mathcal{F}^{CC}$) are defined to be any situation other than the above four ones.

$$\mathcal{F}^{CC} = \{f : f \notin \mathcal{F}^{NL} \cup \mathcal{F}^{IS} \cup \mathcal{F}^{CP} \cup \mathcal{F}^{CL}\}. \tag{17.5}$$

First, the term $\partial \int_{x' \in \mathcal{X}} R^f_{x'} dx' / \partial \pi_x = \mathbf{0}$ in (17.3) can be written as $\int_{x' \in \mathcal{X}} \partial R^f_{x'} / \partial R^f_x dx' = 0$ (proof is not provided here, which refers to original paper), which makes an alternative (17.3). Considering $\mathcal{F}^{CP}$ in this alternative (17.3) and $\mathcal{F}^{CL}$ in (17.5), an intuitive explanation of $\mathcal{F}^{CC}$ is that there exist circumstances when $\partial R^f_{x'} / \partial R^f_x < 0$, meaning that the agents are competitive at this point. But the derivative of total interest $\int_{x' \in \mathcal{X}} \partial R^f_{x'} / \partial R^f_x dx'$ is not always 0, therefore, the total interest can be maximized with specific policies, meaning that the agents are collaborative at this point.

Apart from providing several practical f in each BMaRS, *Arena* also provides a verification option for each BMaRS, meaning that one can customize a f and use this verification option to make sure that the programmed f lies in a specific BMaRS.

The above contents provide the theory about how to use different kinds of reward functions to define social relationships. Moreover, the reward functions should be defined with respect to the categories of the above definitions to achieve the expected social relationships in the population. In practice, the reward functions have some specific formats like those we mentioned in previous sections. The *Arena* framework usually defines the **Collaborative** and **Competitive** reward functions in the **Arena Node** of **GlobalManager**, and the **Isolated** reward functions are defined in the **Arena Node** of agents like **BasicAgent**.

Here is an example for understanding the social tree relationship using different BMaRs for each **Arena Node** as shown in Fig. 17.18.[3] The reward schemes are assigned at each **Arena Node** to define the social relationships of its sub-nodes. The graphical user interface (GUI) in Fig. 17.18a defines a tree structure in Fig. 17.18b, representing a population of 4 agents. The tree structure can be easily reconfigured through dragging, duplicating, or deleting in the GUI in Fig. 17.18a. In this example,

[3]Figure source: Song, Yuhang, et al. "Arena: A General Evaluation Platform and Building Toolkit for Multi-Agent Intelligence." arXiv preprint arXiv:1905.08085 (2019).

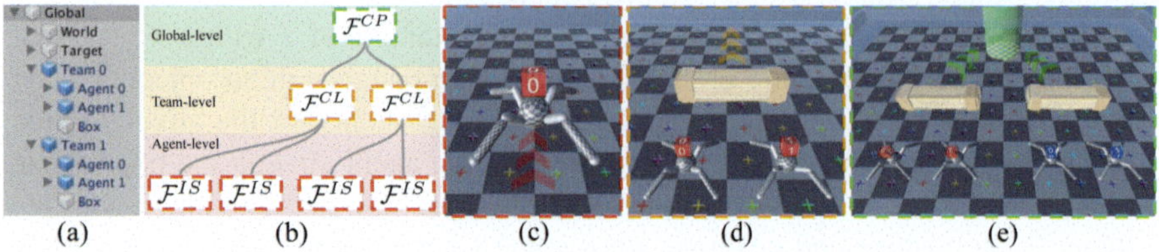

Fig. 17.18 The social tree defined in *Arena* using different BMaRs for each **Arena Node**

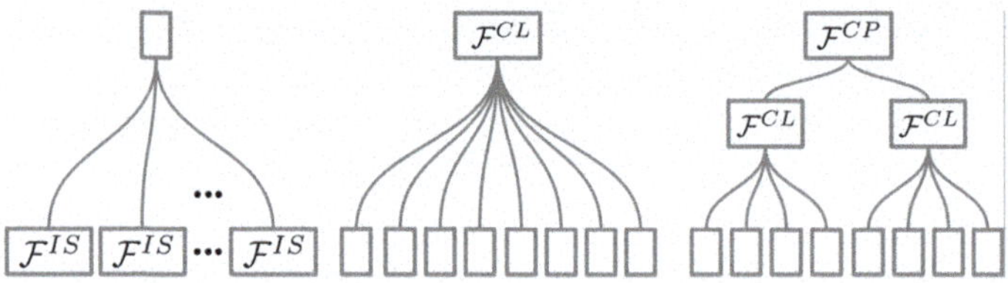

Fig. 17.19 Common social paradigms defined in *Arena* framework

each agent has an agent-level BMaRS. The agent is a robot ant, so that the agent-level BMaRSs are $\mathcal{F}^{IS}$, specifically, the option of *ant-motion* that directs the learning towards basic motion skills such as moving forward, as shown in Fig. 17.18c. Each two agents form a *team* (which is a set of agents or teams), and the two agents have team-level BMaRSs. In this example, the two robot ants collaborate with each other to push a box forward, as shown in Fig. 17.18d. Therefore, the team-level BMaRSs are $\mathcal{F}^{CL}$, specifically, the moving distance of the box. On the two teams, *Arena* has global-level BMaRSs. In this example, the two teams are set to have a match regarding which team pushes its box to the target point first, as shown in Fig. 17.18e. Therefore, the global-level BMaRSs are $\mathcal{F}^{CP}$, specifically, the ranking of the box reaching the target. The final reward function applied to each agent is a weighted sum of the above three BMaRSs at three levels. One can imagine defining a social tree of more than three levels, where small teams form into bigger teams, and BMaRSs are defined at each node to give more complex and structured social problems. After defining the social tree and applying BMaRSs on each node, the environment is ready for use with the abstraction layer handling everything else, such as assigning viewports to each agent in the window, applying the team color, displaying the agent ID, and generating a top-down view.

Moreover, we can easily extend the above framework to other common social paradigms, as shown in Fig. 17.19.[4]

[4]Figure source: Song, Yuhang, et al. "Arena: A General Evaluation Platform and Building Toolkit for Multi-Agent Intelligence." arXiv preprint arXiv:1905.08085 (2019).

More Agent Prefabs

Apart from the previous **BasicAgent**, *Arena* has other more advanced agent prefabs for the off-the-shelf usage, as shown in Fig. 17.20. The usage of other agents is mostly like **BasicAgent**, through dragging and attaching it under the **GlobalManager**. The only difference lies in the action space, you need to change a corresponding brain for controlling different agents. For example, the **ArenaCrawlerAgent** in the agent prefabs looks like Fig. 17.21, which has continuous action space for controlling the action values of joints. In order to use this agent properly, we need to change the brain of **ArenaCrawlerAgent** to be **ArenaCrawlerPlayerContinuous (PlayerBrain)** as shown in Fig. 17.22. Then the game can be exported and used for training as general games.

17.2.4 Export Binary Game

After you have tested the game in Unity within the player mode, and make sure that there is no problem in the game settings, you can export the game to be a standalone binary file, and use it for training the MARL algorithms with Python scripts. This section will show you how to export the game.

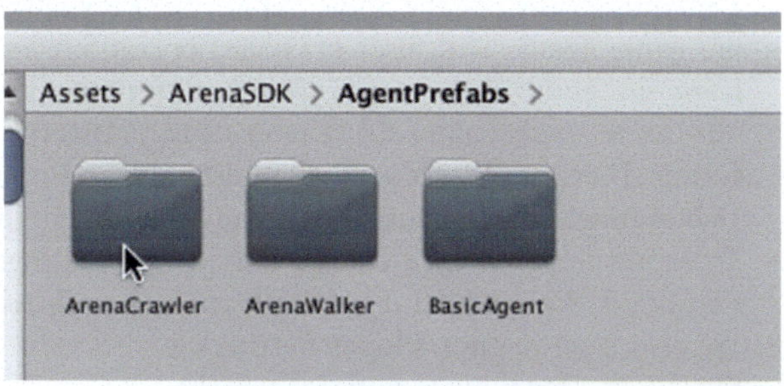

Fig. 17.20 Different agent prefabs in *Arena*

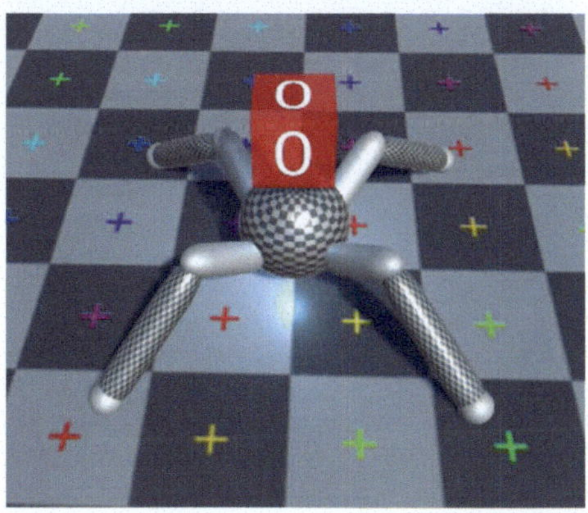

Fig. 17.21 The **ArenaCrawlerAgent** in the scene

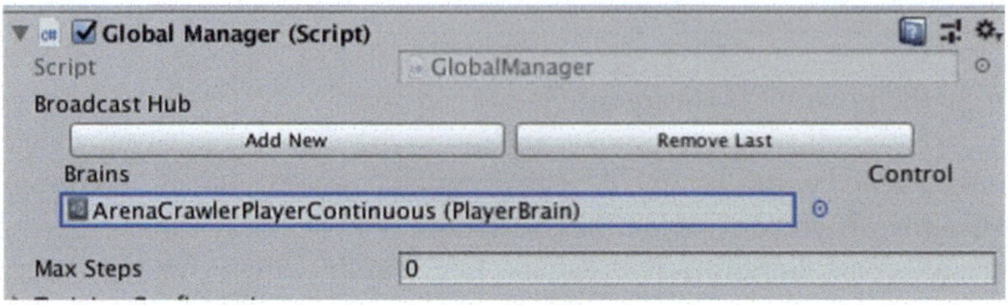

Fig. 17.22 Change the brain for **ArenaCrawlerAgent**

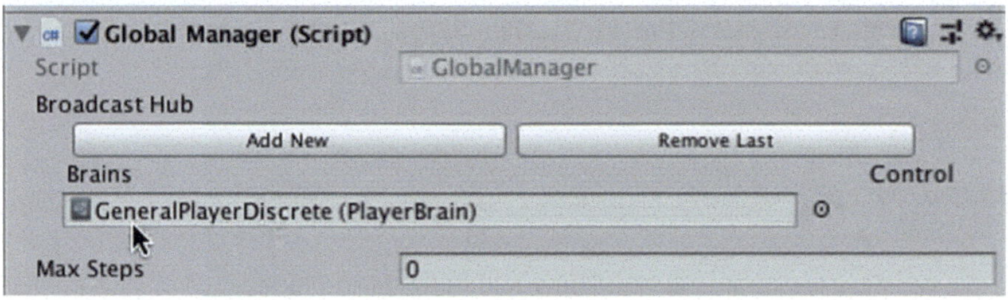

Fig. 17.23 The original brain type in player mode

- First, we need to change the brain type from **PlayerBrain** to a corresponding **LearningBrain** (of the same type), the **PlayerBrain** is used for controlling the game agents with user keyboard operations and the **LearningBrain** is to learn the controlling with learning algorithms. As shown in Fig. 17.23, for this game, we change the **GeneralPlayerDiscrete (PlayerBrain)** to be the **GeneralLearnerDiscrete (LearningBrain)** in Fig. 17.24. We also uncheck the **Debugging** to reduce output information during training.
- To export the game, we choose File->Build Settings, and get a window like in Fig. 17.25. We set the **Target Platform** and **Architecture** accordingly.
- We also need to click **Player Settings** to check the configurations, as shown in Fig. 17.26. One thing to notice is that the **Display Resolution Dialog** needs to be **Disabled** to work correctly. Then we go back to the previous window and click **Build**. We can get the binary file of the game after building.

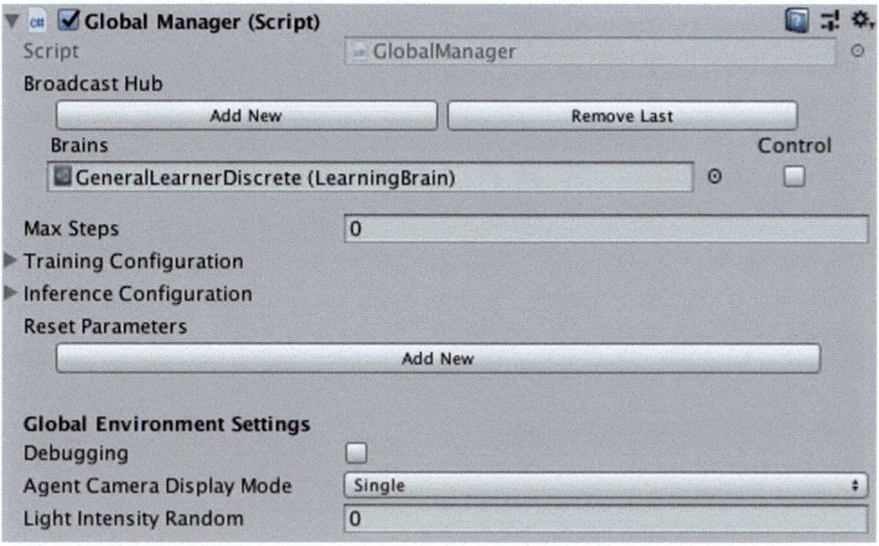

Fig. 17.24 Change the brain type to be **LearningBrain** to export the game for training

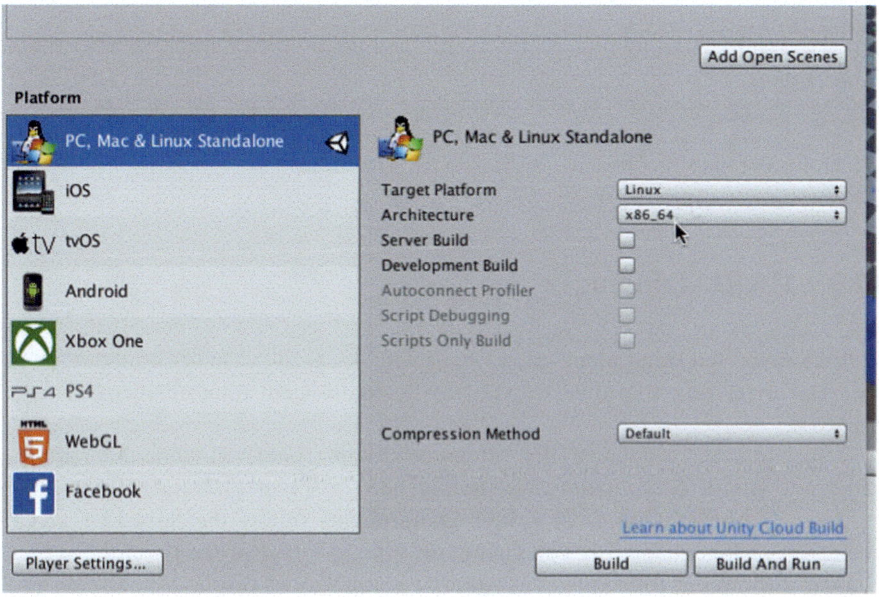

Fig. 17.25 Check the configurations when building the game

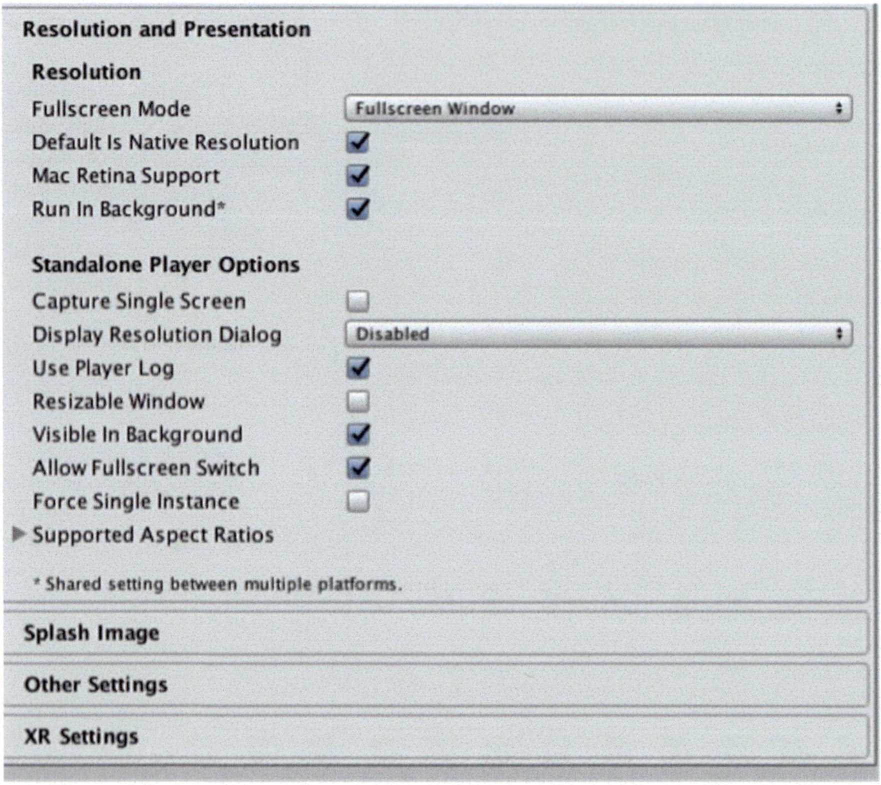

Fig. 17.26 The window of configuring export of the game

17.3 MARL Training

With the exported stand-alone games built with *Arena*, we can set up the training process for investigating diverse problems in multi-agent reinforcement learning (MARL).

Before training, we need to first set up the system. As MARL generally requires large amounts of computation, we usually need a server to handle the training process. The basic settings of *Arena* environments follow the Sect. 17.1 at the beginning of this chapter. If you cannot use the X-Server properly on the server, you can follow the following subsection to setup virtual display; otherwise, just jump to the training sections.

17.3.1 Setup X-Server

The basic settings of using virtual display are as follows:

```
# Install Xorg
sudo apt-get update
sudo apt-get install -y xserver-xorg mesa-utils
sudo nvidia-xconfig -a --use-display-device=None
    --virtual=1280x1024

# Get the BusID information
nvidia-xconfig --query-gpu-info

# Add the BusID information to your /etc/X11/xorg.conf file
sudo sed -i 's/ BoardName "GeForce GTX TITAN X"/ BoardName
    "GeForce GTX TITAN X"\n BusID "0:30:0"/g' /etc/X11/xorg.conf

# Remove the Section "Files" from the /etc/X11/xorg.conf file
# And remove two lines that contain Section "Files" and
    EndSection
sudo vim /etc/X11/xorg.conf

# Download and install the latest Nvidia driver for ubuntu
# Please refer to
    http://download.nvidia.com/XFree86/Linux-#x86_64/latest.txt
wget http://download.nvidia.com/XFree86/Linux-x86_64/390.87/
    NVIDIA-Linux-x86_64-390.87.run
sudo /bin/bash ./NVIDIA-Linux-x86_64-390.87.run --accept-license
    --no-questions --ui=none

# Disable Nouveau as it will clash with the Nvidia driver
sudo echo 'blacklist nouveau' | sudo tee -a
    /etc/modprobe.d/blacklist.conf
sudo echo 'options nouveau modeset=0' | sudo tee -a
    /etc/modprobe.d/blacklist.conf
sudo echo options nouveau modeset=0 | sudo tee -a
    /etc/modprobe.d/nouveau-kms.conf
sudo update-initramfs -u

sudo reboot now
```

Kill Xorg using one of the following three ways (different ways may work on different Linux versions):

```
# approach 1: run following and then run the output of this
    command
ps aux | grep -ie Xorg | awk '{print "sudo kill -9 " $2}'
# approach 2: run following
sudo killall Xorg
# approach 3: run following
sudo init 3
```

Start vitual display with:

```
sudo ls
sudo /usr/bin/X :0 &
```

You should see the virtual display starts successfully with the output as follows:

```
X.Org X Server 1.19.5
Release Date: 2017-10-12
X Protocol Version 11, Revision 0
Build Operating System: Linux 4.4.0-97-generic x86_64 Ubuntu
Current Operating System: Linux W5 4.13.0-46-generic #51-Ubuntu
    SMP Tue Jun 12 12:36:29 UTC 2018 x86_64
Kernel command line:
    BOOT_IMAGE=/boot/vmlinuz-4.13.0-46-generic.efi.signed
    root=UUID=5fdb5e18-f8ee-4762-a53b-e58d2b663df1 ro quiet
    splash nomodeset acpi=noirq thermal.off=1 vt.handoff=7
Build Date: 15 October 2017 05:51:19PM
xorg-server 2:1.19.5-0ubuntu2 (For technical support please see
    http://www.ubuntu.com/support)
Current version of pixman: 0.34.0
   Before reporting problems, check http://wiki.x.org
   to make sure that you have the latest version.
Markers: (--) probed, (**) from config file, (==) default
    setting,
   (++) from command line, (!!) notice, (II) informational,
   (WW) warning, (EE) error, (NI) not implemented, (??) unknown.
(==) Log file: "/var/log/Xorg.0.log", Time: Fri Jun 14 01:18:40
    2019
(==) Using config file: "/etc/X11/xorg.conf"
(==) Using system config directory "/usr/share/X11/xorg.conf.d"
```

If you are seeing errors, go back to "kill Xorg using one of following three way" and try another way.

Before running "Arena-Baselines" in a new window, run following command to attach a virtual display port to the window:

```
export DISPLAY=:0
```

17.3.2 *Run Training*

Create TMUX session (if the machine is a server you connect via SSH) and enter virtual environment:

```
tmux new-session -s Arena
source activate Arena
```

Continuous Action Space

List of games with continuous action space in *Arena*:

- ArenaCrawler-Example-v2-Continuous
- ArenaCrawlerMove-2T1P-v1-Continuous
- ArenaCrawlerRush-2T1P-v1-Continuous
- ArenaCrawlerPush-2T1P-v1-Continuous
- ArenaWalkerMove-2T1P-v1-Continuous
- Crossroads-2T1P-v1-Continuous
- Crossroads-2T2P-v1-Continuous
- ArenaCrawlerPush-2T2P-v1-Continuous
- RunToGoal-2T1P-v1-Continuous
- Sumo-2T1P-v1-Continuous
- YouShallNotPass-Dense-2T1P-v1-Continuous

Run the training commands, replace **GAME_NAME** with above games and choose proper **num-processes** (with **num-mini-batch** equivalent to **num-processes**) according to your machine,:

```
tmux new-session -s Arena
CUDA_VISIBLE_DEVICES=0 python main.py --mode train --env-name
    GAME_NAME --obs-type visual --num-frame-stack 4
    --recurrent-brain --normalize-obs --trainer ppo --use-gae
    --lr 3e-4 --value-loss-coef 0.5 --ppo-epoch 10
    --num-processes 16 --num-steps 2048 --num-mini-batch 16
    --use-linear-lr-decay --entropy-coef 0 --gamma 0.995 --tau
    0.95 --num-env-steps 100000000
    --reload-playing-agents-principle OpenAIFive --vis
    --vis-interval 1 --log-interval 1 --num-eval-episodes 10
    --arena-start-index 31969 --aux 0
```

Discrete Action Space

List of games with discrete action space in *Arena*:

- Crossroads-2T1P-v1-Discrete
- FighterNoTurn-2T1P-v1-Discrete
- FighterFull-2T1P-v1-Discrete
- Soccer-2T1P-v1-Discrete
- BlowBlow-2T1P-v1-Discrete
- Boomer-2T1P-v1-Discrete
- Gunner-2T1P-v1-Discrete
- Maze2x2Gunner-2T1P-v1-Discrete
- Maze3x3Gunner-2T1P-v1-Discrete
- Maze3x3Gunner-PenalizeTie-2T1P-v1-Discrete

- Barrier4x4Gunner-2T1P-v1-Discrete
- Soccer-2T2P-v1-Discrete
- BlowBlow-2T2P-v1-Discrete
- BlowBlow-Dense-2T2P-v1-Discrete
- Tennis-2T1P-v1-Discrete
- Tank-FP-2T1P-v1-Discrete
- BlowBlow-Dense-2T1P-v1-Discrete

Run the training commands, replace **GAME_NAME** with above games and choose proper **num-processes** (with **num-mini-batch** equivalent to **num-processes**) according to your machine,:

```
CUDA_VISIBLE_DEVICES=0 python main.py --mode train --env-name
    GAME_NAME --obs-type visual --num-frame-stack 4
    --recurrent-brain --normalize-obs --trainer ppo --use-gae
    --lr 2.5e-4 --value-loss-coef 0.5 --ppo-epoch 4
    --num-processes 16 --num-steps 1024 --num-mini-batch 16
    --use-linear-lr-decay --entropy-coef 0.01 --clip-param 0.1
    --num-env-steps 100000000 --reload-playing-agents-principle
    OpenAIFive --vis --vis-interval 1 --log-interval 1
    --num-eval-episodes 10 --arena-start-index 31569 --aux 0
```

You can also change other MARL algorithms instead of the PPO above to test the games you build.

17.3.3 Visualization

To visualize the learning curves for analyzing the training process with Tensorboard, run:

```
source activate Arena && tensorboard --logdir ../results/ --port
    8888
```

and visit http://localhost:4253 for visualization with tensorboard.

Acknowledgments We give special thanks to Yuhang Song, Prof. Zhenghua Xu, Prof. Thomas Lukasiewicz, et al., for their great contributions in the *Arena* project.

References

Balduzzi D, Garnelo M, Bachrach Y, Czarnecki WM, Perolat J, Jaderberg M, Graepel T (2019) Open-ended learning in symmetric zero-sum games. arXiv:190108106

Bansal T, Pachocki J, Sidor S, Sutskever I, Mordatch I (2018) Emergent complexity via multi-agent competition. In: ICLR. In: International Conference on Learning Representations. https://openreview.net/forum?id=Sy0GnUxCb

Cai Y, Daskalakis C (2011) On minmax theorems for multiplayer games. In: Proceedings of the twenty-second annual ACM-SIAM symposium on discrete algorithms. Society for Industrial and Applied Mathematics, Philadelphia

Heess N, Sriram S, Lemmon J, Merel J, Wayne G, Tassa Y, Erez T, Wang Z, Eslami S, Riedmiller M, et al. (2017) Emergence of locomotion behaviours in rich environments. arXiv:170702286

Hendtlass T (2004) An introduction to collective intelligence. In: Applied intelligent systems. Springer, Berlin,

Jaderberg M, Czarnecki WM, Dunning I, Marris L, Lever G, Castaneda AG, Beattie C, Rabinowitz NC, Morcos AS, Ruderman A, et al. (2018) Human-level performance in first-person multi-player games with population-based deep reinforcement learning. Computing Res Repository. http://arxiv.org/abs/1807.01281

Myerson RB (2013) Game theory. Harvard University Press, Cambridge

Singh S, Lewis RL, Barto AG (2009) Where do rewards come from. In: Proceedings of the annual conference of the cognitive science society

Singh S, Lewis RL, Barto AG, Sorg J (2010) Intrinsically motivated reinforcement learning: an evolutionary perspective. IEEE Trans Auton Ment Dev 2(2):70–82

Song Y, Wang J, Lukasiewicz T, Xu Z, Xu M, Ding Z, Wu L (2019) Arena: a general evaluation platform and building toolkit for multi-agent intelligence. arXiv:190508085

Chapter 18
Tricks of Implementation

Zihan Ding and Hao Dong

Abstract Previous chapters have provided the readers the main knowledge of deep reinforcement learning, main categories of reinforcement learning algorithms as well as their code implementations, and several practical projects for better understanding deep reinforcement learning in practice. However, due to the aforementioned challenges like low sample efficiency, instability, and so on, it may still be hard for the novices to employ those algorithms well in their own applications. So in this chapter, we summarize some common tricks and methods in detail, either mathematically or empirically for deep reinforcement learning applications in practice. The methods and tips are provided from both the stage of algorithm implementation and the stage of training and debugging, to avoid the readers from getting trapped in some practical dilemmas. These empirical tricks can be significantly effective in some cases, but not always. This is due to the complexity and sensitivity of deep reinforcement learning models, where sometimes an ensemble of tricks needs to be applied. People can also refer to this chapter to get some enlightenment of solutions when getting stuck on the projects.

Keywords Deep reinforcement learning · Application · Implementation · Reward engineering · Neural network · Normalization · Efficiency · Stability

Z. Ding (✉)
Imperial College London, London, UK
e-mail: zhding@mail.ustc.edu.cn

H. Dong
Peking University, Beijing, China
e-mail: hao.dong@pku.edu.cn

H. Dong et al. (eds.), *Deep Reinforcement Learning*,
https://doi.org/10.1007/978-981-15-4095-0_18

18.1 Overview: How to Apply Deep Reinforcement Learning?

Deep learning is often referred to as a "black-box" method. However, it is not actually "black-box," but can sometimes still be unstable and provide unpredictable results. In deep reinforcement learning, this problem is even worse as the basic learning process of reinforcement learning requires the agent to learn in a dynamic process with reward signals instead of labels. This is in contrast to supervised learning. Rewards in reinforcement learning settings may also contain incomplete or local information, and the agent needs to chase a changing goal when using bootstrapping learning methods. Moreover, an algorithm with more than one deep neural network is common in deep reinforcement learning, especially for more advanced and recent proposed methods, which also makes deep reinforcement learning algorithms unstable and potentially sensitive to hyperparameters. The above problems make research or applications with deep reinforcement learning hard to work in practice. Because of this, here we described some common tricks and suggestions for implementing deep reinforcement learning.

First of all, you need to know whether a reinforcement learning algorithm should be used for a certain problem or not, and it definitely does not work for all tasks. We usually need to carefully consider if reinforcement learning is suitable for solving a task. Generally, reinforcement learning works for sequential decision-making problems, which can be estimated via a Markov decision process. A prediction task with a labeled data set does not require reinforcement learning algorithms, and a supervised-learning method can be more straightforward and effective. Reinforcement learning tasks always contain at least two key components: (1) the environment, which provides the dynamic process and reward signals and (2) the agent, which is controlled by a policy learned with reinforcement learning. The reinforcement learning algorithms implemented in previous chapters are used for solving tasks like OpenAI Gym environments. In these experiments, you do not need to care more about the environments as they are mostly already standardized and normalized. However, the environments in some projects described in the application chapters need to be defined manually, as well as applying reinforcement learning algorithms for the agent to make it work.

Generally, there are several stages for applying deep reinforcement learning in applications:

1. Toy-test stage: you should use the simple models including the reinforcement learning algorithms with high confidence of the accuracy and correctness, to explore the environments (even using random policy) if it is a new task, or to testify the extensions you are going to apply on the final model step by step, instead of applying the complicated models all at once. You can run these experiments very quickly to see if there are any problems with the environment or general model settings, or at least you can familiarize yourself with the tasks

you are going to handle, which could give you some useful intuitions afterward and show you some corner cases for consideration.
2. Fast-configuration stage: you should take fast tests of model settings to evaluate the potential of success. Visualize the learning process as much as possible if there is any bug, and use statistical variables if you cannot get the underlying relationships from the numbers directly. This step should start from the toy test in the first stage and gradually increases the complexity of your new model. Every attempt should be tested if you are not 100% certain about its effectiveness.
3. Deploy-training stage: after you have carefully configured the correctness of the model, you can start deploying the training in a large scale. As deep reinforcement learning usually needs a large number of samples to train for a long time, you will always be encouraged to use parallel training, cloud computing, etc., to speed up the large-scale training of the final model. Sometimes this stage has to be mixed with the second stage and could potentially take a long time in practice.

In the following sections the tricks of applying deep reinforcement learning will be described in several parts.

18.2 Implementation

- **Implement some fundamental reinforcement learning algorithms from scratch.** For a starter in the deep reinforcement learning area, it is a good practice to implement some fundamental reinforcement learning algorithms from scratch, debug it, and make it work finally. The Deep Q-Network is worth implementing as it is the foundation of most value-based algorithms. Policy gradient and actor-critic algorithms are good choice to start with for continuous action space. This process will require you to understand each line in the implemented reinforcement learning algorithms and provide you a rough idea of reinforcement learning process. At the beginning, you do not need a complicated large-scale task, but a relatively simple one with a fast learning process, like those in OpenAI Gym environments. You should follow a common structure as well as using one type of deep learning framework (i.e., *TensorFlow*, *PyTorch*, etc.) when implementing those basic methods, and gradually scale it up to more complicated tasks as well as more advanced techniques (e.g., prioritized experience replay, etc.). This will significantly accelerate the process when you work on other projects with different deep reinforcement learning algorithms later. If you meet some problems during the implementation process, you can refer to others' implementation (e.g., the tutorial reinforcement learning algorithms implementation[1] provided together with the book) or search the problem on the internet. Most of the problems are solved by others.

- **Moderately implement the details in papers.** After you get familiar with those fundamental reinforcement learning algorithms, you can start to implement and test some methods in the literature. Usually papers in the deep reinforcement learning research area contain lots of implementation details, and sometimes they are not even consistent across papers. So, when you implement those methods, do not overfit to the details of the paper, but get a general understanding about why the authors chose to apply the tricks in a specific situation. A typical example is that, in most papers, the detailed structures of the neural networks used in the experimental tests including even the dimensions and numbers of hidden layers, values of each hyperparameter, etc., are provided in the body of the paper or the supplementary materials. You do not need to strictly follow those details during the implementation of your own version, and you may not even test the method with exactly the same environment as the original paper does. For example, in the original paper of the deep deterministic policy gradient algorithm, the authors recommend to use the Ornstein–Uhlenbeck (OU) noise for exploration. However, it is sometimes hard to say if the OU noise is better than the Gaussian noise in practice, which depends on the specific tasks to a great extent. Another example is that, in the work of AlphaStar by Vinyals et al. (2019), the vanilla TD(λ) method is found to be better than the more advanced off-policy correction method V-trace (Espeholt et al. 2018) in practice. Therefore, if those tricks are not general enough to work well, it may not even be worth your effort on implementing them. Some fine-tuning methods for a specific task may produce better results. However, as mentioned above, understanding why the tricks can work for the cases by the authors is more critical and meaningful. These suggestions are more useful when you try to borrow an idea from a paper and apply it in your own method, because sometimes it is not the main idea in the paper but a specific trick or operation helps the most in your cases.
- **Explore the environment if you are working on solving a specific task.** You should check the environment details including properties of the observation and actions like dimension, value range, property of continuous or discrete value, etc. You should normalize values like observations of environments if they are unknown or within a large valid range. For example, if you are using tanh or sigmoid as activation functions, large values of inputs will have the chance to saturate the nodes in the first hidden layer, which can lead to a slow learning process with small gradients at the beginning of training. Moreover, you should choose good input features for reinforcement learning, which contain more useful information for the environment. You can also use the agent with random actions to explore the environment and visualize it to see if there are any corner cases. This step is more important if the environment is built by yourself.

[1]https://github.com/tensorlayer/tensorlayer/tree/master/examples/reinforcement_learning.

- **Choose a proper output activation function for each network**. You should apply the proper output activation function for the actor network according to the environment. For example, ReLU could work quite well for the hidden layer, for both computation time and convergence performance; but may not be proper for output action range with possible negative values. It is better to fit the output range of policy with the action value range for the environment, like using tanh activation as output layers for action value range of $(-1, 1)$.
- **Start with the toy example and gradually increase the complexity.** You should begin with the models or environments you are very clear about for tests, and go step by step with new components instead of assembling all modules before testing and debugging. Keep taking tests during implementation. You should never expect a complicated model can be implemented once and work directly unless you are an expert and lucky.
- **Start with dense reward functions.** The reward function can affect the convexity of the optimization in the learning process, so a smooth and dense reward function is always what you should try at the beginning stage. For example, in the robotic reaching task defined in the application Chap. 16, we start the robot learning with a dense reward, which is defined to be the negative distance from the robot gripper to the target object. This will ensure a smooth hyperplane for the value networks and policy networks to be optimized upon, and, therefore, speed up the learning process significantly. The sparse reward can be defined to be a simple binary value indicating whether the robot has reached the target object or not, which is hard to explore and learn for the robot without leveraging additional information.
- **Choose proper network architectures.** Although in deep learning it is common to see a network with dozens of layers and billions of parameters, especially in fields like computer vision (He et al. 2016) and natural language processing (Jaderberg et al. 2015), in deep reinforcement learning generally a "deep" network indicates its depth of more than five layers, and it is quite common to see the networks in deep reinforcement learning algorithms with only 2–5 layers, depending on the environment. Therefore, unless the environment is really in a huge scale and you have hundreds or thousands of GPUs or TPUs, you are not supposed to apply a network with ten layers or more in deep reinforcement learning. This is not only limited by the training resources, but also related to the instability and non-monotonic improvement of deep reinforcement learning due to a lack of supervisory signals. The network can overfit the data set if the network capacity is too large compared to the data in supervised learning, while in deep reinforcement learning it may just converge slowly or even diverge due to the strong correlations between the exploration and exploitation. For choosing the size of the network, it usually depends on the state space and action space. A discrete environment with dozens of state-action combinations may be solved with a tabular method, or a single- or two-layer neural network. More complicated cases like applications in Chaps. 13 and 16 with dozens of continuous dimensions for states and actions may require more than three layers,

but still on a small scale compared with gigantic networks in other deep learning fields.

As for the structure of the networks, it is common to see plain multi-layer perceptrons (MLPs), convolutional neural networks (CNNs), and recurrent neural networks (RNNs) in literature. Advanced and sophisticated network structures are rarely applied unless there are specific requirements for fine-tuning the models and other special cases. A low-dimensional vectorized input can be handled with MLPs, and vision-based policy learning usually requires a CNN backbone to extract information beforehand, either trained together with reinforcement learning loss or pre-trained with other computer vision methods. There are also cases when both vectorized low-dimensional inputs and image-based high-dimensional inputs are used at the same time; a backbone for feature extraction on high-dimensional inputs and concatenation with the rest low-dimensional inputs are usually employed in practice. RNNs can be applied when the environment is not fully observable or non-Markovian, where the optimal action choice may depend on previous states other than the current state. The above empirical instructions work for both the policy and the value networks in practice. Sometimes the policy and the value may take different inputs as states to form an asymmetric actor-critic structure, which can be applied when the value network is only a guidance for the policy network during training and not accessible anymore in action prediction.

- **Get familiar with the properties of the reinforcement learning algorithm you use.** For example, the trust-region based methods like PPO or TRPO may need large batch size to ensure safe improvement. For these trust-region methods, we would expect a steady improvement in policy performance, instead of a sudden dramatic decrease at some points on the learning curve. The reason for the trust-region methods like TRPO to use a larger batch size is that it applies the conjugate gradient to approximate the Fisher information matrix, with respect to current sampled batch. This can be problematic when the batch is too small or biased, which will end up with an inaccurate Fisher information matrix (or inverse Hessian product) approximation and therefore a decrease in learning performances. So, in practice, the batch size of algorithms TRPO and PPO should be enlarged until the agent can learn with a steadily improved performance. Therefore, TRPO sometimes cannot scale up to large-scale networks or deep CNNs and RNNs well. DDPG is usually considered to be sensitive to hyperparameters although it has been shown to be effective on tasks with continuous action space. The sensitivity can be more significant when it is applied on large-scale or real-world tasks (Mahmood et al. 2018). For example, although a thorough hyperparameter searching process can provide an optimal performance on simulated toy tests ultimately, learning in the real-world may not allow this kind of hyperparameter searching due to the time and resources limitation, and, therefore, DDPG may lead to a worse learning performance compared with other algorithms like TRPO and SAC. Also, although DDPG algorithms are originally designed for tasks with continuous-valued actions, it does not mean that it cannot work for discrete-action cases. When you try to apply it on tasks with discrete-valued actions, some additional tricks can be applied for solving this, like a sigmoid(tx) output activation with a large t value can be added and then clipped for binary-valued output with small truncated errors, or you can directly apply the *Gumbel–Softmax* trick and change the deterministic output to be a categorical distribution. Other algorithms can be processed similarly.

- **Normalize the values.** Generally, you should normalize the reward values with rescaling but not shifting the mean, and standardize the prediction targets for the value functions as well in the same manner. Reward scaling is conducted on the sample batch for training. The reason for only taking the value scaling (e.g., divided by the standard deviation value) but no mean shifting (e.g., zero mean) is that mean shifting may hurt the living will of the agent. This is actually related to the signs of general reward values in the reward function and only holds for the cases when you are using the "done" signals, which means you can take a mean shifting when not using "done" signals for terminating the episode in advance. Considering the agent is going through an episode with a "done=True" happened before the maximum length of the episode. Then the reward values after this "done" signal are actually zeros if we still pretend their existence. If these zero values are generally higher than previous reward values (i.e., previous reward values are generally negative), then the agent will tend to terminate the episode as early as possible, to maximize its episode reward. Otherwise the agent will "live" longer if its reward values are generally positive. If we take the mean shifting in the reward values, it will break the living will described in the above scenarios, which may not make the agent live longer even when reward values are generally positive. So this could hurt the performance in training. Standardizing the target value function is similar. For example, the average Q-value in some DQN-based algorithms can keep increasing during learning in an unexpected mode, which is caused by overestimation of Q-value with the maximized optimization formula. A normalized target Q-value could alleviate the problem, as well as applying tricks like double Q-learning.
- **A tip of the discount factor.** You can have a rough sense of the effective time horizon for evaluating the single-step action choice according to the discount factor γ: $1 + \gamma + \gamma^2 + \cdots = 1/(1 - \gamma)$. For $\gamma = 0.99$, we usually neglect the rewards after 100 time steps, and this can speed up the process when you set the parameter.
- **Done signal is true only for terminal states.** There are some nuances in deep reinforcement learning which are easy to ignore for novices, and the "done" signal in episodic reinforcement learning is one of them. These nuances make different implementations of even the same algorithms display totally different performances in practice. The "done" signal is commonly applied in episodic reinforcement learning when trying to finish the episodes, and it is a function of the state of the environment, set to be true whenever the terminal state is reached by the agent. Note that here the terminal state is defined to be a state indicating the agent has finished the episode, either in success or failure cases, rather than an arbitrary state when the time limit or maximal episode length is reached. It is not trivial to require that the "done" signal is set to be true only when the state is

a terminal state. For example, if the task is to manipulate the robot arm to reach a specific point in the space, the "done" signal is supposed to be true only when the robot has indeed reached the point, but not for the cases when a default maximal length of episode is reached. To understand the differences, we need to know that in reinforcement learning some environments are infinite and some are finite. But in the sampling process, the algorithms usually handle the trajectories in a finite length. Two common ways of achieving that are either setting a maximal episode length or using the "done" signals as feedback from the environment to finish the episode with a break in loops. When applying the "done" signal as a breakpoint in sampling, it should not be set to be true when the episode is finished due to the accomplishment of the maximal length, but only when the terminal state is reached. If the robot finishes its trajectory somewhere else rather than the target point with the done signal set to be true, it may negatively affect the learning process. Specifically, take the example of PPO algorithm, cumulative rewards starting from states S_t are used to estimate the value of states $V(S_t)$, while a terminal state has the value zero. If the "done" signal is true for a non-terminal state, the state value is set to be zero although it should not be, it will confuse the value network when estimating its previous states and therefore hinder the learning process.

- **Prevent numerical problem.** For the division case, it is possible to have infinite value as outcome without getting an error if not used properly. Two tricks can alleviate these problems: the first solution is to use exponential scale for positive-value cases like $a/b = \exp(\log(a) - \log(b))$; another solution is to add small value to the denominator if it is non-negative, like $a/b \approx a/(b + 10^{-6})$.
- **Be aware of the divergence between reward functions and final objectives.** Reinforcement learning is usually applied on a specific task with a final goal, and a reward function is usually manually designed to align with the goal and make it easier for the reinforcement learning agent to learn. In this sense, the reward function is usually a quantitative form of the goal, which also means they are two different things. Divergence could happen in these cases. As a reinforcement learning agent is able to overfit to the reward function you set for the task, you may notice that the final policy provides a result different from what you expect, for achieving the final goal. One of the most probable reasons for this is the divergence of the reward function and the final objective. In most cases, the reward function is easy to be chosen to bias towards the final task objective, but it is not trivial to design a reward function consistent with the objective all the time and for all corner cases. What you should do is to try to reduce this kind of divergence, to make sure your reward function can smoothly help the agent to achieve the true objective.
- **Reward may not always be a good way of displaying learning performances** People usually display the reward values (with moving average or without) through the learning process as representations of an algorithm's capability. However, as said in the tip above, there can be a divergence between the final goal and the way you define the reward function, which makes it possible for a higher-reward state to correspond with a worse case towards the final goal, or at least

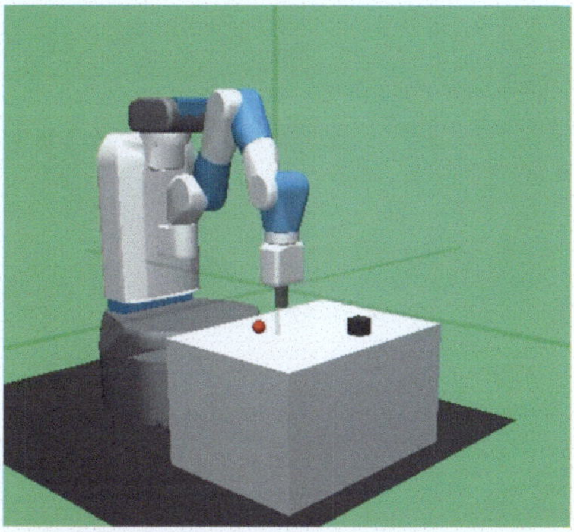

Fig. 18.1 For the robot learning task *FetchPush* environment in OpenAI Gym, it may be better to use the final distance from the object to the target position instead of the reward value to evaluate the performance level of the learned policy, because it is the most straightforward representation of the overall goal for the task. The reward function, however, may be engineered to have some other factors like minimizing the distance of the gripper to the object

not explicitly displaying the relationship of the state with an optimal state. For this reason, we always need to consider the possibility of this divergence when applying reinforcement learning and displaying results. Therefore, it is common to see in the literature (Fu et al. 2018) that sometimes the learning performances are not evaluated and displayed with the smoothed episode reward (depending on the design of reward function as well), but with more specific metric for the task, like the final distance from the robotic gripper to the object for the reaching manipulation or the distance from the object to the target for pushing manipulations in robotic learning tasks as shown in Fig. 18.1. The distance to the object, or whether or not the object is grasped, is the true evaluation metric for the goal of task. Therefore, these can be used for displaying the learning performances. This can be useful when there are indeed divergences between the final goals and the manually designed reward functions, and even critical if you are trying to compare different reward schemes.

- **Non-Markovian cases.** As introduced in previous chapters, most theoretical results introduced in this book are dependent on the Markov process assumption or Markov property of the state. The Markov property not only simplifies the problem and derivations, but also makes the problems describable and solvable with compact representation of solutions using iterative methods. However, in practice, the Markov process assumption does not hold all the time. For example, as shown in Fig. 18.2, the Pong game in the Gym environments does not satisfy the Markov process assumption of the state for the agent to take optimal actions. We need to keep in mind the Markov property is a property of the state, or the environment, and therefore determined by the definition of the state. The

Fig. 18.2 The Pong-v0 game in Gym. The speed of the ball cannot be captured in a single frame, which makes the problem to be non-Markovian. Stacked frames are used as observations to solve it

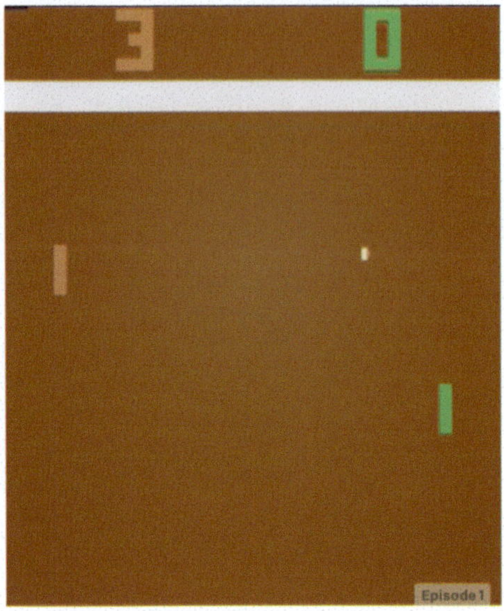

difference of the non-Markov decision process and partially observed Markov decision process (POMDP) is sometimes subtle. For example, if the state is defined to contain both the position and the velocity (no acceleration as an assumption) of the ball, but the observation contains the position only, then it is POMDP instead of non-Markov process. However, the state in the Pong game is usually assumed to be the static frame of each time step. The current state containing the position of the ball does not provide all the information the agent will need to make an optimal action choice, based on the fact that the velocity and moving direction of the ball will also affect the optimal actions. So it is a non-Markov environment in this sense. One way of providing the velocity and moving direction information is to use the historical states, which violates the Markov process. Therefore, in original paper of DQN (Mnih et al. 2015), stacked frames are used for solving the task of Pong with an approximated MDP. On the other hand, if we take the stacked frames as a single state all the time, and assume that the stacked frames always contain complete information for making the optimal action choice, then it still follows the Markov process assumption actually. As in simulation all processes are discrete after all, rather than a continuous sequence of time in the real-world, we can usually take this kind of transformation to regard a non-Markov process as a Markov one. Apart from using the stacked frames as in original DQN, the recurrent neural network (RNN) (Heess et al. 2015) or more advanced long short-term memory (LSTM) can be applied for control with memory of history, to solve non-Markovian problems as well.

18.3 Training and Debugging

- **Initialization matters.** Deep reinforcement learning methods usually update the policy either in on-policy manner, with samples for each episode, or using a dynamic replay buffer, which contains various samples through time. This makes deep reinforcement learning different from supervised learning, which learns from a fixed data set and the order of learning samples will not matter too much. However, in deep reinforcement learning, the initialization of policy can affect the possible exploration range afterwards and determines the samples sent into the buffer or directly used for updating, and, therefore, affect the overall learning performances. Starting with a random policy can result in larger chances to have more various samples, which is great at the beginning of the training stage. But with the convergence and improvement of the policy, the explored areas are restricted, which are close to the trajectories generated by current policy. For the initialization of weight parameters, it is generally better to use more advanced methods like Xavier initialization (Glorot and Bengio 2010) or orthogonal initialization (Saxe et al. 2013), which can help with avoiding vanishing or exploding gradients and provide more robust learning performances in general deep learning cases.
- **Add useful probes to the program.** As deep learning deals with a large amount of data, there could be some hidden operations we are not always aware of, especially when we are not familiar with the model. The error reports may be not designed for some of those mistakes. It could be dangerous to have these kinds of underlying problems in the model. For example, in deep reinforcement learning tasks you may only care about the reward functions, but you should also visualize the value of loss functions to know about how good the value functions are approximated, and the entropy of the stochastic policy to know the exploration status. If there is a premature drop of the policy entropy, it basically indicates the agent cannot explore more useful samples with current policy. This can be alleviated through using entropy bonus or KL divergence penalty. Algorithms like soft actor-critic (SAC) use adaptive entropy for solving this problem automatically. For trust-region based methods, you will need indicators to know the value of KL divergence of old and new policies, which tells you if the model is working healthily. Sometimes you need to output the gradients value in your network if it does not work. The gradient values for normal layers should not be too large or all zeros, which could indicate either an abnormal gradient or no gradient flow. Other useful indicators include update size in output space and parameter space, and standard diagnostics for deep neural networks. For above cases, *Tensorboard*[2] module can be a powerful tool, which is originally implemented for *TensorFlow* but also has support for *PyTorch*. It simplifies the

[2]Details of tensorboard: https://www.tensorflow.org/tensorboard.

visualization process of variables and graphs in neural network methods, which helps to achieve those probes in practice.

- **Apply several random seeds and take an average to reduce randomness.** Deep reinforcement learning methods have typical properties of unstable training processes, as described in Chap. 7. Even the random seeds can affect the learning performances a lot. There are random seeds for *Numpy*, for *TensorFlow* or *PyTorch*, for environments, and so on. When randomizing the seeds, all of those seeds need to be randomized properly as a default setting. You can set fixed seeds first to see if there are any differences along the sampled trajectories, and other sources of randomness can exist if it still contains randomness. Fixed seeds can be used for reproducing the learning process. Taking random seeds and averaging the learning curves can reduce the chances of deriving wrong conclusions from experimental comparisons due to the randomness of deep reinforcement learning. Usually the larger number of the random seeds is taken, the more reliable the results are, but at the same time with the increase of experimental time. An empirical setting with three to five runs using different random seeds can have a relatively solid result, but more is better.
- **Balance the CPU and GPU resources to speed up training.** This tip is actually about finding and solving the bottlenecks of the speed in your training process. The problem of better leveraging the computational resources on the limited machines is more complicated in reinforcement learning than supervised learning. In supervised learning, the CPUs are usually used for data pre-processing, reading and writing, while GPUs are conducting the forward inference and back-propagation process. However, as in reinforcement learning the inference process always contains interactions with the environment, the devices for gradients calculation need to match with the ones handling the environment interaction in computational capability, otherwise it is a waste of exploration or a waste of exploitation. In reinforcement learning, the CPUs are usually used for sampling through interactions with the environment, which can involve a large amount of computation for some complicated simulation systems. The GPUs are used for forward inference and backward updating with back-propagation. You should check the occupancy of both CPU and GPU resources when deploying a large-scale training process and prevent thread/process sleeping. This is important especially when you distribute your program on a large-scale parallel computing system. For the case of overusing the GPUs, more threads or processes for sampling through interaction with environments can be employed. For the overusing of CPUs, you can reduce the distributed sampling workers, or increase the number of parallel updating workers, choose a larger update iteration number, or increase the batch size for off-line updating, depending on the way in which you manage the parallel threads/processes. Note that the above is only about how to best leverage your computational resources, you also need to consider the trade-off between exploration and exploitation, as well as different levels of sample efficiency for various reinforcement learning tasks.

  To solve the balancing problem between CPU and GPU resources, you will usually need parallel computing with multi-threading or multi-processing for both sampling and training to make full use of your available machines. The design of parallel training framework for running both threads/processes for sampling and threads/processes for updating the networks needs to be carefully considered. Locks and pipes are usually needed in these kinds of framework to make it work smoothly. Redundant processes can be created to save the waiting time. It could be different for on-policy and off-policy learning, the settings of off-policy training in parallel are usually more flexible than on-policy ones, because you can update the policy whenever you want instead of only at each final step of episodes. A typical usage of distributional training with multiple GPUs with PyTorch framework is shown in Fig. 18.3.[3] When PyTorch is handling the multiple-GPU process, a model replication process and a gathering process of inferred results are employed in the forward inference process, and parallel gradients back-propagation with gradients reduction are employed in the backward updating process. More details are discussed in chapters of applications with deep reinforcement learning and provided code examples in our repository.

- **Visualization.** Try to visualize the data if you cannot see the underlying relationships from the numerical values directly. For example, sometimes if the reward values are very shaky due to the unstable property of deep reinforcement learning, then you may need to plot the running average of the values to see if the agent is improving through training.
- **Smooth the learning curves.** The learning process of reinforcement learning can be very unstable, as shown in Fig. 18.4. Directly giving conclusions from raw learning curves is usually not reliable, as the unsmoothed learning curve shows in the graph. We usually smooth the learning curves with moving averages, convolutional kernels, etc., using a proper window length. The increasing/decreasing learning performance can be displayed more explicitly in this approach. This can be critical when you are handling with a complicated reinforcement learning task with a long period of training time and slow performance improvement.
- **Understand exploration and exploitation** From the Fig. 18.4, we can also see that there is a plateau in the learning curve at the early stage of training. This is actually not rare but a common case in the reinforcement learning training process. This is because samples for reinforcement learning to learn from are not pre-prepared like in supervised-learning settings, but explored with the policy that the agent is applying. Therefore, whether or not current policy is able to explore high-reward trajectories can be critical in reinforcement learning training. This is the problem of exploration, and we need to make sure our policy can gradually explore near-optimal trajectories. When the reinforcement learning algorithms do not work on a specific task, you need to investigate whether the agent has explored those good trajectories or not. If not, at least there is the problem of current exploration strategy. However, if a current policy is able to explore good trajectories, but it still does not converge to great

[3]Figure from PyTorch Forum.

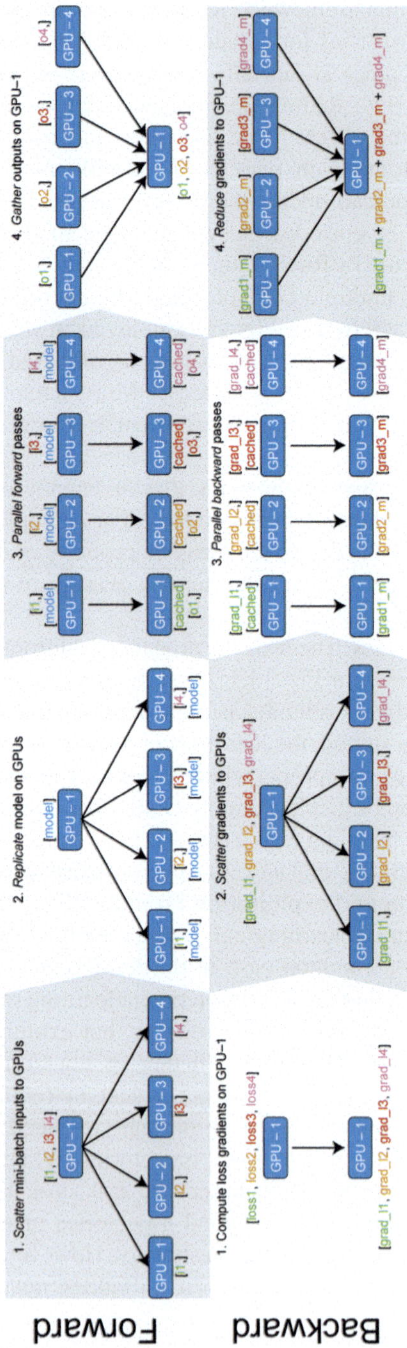

Fig. 18.3 The forward and backward passes of torch.nn.DataParallel

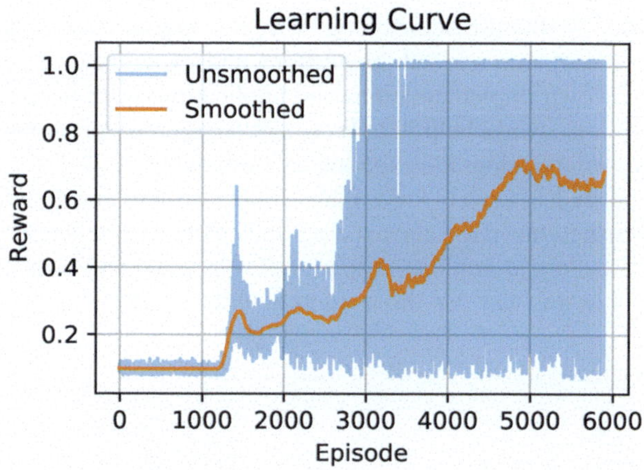

Fig. 18.4 The smoothed and unsmoothed learning curves in reinforcement learning

actions, then it can be the problem of exploitation. This means the policy cannot learn from good trajectories well. The problem of exploitation can be caused by low sample efficiency, poor value function approximation, low learning rate of value function, poor learning performance of policy networks, etc. The learning curves in Fig. 18.4 show a healthy learning improvement: once the good samples are explored (during the plateau), the learning performance will increase dramatically (after the plateau) with the policy improved.

- **Question the correctness of your algorithm implementations first.** It is very common to see the code implementation not work immediately after you finish it, so be patient at debugging the code is important. The correctness of algorithm implementation always takes precedence over the fine-tuned relatively good results. So making sure the implementation is correct should be always before fine-tuning the hyperparameters. This is actually just the procedure for reinforcement learning applications mentioned at the beginning of this chapter: testing on small-scale cases and making sure the method is implemented correctly, then scaling it up to large-scale environments and fine-tuning with distributed training process. The bad learning performances can be caused by lots of different factors, the insufficient learning time, the poor choice of hyperparameters, the unnormalized input data, etc., but the most common reason is the mistakes in implementations.

In order to provide a more complete guidance for the readers to apply the reinforcement learning algorithms in their projects, we also refer to some external resources when writing up this chapter, including OpenAI Spinning Up,[4] John

[4]OpenAI Spinning Up: https://spinningup.openai.com/en/latest/index.html.

Schulman's slide,[5] William Falcon's blog,[6] etc. We would recommend the readers to refer to those well-summarized suggestions and experiences from researchers to help with their own implementations and applications with deep reinforcement learning as well. It is always helpful to look at other people's previous works which are similar to yours and adopt the experiences from them.

Moreover, the readers need to know that it is not useful to merely read through all the empirical instructions in the above paragraphs without practice. So, we strongly recommend the readers to implement some codes by hand and gain some practical experience. Only in this way can those tricks help the most.

References

Espeholt L, Soyer H, Munos R, Simonyan K, Mnih V, Ward T, Doron Y, Firoiu V, Harley T, Dunning I, et al. (2018) IMPALA: Scalable distributed deep-RL with importance weighted actor-learner architectures. arXiv:180201561

Fu J, Singh A, Ghosh D, Yang L, Levine S (2018) Variational inverse control with events: a general framework for data-driven reward definition. In: Advances in neural information processing systems, pp 8538–8547

Glorot X, Bengio Y (2010) Understanding the difficulty of training deep feedforward neural networks. In: Proceedings of the thirteenth international conference on artificial intelligence and statistics, pp 249–256

He K, Zhang X, Ren S, Sun J (2016) Deep residual learning for image recognition. In: Proceedings of the IEEE conference on computer vision and pattern recognition (CVPR)

Heess N, Hunt JJ, Lillicrap TP, Silver D (2015) Memory-based control with recurrent neural networks. arXiv:151204455

Jaderberg M, Simonyan K, Zisserman A, Kavukcuoglu K (2015) Spatial transformer networks. In: Proceedings of the Neural Information Processing Systems (Advances in neural information processing systems) conference, pp 2017–2025

Mahmood AR, Korenkevych D, Vasan G, Ma W, Bergstra J (2018) Benchmarking reinforcement learning algorithms on real-world robots. arXiv:1809.07731

Mnih V, Kavukcuoglu K, Silver D, Rusu AA, Veness J, Bellemare MG, Graves A, Riedmiller M, Fidjeland AK, Ostrovski G, Petersen S, Beattie C, Sadik A, Antonoglou I, King H, Kumaran D, Wierstra D, Legg S, Hassabis D (2015) Human-level control through deep reinforcement learning. Nature 518(7540):529–533

Saxe AM, McClelland JL, Ganguli S (2013) Exact solutions to the nonlinear dynamics of learning in deep linear neural networks. arXiv:13126120

Vinyals O, Babuschkin I, Czarnecki WM, Mathieu M, Dudzik A, Chung J, Choi DH, Powell R, Ewalds T, Georgiev P, et al (2019) Grandmaster level in StarCraft II using multi-agent reinforcement learning. Nature 575(7782):350–354

[5]The Nuts and Bolts of Deep RL Research. John Schulman: http://joschu.net/docs/nuts-and-bolts.pdf.

[6]Deep RL Hacks: https://github.com/williamFalcon/DeepRLHacks.

Part IV
Summary

Hao Dong
e-mail: hao.dong@pku.edu.cn

To help the readers compare and check different algorithms easier, we summarized the papers for the algorithms we introduced, as listed in Chap. 19 and summarized the pseudocode of all important algorithms in Chap. 20.

Chapter 19
Algorithm Table

Zihan Ding

Abstract In this chapter, we summarize the references of some important reinforcement learning algorithms introduced in the book as a table.

Keywords Reinforcement learning · Algorithm · On-policy · Off-policy · Action space

In this chapter, Table 19.1 containing the most popular reinforcement learning algorithms is summarized, especially for those introduced in this book. We hope this will help the readers to refer to the original papers.

Z. Ding (✉)
Imperial College London, London, UK
e-mail: zhding@mail.ustc.edu.cn

H. Dong et al. (eds.), *Deep Reinforcement Learning*,
https://doi.org/10.1007/978-981-15-4095-0_19

Table 19.1 Reinforcement learning algorithms

RL algorithms	Policy	Action space	Year	Paper	Authors
Q-learning	Off-policy	Discrete	1992	Q-learning (Watkins and Dayan 1992)	Cristopher J.C.H Watkins and Peter Dayan
SARSA	On-policy	Discrete	1994	Online Q-learning using connectionist systems (Rummery and Niranjan 1994)	G.A. Rummery and M. Niranjan
DQN	Off-policy	Discrete	2015	Human-level control through deep reinforcement learning (Mnih et al. 2015)	Volodymyr Mnih, et al.
Dueling DQN	Off-policy	Discrete	2015	Dueling network architectures for deep reinforcement learning (Wang et al. 2015)	Ziyu Wang, et al.
Double DQN	Off-policy	Discrete	2016	Deep reinforcement learning with double Q-learning (Van Hasselt et al. 2016)	Hado van Hasselt, et al.
Noisy DQN	Off-policy	Discrete	2017	Noisy networks for exploration (Fortunato et al. 2017)	Meire Fortunato, et al.
Distributed DQN	Off-policy	Discrete	2017	A distributional perspective on reinforcement learning (Bellemare et al. 2017)	Marc G. Bellemare, et al.
Actor-critic (QAC)	On-policy	Discrete or continuous	2000	Actor-critic algorithms (Konda and Tsitsiklis 2000)	Vijay R. Konda and John N. Tsitsiklis
A3C	On-policy	Discrete or continuous	2016	Asynchronous methods for deep reinforcement learning (Mnih et al. 2016)	Volodymyr Mnih, et al.
REINFORCE	On-policy	Discrete or continuous	1988	On the use of backpropagation in associative reinforcement learning (Williams 1988)	Ronald J. Williams

(continued)

Table 19.1 continued

RL algorithms	Policy	Action space	Year	Paper	Authors
DDPG	Off-policy	Continuous	2016	Continuous control with deep reinforcement learning (Lillicrap et al. 2015)	Timothy P. Lillicrap, et al.
TD3	Off-policy	Continuous	2018	Addressing function approximation error in actor-critic methods (Fujimoto et al. 2018)	Scott Fujimoto, et al.
SAC	Off-policy	Discrete or continuous	2018	Soft actor-critic algorithms and applications (Haarnoja et al. 2018)	Tuomas Haarnoja, et al.
TRPO	On-policy	Discrete or continuous	2015	Trust region policy optimization (Schulman et al. 2015)	John Schulman, et al.
PPO	On-policy	Discrete or continuous	2017	Proximal policy optimization algorithms (Schulman et al. 2017)	John Schulman, et al.
DPPO	On-policy	Discrete or continuous	2017	Emergence of locomotion behaviours in rich environments (Heess et al. 2017)	Nicolas Heess, et al.
ACKTR	On-policy	Discrete or continuous	2017	Scalable trust-region method for deep reinforcement learning using Kronecker-factored approximation Wu et al. (2017)	Yuhuai Wu, et al.
CE method	On-policy	Discrete or continuous	2004	The cross-entropy method: A unified approach to Monte Carlo simulation, randomized optimization and machine learning (Rubinstein and Kroese 2004)	R. Rubinstein and D. Kroese

References

Bellemare MG, Dabney W, Munos R (2017) A distributional perspective on reinforcement learning. In: Proceedings of the 34th international conference on machine learning, vol 70, pp 449–458. JMLR.org

Fortunato M, Azar MG, Piot B, Menick J, Osband I, Graves A, Mnih V, Munos R, Hassabis D, Pietquin O, et al. (2017) Noisy networks for exploration. arXiv:170610295

Fujimoto S, van Hoof H, Meger D (2018) Addressing function approximation error in actor-critic methods. arXiv:180209477

Haarnoja T, Zhou A, Hartikainen K, Tucker G, Ha S, Tan J, Kumar V, Zhu H, Gupta A, Abbeel P, et al. (2018) Soft actor-critic algorithms and applications. arXiv:181205905

Heess N, Sriram S, Lemmon J, Merel J, Wayne G, Tassa Y, Erez T, Wang Z, Eslami S, Riedmiller M, et al. (2017) Emergence of locomotion behaviours in rich environments. arXiv:170702286

Konda VR, Tsitsiklis JN (2000) Actor-critic algorithms. In: Advances in neural information processing systems, pp 1008–1014

Lillicrap TP, Hunt JJ, Pritzel A, Heess N, Erez T, Tassa Y, Silver D, Wierstra D (2015) Continuous control with deep reinforcement learning. arXiv:150902971

Mnih V, Kavukcuoglu K, Silver D, Rusu AA, Veness J, Bellemare MG, Graves A, Riedmiller M, Fidjeland AK, Ostrovski G, Petersen S, Beattie C, Sadik A, Antonoglou I, King H, Kumaran D, Wierstra D, Legg S, Hassabis D (2015) Human-level control through deep reinforcement learning. Nature 518(7540):529–533

Mnih V, Badia AP, Mirza M, Graves A, Lillicrap T, Harley T, Silver D, Kavukcuoglu K (2016) Asynchronous methods for deep reinforcement learning. In: International conference on machine learning (ICML), pp 1928–1937

Rubinstein RY, Kroese DP (2004) The cross-entropy method: a unified approach to monte carlo simulation, randomized optimization and machine learning (Information science and statistics). Springer, New York

Rummery GA, Niranjan M (1994) On-line Q-learning using connectionist systems, vol 37. University of Cambridge, Department of Engineering Cambridge, Cambridge

Schulman J, Levine S, Abbeel P, Jordan M, Moritz P (2015) Trust region policy optimization. In: International conference on machine learning (ICML), pp 1889–1897

Schulman J, Wolski F, Dhariwal P, Radford A, Klimov O (2017) Proximal policy optimization algorithms. arXiv:170706347

Van Hasselt H, Guez A, Silver D (2016) Deep reinforcement learning with double Q-learning. In: Thirtieth AAAI conference on artificial intelligence

Wang Z, Schaul T, Hessel M, Van Hasselt H, Lanctot M, De Freitas N (2015) Dueling network architectures for deep reinforcement learning. arXiv:151106581

Watkins CJ, Dayan P (1992) Q-learning. Mach Learn 8(3–4):279–292

Williams RJ (1988) On the use of backpropagation in associative reinforcement learning. In: Proceedings of the IEEE international conference on neural networks, vol 1, San Diego, pp 263–270

Wu Y, Mansimov E, Grosse RB, Liao S, Ba J (2017) Scalable trust-region method for deep reinforcement learning using Kronecker-factored approximation. In: Advances in neural information processing systems, pp 5279–5288

Chapter 20
Algorithm Cheatsheet

Zihan Ding

Abstract In this chapter, we summarized the algorithms introduced throughout the book, which are categorized into four sections of deep learning, reinforcement learning, deep reinforcement learning, and advanced deep reinforcement learning. The pseudo-code is provided for each algorithm to facilitate the learning process of readers.

Keywords Deep learning · Reinforcement learning · Pseudo-code · Algorithm

This chapter provides a summary of algorithms and key concepts in (deep) reinforcement learning here. We also try to keep the mathematical notations and terminology consistent with the rest of the book, which can be referred to the section of mathematical notation at the beginning of the book and Chap. 2.

20.1 Deep Learning

20.1.1 Stochastic Gradient Descent

Algorithm 1 The training process of stochastic gradient descent (SGD)

Input: Parameters $\boldsymbol{\theta}$, learning rate α, number of training steps/iterations S

for $i = 0$ **to** S **do**
    Compute $\mathcal{L}$ of a mini-batch;
    Compute $\frac{\partial \mathcal{L}}{\partial \theta}$ by back-propagation;
    $\nabla\boldsymbol{\theta} \leftarrow -\alpha * \frac{\partial \mathcal{L}}{\partial \boldsymbol{\theta}}$;
    $\boldsymbol{\theta} \leftarrow \boldsymbol{\theta} + \nabla\boldsymbol{\theta}$; update the parameters
end for
return $\boldsymbol{\theta}$; return the trained parameters;

Z. Ding (✉)
Imperial College London, London, UK
e-mail: zhding@mail.ustc.edu.cn

H. Dong et al. (eds.), *Deep Reinforcement Learning*,
https://doi.org/10.1007/978-981-15-4095-0_20

20.1.2 Adam Optimizer

Algorithm 2 The training process of Adam optimization

Input: parameters $\boldsymbol{\theta}$, learning rate α, number of training steps/iterations S, $\beta_1 = 0.9$, $\beta_2 = 0.999$, $\epsilon = 10^{-8}$

$\boldsymbol{m}_0 \leftarrow 0$; initialize the first moment vector

$\boldsymbol{v}_0 \leftarrow 0$; initialize the second moment vector

for $t = 1$ **to** S **do**

  $\frac{\partial \mathcal{L}}{\partial \boldsymbol{\theta}}$; compute the gradient using a random mini-batch

  $\boldsymbol{m}_t \leftarrow \beta_1 * \boldsymbol{m}_{t-1} + (1 - \beta_1) * \frac{\partial \mathcal{L}}{\partial \boldsymbol{\theta}}$; update the first moment

  $\boldsymbol{v}_t \leftarrow \beta_2 * \boldsymbol{v}_{t-1} + (1 - \beta_2) * (\frac{\partial \mathcal{L}}{\partial \boldsymbol{\theta}})^2$; update the second moment

  $\hat{\boldsymbol{m}}_t \leftarrow \frac{\boldsymbol{m}_t}{1-\beta_1^t}$; compute the running average of the first moment

  $\hat{\boldsymbol{v}}_t \leftarrow \frac{\boldsymbol{v}_t}{1-\beta_2^t}$; compute the running average of the second moment

  $\nabla \boldsymbol{\theta} \leftarrow -\alpha * \frac{\hat{\boldsymbol{m}}_t}{\sqrt{\hat{\boldsymbol{v}}_t + \epsilon}}$;

  $\boldsymbol{\theta} \leftarrow \boldsymbol{\theta} + \nabla \boldsymbol{\theta}$; update parameters

end for

return $\boldsymbol{\theta}$; return the trained parameters

20.2 Reinforcement Learning

20.2.1 Bandit

Stochastic Multi-Armed Bandit

Algorithm 3 Multi-armed bandit learning

Initialize K arms;

Number of time steps T;

Each arm is associated with $v_i \in [0, 1]$. The reward being returned is drawn i.i.d from v_i

for $t = 1, 2, \ldots, T$ **do**

  The agent selects $A_t = i$ from the K arms.

  The environment returns the reward vector $R_t = (R_t^1, R_t^2, \cdots, R_t^K)$.

  The agent observes reward R_t^i.

end for

Adversarial Multi-Armed Bandit

Algorithm 4 Adversarial multi-armed bandit

Initialize K arms;
for $t = 1, 2, \ldots, T$ **do**
    The agent selects I_t from the K arms.
    The adversary selects a reward vector $\boldsymbol{R}_t = (R_t^1, R_t^2, \ldots, R_t^K) \in [0, 1]^K$
    The agent observes reward $R_t^{I_t}$ and maybe also observes the rest of the reward vector depending on the specific problem set up.
end for

Algorithm 5 Hedge for adversarial multi-armed bandit

Initialize K arms;
$G_i(0)$ for $i = 1, 2, \ldots, K$;
for $t = 1, 2, \ldots, T$ **do**
    The agent selects $A_t = i_t$ from the distribution $p(t)$, where

$$p_i(t) = \frac{\exp(\eta G_i(t-1))}{\sum_j^K \exp(\eta G_j(t-1))}$$

    The agent observes reward vector g_t.
    Let $G_i(t) = G(t-1) + g_t^i, \ \forall i \in [1, K]$.
end for

20.2.2 Dynamic Programming (DP)

Policy Iteration

Algorithm 6 Policy iteration

Initialize V and π for all states
repeat
  // Do policy evaluation
  repeat
    $\delta \leftarrow 0$
    for $s \in \mathcal{S}$ **do**
      $v \leftarrow V(s)$
      $V(s) \leftarrow \sum_{r,s'}(r + \gamma V(s'))P(r, s'|s, \pi(s))$
      $\delta \leftarrow \max(\delta, |v - V(s)|)$
    end for
  until δ is smaller than a positive threshold
  // Do policy improvement
  $stable \leftarrow true$
  for $s \in \mathcal{S}$ **do**
    $a \leftarrow \pi(s)$
    $\pi(s) \leftarrow \arg\max_a \sum_{r,s'}(r + \gamma V(s'))P(r, s'|s, a)$
    if $a \neq \pi(s)$ **then**
      $stable \leftarrow false$
    end if
  end for
until $stable = true$
return policy π

Value Iteration

Algorithm 7 Value iteration

Initialize V for all states
repeat
  $\delta \leftarrow 0$
  for $s \in \mathcal{S}$ **do**
    $u \leftarrow V(s)$
    $V(s) \leftarrow \max_a \sum_{r,s'} P(r, s'|s, a)(r + \gamma V(s'))$
    $\delta \leftarrow \max(\delta, |u - V(s)|)$
  end for
until δ is smaller than a positive threshold
Output greedily policy $\pi(s) = \arg\max_a \sum_{r,s'} P(r, s'|s, a)(r + \gamma V(s'))$

20.2.3 *Monte Carlo (MC)*

MC Prediction

Algorithm 8 First-visit MC prediction

Input: Initialize policy π
Initialize $V(s)$ for all states
Initialize a list of returns: $Returns(s)$ for all states
repeat
  Generate an episode under π: $S_0, A_0, R_0, S_1, \cdots, S_{T-1}, A_{T-1}, R_t$
  $G \leftarrow 0$
  $t \leftarrow T - 1$
  for $t >= 0$ **do**
    $G \leftarrow \gamma G + R_{t+1}$
    if $S_0, S_1, \cdots, S_{t-1}$ does not have S_t **then**
      $Returns(S_t)$.append(G)
      $V(S_t) \leftarrow$ mean($Returns(S_t)$)
    end if
    $t \leftarrow t - 1$
  end for
until convergence

MC Control

Algorithm 9 MC exploring starts

Initialize $\pi(s)$ for all states
Initialize $Q(s, a)$ and $Returns(s, a)$ for all state-action pairs
repeat
  Randomly select S_0 and A_0 s.t. all state-action pairs' probabilities are nonzero.
  Generate an episode from S_0, A_0 under π: $S_0, A_0, R_0, S_1, \cdots, S_{T-1}, A_{T-1}, R_T$
  $G \leftarrow 0$
  $t \leftarrow T - 1$
  for $t >= 0$ **do**
    $G \leftarrow \gamma G + R_{t+1}$
    if $S_0, A_0, S_1, A_1 \cdots, S_{t-1}, A_{t-1}$ does not have S_t, A_t **then**
      $Returns(S_t, A_t)$.append(G)
      $Q(S_t, A_t) \leftarrow$ mean($Returns(S_t, A_t)$
      $\pi(S_t) \leftarrow \arg\max_a Q(S_t, a)$
    end if
    $t \leftarrow t - 1$
  end for
until convergence

Temporal Different (TD)

Algorithm 10 TD(0) for state-value estimation

Input policy π
Initialize $V(s)$ and step size $\alpha \in (0, 1]$
for each episode **do**
  Initialize S_0
  for Each step S_t in the current episode **do**
    $A_t \leftarrow \pi(S_t)$
    $R_{t+1}, S_{t+1} \leftarrow Env(S_t, A_t)$
    $V(S_t) \leftarrow V(S_t) + \alpha[R_{t+1} + \gamma V(S_{t+1}) - V(S_t)]$
  end for
end for

TD(λ)

Algorithm 11 Semi-gradient TD(λ) for state-value

Input: policy π
Initialize a differentiable state function v, step size α and value function weight **w**
for each episode **do**
  Initialize S_0
  $z \leftarrow 0$
  for Each step S_t in the current episode **do**
    Select A_t using policy that is based on π
    $R_{t+1}, S_{t+1} \leftarrow Env(S_t, A_t)$
    $z \leftarrow \gamma\lambda z + \nabla V(S_t, \boldsymbol{w}_t)$
    $\delta \leftarrow R_{t+1} + \gamma V(S_{t+1}, \boldsymbol{w}_t) - V(S_t, \boldsymbol{w}_t)$
    $\boldsymbol{w} \leftarrow w + \alpha\delta z$
  end for
end for

Sarsa: On-Policy TD Control

Algorithm 12 Sarsa (on-policy TD control)

Initialize $Q(s, a)$ for all state-action pairs.
for each episode **do**
  Initialize S_0
  Select A_0 using policy that is based on Q
  for Each step S_t in the current episode **do**
    Select A_t from S_t using policy that is based on Q
    $R_{t+1}, S_{t+1} \leftarrow Env(S_t, A_t)$
    Select A_{t+1} from S_{t+1} using policy that is based on Q
    $Q(S_t, A_t) \leftarrow Q(S_t, A_t) + \alpha[R_{t+1} + \gamma Q(S_{t+1}, A_{t+1}) - Q(S_t, A_t)]$
  end for
end for

N-Step Sarsa

Algorithm 13 n-step Sarsa

Initialize $Q(s, a)$ for all state-action pairs.
Initialize step-size $\alpha \in (0, 1]$.
Determine a fixed policy π or use ϵ-greedy.
for each episode **do**
  Initialize S_0
  Select A_0 using $\pi(S_0, A)$
  $T \leftarrow$ INTMAX (the length of an episode)
  $\gamma \leftarrow 0$
  for $t \leftarrow 0, 1, 2, \ldots$ until $\gamma - T - 1$ **do**
    if $t < T$ **then**
      $R_{t+1}, S_{t+1} \leftarrow Env(S_t, A_t)$
      if S_{t+1} is terminal **then**
        $T \leftarrow t + 1$
      else
        Select A_{t+1} using $\pi(S_t, A)$
      end if
    end if
    $\tau \leftarrow t - n + 1$ (the time step to update. This is an n-step Sarsa, so we will only update the estimate that is $n + 1$ steps ago and we will continue to do so until all the eligible states have been updated.
    if $\tau \geq 0$ **then**
      $G \leftarrow \sum_{i=\tau+1}^{min(r+n,T)} \gamma^{i-\gamma-1} R_i$
      if $\gamma + n < T$ **then**
        $G \leftarrow G + \gamma^n Q(S_{t+n}, A_{\gamma+n})$
      end if
      $Q(S_\gamma, A_\gamma) \leftarrow Q(S_\gamma, A_\gamma) + \alpha[G - Q(S_\gamma, A_\gamma)]$
    end if
  end for
end for

Q-Learning (Off-Policy TD Control)

Algorithm 14 Q-learning (off-policy TD control)

Initialize $Q(s, a)$ for all state-action pairs and step size $\alpha \in (0, 1]$
for each episode **do**
  Initialize S_0
  for Each step S_t in the current episode **do**
    Select A_t using policy that is based on Q
    $R_{t+1}, S_{t+1} \leftarrow Env(S_t, A_t)$
    $Q(S_t, A_t) \leftarrow Q(S_t, A_t) + \alpha[R_{t+1} + \gamma \max_a Q(S_{t+1}, a) - Q(S_t, A_t)]$
  end for
end for

20.3 Deep Reinforcement Learning

Deep Q-Network (DQN) is an extension of Q-learning to high-dimensional cases with deep neural network for value function approximation, employing a target action-value network and a replay buffer for updating.

Key Ideas

- Neural networks for Q-value function approximation;
- Replay buffer for off-line updating;
- Target network and delayed update;
- Mean Square Error/Huber loss for minimizing the temporal-difference (TD) error;

Algorithm 15 DQN

1: **Hyperparameters**: replay buffer capacity N, reward discount factor γ, delayed steps C for target action-value function update, ϵ-greedy factor ϵ
2: **Input**: empty replay buffer $\mathcal{D}$, initial parameters θ of action-value function Q
3: Initialize target action-value function $\hat{Q}$ with parameter $\hat{\theta} \leftarrow \theta$
4: **for** episode = 0, 1, 2, ... **do**
5:     Initialize environment and get observation O_0
6:     Initialize sequence $S_0 = \{O_0\}$ and preprocess sequence $\phi_0 = \phi(S_0)$
7:     **for** t = 0, 1, 2, ... **do**
8:         With probability ϵ select a random action A_t, otherwise select $A_t = \arg\max_a Q(\phi(S_t), a; \theta)$
9:         Execute action A_t and observe O_{t+1} and reward R_t
10:         If the episode has ended, set $D_t = 1$. Otherwise, set $D_t = 0$
11:         Set $S_{t+1} = \{S_t, A_t, O_{t+1}\}$ and preprocess $\phi_{t+1} = \phi(S_{t+1})$
12:         Store transition $(\phi_t, A_t, R_t, D_t, \phi_{t+1})$ in $\mathcal{D}$
13:         Sample random minibatch of transitions $(\phi_i, A_i, R_i, D_i, \phi_i')$ from $\mathcal{D}$
14:         If $D_i = 0$, set $Y_i = R_i + \gamma \max_{a'} \hat{Q}(\phi_i', a'; \hat{\theta})$. Otherwise, set $Y_i = R_i$
15:         Perform a gradient descent step on $(Y_i - Q(\phi_i, A_i; \theta))^2$ with respect to θ
16:         Synchronize the target $\hat{Q}$ every C steps
17:         If the episode has ended, break the loop
18:     **end for**
19: **end for**

Double Deep Q-Network is a modified version of DQN for solving the overestimation problem.

Key Ideas

- Double Q-networks in an embedded manner for target value estimation. Change line 14 in DQN Algorithm 15 to be: Set $Y_j = R_j + \gamma(1 - D_j)\hat{Q}(\phi_{j+1}, \arg\max_{a'} Q(\phi_{j+1}, a'; \theta_j); \hat{\theta})$.

Dueling Deep Q-Network is a modified version of DQN with the action-value function decomposed into one state-value function and one state-dependent action advantage function.

Key Ideas

- Factorize the action-value function Q to be value function V and advantage function A.

Change the way of parameterization for action-value function Q (as well as its target $\hat{Q}$) in DQN as:

$$Q(s, a; \theta, \theta_v, \theta_a) = V(s; \theta, \theta_v) + (A(s, a; \theta, \theta_a) - \max_{a'} A(s, a'; \theta, \theta_a))$$

or,

$$Q(s, a; \theta, \theta_v, \theta_a) = V(s; \theta, \theta_v) + (A(s, a; \theta, \theta_a) - \frac{1}{|\mathcal{A}|} A(s, a'; \theta, \theta_a))$$

REINFORCE is an algorithm using policy-based optimization and on-policy update.

Algorithm 16 REINFORCE

1: **Input**: initial policy parameters θ
2: **for** k = 0, 1, 2, ... **do**
3: Initialize environment.
4: Collect set of trajectories $\mathcal{D}_k = \{\tau_i = \{(S_t, A_t, R_t)|t = 0, 1, \ldots, T\}\}$ by running policy $\pi_k = \pi(\theta_k)$ in the environment.
5: Compute cumulative return G_t
6: Estimate policy gradient as

$$g_k = \frac{1}{|\mathcal{D}_k|} \sum_{\tau \in \mathcal{D}_k} \sum_{t=0}^{T} \nabla_\theta \log \pi_\theta(A_t|S_t)|_{\theta_k} G_t$$

7: Update the policy using gradient ascent,

$$\theta_{k+1} = \theta_k + \alpha_k g_k$$

8: **end for**

REINFORCE with Baseline/Vanilla Policy Gradient is another version of REINFORCE with the policy gradient estimated with action advantage function instead of the cumulative return.

Algorithm 17 REINFORCE with baseline

Hyperparameters: step size η_θ, reward discount factor γ, number of time steps L, batch size B, baseline value b
Input: initial policy parameters θ_0
Initialize $\theta = \theta_0$
for $k = 1, 2, \ldots,$ **do**
  Run policy π_θ for B trajectories, each one with L time steps, and collect $\{S_{t,\ell}, A_{t,\ell}, R_{t,\ell}\}$
  $\hat{A}_{t,\ell} = \sum_{\ell'=\ell}^{L} \gamma^{\ell'-\ell} R_{t,\ell} - b(S_{t,\ell})$
  $J(\theta) = \frac{1}{B}\sum_{t=1}^{B}\sum_{\ell=0}^{L} \log \pi_\theta(A_{t,\ell}|S_{t,\ell})\hat{A}_{t,\ell}$
  $\theta = \theta + \eta_\theta \nabla J(\theta)$
  Update $b(S_{t,\ell})$ by $\{S_{t,\ell}, A_{t,\ell}, R_{t,\ell}\}$
end for
Return θ

Actor-Critic is modified from REINFORCE algorithm with value function approximation.

Algorithm 18 Actor-Critic

Hyperparameters: step size η_θ and η_ψ, reward discount factor γ
Input: initial policy parameters θ_0, initial value function parameters ψ_0
Initialize $\theta = \theta_0$ and $\psi = \psi_0$
for $t = 0, 1, 2, \ldots$ **do**
  Run policy π_θ for one step, collection $\{S_t, A_t, R_t, S_{t+1}\}$
  Estimate advantages $\hat{A}_t = R_t + \gamma V_\psi^{\pi_\theta}(S_{t+1}) - V_\psi^{\pi_\theta}(S_t)$
  $J(\theta) = \sum_t \log \pi_\theta(A_t|S_t)\hat{A}_t$
  $J_{V_\psi^{\pi_\theta}}(\psi) = \sum_t \hat{A}_t^2$
  $\psi = \psi + \eta_\psi \nabla J_{V_\psi^{\pi_\theta}}(\psi), \theta = \theta + \eta_\theta \nabla J(\theta)$
end for
Return (θ, ψ)

***Q*-value Actor-Critic (QAC)** is another version of actor-critic algorithm, as a combination of value-based (e.g., Q-learning) and policy-based (e.g., REINFORCE) optimization using on-policy update.

Key Ideas

- Combination of DQN and REINFORCE.

Algorithm 19 QAC

1: **Input**: initial policy parameters θ, initial parameters ω of action-value function Q, discounted factor γ
2: **for** k = 0, 1, 2, ... **do**
3: Initialize environment.
4: Collect set of trajectories $\mathcal{D}_k = \{\tau_i = \{(S_t, A_t, R_t, D_t)|t = 0, 1, \ldots, T\}\}$ by running policy $\pi_k = \pi(\theta_k)$ in the environment.
5: Compute TD-error $\delta_t = R_t + \gamma \max_{a'} Q_\omega(S_{t+1}, a') - Q_\omega(S_t, A_t)$
6: Estimate policy gradient as

$$g_k = \frac{1}{|\mathcal{D}_k|} \sum_{\tau \in \mathcal{D}_k} \sum_{t=0}^{T} \nabla_\theta \log \pi_\theta(A_t|S_t)|_{\theta_k} Q_\omega(S_t, A_t)$$

7: Update the policy using gradient ascent,

$$\theta_{k+1} = \theta_k + \alpha_k g_k$$

8: Update the action-value function with mean-squared error:

$$\phi_{k+1} = \arg\min_\phi \frac{1}{|\mathcal{D}_k|T} \sum_{\tau \in \mathcal{D}_k} \sum_{t=0}^{T} \delta_t^2$$

via gradient descent algorithm.
9: **end for**

Advantage Actor-Critic (A2C) is a modified version of actor-critic algorithm, which applies the trick of REINFORCE with baseline instead of vanilla REINFORCE for policy optimization, also using on-policy update.

Key Ideas

- Combination of DQN and REINFORCE with baseline.

Algorithm 20 A2C

Master:
Hyperparameters: step size η_ψ and η_θ, set of workers $\mathcal{W}$
Input: initial policy parameters θ_0, initial value function parameters ψ_0
Initialize $\theta = \theta_0$ and $\psi = \psi_0$
for $k = 0, 1, 2, \ldots$ **do**
  $(g_\psi, g_\theta) = 0$
  for worker in $\mathcal{W}$ **do**
    $(g_\psi, g_\theta) = (g_\psi, g_\theta) + \textbf{worker}(V_\psi^{\pi_\theta}, \pi_\theta)$
  end for
  $\psi = \psi - \eta_\psi g_\psi; \theta = \theta + \eta_\theta g_\theta.$
end for

Worker:
Hyperparameters: reward discount factor γ, the length of trajectory L
Input: value function $V_\psi^{\pi_\theta}$, policy π_θ
Run policy π_θ for L time steps, collection $\{S_t, A_t, R_t, S_{t+1}\}$
Estimate advantages $\hat{A}_t = R_t + \gamma V_\psi^{\pi_\theta}(S_{t+1}) - V_\psi^{\pi_\theta}(S_t)$
$J(\theta) = \sum_t \log \pi_\theta(A_t|S_t)\hat{A}_t$
$J_{V_\psi^{\pi_\theta}}(\psi) = \sum_t \hat{A}_t^2$
$(g_\psi, g_\theta) = (\nabla J_{V_\psi^{\pi_\theta}}(\psi), \nabla J(\theta))$
Return (g_ψ, g_θ)

Asynchronous Advantage Actor-Critic (A3C) is a modified version of A2C with asynchronous gradient update for large-scale parallel computation.

Key Ideas

- Asynchronous updating of the policy.

Algorithm 21 A3C

Master:
Hyperparameters: step size η_ψ and η_θ, current policy π_θ, value function $V_\psi^{\pi_\theta}$
Input: gradients g_ψ, g_θ
$\psi = \psi - \eta_\psi g_\psi; \theta = \theta + \eta_\theta g_\theta.$
Return $(V_\psi^{\pi_\theta}, \pi_\theta)$

Worker:
Hyperparameters: reward discount factor γ, the length of trajectory L
Input: value function $V_\psi^{\pi_\theta}$, policy π_θ
$(g_\theta, g_\psi) = (0, 0)$
for $k = 1, 2, \ldots,$ **do**
  $(\theta, \psi) = \textbf{Master}(g_\theta, g_\psi)$
  Run policy π_θ for L time steps, collection $\{S_t, A_t, R_t, S_{t+1}\}$
  Estimate advantages $\hat{A}_t = R_t + \gamma V_\psi^{\pi_\theta}(S_{t+1}) - V_\psi^{\pi_\theta}(S_t)$
  $J(\theta) = \sum_t \log \pi_\theta(A_t|S_t)\hat{A}_t$
  $J_{V_\psi^{\pi_\theta}}(\psi) = \sum_t \hat{A}_t^2$
  $(g_\psi, g_\theta) = (\nabla J_{V_\psi^{\pi_\theta}}(\psi), \nabla J(\theta))$
end for

Deep Deterministic Policy Gradient (DDPG) is a combination of DQN and QAC with deterministic policy, with off-policy update using a replay buffer.

Key Ideas

- Deterministic policy as an approximate Q-value maximizer over action space;
- Ornstein–Uhlenbeck/Gaussian noise for stochastic actions in exploration;
- Target networks and delayed update.

Algorithm 22 DDPG

Hyperparameters: soft update factor ρ, reward discount factor γ
Input empty replay buffer $\mathcal{D}$, initialize parameters θ^Q of critic network $Q(s, a|\theta^Q)$ and parameters θ^π of actor network $\pi(s|\theta^\pi)$, target network Q' and π'
Initialize target network Q' and π' with weights $\theta^{Q'} \leftarrow \theta^Q, \theta^{\pi'} \leftarrow \theta^\pi$
for episode $= 1, M$ **do**
  Initialize a random process $\mathcal{N}$ for action exploration
  Receive initial observation state s_1
  for t $= 1, T$ **do**
    Selection action $A_t = \pi(S_t|\theta^\pi) + \mathcal{N}_t$
    Execute action A_t and observe reward R_t, and observe new state S_{t+1}
    Store transion $(S_t, A_t, R_t, D_t, S_{t+1})$ in $\mathcal{D}$
    Set $Y_i = R_i + \gamma(1 - D_t)Q'(S_{t+1}, \pi'(S_{t+1}|\theta^{\pi'})|\theta^{Q'})$
    Update critic by minimizing the loss:

$$L = \frac{1}{N}\sum_i (Y_i - Q(S_i, A_i|\theta^Q))^2$$

    Update the actor policy using the sampled policy gradient:

$$\nabla_{\theta^\pi} J \approx \frac{1}{N}\sum_i \nabla_a Q(s, a|\theta^Q)|_{s=S_i, a=\pi(S_i)} \nabla_{\theta^\pi} \pi(s|\theta^\pi)|_{S_i}$$

    Update the target networks:
    $\theta^{Q'} \leftarrow \rho\theta^Q + (1 - \rho)\theta^{Q'}$
    $\theta^{\pi'} \leftarrow \rho\theta^\pi + (1 - \rho)\theta^{\pi'}$
  end for
end for

Twin Delayed DDPG (TD3) is a more advanced algorithm based on DDPG with twin action-value networks and delayed updating for the policy and the target networks.

Key Ideas

- Double Q-Learning;
- Delayed update for the target networks and the policy;
- Smoothing regularization for the target policy.

Algorithm 23 TD3

Hyperparameters: soft update factor ρ, reward discount factor γ, clip factor c
Input: empty replay buffer $\mathcal{D}$, initial parameters θ_1, θ_2 of critic networks $Q_{\theta_1}, Q_{\theta_2}$, initial parameters ϕ of actor network π_ϕ
Initialize target networks $\hat{\theta}_1 \leftarrow \theta_1, \hat{\theta}_2 \leftarrow \theta_2, \hat{\phi} \leftarrow \phi$
for $t = 1$ to T do **do**
  Select action with exploration noise $A_t \sim \pi_\phi(S_t) + \epsilon, \epsilon \sim \mathcal{N}(0, \sigma)$
  Observe reward R_t and new state S_{t+1}
  Store transition tuple $(S_t, A_t, R_t, D_t, S_{t+1})$ in $\mathcal{D}$
  Sample mini-batch of N transitions $(S_t, A_t, R_t, D_t, S_{t+1})$ from $\mathcal{D}$
  $\tilde{a}_{t+1} \leftarrow \pi_{\phi'}(S_{t+1}) + \epsilon, \epsilon \sim clip(\mathcal{N}(0, \tilde{\sigma}, -c, c))$
  $y \leftarrow R_t + \gamma(1 - D_t) \min_{i=1,2} Q_{\theta_i'}(S_{t+1}, \tilde{a}_{t+1})$
  Update critics $\theta_i \leftarrow \arg\min_{\theta_i} N^{-1} \sum (y - Q_{\theta_i}(S_t, A_t))^2$
  if t mod d **then**
    Update ϕ by the deterministic policy gradient:
    $\nabla_\phi J(\phi) = N^{-1} \sum \nabla_a Q_{\theta_1}(S_t, A_t)|_{A_t = \pi_\phi(S_t)} \nabla_\phi \pi_\phi(S_t)$
    Update target networks:
    $\hat{\theta}_i \leftarrow \rho\theta_i + (1 - \rho)\hat{\theta}_i$
    $\hat{\phi} \leftarrow \rho\phi + (1 - \rho)\hat{\phi}$
  end if
end for

Soft Actor-Critic (SAC) is a more advanced algorithm based on DDPG using an additional soft entropy term for boosting exploration.

Key Ideas

- Entropy regularization for boosting exploration;
- Double Q-Learning;
- Reparameterization trick for making stochastic policy differentiable and update with deterministic policy gradient;
- Tanh Gaussian action distribution.

Algorithm 24 Soft actor-critic (SAC)

Hyperparameters: target entropy κ, step sizes $\lambda_Q, \lambda_\pi, \lambda_\alpha$, exponentially moving average coefficient τ
Input: initial policy parameters θ, initial Q value function parameters ϕ_1 and ϕ_2
$\mathcal{D} = \emptyset$; $\tilde{\phi}_i = \phi_i$, for $i = 1, 2$
for $k = 0, 1, 2, \ldots$ **do**
  for $t = 0, 1, 2, \ldots$ **do**
    Sample A_t from $\pi_\theta(\cdot|S_t)$, collect (R_t, S_{t+1})
    $\mathcal{D} = \mathcal{D} \cup \{S_t, A_t, R_t, S_{t+1}\}$
  end for
  Perform multiple step of gradients:
    $\phi_i = \phi_i - \lambda_Q \nabla J_Q(\phi_i)$ for $i = 1, 2$
    $\theta = \theta - \lambda_\pi \nabla_\theta J_\pi(\theta)$
    $\alpha = \alpha - \lambda_\alpha \nabla J(\alpha)$
    $\tilde{\phi}_i = (1 - \tau)\phi_i + \tau\tilde{\phi}_i$ for $i = 1, 2$
end for
Output θ, ϕ_1, ϕ_2

Trust Region Policy Optimization (TRPO) is a trust-region algorithm using the second-order gradient descent and on-policy update.

Key Ideas

- KL-divergence to keep new and old policies close in policy space;
- Second-order method with the constraint;
- Using the conjugate gradient to avoid the computation of inverse matrix.

Algorithm 25 TRPO

Hyperparameters: KL-divergence limit δ, backtracking coefficient α, maximum number of backtracking steps K

Input: empty replay buffer $\mathcal{D}_k$, initial policy parameters θ_0, initial value function parameters ϕ_0

for episode $= 0, 1, 2, \ldots$ **do**

Collect set of trajectories $\mathcal{D}_k = \{\tau_i\}$ by running policy $\pi_k = \pi(\theta_k)$ in the environment

Compute rewards-to-go $\hat{G}_t$

Compute advantage estimates, $\hat{A}_t$ (using any method of advantage estimation) based on the current value function V_{ϕ_k}

Estimate policy gradient as

$$\hat{g}_k = \frac{1}{|\mathcal{D}_k|} \sum_{\tau \in \mathcal{D}_k} \sum_{t=0}^{T} \nabla_\theta \log \pi_\theta(A_t|S_t)|_{\theta_k} \hat{A}_t \tag{20.1}$$

Use the conjugate gradient algorithm to compute

$$\hat{x}_k \approx \hat{H}_k^{-1} \hat{g}_k \tag{20.2}$$

where $\hat{H}_k$ is the Hessian of the sample average KL-divergence

Update the policy by backtracking line search with

$$\theta_{k+1} = \theta_k + \alpha^j \sqrt{\frac{2\delta}{\hat{x}_k^T \hat{H}_k \hat{x}_k}} \hat{x}_k \tag{20.3}$$

where $j \in \{0, 1, 2, \ldots K\}$ is the smallest value which improves the sample loss and satisfies the sample KL-divergence constraint

Fit value function by regression on mean-squared error:

$$\phi_{k+1} = \arg\min_\phi \frac{1}{|\mathcal{D}_k|T} \sum_{\tau \in \mathcal{D}_k} \sum_{t=0}^{T} \left(V_\phi(S_t) - \hat{G}_t\right)^2 \tag{20.4}$$

typically via some gradient descent algorithm

end for

Proximal Policy Optimization (PPO-Penalty) is a trust-region algorithm based on TRPO using the first-order gradient, with the trust-region constraint achieved with an adaptive penalty term.

Key Ideas

- KL-divergence to keep new and old policies close in policy space;
- Transferring the constrained optimization problem into unconstrained one;
- First-order method to avoid the computation of Hessian Matrix;
- Adjusting the penalty coefficient adaptively.

Algorithm 26 PPO-Penalty

Hyperparameters: reward discount factor γ, KL penalty coefficient λ, adaptive parameters $a = 1.5$, $b = 2$, the number of sub-iterations M, B

Input: initial policy parameters θ, initial value function parameters ϕ

for k = 0, 1, 2, ... **do**

  Run policy π_θ for T time steps, collection $\{S_t, A_t, R_t\}$

  Estimate advantages $\hat{A}_t = \sum_{t'>t} \gamma^{t'-t} R_{t'} - V_\phi(S_t)$

  $\pi_{\text{old}} \leftarrow \pi_\theta$

  for $m \in \{1, \ldots, M\}$ **do**

    $J_{\text{PPO}}(\theta) = \sum_{t=1}^{T} \frac{\pi_\theta(A_t|S_t)}{\pi_{\text{old}}(A_t|S_t)} \hat{A}_t - \lambda \hat{\mathbb{E}}_t \left[D_{\text{KL}}(\pi_{\text{old}}(\cdot|S_t) \| \pi_\theta(\cdot|S_t))\right]$

    Update θ by a gradient method w.r.t $J_{\text{PPO}}(\theta)$

  end for

  for $b \in \{1, \ldots, B\}$ **do**

    $L_{BL}(\phi) = -\sum_{t=1}^{T} \left(\sum_{t'>t} \gamma^{t'-t} R_{t'} - V_\phi(S_t)\right)^2$

  end for

  Compute $d = \hat{\mathbb{E}}_t \left[D_{\text{KL}}(\pi_{\text{old}}(\cdot|S_t) \| \pi_\theta(\cdot|S_t))\right]$

  if $d < d_{\text{target}}/a$ **then**

    $\lambda \leftarrow \lambda/b$

  else if $d > d_{\text{target}} \times a$ **then**

    $\lambda \leftarrow \lambda \times b$

  end if

end for

Proximal Policy Optimization (PPO-Clip) is a trust-region algorithm based on TRPO using the first-order gradient, with the trust-region constraint achieved with a clipping method on the gradients.

Key Ideas

- Replacing KL-divergence with clipping method in the objective function.

Algorithm 27 PPO-Clip

Hyperparameters: clip factor ϵ, the number of sub-iterations M, B
Input: initial policy parameters θ, initial value function parameters ϕ
for k = 0, 1, 2, ... **do**
Collect set of trajectories $\mathcal{D}_k = \{\tau_i\}$ by running policy π_{θ_k} in the environment
Compute rewards-to-go $\hat{G}_t$
Compute advantage estimates, $\hat{A}_t$(using any method of advantage estimation) based on the current value function V_{ϕ_k}
for $m \in \{1, \ldots, M\}$ **do**

$$\ell_t(\theta') = \frac{\pi_\theta(A_t|S_t)}{\pi_{\theta_{\text{old}}}(A_t|S_t)} \tag{20.5}$$

Update the policy by maximizing the PPO-Clip objective:

$$\theta_{k+1} = \arg\max_\theta \frac{1}{|\mathcal{D}_k|T} \sum_{\tau \in D_k} \sum_{t=0}^{T} \min(\ell_t(\theta')A^{\pi_{\theta_{\text{old}}}}(S_t, A_t), \tag{20.6}$$

$$\text{clip}(\ell_t(\theta'), 1-\epsilon, 1+\epsilon)A^{\pi_{\theta_{\text{old}}}}(S_t, A_t)) \tag{20.7}$$

typically via stochastic gradient ascent with Adam
end for
for $b \in \{1, \ldots, B\}$ **do**
Fit value function by regression on mean-squared error:

$$\phi_{k+1} = \arg\min_\phi \frac{1}{|\mathcal{D}_k|T} \sum_{\tau \in \mathcal{D}_k} \sum_{t=0}^{T} \left(V_\phi(S_t) - \hat{G}_t\right)^2$$

typically via some gradient descent algorithm
end for
end for

Actor Critic using Kronecker-Factored Trust Region (ACKTR) is trust-region on-policy algorithm using Kronecker-factored approximation for the second-order natural gradient computation.

Key Ideas

- Second-order optimization with natural gradient;
- K-FAC approximation for natural gradient.

Algorithm 28 ACKTR

Hyperparameters: learning rate η_{max}, trust region radius δ

Input: empty replay buffer $\mathcal{D}$, initial policy parameters θ_0, initial value function parameters ϕ_0

for k = 0, 1, 2, ... **do**

  Collect set of trajectories $\mathcal{D}_k = \{\tau_i | i = 0, 1, \ldots\}$ by running policy $\pi_k = \pi(\theta_k)$ in the environment

  Compute cumulative return G_t

  Compute advantage estimates, $\hat{A}_t$ (using any method of advantage estimation) based on the current value function V_{ϕ_k}

  Estimate policy gradient as

$$\hat{g}_k = \frac{1}{|\mathcal{D}_k|} \sum_{\tau \in \mathcal{D}_k} \sum_{t=0}^{T} \nabla_\theta \log \pi_\theta(A_t|S_t)|_{\theta_k} \hat{A}_t \tag{20.8}$$

  for $l = 0, 1, 2, \ldots$ **do**

$$\text{vec}(\Delta\theta_k^l) = \text{vec}(A_l^{-1} \nabla_{\theta_k^l} \hat{g}_k S_l^{-1}) \tag{20.9}$$

   where $A_l = \mathbb{E}[a_l a_l^T]$, $S_l = \mathbb{E}[(\nabla_{s_l} \hat{g}_k)(\nabla_{s_l} \hat{g}_k)^T]$ (A_l, S_l are calculated with running average among episodes), a_l is the input activation vector and $s_l = W_l a_l$ with the weight matrix W_l of lth layer, and vec($\cdot$) operation is used for transforming the matrix into one-dimensional vector

  end for

  Update the policy by K-FAC approximated natural gradient as

$$\theta_{k+1} = \theta_k + \eta_k \Delta\theta_k \tag{20.10}$$

  where $\eta_k = \min(\eta_{max}, \sqrt{\frac{2\delta}{\theta_k^T \hat{H}_k \theta_k}})$ and $\hat{H}_k^l = A_l \otimes S_l$ for l^{th} layer.

  Fit value function by regression on mean-squared error:

$$\phi_{k+1} = \arg\min_\phi \frac{1}{|\mathcal{D}_k|T} \sum_{\tau \in \mathcal{D}_k} \sum_{t=0}^{T} (V_\phi(S_t) - G_t)^2 \tag{20.11}$$

  via Gauss-Newton second-order gradient descent algorithm (also uses K-FAC approximation)

end for

20.4 Advanced Deep Reinforcement Learning

20.4.1 Imitation Learning

DAgger

Algorithm 29 DAgger

Initialize $\mathcal{D} \leftarrow \emptyset$.
Initialize the policy $\hat{\pi}_1$ to any policy in policy set Π.
for i = $1, 2, \ldots, N$ **do**
    $\pi_i \leftarrow \beta_i \pi^* + (1 - \beta_i)\hat{\pi}_i$.
    Sample several T-step trajectories using π_i.
    Get dataset $\mathcal{D}_i = \{(s, \pi^*(s))\}$ of visited states by π_i and actions given by the expert.
    Aggregate datasets: $\mathcal{D} \leftarrow \mathcal{D} \cup \mathcal{D}_i$. Train current policy $\hat{\pi}_{i+1}$ on $\mathcal{D}$.
end for
Return policy $\hat{\pi}_{N+1}$.

20.4.2 Model-Based Reinforcement Learning

Dyna-Q

Algorithm 30 Dyna-Q

Initialize $Q(s, a)$ and $Model(s, a)$ for all $s \in \mathcal{S}$ and $a \in \mathcal{A}(s)$
Do forever:
  (a) $S \leftarrow$ current (non-terminal) state
  (b) $A \leftarrow \epsilon$-greedy(S, Q)
  (c) Execute action A; Observe resultant reward R, Get next state S'
  (d) $Q(S, A) \leftarrow Q(S, A) + \alpha \left[R + \gamma \max_a Q(S', a) - Q(S, A)\right]$
  (e) $Model(S, A) \leftarrow R, S'$
  (f) Repeat n times:
    $S \leftarrow$ random previously observed state
    $A \leftarrow$ random action previously taken in S
    $R, S' \leftarrow Model(S, A)$ random action previously taken in S
    $Q(S, A) \leftarrow Q(S, A) + \alpha \left[R + \gamma \max_a Q(S', a) - Q(S, A)\right]$

Simple Monte Carlo Search

Algorithm 31 Simple Monte Carlo search

Provided the model $\mathcal{M}$ and simulation policy π
for each action $a \in \mathcal{A}$ **do**
  for each episode $k \in \{1, 2, \ldots, K\}$ **do**
    Following the model $\mathcal{M}$ and simulation policy π, roll out in the environment started from current state S_t
    Record the trajectory as $\{S_t, a, R_{t+1}^k, S_{t+1}^k, A_{t+1}^k, R_{t+2}^k, \ldots S_T^k\}$
  end for
  Evaluate actions by mean return. $Q(S_t, a) = \frac{1}{K} \sum_{k=1}^{K} G_t^k$
end for
The learnt policy is to select current action with maximum Q value $A_t = \arg\max_{a \in \mathcal{A}} Q(S_t, a)$

Monte Carlo Tree Search

Algorithm 32 Monte Carlo tree search

Provided the model $\mathcal{M}$
Initialize simulation policy π
for each action $a \in \mathcal{A}$ **do**
  for each episode $k \in \{1, 2, \ldots, K\}$ **do**
    Following the model $\mathcal{M}$ and simulation policy π, roll out in the environment started from current state S_t
    Record the trajectory as $\{S_t, a, R_{t+1}, S_{t+1}, A_{t+1}, R_{t+2}, \ldots S_T\}$
    Update the Q value of every $(s_i, a_i), i = t, \ldots, T$ by mean return starting from (s_i, a_i) with $A_t = a$
    Update the simulation policy π according to the current Q values
  end for
end for
Output the action with maximum Q value at the current state, $A_t = \arg\max_{a \in \mathcal{A}} Q(S_t, a)$

Dyna-2

Algorithm 33 Dyna-2

```
function LEARNING
    Initialize F_s and F_r
    θ ← 0            # Initialize the weights of long-term memory
    loop
        s ← s_0
        θ̄ ← 0        # Initialize the weights of short-term memory
        z ← 0        # Initialize eligibility trace
        SEARCH(s)
        a ← π(s; Q̄)          # Choose action based on polity related with Q̄
        while s is not terminal do
            Execute a, observe reward r and next state s'
            (F_s, F_r) ← UpdateModel(s, a, r, s')
            SEARCH(s')
            a' ← π(s'; Q̄)          # Choose action applied in the next state s'
            δ ← r + Q(s', a') − Q(s, a)          # Calculate TD-error
            θ ← θ + α(s, a)δz          # Update weights of long-term memory
            z ← λz + φ          # Update eligibility trace
            s ← s', a ← a'
        end while
    end loop
end function

function SEARCH(s)
    while time available do
        z̄ ← 0          # Clear eligibility trace
        a ← pi(s; Q̄)          # Choose action based on polity related with Q̄
        while s is not terminal do
            s' ← F_s(s, a)          # Sample transition
            r ← F_r(s, a)          # Sample reward
            a' ← π̄(s'; Q̄
            δ̄ ← r + Q̄(s', a') − Q̄(s, a)          # Calculate TD-error
            θ̄ ← θ̄ + ᾱ(s, a)δ̄z̄          # Update weights of short-term memory
            z̄ ← λ̄z̄ + φ̄          # Update eligibility trance in short-term memory
            s ← s', a ← a'
        end while
    end while
end function
```

20.4.3 *Hierarchical Reinforcement Learning*

STRategic Attentive Write (STRAW)

Algorithm 34 Plans update in STRAW

if $g_t = 1$ **then**
  Compute attention parameter of action-plan $\psi_t^A = f^{\psi}(z_t)$
  Apply attentive read: $\beta_t = read(\mathbf{A}^{t-1}, \psi_t^A)$
  Compute intermediate representation $\epsilon_t = h(\text{concat}(\beta_t, z_t))$
  Compute attention parameter of commitment-plan $\psi_t^c = f^c(\text{concat}(\psi_t^A, \epsilon_t))$
  Update $\boldsymbol{A}^t = \rho(\mathbf{A}^{t-1}) + \text{write}(f^A(\epsilon_t), \psi_t^A)$
  Update $\mathbf{c}_t = Sigmoid(\boldsymbol{b} + \text{write}(e, \psi_t^c))$
else
  Update $\boldsymbol{A}^t = \rho(\mathbf{A}^{t-1})$
  Update $\mathbf{c}_t = \rho(\mathbf{c}_{t-1})$
end if

20.4.4 *Multi-Agent Reinforcement Learning*

Multi-Agent Q Learning

Algorithm 35 Multi-agent general Q learning

Set initial values for Q table $Q_i(s, a_i, \mathbf{a}_{-i}) = 1, \forall i \in \{1, 2, \ldots, m\}$.
for episode = 1 to M **do**
  Set initial state $s = S_0$.
  for step = 1 to T **do**
    Each agent i chooses action a_i based on $\pi_i(s)$, which is a mixed Nash equilibrium strategy based on **Q** values of all agents.
    Observe experience $(s, a_i, \mathbf{a}_{-i}, r_i, s')$ and apply it to update Q_i value
    Update the state $s = s'$.
  end for
end for

Multi-Agent Deep Deterministic Policy Gradient (MADDPG)

Algorithm 36 Multi-agent deep deterministic policy gradient (MADDPG)

for episode = 1 to M **do**
  Set initial state $s = S_0$.
  for step = 1 to T **do**
    Each agent i chooses action a_i based on current policy π_{θ_i}.
    The actions of all agents $\mathbf{a} = (a_1, a_2, \ldots, a_m)$ are executed simultaneously.
    Store $(s, \mathbf{a}, r, s')$ in replay buffer $\mathcal{M}$
    Update the state $s = s'$.
    for agent i = 1 to m **do**
      Sample a batch of previous experience from replay buffer $\mathcal{M}$.
      Calculate the gradients and update the weights for both actor and critic network.
    end for
  end for
end for

20.4.5 *Parallel Computing*

Asynchronous Advantage Actor-Critic (A3C)

Algorithm 37 Asynchronous advantage actor-critic (Actor-Learner)

Hyperparameters: Total number of steps T_{max}. Maximum steps for each episode t_{max}.
Initialize step counter $t = 1$.
while $T \leq T_{max}$ **do**
  Reset gradients: $d\theta = 0$ and $d\theta_v = 0$.
  Sync with parameter server to obtain network parameters $\theta' = \theta$ and $\theta'_v = \theta_v$.
  $t_{start} = t$
  Set starting state S_t for the episode
  while Reach terminal state **or** $t - t_{start} == t_{max}$ **do**
    Choose action A_t based on policy $\pi(S_t|\theta')$
    Act in the environment and receive rewards R_t and next state S_{t+1}
    $t = t + 1, T = T + 1$
  end while
  if Reach terminal state **then**
    $R = 0$
  else
    $R = V(S_t|\theta'_v)$
  end if
  for $i = t - 1, t - 2, \ldots, t_s tart$ **do**
    Update discounted rewards $R = R_i + \gamma R$
    Accumulate gradients wrt θ', $d\theta = d\theta + \nabla_{\theta'} \log \pi(S_i|\theta')(R - V(S_i|\theta'_v))$
    Accumulate gradients wrt θ'_v, $d\theta_v = d\theta_v + \partial(R - V(S_i|\theta'_v))^2/\partial\theta'_v$
  end for
  Asynchronously update θ with $d\theta$ and θ_v with $d\theta_v$.
end while

Distributed Proximal Policy Optimization

Algorithm 38 DPPO (chief)

Hyperparameters: the number of workers W, threshold for numbers of gradients available workers D, the number of sub-iterations M, B
Input: initial global policy parameters θ, initial global value function parameters ϕ
for k = 0, 1, 2, ... **do**
  for $m \in \{1, \dots, M\}$ **do**
    Wait until at least $W - D$ gradients wrt. θ are available average gradients and update global θ
  end for
  for $b \in \{1, \dots, B\}$ **do**
    Wait until at least $W - D$ gradients wrt. ϕ are available average gradients and update global ϕ
  end for
end for

Algorithm 39 DPPO (PPO-Penalty worker)

Hyperparameters: KL penalty coefficient λ, adaptive parameters $a = 1.5, b = 2$, the number of sub-iterations M, B
Input: initial local policy parameters θ, initial local value function parameters ϕ
for k = 0, 1, 2, ... **do**
  Collect set of trajectories $\mathcal{D}_k = \{\tau_i\}$ by running policy π_θ in the environment
  Compute rewards-to-go $\hat{G}_t$
  Compute advantage estimates, $\hat{A}_t$(using any method of advantage estimation) based on the current value function V_{ϕ_k}
  Store partial trajectory information
  $\pi_{old} \leftarrow \pi_\theta$
  for $m \in \{1, \dots, M\}$ **do**

$$J_{PPO}(\theta) = \sum_{t=1}^{T} \frac{\pi_\theta(A_t|S_t)}{\pi_{old}(A_t|S_t)} \hat{A}_t - \lambda \mathrm{KL}[\pi_{old}|\pi_\theta] - \xi \max(0, \mathrm{KL}[\pi_{old}|\pi_\theta] - 2\mathrm{KL}_{target})^2$$

    if $\mathrm{KL}[\pi_{old}|\pi_\theta] > 4\mathrm{KL}_{target}$ **then**
      break and continue with next outer iteration $k + 1$
    end if
    Compute $\nabla_\theta J_{PPO}$
    send gradient wrt. θ to chief
    wait until gradient accepted or dropped; update parameters
  end for
  for $b \in \{1, \dots, B\}$ **do**
    $L_{BL}(\phi) = -\sum_{t=1}^{T}(\hat{G}_t - V_\phi(S_t))^2$
    Compute $\nabla_\phi L_{BL}$
    send gradient wrt. ϕ to chief
    wait until gradient accepted or dropped; update parameters
  end for
  Compute $d = \hat{\mathbb{E}}_t\,[\mathrm{KL}[\pi_{old}(\cdot|S_t), \pi_\theta(\cdot|S_t)]]$
  if $d < d_{target}/a$ **then**
    $\lambda \leftarrow \lambda/b$
  else if $d > d_{target} \times a$ **then**
    $\lambda \leftarrow \lambda \times b$
  end if
end for

Algorithm 40 DPPO (PPO-Clip worker)

Hyperparameters: clip factor ϵ, the number of sub-iterations M, B
Input: initial local policy parameters θ, initial local value function parameters ϕ
for k = 0, 1, 2, ... **do**
  Collect set of trajectories $\mathcal{D}_k = \{\tau_i\}$ by running policy π_θ in the environment
  Compute rewards-to-go $\hat{G}_t$
  Compute advantage estimates, $\hat{A}_t$(using any method of advantage estimation) based on the current value function V_{ϕ_k}
  Store partial trajectory information
  $\pi_{old} \leftarrow \pi_\theta$
  for $m \in \{1, \dots, M\}$ **do**
    Update the policy by maximizing the PPO-Clip objective:

$$J_{PPO}(\theta) = \frac{1}{|\mathcal{D}_k|T} \sum_{\tau \in D_k} \sum_{t=0}^{T} \min\left(\frac{\pi_\theta(A_t|S_t)}{\pi_{old}(A_t|S_t)}\hat{A}_t, clip(\frac{\pi(A_t|S_t)}{\pi_{old}(A_t|S_t)}, 1-\epsilon, 1+\epsilon)\hat{A}_t\right)$$

    Compute $\nabla_\theta J_{PPO}$
    send gradient wrt. θ to chief
    wait until gradient accepted or dropped; update parameters
  end for
  for $b \in \{1, \dots, B\}$ **do**
    Fit value function by regression on mean-squared error:

$$L_{BL}(\phi) = -\frac{1}{|\mathcal{D}_k|T} \sum_{\tau \in \mathcal{D}_k} \sum_{t=0}^{T} \left(V_\phi(S_t) - \hat{G}_t\right)^2$$

    typically via some gradient descent algorithm
    send gradient wrt. ϕ to chief
    wait until gradient accepted or dropped; update parameters
  end for
end for

Ape-X

Algorithm 41 Ape-X (Actor)

Hyperparameters: Send to replay with batch size B in local buffer. Number of iterations T
Sync with learner to obtain latest network parameters θ_0.
Get initial state S_0 from environment.
for t = 0, 1, 2, ..., $T-1$ **do**
  Choose action a_t based on policy $\pi(S_t|\theta_t)$
  Add experience (S_t, A_t, R_t, S_{t+1}) to the local buffer
  if The local buffer reaches its size requirements B **then**
    Get buffered data with batch size B
    Calculate the priorities p of the buffered data.
    Send the batched buffered data and its priorities to the replay
  end if
  Periodically sync and update the latest network parameters θ_t
end for

Algorithm 42 Ape-X (Learner)

Hyperparameters: Number of learning episodes T.
Initialize the network parameters θ_0.
for $t = 1, 2, 3, \ldots, T$ **do**
    Sample a prioritized batch of data (i, d) from replay
    Applying training with the batched data
    Update network parameters to θ_t
    Calculate the priorities p for batched data d
    Update the priorities p for data with index i on replay
    Periodically remove data with low priorities in replay
end for

The manufacturer's authorised representative in the EU is Springer Nature Customer Service Centre GmbH, Europaplatz 3, 69115 Heidelberg, Germany. If you have any concerns regarding our products, please contact ProductSafety@springernature.com

Printed and bound by CPI Group (UK) Ltd, Croydon, CR0 4YY
30/09/2026
02225299-0012